Lecture Notes in Computer Science 11396

Commenced Publication in 1973
Founding and Former Series Editors:
Gerhard Goos, Juris Hartmanis, and Jan van Leeuwen

More information about this series at http://www.springer.com/series/7410

Kwangsu Lee (Ed.)

Information Security and Cryptology – ICISC 2018

21st International Conference
Seoul, South Korea, November 28–30, 2018
Revised Selected Papers

 Springer

Editor
Kwangsu Lee (iD)
Sejong University
Seoul, South Korea

ISSN 0302-9743   ISSN 1611-3349 (electronic)
Lecture Notes in Computer Science
ISBN 978-3-030-12145-7   ISBN 978-3-030-12146-4 (eBook)
https://doi.org/10.1007/978-3-030-12146-4

Library of Congress Control Number: 2018968329

LNCS Sublibrary: SL4 – Security and Cryptology

This Springer imprint is published by the registered company Springer Nature Switzerland AG
The registered company address is: Gewerbestrasse 11, 6330 Cham, Switzerland

Preface

ICISC 2018, the 21st International Conference on Information Security and Cryptology, was held in Seoul, South Korea, during November 28–30, 2018. This year the conference was hosted by the KIISC (Korea Institute of Information Security and Cryptology).

The aim of this conference is to provide an international forum for the latest results of research, development, and applications in the field of information security and cryptology. This year we received 49 submissions and were able to accept 21 papers. The review and selection processes were carried out by the Program Committee (PC) members via the EasyChair review system. First, each paper was blind reviewed, by at least three PC members for most cases. Second, to resolve conflicts in the reviewers' decisions, the individual review reports were open to all PC members, and detailed interactive discussions on each paper followed.

The conference featured four invited talks given by Hoeteck Wee, Jooyoung Lee, Aaram Yun, and Katsuyuki Takashima and one additional invited tutorial given by Katsuyuki Takashima. We thank the invited speakers for their kind acceptance and nice presentations. We would like to thank all authors who submitted their papers to ICISC 2018 and all PC members. It was a truly rewarding experience to work with such talented and hard-working researchers. We also appreciate the external reviewers for assisting the PC members.

Finally, we would like to thank all attendees for their active participation and the Organizing Committee members who expertly managed this conference. We look forward to seeing you again at next year's ICISC.

November 2018   Kwangsu Lee

Organization

Program Committee

Joonsang Baek	University of Wollongong, Australia
Lynn Batten	Deakin University, Australia
Olivier Blazy	XLim, Université de Limoges, France
Zhenfu Cao	East China Normal University, China
Donghoon Chang	IIIT-Delhi, India
Jie Chen	East China Normal University, China
Keita Emura	NICT, Japan
David Jao	University of Waterloo, Canada
Dong Seong Kim	The University of Queensland, Australia
Dong-Chan Kim	Kookmin University, South Korea
Huy Kang Kim	Korea University, South Korea
Taekyoung Kwon	Yonsei University, South Korea
Hyang-Sook Lee	Ewha Womans University, South Korea
Hyung Tae Lee	Chonbuk National University, South Korea
Jooyoung Lee	Korea Advanced Institute of Science and Technology, South Korea
Kwangsu Lee	Sejong University, South Korea
Moon Sung Lee	UNIST, South Korea
Joseph Liu	Monash University, Australia
Koji Nuida	The University of Tokyo, Japan
Daehun Nyang	Inha University, South Korea
Jong Hwan Park	Sangmyung University, South Korea
Josef Pieprzyk	Queensland University of Technology, Australia
Kui Ren	State University of New York at Buffalo, USA
Kouichi Sakurai	Kyushu University, Japan
Jae Hong Seo	Hanyang University, South Korea
Wenling Wu	Chinese Academy of Sciences, China
Dae Hyun Yum	Myongji University, South Korea
Aaram Yun	UNIST, South Korea

Additional Reviewers

Abuhmed, Tamer
Chang, Seunghwan
Chen, Hua
Chen, Zhenhua
Cheng, Chen-Mou
Choi, Wonseok
Dagvatur, Zayabaatar
Dutta, Sabyasachi
Eom, Jieun
Esser, Andre
Hasan, Munawar
Hayashi, Takuya
Jati, Arpan
Kang, Jeonil
Kim, Jeongsu
Kim, Jonghyun
Kim, Junsik
Kim, Seongkwang
Kim, Younjin
Lee, Byeonghak
Lee, Dong Hoon

Lee, Jinwoo
Lee, Young Kyung
Lim, Seongan
Maeng, Youngjae
Mishina, Ibuki
Ohata, Satsuya
Ouaddah, Aafaf
Roy, Partha Sarathi
Seo, Minhye
Shinagawa, Kazumasa
Singh, Ajit Pratap
Su, Chunhua
Takashima, Katsuyuki
Takayasu, Atsushi
Tian, Miaomiao
Une, Masashi
Wang, Luping
Wu, Longfei
Zeng, Ming
Zhao, Qian

Abstracts of Invited Talks

Tweakable Block Ciphers: Construction and Applications

Jooyoung Lee

KAIST, Korea
hicalf@kaist.ac.kr

A tweakable block cipher is a block cipher that accepts additional inputs called tweaks to provide variability to encryption. Recently, this primitive is widely used in various cryptographic schemes such as (authenticated) encryption and message authentication. In the first part of this talk, we survey recent results on construction and application of tweakable block ciphers. We also compare security notions between block ciphers and tweakable block ciphers in the standard/ideal primitive models.

In the second part, we propose a new construction of tweakable block ciphers from standard block ciphers. Our construction, dubbed XHX2, is the cascade of two independent XHX constructions, so it makes two calls to the underlying block cipher using tweak-dependent keys. We prove the security of XHX2 up to $\min\{2^{2(n+m)/3}, 2^{n+m/2}\}$ queries (ignoring logarithmic factors) in the ideal cipher model, when the block cipher operates on n-bit blocks using m-bit keys. The XHX2 tweakable block cipher is the first construction that achieves beyond-birthday-bound security with respect to the input size of the underlying block cipher in the ideal cipher model.

Security Against Quantum Superposition Attacks

Aaram Yun

School of Electrical and Computer Engineering,
Ulsan National Institute of Science and Technology (UNIST), Ulsan, Korea
aaramyun@unist.ac.kr

Abstract. In the post-quantum cryptography, we usually consider adversaries who have quantum computational capabilities. Such an adversary has a quantum computer 'at home', and is able to carry out arbitrary polynomial-time quantum computation privately, but all the interaction between the adversary and its environment (including the legitimate users) are classical. In other words, such an adversary can compute quantumly, but all of its oracle queries are classical.

On the other hand, there are situations where we also consider adversaries who not only have quantum computational capabilities, but also have ability to make its oracle queries in quantum superposition. One obvious example is the quantum random oracle model, where the adversary can make superposition queries to the random oracle, since such a query models adversary's private hash function evaluation, which can be carried out in superposition. There are also works which consider adversaries making quantum superposition queries not only to the random oracle but to all oracles.

In this talk, I would like to discuss the above adversarial model, and make a survey of some of the works in that model, both cryptanalytic attacks and constructions secure against quantum superposition attacks.

Contents

Invited Talk

Revised Talk.

New Assumptions on Isogenous Pairing Groups with Applications to Attribute-Based Encryption

Takeshi Koshiba[1] and Katsuyuki Takashima[2(✉)]

[1] Waseda University, Tokyo, Japan
tkoshiba@waseda.jp
[2] Mitsubishi Electric, Kanagawa, Japan
Takashima.Katsuyuki@aj.MitsubishiElectric.co.jp

Abstract. We introduce new isogeny-related assumptions called Isog-DDH and Isog-DBDH assumptions. By using the assumptions, we reinforce security of several existing (hierarchical) identity-/attribute-based encryption (HIBE/ABE) schemes. While the existing schemes are proven from the standard DBDH assumption, our reinforced secure ones have *two incomparable* security proofs: one is proven from the DBDH as well and another is proven from the Isog-DDH assumption which is incomparable with DBDH. As a result, if *either DBDH or Isog-DDH* assumption holds, the proposed HIBE/ABE schemes are secure. For obtaining our (H)IBE secure in the standard model, we assign a *unique (product) group* called ID-group to each (H)ID, and introduce a new proof technique, i.e., *ID-group partitioning* by using isogenies as trapdoors.

Keywords: Isogenous pairing groups · Identity-based encryption · Attribute-based encryption · Security reinforcement

1 Introduction

1.1 Background

Since the National Institute of Standards and Technology (NIST) has initiated a standardization process for post-quantum public-key cryptographic algorithms [23], studying such cryptosystems is a hot research area. Aside from lattice-based, code-based, and multivariate cryptography, isogeny-based cryptography is (relatively) newly entered as a candidate post-quantum primitive.

Very recently, Boneh et al. [8] suggest another research direction for using isogenies on elliptic curves in *applied* cryptography which includes n-way non-interactive key exchange, verifiable random functions, constrained pseudorandom functions, broadcast encryption, and witness encryption. We also seek such applications of isogenies in (hierarchical) identity-/attribute-based encryption ((H)IBE/ABE). For the purpose, we use a unified framework for operations on

© Springer Nature Switzerland AG 2019
K. Lee (Ed.): ICISC 2018, LNCS 11396, pp. 3–19, 2019.
https://doi.org/10.1007/978-3-030-12146-4_1

elliptic curves (i.e., scalar multiplication, pairing and isogeny) called isogenous pairing groups (IPG) by Koshiba and Takashima [21].

(Key-policy) attribute-based encryption (KP-ABE) is a powerful and useful generalization of identity-based encryption (IBE). In a KP-ABE system, ciphertexts are associated with sets of attributes and user secret keys distributed by an authority are associated with formulas over attributes. A user should be able to decrypt a ciphertext if and only if the secret key formula is satisfied by the ciphertext attributes. Previously, IBE schemes were constructed on three mathematical primitives, i.e., pairing [4,5,28], factoring [7,12] and lattices [1,16] and ABE schemes were constructed on only two mathematical primitives, pairing [18,24] and lattices [6,17]. Through reformulations of previous pairing based IBE/ABE on the IPG framework, we obtain several IBE/ABE schemes based on another mathematical primitive, i.e., *system of pairings and isogenies*, in this paper (as in [21]). (Refer to Remark 1 in Sect. 2.1 for the concrete instantiation of IPG by using supersingular elliptic curves.)

In constructing IBE/ABE schemes, *quality of underlying assumptions* is an important factor for assessing the schemes. Previous works in pairing cryptography realized a highly secure IBE/ABE (e.g., adaptive security) under non-standard assumptions like *q-type* ones for the first time. However, the q-type assumptions (and also the associated schemes) suffered a special attack which was presented by Cheon [11] at Eurocrypt 2006. Consequently, it is very desirable that the non-standard q-type assumption should be replaced by a *static* (non-q type) assumption. Hence, in subsequent works, the same security level can be accomplished under standard (or better) assumptions. For example, many high secure pairing cryptosystems have been established under one of standard assumptions like DBDH and DLIN assumptions.[1] Our starting point in this work is the following question.

> *For pairing cryptosystems with some desirable security which is proven under the standard DBDH or DLIN assumption, can we add other incomparable security proofs for them by using isogenies?*

We propose new methodologies for evolving pairing systems further into the desirable directions. The key for the step comes from rich algebraic structures on elliptic curves, and specifically from isogeny operations, which have attracted notice in the context of post-quantum cryptography. We extend Koshiba-Takashima's work [21], however, *our aim is not post-quantum* schemes, but, we *reinforce (classical) security* of pairing-based IBE/ABE schemes by using isogenies as *trapdoor homomorphisms*.

For example, we reformulate the basic Boneh-Franklin (BF) IBE [5] in the setting of IPG with minor modifications. The reformulated IBE has two *incomparable* security proofs, one is from DBDH (as before) and the other is from a new, isogeny-related Isog-DDH assumption. The additional Isog-DDH based

---

[1] Here, DLIN (resp. DBDH) represents a family of progressively weaker k-Lin (resp. k-BDH) assumptions, e.g., [3,10,25] since DLIN = 2-Lin and DBDH is equivalent to 1-BDH.

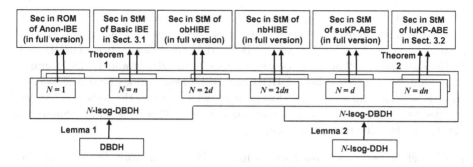

Fig. 1. Reductions between problems on IPG and security of proposed IBE/ABE, where n is the bit length of identities and attributes, and d is the maximum hierarchy length (depth) in HIBE and the maximum size of small universe (and the set of categories) in KP-ABE, and ROM, StM, obHIBE, nbHIBE, suKP-ABE and luKP-ABE stand for Random Oracle Model, Standard Model, one-bit (resp. n-bit) ID-space HIBE and small universe (resp. large universe) KP-ABE, respectively. The proposed anonymous IBE, obHIBE, nbHIBE and suKP-ABE schemes will be given in the full version of this paper.

security proof *reinforces* the standard DBDH based security. Such kind of security reinforcement was used in a variant of Cramer-Shoup encryption scheme [13, 26]. The CS3b encryption scheme in [13] has two *incomparable* IND-CCA security proofs: one is proven from the DDH assumption (and standard assumptions for hash and key derivation functions) in the standard model, and another is proven from the CDH assumption (and a standard assumption for hash) in the random oracle model.

In our ABE, by using *randomly chosen multiple pairing groups* and encoding various attributes of some user, i.e., his/her name, age, affiliate, gender, etc. onto *the different groups*, we have more independence among the encoded attributes since all previous ABE schemes encode various types of attributes *onto one same pairing group*. This intuition for security reinforcement is captured by new isogeny-related assumptions which give incomparable security proofs as well as previous ones. Furthermore, decryption executes pairing operations between key and ciphertext components and computes decryption results on the target group $\mathbb{G}_T$ (refer to Fig. 2).

1.2 Our Results

We define two new assumptions on IPG, i.e., Isog-DDH and Isog-DBDH assumptions, and obtain security reinforced several IBE/ABE schemes, which have two security proofs, one is reduced from DBDH and another is reduced from $(N\text{-})$Isog-DDH assumption (Fig. 1). It implies that if either DBDH *or* Isog-DDH assumption holds, our reinforced IBE/ABE schemes are secure.[2] We stress that our situation is *different* from several previous works. For example, in [28], Waters proposed celebrated adaptive secure IBE and HIBE schemes from DBDH and DLIN assumptions. The security proof needs both DBDH and DLIN assumptions, and implies that if both DBDH *and* DLIN assumptions hold, the Waters (H)IBE schemes are secure (see footnote 2), which means the logical structure is different from ours. We summarize our Lemmas, Theorems and Corollary in Fig. 1, and describe our results below.

1. We define two new assumptions in the IPG framework for proving the reinforced security of our IBE and ABE schemes: the (N-)Isog-DDH and (N-)Isog-DBDH assumptions (For the terminology, see the footnote for Definition 3). Then we show two reductions in Lemmas 1 and 2: the Isog-DBDH assumption is reduced to the DBDH assumption (resp. the Isog-DDH assumption). We stress that the two underlying assumptions are incomparable, that means that even if the DBDH assumption would be broken by some future progress of the attack techniques, the scheme possibly survives because of the isogeny-related assumption. As another interpretation, while someone may find a new attack in future since isogeny problems have not yet mature history of scrutiny, but, our schemes are secure based on the traditional DBDH, which means that our Isog-DBDH based schemes are at least as secure as the underlying DBDH based ones. For the details, see Sect. 2.

2. We present an anonymous IBE construction adaptive-ID secure in the random oracle model (ROM) whose security is proven from Isog-DBDH (in the full version of this paper). Our IBE scheme is based on the Boneh-Franklin IBE (BF-IBE), and has an efficiency (or practicality) comparable to the BF-IBE. One of the main differences is that a public master key has two elliptic curve parameters for using isogeny in the cryptographic construction. Since BF-IBE was adopted as an international standard [20], our IBE is quite practical with respect to the data sizes, encryption and decryption times.

3. We also present (H)IBE constructions which are selective-ID secure from Isog-DBDH in the standard model (Theorem 1 in Sect. 3.1). Our first IBE scheme (basic IBE) is a special form of our KP-ABE in Sect. 3.2. The scheme is conceptually simple and can be extended to a hierarchical IBE, but the security of the HIBE is not easy to be based on Isog-DBDH since we should publish auxiliary group elements for delegation. Therefore, we modify it and obtain our one-bit (resp. n-bit) ID-space HIBE scheme based on Boneh-Boyen HIBE [4], whose security is proven from the Isog-DBDH assumption (in the full version of this paper).

4. We construct small and large universe KP-ABE schemes which are selective-attribute secure from Isog-DBDH in the standard model (see Theorem 2 in

---

[2] Since logical equivalence $A \rightarrow C \equiv \neg A \vee C$ holds for truth variables A, B, C, it holds that $(A \vee B) \rightarrow C \equiv \neg(A \vee B) \vee C \equiv (\neg A \vee C) \wedge (\neg B \vee C) \equiv (A \rightarrow C) \wedge (B \rightarrow C)$. We have $(A \wedge B) \rightarrow C \equiv (A \rightarrow C) \vee (B \rightarrow C)$ in a similar manner (in the case [28]). In our case, $A =$ "DBDH assumptions holds", $B =$ "N-Isog-DDH assumptions hold", $C =$ "our schemes are secure".

Sect. 3.2 for the large universe one). First, we construct a small universe KP-ABE and then obtain the large universe construction by hierarchically combining the small universe ABE and the basic IBE as underlying ones. The proposed KP-ABE schemes are based on the GPSW06 KP-ABE [18]. We note that all sizes of secret keys and ciphertxts of our small universe KP-ABE are *quite comparable* to those of GPSW06 small universe KP-ABE (and public parameter sizes of the two schemes are of asymptotically same order). Since there exists a generic conversion from KP-ABE to CP-ABE and vice versa [2] (with large overheads and some restrictions), we obtain CP-ABE schemes from our KP-ABE schemes, whose security is also proven from Isog-DBDH.

We point out a comparison with existing works. While there exist some generic combiners for encryption, e.g., [15,19], which need two different proofs for the reinforced security, our schemes have two incomparable proofs just by proving security from one simple assumption (i.e., Isog-DBDH assumption). Therefore, our constructions have efficiency advantages over generic combiner constructions (see our anonymous IBE and small universe KP-ABE, for example).

1.3 Key Techniques

Underlying Mathematical Framework of IPG [21]. For most pairing cryptosystems, the bilinear property $e(g_0, \hat{g}_0^\alpha) = e(g_0, \hat{g}_0)^\alpha$ is a key point, which is considered as a compatibility condition on pairing with (public key) $\hat{g}_0^\alpha$ and scalar multiplication with (secret key) α. Based on the above similarity, for our IBE and ABE, a compatibility of pairing and isogeny, e.g., $e_0(g_0, \hat{g}_0) = e_1(\phi(g_0), \phi(\hat{g}_0))$, is required. Note that since we use multiple elliptic curves, pairings e_0 and e_1 are defined on different curves E_0 and $E_1 := \phi(E_0)$, respectively. Based on the compatibility, the notion of isogenous pairing groups (IPG), an extension of that of pairing groups, was formulated. In the system, multiple pairing groups of the same prime order are employed, where efficient homomorphisms between them are hidden from adversaries. It is schematically presented in Fig. 2 in Sect. 2.1.

New Assumptions on IPG: Isog-DDH and Isog-DBDH. We formulate new assumptions for our proposals, namely, N-Isog-DDH and N-Isog-DBDH assumptions. An instance for the 1-Isog-DDH problem consists of $(g, \hat{g}, g^\alpha, \phi(\hat{g}), \phi(\hat{g})^\beta, h_T)$ for distinguishing whether h_T is $g_T^{\alpha\beta}$ or random, where $g \in \mathbb{G}_0 \subset E_0$ and $\phi(\hat{g}) \in \hat{\mathbb{G}}_1 \subset E_1$ and then secret scalars α and β are encoded on different (isogenous) elliptic curves E_0 and E_1, and $g_T := e_0(g, \hat{g})$. Therefore, informally, an adversary without knowing ϕ cannot obtain a meaningful pairing value $g_T^{\alpha\beta}$ from g^α and $\phi(\hat{g})^\beta$. We extend it to 1-Isog-DBDH, whose (simple form) instance consists of $(g, \hat{g}, g^\alpha, \phi(\hat{g}), \phi(\hat{g})^\beta, \phi(g)^\gamma, \phi(\hat{g})^\gamma, h_T)$ for distinguishing whether h_T is $g_T^{\alpha\beta\gamma}$ or random. The intractability of N-Isog-DBDH on $N+1$ multiple groups is reduced from that of N-Isog-DDH (resp. the standard DBDH). Our basic IBE (resp., small, large universe ABE) scheme is secure under the n-Isog-DBDH (resp., d-Isog-DBDH, dn-Isog-DBDH) assumption, where d is small universe size and n is bitlength of an identity or attribute.

First Step for Isog-DBDH-Based IBE: Partitioning via Isogeny in ROM. By using the above similarity, we replace a part of master key pair of BF-IBE, ($\mathsf{pk} := (\hat{g}, \hat{g}^\gamma)$, $\mathsf{sk} := \gamma$) for a group element $\hat{g} \in \hat{\mathbb{G}}_0$ and a random scalar γ, by ($\mathsf{pk} := (\hat{g}, \phi(\hat{g}^\gamma))$, $\mathsf{sk} := (\phi, \gamma)$) for a randomly chosen isogeny ϕ. The important difference is that further randomization is applied by isogeny ϕ as well as random scalar γ. The difference leads to security under incomparable DBDH and 1-Isog-DDH. In our IBE in the random oracle model, the target ID^* is assigned as $H(\mathsf{ID}^*) := g^\alpha$ and secret key $\mathsf{sk}_{\mathsf{ID}^*} := \phi(H(\mathsf{ID}^*))^\gamma = \phi(g^\alpha)^\gamma$, which cannot be computed by adversary. However, for other $\mathsf{ID} \neq \mathsf{ID}^*$ and $H(\mathsf{ID}) := g^{\tau_{\mathsf{ID}}}$ with simulated $\tau_{\mathsf{ID}} \xleftarrow{\mathsf{U}} \mathbb{F}_q$, we can compute isogeny image $\mathsf{sk}_{\mathsf{ID}} := \phi(H(\mathsf{ID}))^\gamma = \phi(g^{\tau_{\mathsf{ID}}})^\gamma = (\phi(g)^\gamma)^{\tau_{\mathsf{ID}}}$ since simulator knows $\phi(g)^\gamma$ and τ_{ID}. We should extend this partitioning in ROM to more complicated setting like the standard model, hierarchical or attribute-based settings via trapdoor isogenies.

New Proof Technique: ID-Group Partitioning via Trapdoor Isogenies. For properly partitioning IDs in security proofs in the standard model, in our (hierarchical) IBE schemes, n-bit identity $\mathsf{ID} := (\mathsf{ID}_j)_{j \in [n]}$ is encoded as a *unique group* $\mathbb{G}_{\mathsf{ID}}$ (resp. $\hat{\mathbb{G}}_{\mathsf{ID}}$) called ID-groups in secret key (resp. ciphertext) such that $\mathbb{G}_{\mathsf{ID}} = \mathbb{G}_{\mathsf{ID}'}$ (resp. $\hat{\mathbb{G}}_{\mathsf{ID}} = \hat{\mathbb{G}}_{\mathsf{ID}'}$) *if and only if* $\mathsf{ID} = \mathsf{ID}'$. The ID-group is given as a direct product of n isogenous pairing groups, i.e., $\mathbb{G}_{\mathsf{ID}} := \mathbb{G}_{1,\mathsf{ID}_1} \times \cdots \times \mathbb{G}_{n,\mathsf{ID}_n}$ (resp. $\hat{\mathbb{G}}_{\mathsf{ID}} := \hat{\mathbb{G}}_{1,\mathsf{ID}_1} \times \cdots \times \hat{\mathbb{G}}_{n,\mathsf{ID}_n}$). In our IBE, public parameters include $2n$ isogenous pairing groups $(\mathbb{G}_{j,\iota}, \hat{\mathbb{G}}_{j,\iota})_{j \in [n]; \iota = 0,1}$. In selective-ID security proof, if adversary declares challenge identity $\mathsf{ID}^* := (\mathsf{ID}_j^*)$, then n groups $(\hat{\mathbb{G}}_j)_{j \in [n]}$ given in the n-Isog-DBDH instance are used for the challenge ID-group $\hat{\mathbb{G}}_{\mathsf{ID}^*}$, i.e., $\hat{\mathbb{G}}_{j,\mathsf{ID}_j^*} := \hat{\mathbb{G}}_j$ for $j \in [n]$. Other n isogenous pairing groups $(\mathbb{G}_{j,\overline{\mathsf{ID}_j^*}}, \hat{\mathbb{G}}_{j,\overline{\mathsf{ID}_j^*}})_{j \in [n]}$ in public parameters are newly generated together with *trapdoor isogenies* $\phi_j : \mathbb{G}_0 \to \mathbb{G}_{j,\overline{\mathsf{ID}_j^*}}$ where $\mathbb{G}_0$ is given in the n-Isog-DBDH instance. For a key queried identity $\mathsf{ID} := (\mathsf{ID}_j)$, we have at least one index j such that $\mathsf{ID}_j \neq \mathsf{ID}_j^*$ (i.e., $\mathsf{ID}_j = \overline{\mathsf{ID}_j^*}$) by definition. Therefore, we can use the above isogeny ϕ_j for the key generation simulation on the ID-group $\mathbb{G}_{\mathsf{ID}}$. We call the technique *ID-group partitioning* since we partition 2^n ID-groups into key-simulatable $2^n - 1$ ones (other than for challenge ID^*) and one ciphertext-simulatable ID-group for challenge ID^* in the setup phase via the above trapdoor isogenies $(\phi_j)_{j \in [n]}$. For the details of the idea, see Sect. 3.1. Moreover, this ID-group partitioning technique is used for a conversion from small universe KP-ABE to large universe one in a modular manner (Theorem 2). We stress that our schemes which are secure in the standard model *do not use isogenies in real operations* except for master key generation, but *use them in simulation* in a crucial manner as we see above.

1.4 Notations

When A is a random variable or distribution, $y \xleftarrow{\mathsf{R}} A$ denotes that y is randomly selected from A according to its distribution. When A is a set, $y \xleftarrow{\mathsf{U}} A$ denotes that y is uniformly selected from A. We denote the finite field of order q by $\mathbb{F}_q$. Let $[n] := \{1, .., n\}$ and $[0, n] := \{0\} \cup [n] := \{0, .., n\}$ for any $n \in \mathbb{Z}_{>0}$. For two vectors $\vec{y} = (y_i)_{i \in [r]}$ and $\vec{v} = (v_i)_{i \in [r]}$, $\vec{y} \cdot \vec{v}$ denotes the inner-product $\sum_{i=1}^r y_i v_i$. For an element $\mathfrak{g} := (\mathfrak{g}_i)$ (resp. $\hat{\mathfrak{g}} := (\hat{\mathfrak{g}}_i)$) in product group $\mathbb{K} := \mathbb{G}_1 \times \cdots \times \mathbb{G}_r$ (resp. $\hat{\mathbb{K}} := \hat{\mathbb{G}}_1 \times \cdots \times \hat{\mathbb{G}}_r$) and vector $\vec{y} = (y_i)_{i \in [r]} \in \mathbb{F}_q^r$, $\mathfrak{g}^{\vec{y}}$ denotes the group element $(\mathfrak{g}_i^{y_i})_{i \in [r]}$. For scalar $\zeta \in \mathbb{F}_q$, $\mathfrak{g}^\zeta$ denotes scalar exponentiation $(\mathfrak{g}_i^\zeta)_{i \in [r]}$. For two elements $\mathfrak{g} := (\mathfrak{g}_i)$ and $\hat{\mathfrak{g}} := (\hat{\mathfrak{g}}_i)$ in the above product group $\mathbb{K}$ and $\hat{\mathbb{K}}$, pairing $e_{\mathbb{K}}$ is defined as $e_{\mathbb{K}}(\mathfrak{g}, \hat{\mathfrak{g}}) := \prod_{i \in [r]} e_i(\mathfrak{g}_i, \hat{\mathfrak{g}}_i)$ if the pairing e_i is defined on $\mathbb{G}_i \times \hat{\mathbb{G}}_i$ for $i \in [r]$.

2 Isogenous Pairing Groups (IPG) and Assumptions

We review the notion of isogenous pairing groups (IPG) [21] which models isogenous (supersingular) elliptic curves and related operations on them (Sect. 2.1). For mathematical background on IPG, refer to Appendix A in [21]. We then give definitions of key assumptions, i.e., Isog-DDH, DBDH, Isog-DBDH assumptions, and their relations (Lemmas 1 and 2) in Sect. 2.2.

2.1 Definition of Isogenous Pairing Groups (IPG)

We first define trapdoor homomorphisms (TH) for unifying operations on elliptic curves. Three THs, i.e., exponentiation (scalar multiplication), pairing and isogeny, give a rich algebraic structure.

Definition 1 (Trapdoor Homomorphism (TH)). *A (randomly chosen) function $\phi := \phi_\xi : G_0 \to G_1$ with two (randomly chosen) cyclic groups G_0, G_1 of a prime order q is called a* trapdoor homomorphism *if the following conditions hold.*

- *(Homomorphism) ϕ is a non-trivial (e.g., non-zero for an additive group) homomorphism.*
- *(TH-DH (TH-Diffie-Hellman) intractability assumption) Any probabilistic polynomial-time (ppt) machine $\mathcal{B}$ computes $\phi(g)$ only with a negligible probability when given $(g_0, \phi(g_0), g)$ for randomly chosen ϕ and $g_0, g \xleftarrow{\mathsf{U}} G_0$.*
- *(Polynomial-size trapdoor) There exists a ppt machine $\mathcal{B}$ which computes $\phi(g)$ for any $g \in G_0$ given a polynomial-size trapdoor ξ for $\phi := \phi_\xi$.*

Examples. By using elliptic curves, we have three examples of THs.

1. (Exponentiation) $G_0 := G_1 := \mathbb{G}$ is an elliptic curve cyclic group and $\phi := \phi_\xi$ is an exponentiation on $\mathbb{G}$ (i.e., scalar multiplication on the curve), i.e., $\phi_\xi : g \mapsto g^\xi$, where ξ is a scalar. TH-DH input and output are given as $(g_0, \phi(g_0), g) = (g_0, g_0^\xi, g)$ and $\phi(g) = g^\xi$, respectively, and then TH-DH intractability is the same as the usual computational DH assumption.

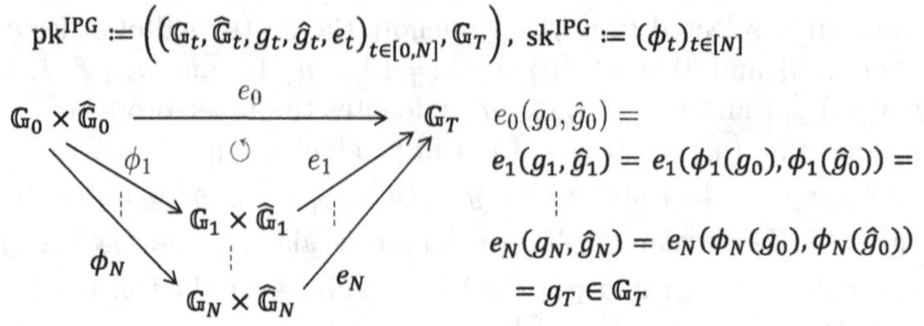

$$\mathsf{pk}^{\mathsf{IPG}} := \left(\left(\mathbb{G}_t, \widehat{\mathbb{G}}_t, g_t, \hat{g}_t, e_t \right)_{t \in [0,N]}, \mathbb{G}_T \right), \quad \mathsf{sk}^{\mathsf{IPG}} := (\phi_t)_{t \in [N]}$$

$$e_0(g_0, \hat{g}_0) =$$
$$e_1(g_1, \hat{g}_1) = e_1(\phi_1(g_0), \phi_1(\hat{g}_0)) =$$
$$\vdots$$
$$e_N(g_N, \hat{g}_N) = e_N(\phi_N(g_0), \phi_N(\hat{g}_0))$$
$$= g_T \in \mathbb{G}_T$$

Fig. 2. Compatibility of IPG

2. (Pairing) $G_0 := \mathbb{G}, G_1 := \mathbb{G}_T$ is a part of asymmetric pairing groups and $\phi := \phi_\xi$ is a pairing operation on $\mathbb{G} \times \widehat{\mathbb{G}}$, i.e., $\phi_\xi : g \mapsto e(g, \xi)$, where ξ is an element in $\widehat{\mathbb{G}}$, TH-DH input and output are given as $(g_0, \phi(g_0), g) = (g_0, e(g_0, \xi), g)$ and $\phi(g) = e(g, \xi)$, respectively, and then TH-DH intractability is reduced to the computational BDH (CBDH) assumption.

3. (Isogeny) $G_0 := \mathbb{G}_0, G_1 := \mathbb{G}_1$ are two different elliptic curve cyclic groups obtained from two curves E, E', respectively, and $\phi := \phi_\xi$ is an isogeny from $\mathbb{G}_0$ to $\mathbb{G}_1$, i.e., $\phi_\xi : E \to E' := E/C$, where $\xi := C$ is a (cyclic) subgroup in E. The TH-DH intractability is another kind of natural extensions of the DH assumption obtained by using isogenies other than that given in [14].

Combining the three trapdoor homomorphic structures, we propose a useful cryptographic framework called *Isogenous Pairing Groups (IPG)*. When applying IPG framework and Isog-DBDH to crypto constructions, *compatibility* of pairings on different groups (or elliptic curves) and isogenies is a main ingredient. For asymmetric pairing groups $(\mathbb{G}_0, \widehat{\mathbb{G}}_0)$ and $(\mathbb{G}_1, \widehat{\mathbb{G}}_1)$ with TH $\phi : \mathbb{G}_0 \times \widehat{\mathbb{G}}_0 \to \mathbb{G}_1 \times \widehat{\mathbb{G}}_1$ (given by an isogeny), it is informally described as

$$e_0(g, \hat{g}) = e_1(\phi(g), \phi(\hat{g})), \tag{1}$$

where e_0 (resp., e_1) is an efficiently computable pairing on $\mathbb{G}_0 \times \widehat{\mathbb{G}}_0$ (resp., $\mathbb{G}_1 \times \widehat{\mathbb{G}}_1$), i.e., on the curve E_0 (resp., E_1). For the correctness, Theorem 6.1 and Proposition 8.2 in chapter III of [27] show that $e_{\mathsf{weil},0}(g, \hat{g})^{\deg \phi} = e_{\mathsf{weil},1}(\phi(g), \phi(\hat{g}))$, where $e_{\mathsf{weil},0}$ (resp. $e_{\mathsf{weil},1}$) is the Weil pairing on E_0 (resp. E_1), and then we have Eq. (1) for $e_0 := e_{\mathsf{weil},0}(\cdot, \cdot)^{\deg \phi}$ and $e_1 := e_{\mathsf{weil},1}(\cdot, \cdot)$ when $\deg \phi$ is fixed. The above property (1) is extended to among multiple curves $\{E_t\}_{t \in [0,N]}$ or multiple asymmetric pairing groups $(\mathbb{G}_t, \widehat{\mathbb{G}}_t)_{t \in [0,N]}$ as is given in Eq. (2) below.

Definition 2 (Isogenous Pairing Groups (IPG)). *Isogenous Pairing Groups (IPG) generator generates a random instance as follows:*

$$\mathsf{Gen}^{\mathsf{IPG}}(1^\lambda, N) \xrightarrow{\mathsf{R}} (\, \mathsf{pk}^{\mathsf{IPG}} := ((\mathbb{G}_t, \widehat{\mathbb{G}}_t, g_t, \hat{g}_t, e_t)_{t \in [0,N]}, \mathbb{G}_T), \ \mathsf{sk}^{\mathsf{IPG}} := (\phi_t)_{t \in [N]} \,),$$

where $(\mathbb{G}_t, \widehat{\mathbb{G}}_t, e_t, \mathbb{G}_T)$ are asymmetric pairing groups of a prime order q with pairings $e_t : \mathbb{G}_t \times \widehat{\mathbb{G}}_t \to \mathbb{G}_T$, trapdoor homomorphisms $\phi_t : \mathbb{G}_0 \times \widehat{\mathbb{G}}_0 \to \mathbb{G}_t \times \widehat{\mathbb{G}}_t$

such that $\mathbb{G}_t = \phi_t(\mathbb{G}_0)$ *and* $\hat{\mathbb{G}}_t = \phi_t(\hat{\mathbb{G}}_0)$ *under natural identifications* $\mathbb{G} = \mathbb{G} \times 1_{\hat{\mathbb{G}}}$ *and* $\hat{\mathbb{G}} = 1_{\mathbb{G}} \times \hat{\mathbb{G}}$ *(given by isogenies between different elliptic curves), and* $g_t = \phi_t(g_0) \in \mathbb{G}_t, \hat{g}_t = \phi_t(\hat{g}_0) \in \hat{\mathbb{G}}_t$. *The isogenous pairing groups satisfy.*

Compatibility: $e_0(g_0, \hat{g}_0) = e_t(g_t, \hat{g}_t) = e_t(\phi_t(g_0), \phi_t(\hat{g}_0))$ for any $t \in [N]$.   (2)

We denote the common non-trivial pairing value by g_T, *i.e.,* $g_T = e_0(g_0, \hat{g}_0) \neq 1$. *See Fig. 2. Moreover, we require that* $\mathbb{G}_t \neq \hat{\mathbb{G}}_t$. *(Namely, all the points in* $\mathbb{G}_t$ *and* $\hat{\mathbb{G}}_t$ *generate the group of* q-*torsion points on the* t-*th elliptic curve.)*

Remark 1 (Instantiation of Gen^{IPG} by Elliptic Curves). A trapdoor (homomorphic) isogeny generation algorithm THGen$^{\text{IPG}}(\mathbb{G}_0, \hat{\mathbb{G}}_0, g_0, \hat{g}_0, e_0)$, which is used as a main ingredient of Gen^{IPG} and also used in simulation in security proofs, outputs a newly isogenous group $(\mathbb{G}, \hat{\mathbb{G}}, g, \hat{g}, e, \phi)$ such that trapdoor $\phi : \mathbb{G}_0 \times \hat{\mathbb{G}}_0 \to \mathbb{G} \times \hat{\mathbb{G}}$ with $\mathbb{G} = \phi(\mathbb{G}_0)$, $\hat{\mathbb{G}} = \phi(\hat{\mathbb{G}}_0)$ is an efficiently computable group isomorphism, $g = \phi(g_0)$, $\hat{g} = \phi(\hat{g}_0)$ and the compatibility holds, i.e., $e_0(g_0, \hat{g}_0) = e(\phi(g_0), \phi(\hat{g}_0)) = e(g, \hat{g})$. If we use algorithm THGen^{IPG} as an ingredient, a concrete instantiation of IPG generation algorithm Gen$^{\text{IPG}}(1^\lambda, N)$ is obtained as follows: Generate a (random) supersingular elliptic curve $E_0/\mathbb{F}_{p^2}$ with a sufficiently large, odd prime p and large prime $q \mid p \pm 1$, then we have $\mathbb{G}_0 \times \hat{\mathbb{G}}_0 \subset E(\mathbb{F}_{p^2})$ where $\mathbb{G}_0, \hat{\mathbb{G}}_0$ are cyclic groups of order q since $q^2 \mid \sharp E_0(\mathbb{F}_{p^2}) = (p \pm 1)^2$. Generate bases $g_0 \xleftarrow{\text{U}} \mathbb{G}_0$, $\hat{g}_0 \xleftarrow{\text{U}} \hat{\mathbb{G}}_0$ and the pairing e_0 is defined by $e_0(h_0, \hat{h}_0) := e_{\text{weil},0}(h_0, \hat{h}_0)^{\ell^\kappa}$ for any $h_0 \in \mathbb{G}_0, \hat{h}_0 \in \hat{\mathbb{G}}_0$ from the Weil pairing $e_{\text{weil},0}$ on E_0 and suitably chosen (ℓ, κ). Then we set the 0-th pairing group $(\mathbb{G}_0, \hat{\mathbb{G}}_0, g_0, \hat{g}_0, e_0)$. For a concrete instantiation of the 0-th pairing group generation by using elliptic curves, see the full version.

For each $t \in [N]$, $(\mathbb{G}_t, \hat{\mathbb{G}}_t, g_t, \hat{g}_t, e_t, \phi_t) \xleftarrow{\text{R}} \text{THGen}^{\text{IPG}}(\mathbb{G}_0, \hat{\mathbb{G}}_0, g_0, \hat{g}_0, e_0)$. Then, output (pk$^{\text{IPG}} := ((\mathbb{G}_t, \hat{\mathbb{G}}_t, g_t, \hat{g}_t, e_t)_{t \in [0,N]}, \mathbb{G}_T)$, sk$^{\text{IPG}} := (\phi_t)_{t \in [N]}$).

Below, we give a concrete instantiation of THGen^{IPG} for a supersingular EC E_0 for $(\mathbb{G}_0, \hat{\mathbb{G}}_0, g_0, \hat{g}_0, e_0)$. Namely, generate a random supersingular EC which is ℓ^κ-isogenous to E_0 (given in [9,29]) for $\ell = 2$. For the details of the main subroutine of THGen^{IPG} (Algorithm Isog$^{\text{clg}}_{\ell,\kappa}$), see the full version of this paper.

Concrete instantiation of THGen^{IPG} by using supersingular elliptic curves

Input : An initial elliptic curve E_0.

Output : An isogenous E and all the selector bits $\omega := \{\omega_i\}_{0 \le i < \kappa}$, that is, a trapdoor ξ for computing the isogeny $\phi := \phi_\xi : E_0 \to E$.

for $0 \le i < \kappa$ **do**

generate a random bit $\omega_i \in \{0,1\}$ for selecting a next kernel point R_i, which is either of two points in $K_i := E_i[\ell] \setminus \psi_{i-1}(E_{i-1}[\ell])$ if $i \neq 0$ (resp., in $K_i := \{\text{some fixed two points in } E_i[\ell] \setminus \{\mathcal{O}_{E_i}\}\}$ if $i = 0$) since $\ell = 2$. R_i is determined from ω_i by a lexicographic order in $\mathbb{F}_{p^2}$.

compute $\psi_i : E_i \to E_{i+1} := E_i/\langle R_i \rangle$ for the selected R_i.

end for

we set the composition $\phi := \psi_{\kappa-1} \cdots \psi_0 : E_0 \to E_\kappa$.

return $E := E_\kappa$ (or j-inv. $j(E_\kappa)$) and all the selector bits $\xi := \omega := \{\omega_i\}_{0 \le i < \kappa}$.

After obtaining the above output $E := E_\kappa$ and ξ, $\mathsf{THGen}^{\mathsf{IPG}}$ outputs isogenous pairing group $(\mathbb{G}, \hat{\mathbb{G}}, g, \hat{g}, e, \phi)$, where $\mathbb{G} := \phi(\mathbb{G}_0), \hat{\mathbb{G}} := \phi(\hat{\mathbb{G}}_0), g := \phi(g_0), \hat{g} := \phi(\hat{g}_0)$ using $\phi := \phi_\xi : E_0 \to E$, and $e := e_{\mathsf{weil}}$ on E.

2.2   Assumptions on IPGs and Their Relationships

Definition 3 (N-Isog-DDH Assumption (on IPG)).[3] *Let $\mathcal{B}$ be a ppt adversary. For* $(\mathsf{pk}^{\mathsf{IPG}} := ((\mathbb{G}_t, \hat{\mathbb{G}}_t, g_t, \hat{g}_t, e_t)_{t\in[0,N]}, \mathbb{G}_T), \mathsf{sk}^{\mathsf{IPG}} := (\phi_t)_{t\in[N]}) \xleftarrow{\mathsf{R}}$ $\mathsf{Gen}^{\mathsf{IPG}}(1^\lambda, N)$ *and* $\alpha, \beta, \delta \xleftarrow{\mathsf{U}} \mathbb{F}_q$, $\mathcal{B}$ *receives* $\mathcal{X}_\mathfrak{b}$ *for* $\mathfrak{b} \xleftarrow{\mathsf{U}} \{0,1\}$, *that is defined by*

$$\mathcal{X}_0 := (\ \mathsf{pk}^{\mathsf{IPG}}, \ g_0^\alpha, \ (\hat{g}_t^\beta)_{t\in[N]}, \ g_T^{\alpha\beta}\ ) \text{ and } \mathcal{X}_1 := (\ \mathsf{pk}^{\mathsf{IPG}}, \ g_0^\alpha, \ (\hat{g}_t^\beta)_{t\in[N]}, \ g_T^\delta\ ).$$

$\mathcal{B}$ *outputs a guess bit* $\mathfrak{b}'$. *If* $\mathfrak{b} = \mathfrak{b}'$, $\mathcal{B}$ *wins. The advantage of adversary* $\mathcal{B}$ *is defined as* $\mathsf{Adv}_{\mathcal{B}}^{N\text{-}\mathsf{Isog\text{-}DDH}}(\lambda) := \Pr[\mathfrak{b}' = \mathfrak{b}] - 1/2$ *for any security parameter* λ. *The N-Isog-DDH assumption is: For any ppt adversary $\mathcal{B}$, the advantage of $\mathcal{B}$ for the N-Isog-DDH problem is negligible in λ.*

Our aim is to obtain HIBE and ABE secure under Isog-DDH and also *secure under DBDH*. For that, we define the notion of Isog-DBDH assumption, which is reduced from the Isog-DDH and also from the DBDH (Lemmas 1 and 2 and Fig. 1). First, we define (a form of) the standard DBDH assumption on the 0-th (asymmetric) pairing group $(\mathbb{G}_0, \hat{\mathbb{G}}_0)$ as follows. We then define N-Isog-DBDH assumption on the IPG $(\mathbb{G}_t, \hat{\mathbb{G}}_t)_{t\in[0,N]}$.

Definition 4 (DBDH Assumption (on $(\mathbb{G}_0, \hat{\mathbb{G}}_0)$)). *Let $\mathcal{B}$ be a ppt machine adversary. For* $(\mathsf{pk}^{\mathsf{IPG}} := ((\mathbb{G}_0, \hat{\mathbb{G}}_0, g_0, \hat{g}_0, e_0), \mathbb{G}_T), \mathsf{sk}^{\mathsf{IPG}} := \emptyset) \xleftarrow{\mathsf{R}} \mathsf{Gen}^{\mathsf{IPG}}(1^\lambda, 0)$ *and* $\alpha, \beta, \gamma, \delta \xleftarrow{\mathsf{U}} \mathbb{F}_q$, $\mathcal{B}$ *receives* $\mathcal{X}_\mathfrak{b}$ *for* $\mathfrak{b} \xleftarrow{\mathsf{U}} \{0,1\}$, *that is defined by*

$$\mathcal{X}_0 := (\ \mathsf{pk}^{\mathsf{IPG}}, \ g_0^\alpha, \ \hat{g}_0^\beta, \ g_0^\gamma, \hat{g}_0^\gamma, \ g_T^{\alpha\beta\gamma}\ ) \text{ and } \mathcal{X}_1 := (\ \mathsf{pk}^{\mathsf{IPG}}, \ g_0^\alpha, \ \hat{g}_0^\beta, \ g_0^\gamma, \hat{g}_0^\gamma, \ g_T^\delta\ ),$$

where $g_T := e_0(g_0, \hat{g}_0)$. $\mathcal{B}$ outputs a guess bit $\mathfrak{b}'$. If $\mathfrak{b} = \mathfrak{b}'$, $\mathcal{B}$ wins. The advantage of adversary $\mathcal{B}$ is defined as $\mathsf{Adv}_{\mathcal{B}}^{\mathsf{DBDH}}(\lambda) := \Pr[\mathfrak{b}' = \mathfrak{b}] - 1/2$ *for any security parameter λ. The DBDH assumption is defined as in Definition 3.*

Definition 5 (N-Isog-DBDH Assumption (on IPG)). *Let $\mathcal{B}$ be a ppt adversary. For* $(\mathsf{pk}^{\mathsf{IPG}} := ((\mathbb{G}_t, \hat{\mathbb{G}}_t, g_t, \hat{g}_t, e_t)_{t\in[0,N]}, \mathbb{G}_T), \mathsf{sk}^{\mathsf{IPG}} := (\phi_t)_{t\in[N]}) \xleftarrow{\mathsf{R}}$ $\mathsf{Gen}^{\mathsf{IPG}}(1^\lambda, N)$ *and* $\alpha, \beta, \gamma, \delta \xleftarrow{\mathsf{U}} \mathbb{F}_q$, $\mathcal{B}$ *receives* $\mathcal{X}_\mathfrak{b}$ *for* $\mathfrak{b} \xleftarrow{\mathsf{U}} \{0,1\}$, *that is defined by*

$$\mathcal{X}_0 := (\ \mathsf{pk}^{\mathsf{IPG}}, \ g_0^\alpha, \ (\hat{g}_t^\beta)_{t\in[N]}, \ (g_t^\gamma, \hat{g}_t^\gamma)_{t\in[0,N]}, \ g_T^{\alpha\beta\gamma}\ ) \text{ and}$$
$$\mathcal{X}_1 := (\ \mathsf{pk}^{\mathsf{IPG}}, \ g_0^\alpha, \ (\hat{g}_t^\beta)_{t\in[N]}, \ (g_t^\gamma, \hat{g}_t^\gamma)_{t\in[0,N]}, \ g_T^\delta\ ),$$

---

[3] The terminology in [21] and in this paper are slightly different. In [21], our (N-)Isog-DDH assumption is called (N-)Isog-DBDH assumption and our N-Isog-DBDH assumption is not defined.

where $g_T := e_0(g_0, \hat{g}_0)$. $\mathcal{B}$ outputs a guess bit $\mathfrak{b}'$. If $\mathfrak{b} = \mathfrak{b}'$, $\mathcal{B}$ wins. The advantage of adversary $\mathcal{B}$ is defined as $\mathsf{Adv}_{\mathcal{B}}^{N\text{-Isog-DBDH}}(\lambda) := \Pr[\mathfrak{b}' = \mathfrak{b}] - 1/2$ for any security parameter λ. The N-Isog-DBDH assumption is defined as in Definition 3.

Lemma 1. *The DBDH problem is reduced to the N-Isog-DBDH problem. (In other words, the hardness of the N-Isog-DBDH problem is reduced to the hardness of the DBDH problem.) For any adversary $\mathcal{B}$ against N-Isog-DBDH, there exists an adversary $\mathcal{C}$ for the DBDH problem, such that for any security parameter λ, $\mathsf{Adv}_{\mathcal{B}}^{N\text{-Isog-DBDH}}(\lambda) \leq \mathsf{Adv}_{\mathcal{C}}^{DBDH}(\lambda)$.*

Lemma 2. *The N-Isog-DDH problem is reduced to the N-Isog-DBDH problem. (In other words, the hardness of the N-Isog-DBDH problem is reduced to the hardness of the N-Isog-DDH problem.) For any adversary $\mathcal{B}$ against N-Isog-DBDH, there exists an adversary $\mathcal{C}$ for the N-Isog-DDH problem, such that for any security parameter λ, $\mathsf{Adv}_{\mathcal{B}}^{N\text{-Isog-DBDH}}(\lambda) \leq \mathsf{Adv}_{\mathcal{C}}^{N\text{-Isog-DDH}}(\lambda)$.*

Proofs of Lemmas 1 and 2 will be given in the full version of this paper.

2.3 On Validity of the N-Isog-DDH Assumption

All assumptions for pairing cryptography are reduced to the corresponding assumptions on the target group $\mathbb{G}_T$ by the MOV reduction [22], and the security is evaluated by validity of the reduced assumption in practice.

Our N-Isog-DDH assumption is also reduced to the DDH assumption on $\mathbb{G}_T \subset \mathbb{F}_{p^2}^*$ as follows: Given a random instance $\mathcal{X}_\mathfrak{b}$ for $\mathfrak{b} \xleftarrow{\mathsf{U}} \{0,1\}$ of the N-Isog-DDH problem, that is given by $\mathcal{X}_\mathfrak{b} := ($ $\mathsf{pk}^{\mathsf{IPG}}$, g_0^α, $(\hat{g}_t^\beta)_{t \in [N]}$, g_T^θ $)$, where $(\mathsf{pk}^{\mathsf{IPG}} := ((\mathbb{G}_t, \hat{\mathbb{G}}_t, g_t, \hat{g}_t, e_t)_{t \in [0,N]}, \mathbb{G}_T))$, $\theta = \alpha\beta$ if $\mathfrak{b} = 0$ and $\theta \xleftarrow{\mathsf{U}} \mathbb{F}_q$ if $\mathfrak{b} = 1$, a reduction algorithm computes $g_T^\alpha := e_0(g_0^\alpha, \hat{g}_0)$ and $g_T^\beta := e_1(g_1, \hat{g}_1^\beta)$. Then, $(\mathbb{G}_T, g_T, g_T^\alpha, g_T^\beta, g_T^\theta)$ is a random DDH instance on $\mathbb{G}_T$.

Since the discrete logarithm problem in the quadratic extension of a sufficiently large prime field, $\mathbb{F}_{p^2}^*$, is hard, our N-Isog-DDH assumption (for a large p) is considered as reasonable at present.

3 Proposed Basic IBE and KP-ABE

3.1 Proposed Basic IBE in the Standard Model

Key Ideas in Constructing the Proposed IBE Scheme. We first describe a simple version of our basic IBE, which is proven secure directly from Isog-DDH (as the simple BF-type IBE given in the full version). Public parameters include $2n$ IPGs $((\mathbb{G}_{j,\iota}, \hat{\mathbb{G}}_{j,\iota}, g_{j,\iota}, \hat{g}_{j,\iota}, e_{j,\iota})_{j \in [n]; \iota \in [0,1]}, \mathbb{G}_T)$ and $h_T := g_T^{s_0}$ where n is the bit length of ID and a random $s_0 \xleftarrow{\mathsf{U}} \mathbb{F}_q$ and s_0 is the master secret key.

Key generation and encryption algorithms consist of two steps, i.e., ID-group setup and group element encoding. In all previous pairing cryptosystems, these algorithms consist of only the latter step, group element encoding *on the same pairing group*. An ID-group for n-bit identity $\mathsf{ID} := (\mathsf{ID}_j)_{j \in [n]}$ is given by $\mathbb{G}_{\mathsf{ID}} := \mathbb{G}_{1,\mathsf{ID}_1} \times \cdots \times \mathbb{G}_{n,\mathsf{ID}_n}$ where components $\mathbb{G}_{j,\mathsf{ID}_j}$ are selected by n-bits $(\mathsf{ID}_j)_{j \in [n]}$ out of $2n$ IPG groups given in public parameters. The base element in the ID-group is naturally given as $g_{\mathsf{ID}} := (g_{1,\mathsf{ID}_1}, \ldots g_{n,\mathsf{ID}_n}) \in \mathbb{G}_{\mathsf{ID}}$. We note that the bases $(\mathbb{G}_{\mathsf{ID}}, g_{\mathsf{ID}})$ for element encoding are different for all ID. We then generate n-out-of-n shares $\vec{s} := (s_j)_{j \in [n]}$ such that $s_0 = \sum_{j \in [n]} s_j$ where s_0 is the master secret. The secret key is generated as $k_{\mathsf{ID}} := g_{\mathsf{ID}}^{\vec{s}} := (g_{1,\mathsf{ID}_1}^{s_1}, \ldots g_{n,\mathsf{ID}_n}^{s_n})$.

In a dual manner, in encryption, an ID-group for n-bit identity $\mathsf{ID} := (\mathsf{ID}_j)_{j \in [n]}$ is given by $\hat{\mathbb{G}}_{\mathsf{ID}} := \hat{\mathbb{G}}_{1,\mathsf{ID}_1} \times \cdots \times \hat{\mathbb{G}}_{n,\mathsf{ID}_n}$ and the base element is $\hat{g}_{\mathsf{ID}} := (\hat{g}_{1,\mathsf{ID}_1}, \ldots \hat{g}_{n,\mathsf{ID}_n}) \in \hat{\mathbb{G}}_{\mathsf{ID}}$. A ciphertext for message $m \in \mathbb{G}_T$ is given as (c_{ID}, c_T), which consists of $c_{\mathsf{ID}} := \hat{g}_{\mathsf{ID}}^{\zeta} := (\hat{g}_{1,\mathsf{ID}_1}^{\zeta}, \ldots \hat{g}_{n,\mathsf{ID}_n}^{\zeta}) \in \hat{\mathbb{G}}_{\mathsf{ID}}$ and $z := h_T^{\zeta}$, $c_T := z \cdot m \in \mathbb{G}_T$ with a random $\zeta \xleftarrow{\mathsf{U}} \mathbb{F}_q$.

Decryption has as input a secret key $k_{\mathsf{ID}} := g_{\mathsf{ID}}^{\vec{s}}$ and a ciphertext $c_{\mathsf{ID}'} := \hat{g}_{\mathsf{ID}'}^{\zeta}$. If $\mathsf{ID} = \mathsf{ID}'$, decryptor can use the natural direct product pairing e_{ID} on the ID-groups for ID, i.e.,

$$e_{\mathsf{ID}} : \mathbb{G}_{\mathsf{ID}} \times \hat{\mathbb{G}}_{\mathsf{ID}} \ni g_{\mathsf{ID}} \times \hat{g}_{\mathsf{ID}} \mapsto e_{1,\mathsf{ID}_1}(g_{1,\mathsf{ID}_1}, \hat{g}_{1,\mathsf{ID}_1}) \cdots e_{n,\mathsf{ID}_n}(g_{n,\mathsf{ID}_n}, \hat{g}_{n,\mathsf{ID}_n}) \in \mathbb{G}_T,$$

where $g_{\mathsf{ID}} := (g_{1,\mathsf{ID}_1}, \ldots, g_{n,\mathsf{ID}_n})$ and $\hat{g}_{\mathsf{ID}} := (\hat{g}_{1,\mathsf{ID}_1}, \ldots, \hat{g}_{n,\mathsf{ID}_n})$. The decryptor calculates $z' := e_{\mathsf{ID}}(k_{\mathsf{ID}}, c_{\mathsf{ID}}) = e_{\mathsf{ID}}(g_{\mathsf{ID}}^{\vec{s}}, \hat{g}_{\mathsf{ID}}^{\zeta}) = \prod_{j \in [n]} e_{j,\mathsf{ID}_j}(g_{j,\mathsf{ID}_j}, \hat{g}_{j,\mathsf{ID}_j})^{\zeta s_j} = g_T^{\zeta \sum_{j \in [n]} s_j} = (g_T^{s_0})^{\zeta} = h_T^{\zeta}$ and then decryption succeeds as $m := c_T \cdot (z')^{-1}$ since $z' = z$.

For proving standard model (selective-ID) security from n-Isog-DDH (not n-Isog-DBDH) for the above simple IBE, simulator should execute *ID-group partitioning* in setup phase. Let me explain it below. For the obtained n-Isog-DDH instance $\mathcal{X}_{\mathfrak{b}} := (\mathsf{pk}^{\mathsf{IPG}}, g_0^{\alpha}, (\hat{g}_j^{\beta})_{j \in [n]}, g_T^{\delta})$ with $\delta = \alpha\beta$ if $\mathfrak{b} = 0$ and $\delta \xleftarrow{\mathsf{U}} \mathbb{F}_q$ if $\mathfrak{b} = 1$, the simulator should embed $g_0^{\alpha} \in \mathbb{G}_0$ into user secret keys where implicitly master secret s_0 is set as $s_0 := \alpha$, and $(\hat{g}_j^{\beta} \in \hat{\mathbb{G}}_j)_{j \in [n]}$ and $g_T^{\delta} \in \mathbb{G}_T$ into the challenge ciphertext where implicitly challenge ciphertext randomness ζ is set as $\zeta := \beta$. Here, we note that the fact that we cannot take pairing of group elements g_0^{α} and $(\hat{g}_j^{\beta})_{j \in [n]}$ implies that simulated keys cannot decrypt the challenge ciphertext.

In setup simulation, for ID length n, we prepare $2n$ isogenous pairing groups $(\mathbb{G}_{j,\iota}, \hat{\mathbb{G}}_{j,\iota})_{\iota \in [0,1]}^{j \in [n]}$. At the start of simulation, challenger obtains the challenge identity $\mathsf{ID}^* := (\mathsf{ID}_j^*)_{j \in [n]}$ and n-Isog-DDH instance $\mathcal{X}_{\mathfrak{b}} := (\mathsf{pk}^{\mathsf{IPG}}, g_0^{\alpha}, (\hat{g}_j^{\beta})_{j \in [n]}, g_T^{\delta})$ with $\mathsf{pk}^{\mathsf{IPG}} := ((\mathbb{G}_j, \hat{\mathbb{G}}_j, g_j, \hat{g}_j, e_j)_{j \in [0,n]}, \mathbb{G}_T)$. Let the product groups $\mathbb{G}_{\mathcal{X}}$ and $\hat{\mathbb{G}}_{\mathcal{X}}$ (n-Isog-DDH instance groups) be $\mathbb{G}_{\mathcal{X}} := \mathbb{G}_1 \times \cdots \times \mathbb{G}_n$ and $\hat{\mathbb{G}}_{\mathcal{X}} := \hat{\mathbb{G}}_1 \times \cdots \times \hat{\mathbb{G}}_n$, whose components are given in the n-Isog-DDH instance. The simulator executes ID-group simulation as $\mathbb{G}_{\mathsf{ID}^*} := \mathbb{G}_\chi$ and $\hat{\mathbb{G}}_{\mathsf{ID}^*} := \hat{\mathbb{G}}_\chi$, i.e., $\mathbb{G}_{j,\mathsf{ID}_j^*} := \mathbb{G}_j$ and $\hat{\mathbb{G}}_{j,\mathsf{ID}_j^*} := \hat{\mathbb{G}}_j$ for $j \in [n]$ where $\mathbb{G}_j, \hat{\mathbb{G}}_j$ are obtained from the n-Isog-DDH instance. If we set $\hat{g}_{\mathsf{ID}^*} := (\hat{g}_j)_{j \in [n]}$ in $\hat{\mathbb{G}}_{\mathsf{ID}^*} = \hat{\mathbb{G}}_\chi$, the simulator uses $c_{\mathsf{ID}^*} := \hat{g}_{\mathsf{ID}^*}^\beta := (\hat{g}_j^\beta)_{j \in [n]}$ and $z := g_T^\delta$ in the challenge ciphertext where implicitly ciphertext randomness is set as $\zeta := \beta$. I.e., in public parameters,

if $\iota = \mathsf{ID}_j^*$, the simulator sets $(\mathbb{G}_{j,\iota}, \hat{\mathbb{G}}_{j,\iota}, g_{j,\iota}, \hat{g}_{j,\iota}, e_{j,\iota}) := (\mathbb{G}_j, \hat{\mathbb{G}}_j, g_j, \hat{g}_j, e_j)$

which are obtained from the n-Isog-DBDH instance in order to use

all of $(\hat{g}_j^\beta)_{j \in [n]}$ in challenge ciphertext simulation,

and the simulator generates n new isogenous pairing groups $(\mathbb{G}_j', \hat{\mathbb{G}}_j', g_j', \hat{g}_j', e_j', \phi_j')$ $\overset{\mathsf{R}}{\leftarrow} \mathsf{THGen}^{\mathsf{IPG}}(\mathbb{G}_0, \hat{\mathbb{G}}_0, g_0, \hat{g}_0, e_0)$ for $j \in [n]$, and the n pairing groups are included in public parameters for the indices that $\iota \neq \mathsf{ID}_j^*$ for the challenge $\mathsf{ID}^* := (\mathsf{ID}_j^*)_{j \in [n]}$, i.e.,

if $\iota \neq \mathsf{ID}_j^*$, the simulator sets $(\mathbb{G}_{j,\iota}, \hat{\mathbb{G}}_{j,\iota}, g_{j,\iota}, \hat{g}_{j,\iota}, e_{j,\iota}) := (\mathbb{G}_j', \hat{\mathbb{G}}_j', g_j', \hat{g}_j', e_j')$

in order to use one of $(\phi_j'(g_0^\alpha))_{j \in [n]}$ in key generation simulation.

In key generation, a queried $\mathsf{ID} := (\mathsf{ID}_j)_{j \in [n]}$ is not equal to the challenge ID^*, i.e., $\mathsf{ID}_{j_0} \neq \mathsf{ID}_{j_0}^*$ for some $j_0 \in [n]$. Then simulator obtain $g_{j_0,\mathsf{ID}_{j_0}}^\alpha = \phi_{j_0}'(g_0^\alpha)$ since $\mathbb{G}_{j_0,\mathsf{ID}_{j_0}} := \mathbb{G}_{j_0}'$ is generated from $\mathbb{G}_0$ by simulator. The simulator generates n-shares of zero as $(s_j')_{j \in [n]} \overset{\mathsf{U}}{\leftarrow} \{(s_j') \in \mathbb{F}_q^n \mid \sum_{j \in [n]} s_j' = 0\}$, and then simulates the key $k_{\mathsf{ID}} = (k_j)_{j \in [n]}$ as $k_{j_0} := \phi_{j_0}'(g_0^\alpha) \cdot g_{j_0,\mathsf{ID}_{j_0}}^{s_{j_0}'} = g_{j_0,\mathsf{ID}_{j_0}}^{\alpha+s_{j_0}'}$ and $k_j := g_{j,\mathsf{ID}_j}^{s_j'}$ if $j \neq j_0$. Since (s_j') is uniformly distributed in $\{(s_j') \in \mathbb{F}_q^n \mid \sum_{j \in [n]} s_j' = 0\}$, the exponent system $(\alpha + s_{j_0}', (s_j')_{j \neq j_0})$ is uniformly distributed in $\{(s_j) \in \mathbb{F}_q^n \mid \sum_{j \in [n]} s_j = \alpha\}$ as in real generated keys. This completes the overview of construction idea of simple version of basic IBE secure from the n-Isog-DDH assumption.

For obtaining full version of basic IBE secure from the n-Isog-DBDH assumption, we combine the above simple version and (a special form of) Goyal et al.'s KP-ABE construction. In particular, public parameters include bases $\hat{h}_{j,\iota} := \hat{g}_{j,\iota}^{\tau_{j,\iota}}$, $h_{j,\iota} := g_{j,\iota}^{\frac{1}{\tau_{j,\iota}}}$ with $\tau_{j,\iota} \overset{\mathsf{U}}{\leftarrow} \mathbb{F}_q$ for all $j \in [n], \iota = 0, 1$. The secret key and ciphertext elements are encoded on the ID-groups in a similar manner as above with using bases $h_{j,\iota}$ and $\hat{h}_{j,\iota}$, respectively. Therefore, we have the security proof from the above proof and that for Goyal et al.'s KP-ABE. For details, see Theorem 1 and its proof.

Construction.

$$\mathsf{Setup}(1^\lambda, n) : \Big(\mathsf{pk}^{\mathsf{IPG}} := ((\mathbb{G}_0, \hat{\mathbb{G}}_0, g_0, \hat{g}_0, e_0), (\mathbb{G}_{j,\iota}, \hat{\mathbb{G}}_{j,\iota}, g_{j,\iota}, \hat{g}_{j,\iota}, e_{j,\iota})_{\iota \in [0,1]}^{j \in [n]}, \mathbb{G}_T),$$

$$\mathsf{sk}^{\mathsf{IPG}} := (\phi_{j,\iota})_{\iota \in [0,1]}^{j \in [n]}\Big) \xleftarrow{\mathsf{R}} \mathsf{Gen}^{\mathsf{IPG}}(1^\lambda, 2n),$$

$$s_0 \xleftarrow{\mathsf{U}} \mathbb{F}_q, \ h_T := g_T^{s_0}, \ \tau_{j,\iota} \xleftarrow{\mathsf{U}} \mathbb{F}_q, \ \hat{h}_{j,\iota} := \hat{g}_{j,\iota}^{\tau_{j,\iota}}, \ h_{j,\iota} := g_{j,\iota}^{\frac{1}{\tau_{j,\iota}}} \ \text{for all } j, \iota,$$

$$\text{return } \mathsf{pk} := (((\mathbb{G}_{j,\iota}, \hat{\mathbb{G}}_{j,\iota}, \hat{h}_{j,\iota}, e_{j,\iota})_{\iota \in [0,1]}^{j \in [n]}, \mathbb{G}_T, h_T), \ \mathsf{sk} := (s_0, (h_{j,\iota})).$$

$\mathsf{KeyGen}(\mathsf{pk}, \mathsf{sk}, \mathsf{ID} := (\mathsf{ID}_j)) :$

/ ∗ ID-group setup ∗ / $\mathbb{G}_{\mathsf{ID}} := \mathbb{G}_{1,\mathsf{ID}_1} \times \cdots \times \mathbb{G}_{n,\mathsf{ID}_n}, \ h_{\mathsf{ID}} := (h_{j,\mathsf{ID}_j}) \in \mathbb{G}_{\mathsf{ID}},$

/ ∗ Group element encoding ∗ /   choose random $\vec{s} := (s_j)_{j \in [n]} \in \mathbb{F}_q^n$

such that $s_0 = \sum_{j=1}^n s_j, \ k_{\mathsf{ID}} := h_{\mathsf{ID}}^{\vec{s}}, \ \text{return } \mathsf{sk}_{\mathsf{ID}} := k_{\mathsf{ID}}.$

$\mathsf{Enc}(\mathsf{pk}, m, \mathsf{ID} := (\mathsf{ID}_j)) :$

/ ∗ ID-group setup ∗ / $\hat{\mathbb{G}}_{\mathsf{ID}} := \hat{\mathbb{G}}_{1,\mathsf{ID}_1} \times \cdots \times \hat{\mathbb{G}}_{n,\mathsf{ID}_n}, \ \hat{h}_{\mathsf{ID}} := (\hat{h}_{j,\mathsf{ID}_j}) \in \hat{\mathbb{G}}_{\mathsf{ID}},$

/ ∗ Group element encoding ∗ / $\zeta \xleftarrow{\mathsf{U}} \mathbb{F}_q, \ c_{\mathsf{ID}} := \hat{h}_{\mathsf{ID}}^\zeta, \ z := h_T^\zeta,$

$c_T := z \cdot m, \ \text{return } \mathsf{ct}_{\mathsf{ID}} := (c_{\mathsf{ID}}, c_T).$

$\mathsf{Dec}(\mathsf{pk}, \mathsf{sk}_{\mathsf{ID}} := k_{\mathsf{ID}}, \mathsf{ct}_{\mathsf{ID}'} := (c_{\mathsf{ID}'}, c_T)) :$

if $\mathsf{ID} = \mathsf{ID}', \ z' := e_{\mathsf{ID}}(k_{\mathsf{ID}}, c_{\mathsf{ID}}), \ \text{return } m' := c/z', \ \text{otherwise, return } \perp.$

[Correctness]: If $\mathsf{ID} = \mathsf{ID}', \ z' = e_{\mathsf{ID}}(k_{\mathsf{ID}}, c_{\mathsf{ID}}) = \prod_{j \in [n]} e_{j,\mathsf{ID}_j}(h_{j,\mathsf{ID}_j}^{s_j}, \hat{h}_{j,\mathsf{ID}_j}^\zeta)$

$$= \prod_{j \in [n]} e_{j,\mathsf{ID}_j}(g_{j,\mathsf{ID}_j}^{\frac{s_j}{\tau_{j,\mathsf{ID}_j}}}, \hat{g}_{j,\mathsf{ID}_j}^{\zeta \cdot \tau_{j,\mathsf{ID}_j}}) = \prod_{j \in [n]} g_T^{s_j \cdot \zeta} = g_T^{\sum_{j \in [n]} s_j \cdot \zeta} = g_T^{s_0 \zeta} = h_T^\zeta = z$$

Security. The definition of selective-ID security is standard and will be given in the full version of this paper.

Theorem 1. *The proposed IBE scheme is selective-ID secure under the n-Isog-DBDH assumption in the standard model.*

For any adversary $\mathcal{A}$, there exists an adversary $\mathcal{B}$ for the n-Isog-DBDH problem, such that for any security parameter λ, $\mathsf{Adv}_{\mathcal{A}}^{\mathsf{ibe,sel}}(\lambda) \leq \mathsf{Adv}_{\mathcal{B}}^{n\text{-}\mathsf{Isog\text{-}DBDH}}(\lambda)$.

A proof of Theorem 1 will be given in the full version of this paper.

3.2   Large Universe KP-ABE

The proposed large universe KP-ABE scheme is constructed from the basic IBE in Sect. 3.1 and small universe KP-ABE, which will be given in the full version.

Construction. An attribute $x_t := (x_{t,j})_{j \in [n]}$ for any sub-universe id $t \in [d]$ is an element in $\mathcal{U} := \{0,1\}^n$, and our construction has a hierarchical structure for $t \in [d]$ and $j \in [n]$ with two instantiations of the small universe ABE. In the low level instantiation, i.e., basic IBE, a special form of n-out-of-$2n$ secret sharing predicate is used for identity-matching for the n-bit identities x_t. The IPG with $2dn + 1$ pairing groups is used.

$\mathsf{Setup}(1^\lambda, d, n):$

$$\Big(\mathsf{pk}^{\mathsf{IPG}} := ((\mathbb{G}_0, \hat{\mathbb{G}}_0, g_0, \hat{g}_0, e_0), (\mathbb{G}_{t,j,\iota}, \hat{\mathbb{G}}_{t,j,\iota}, g_{t,j,\iota}, \hat{g}_{t,j,\iota}, e_{t,j,\iota})_{\iota\in[0,1]}^{t\in[d],j\in[n]}, \mathbb{G}_T),$$

$$\mathsf{sk}^{\mathsf{IPG}} := (\phi_{t,j,\iota})_{\iota\in[0,1]}^{t\in[d],j\in[n]}\Big) \xleftarrow{\mathsf{R}} \mathsf{Gen}^{\mathsf{IPG}}(1^\lambda, 2dn),$$

$s_0 \xleftarrow{\mathsf{U}} \mathbb{F}_q,\ h_T := g_T^{s_0},\ \tau_{t,j,\iota} \xleftarrow{\mathsf{U}} \mathbb{F}_q,\ \hat{h}_{t,j,\iota} := \hat{g}_{t,j,\iota}^{\tau_{t,j,\iota}},\ h_{t,j,\iota} := g_{t,j,\iota}^{\frac{1}{\tau_{t,j,\iota}}}$ for all t, j, ι,

return $\mathsf{pk} := ((\mathbb{G}_{t,j,\iota}, \hat{\mathbb{G}}_{t,j,\iota}, \hat{h}_{t,j,\iota}, e_{t,j,\iota})_{\iota\in[0,1]}^{t\in[d],j\in[n]}, \mathbb{G}_T, h_T),\ \mathsf{sk} := (s_0, (h_{t,j,\iota})).$

$\mathsf{KeyGen}(\mathsf{pk}, \mathsf{sk}, \mathbb{S} := (M, \rho)):$ choose random $\vec{u} \in \mathbb{F}_q^r$ such that $\vec{1} \cdot \vec{u} = s_0$,

for $i \in [l]$, $s_i := M_i \cdot \vec{u}$, choose random $\vec{s}_i := (s_{i,j})_{j\in[n]} \in \mathbb{F}_q^n$

such that $s_i = \sum_{j=1}^n s_{i,j}$,

if $\rho(i) = (t, v_i := (v_{i,j}) \in \{0,1\}^n)$, $\mathbb{G}_{t,v_i} := \mathbb{G}_{t,1,v_{i,1}} \times \cdots \times \mathbb{G}_{t,n,v_{i,n}}$,

$h_{t,v_i} := (h_{t,j,v_{i,j}})_{j\in[n]},\ k_i := h_{t,v_i}^{\vec{s}_i} \in \mathbb{G}_{t,v_i},\ /* \mathbb{G}_{t,v_i}$ is the t-th base group $*/$,

return $\mathsf{sk}_{\mathbb{S}} := \{k_i \in \mathbb{G}_{t,v_i}\}_{i\in[l]}.$

$\mathsf{Enc}(\mathsf{pk}, m, \Gamma): \zeta \xleftarrow{\mathsf{U}} \mathbb{F}_q,$

for $(t, x_t := (x_{t,j}) \in \{0,1\}^n) \in \Gamma$, $\mathbb{G}_{t,x_t} := \mathbb{G}_{t,1,x_{t,1}} \times \cdots \times \mathbb{G}_{t,n,x_{t,n}}$,

$\hat{h}_{t,x_t} := (\hat{h}_{t,j,x_{t,j}})_{j\in[n]},\ c_t := \hat{h}_{t,x_t}^\zeta \in \mathbb{G}_{t,x_t},\ /* \mathbb{G}_{t,x_t}$ is the t-th base group $*/$,

$z := h_T^\zeta,\ c_T := z \cdot m,$ return $\mathsf{ct}_\Gamma := (\{c_t \in \mathbb{G}_{t,x_t}\}_{(t,\cdot)\in\Gamma}, c_T).$

$\mathsf{Dec}(\mathsf{pk}, \mathsf{sk}_{\mathbb{S}} := \{k_i\}_{i\in[l]}, \mathsf{ct}_\Gamma := (\{c_t\}_{(t,\cdot)\in\Gamma}, c_T)):$

if $\mathbb{S} := (M, \rho)$ accepts $\Gamma := \{(t, x_t)\}$, then compute $\{\sigma_i\}_{\rho(i)\in\Gamma}$

such that $\vec{1} = \sum_{\rho(i)\in\Gamma} \sigma_i M_i$, where M_i is the i-th row of M,

$z' := \prod_{\rho(i)=(t,v_i)\in\Gamma} e_{t,v_i}(k_i, c_t)^{\sigma_i}$, return $m' := c/z'$. otherwise, return $\bot$.

[Correctness]: If $\mathbb{S}$ accepts Γ, $z' = \prod_{\rho(i)=(t,v_i)\in\Gamma} e_{t,v_i}(k_i, c_t)^{\sigma_i}$

$= \prod_{\rho(i)=(t,v_i)\in\Gamma} e_{t,v_i}(h_i^{\sigma_i \vec{s}_i}, \hat{h}_t^\zeta) = g_T^{\zeta \sum_{\rho(i)=(t,v_i)\in\Gamma} \sigma_i s_i} = (g_T^{s_0})^\zeta = h_T^\zeta = z.$

Security. The definition of selective-attribute security is standard and will be given in the full version of this paper.

Theorem 2. *The proposed large universe KP-ABE scheme is selective-attribute secure under the dn-Isog-DBDH assumption in the standard model.*

For any adversary $\mathcal{A}$, there exists an adversary $\mathcal{B}$ for the dn-Isog-DBDH problem, such that for any security parameter λ, $\mathsf{Adv}_{\mathcal{A}}^{\mathsf{kp\text{-}abe,sel}}(\lambda) \leq \mathsf{Adv}_{\mathcal{B}}^{\mathsf{dn\text{-}Isog\text{-}DBDH}}(\lambda).$

A proof of Theorem 2 will be given in the full version of this paper.

References

1. Agrawal, S., Boneh, D., Boyen, X.: Efficient lattice (H)IBE in the standard model. In: Gilbert, H. (ed.) EUROCRYPT 2010. LNCS, vol. 6110, pp. 553–572. Springer, Heidelberg (2010). https://doi.org/10.1007/978-3-642-13190-5_28
2. Attrapadung, N., Hanaoka, G., Yamada, S.: Conversions among several classes of predicate encryption and applications to ABE with various compactness tradeoffs. In: Iwata, T., Cheon, J.H. (eds.) ASIACRYPT 2015, Part I. LNCS, vol. 9452, pp. 575–601. Springer, Heidelberg (2015). https://doi.org/10.1007/978-3-662-48797-6_24
3. Benson, K., Shacham, H., Waters, B.: The k-BDH assumption family: bilinear map cryptography from progressively weaker assumptions. In: Dawson, E. (ed.) CT-RSA 2013. LNCS, vol. 7779, pp. 310–325. Springer, Heidelberg (2013). https://doi.org/10.1007/978-3-642-36095-4_20
4. Boneh, D., Boyen, X.: Efficient selective-ID secure identity-based encryption without random oracles. In: Cachin, C., Camenisch, J.L. (eds.) EUROCRYPT 2004. LNCS, vol. 3027, pp. 223–238. Springer, Heidelberg (2004). https://doi.org/10.1007/978-3-540-24676-3_14
5. Boneh, D., Franklin, M.: Identity-based encryption from the weil pairing. In: Kilian, J. (ed.) CRYPTO 2001. LNCS, vol. 2139, pp. 213–229. Springer, Heidelberg (2001). https://doi.org/10.1007/3-540-44647-8_13
6. Boneh, D., et al.: Fully key-homomorphic encryption, arithmetic circuit ABE and compact garbled circuits. In: Nguyen, P.Q., Oswald, E. (eds.) EUROCRYPT 2014. LNCS, vol. 8441, pp. 533–556. Springer, Heidelberg (2014). https://doi.org/10.1007/978-3-642-55220-5_30
7. Boneh, D., Gentry, C., Hamburg, M.: Space-efficient identity based encryption without pairings. In: FOCS 2007, pp. 647–657 (2007)
8. Boneh, D., et al.: Multiparty non-interactive key exchange and more from isogenies on elliptic curves. In: MATHCRYPT 2018 (2018). https://eprint.iacr.org/2018/665
9. Charles, D., Lauter, K., Goren, E.: Cryptographic hash functions from expander graphs. J. Crypt. **22**(1), 93–113 (2009). Preliminary version: IACR Cryptology eprint Archiv, 2006:021 (2006)
10. Chen, J., Gay, R., Wee, H.: Improved dual system ABE in prime-order groups via predicate encodings. In: Oswald, E., Fischlin, M. (eds.) EUROCRYPT 2015, Part II. LNCS, vol. 9057, pp. 595–624. Springer, Heidelberg (2015). https://doi.org/10.1007/978-3-662-46803-6_20
11. Cheon, J.H.: Security analysis of the strong Diffie-Hellman problem. In: Vaudenay, S. (ed.) EUROCRYPT 2006. LNCS, vol. 4004, pp. 1–11. Springer, Heidelberg (2006). https://doi.org/10.1007/11761679_1
12. Cocks, C.: An identity based encryption scheme based on quadratic residues. In: Honary, B. (ed.) Cryptography and Coding 2001. LNCS, vol. 2260, pp. 360–363. Springer, Heidelberg (2001). https://doi.org/10.1007/3-540-45325-3_32
13. Cramer, R., Shoup, V.: Design and analysis of practical public-key encryption schemes secure against adaptive chosen ciphertext attack. SIAM J. Comput. **33**(1), 167–226 (2003)
14. De Feo, L., Jao, D., Plût, J.: Towards quantum-resistant cryptosystems from supersingular elliptic curve isogenies. J. Math. Crypt. **8**(3), 209–247 (2014). Preliminary version: IACR Cryptology eprint Archiv, 2011:506 (2011)
15. Dodis, Y., Katz, J.: Chosen-ciphertext security of multiple encryption. In: Kilian, J. (ed.) TCC 2005. LNCS, vol. 3378, pp. 188–209. Springer, Heidelberg (2005). https://doi.org/10.1007/978-3-540-30576-7_11

16. Gentry, C., Peikert, C., Vaikuntanathan, V.: Trapdoors for hard lattices and new cryptographic constructions. In: STOC 2008, pp. 197–206 (2008)
17. Gorbunov, S., Vaikuntanathan, V., Wee, H.: Attribute-based encryption for circuits. In: STOC 2013, pp. 545–554 (2013)
18. Goyal, V., Pandey, O., Sahai, A., Waters, B.: Attribute-based encryption for fine-grained access control of encrypted data. In: ACM CCS 2006, pp. 89–98 (2006)
19. Herzberg, A.: Folklore, practice and theory of robust combiners. J. Comput. Secur. **17**(2), 159–189 (2009)
20. ISO/IEC 18033–5:2015: Information technology - Security techniques - Encryption algorithms - Part 5: Identity-based ciphers. ISO/IEC (2015)
21. Koshiba, T., Takashima, K.: Pairing cryptography meets isogeny: a new framework of isogenous pairing groups. IACR Cryptology ePrint Archive 2016:1138 (2016)
22. Menezes, A., Okamoto, T., Vanstone, S.: Reducing elliptic curve logarithms to logarithms in a finite field. IEEE Trans. Inf. Theory **39**(5), 1639–1646 (1993). Preliminary version appeared in STOC 1991
23. National Institute of Standards and Technology: Post-Quantum crypto standardization: Call for Proposals Announcement, December 2016. http://csrc.nist.gov/groups/ST/post-quantum-crypto/cfp-announce-dec2016.html
24. Okamoto, T., Takashima, K.: Fully secure functional encryption with general relations from the decisional linear assumption. In: Rabin, T. (ed.) CRYPTO 2010. LNCS, vol. 6223, pp. 191–208. Springer, Heidelberg (2010). https://doi.org/10.1007/978-3-642-14623-7_11. http://eprint.iacr.org/2010/563
25. Shacham, H.: A cramer-shoup encryption scheme from the linear assumption and from progressively weaker linear variants. IACR Cryptology ePrint Archive 2007:74 (2007). http://eprint.iacr.org/2007/074
26. Shoup, V.: Using hash functions as a hedge against chosen ciphertext attack. In: Preneel, B. (ed.) EUROCRYPT 2000. LNCS, vol. 1807, pp. 275–288. Springer, Heidelberg (2000). https://doi.org/10.1007/3-540-45539-6_19
27. Silverman, J.: The Arithmetic of Elliptic Curves. GTM, vol. 106, 2nd edn. Springer, New York (2009). https://doi.org/10.1007/978-0-387-09494-6
28. Waters, B.: Dual system encryption: realizing fully secure IBE and HIBE under simple assumptions. In: Halevi, S. (ed.) CRYPTO 2009. LNCS, vol. 5677, pp. 619–636. Springer, Heidelberg (2009). https://doi.org/10.1007/978-3-642-03356-8_36
29. Yoshida, R., Takashima, K.: Computing a sequence of 2-isogenies on supersingular elliptic curves. IEICE Trans. Fundam. **96-A**(1), 158–165 (2013). Preliminary version is available in ICISC 2008. LNCS, vol. 5461, pp. 52–65 (2008)

Public-Key Encryption and Implementation

Mitigating the One-Use Restriction in Attribute-Based Encryption

Lucas Kowalczyk$^{(\boxtimes)}$, Jiahui Liu, Tal Malkin, and Kailash Meiyappan

Columbia University, New York, USA
{luke,tal}@cs.columbia.edu,
{jl4161,kkm2142}@columbia.edu

Abstract. We present a key-policy attribute-based encryption scheme that is adaptively secure under a static assumption and is not directly affected by an attribute "one-use restriction". Our construction improves upon the only other such scheme (Takashima '17) by mitigating its downside of a ciphertext size that is dependent on the maximum size of any supported attribute set.

1 Introduction

Attribute-based encryption (ABE) is a type of public key encryption which allows for fine-grained access control to encrypted data. In Key-Policy ABE, ciphertexts are associated with attributes, and secret-keys are associated with Boolean access policies that take in a set of attributes and return True if the key is capable of decrypting ciphertexts associated with that set and return False otherwise. Security guarantees that (potentially colluding) users without an authorized key should not be able to learn anything about an encrypted message. (A dual variant called Ciphertext-Policy ABE swaps the roles of attributes and access policies to be associated with the secret keys and ciphertexts respectively).

One way to make security proofs for ABE more attainable is to consider restricted notions of security. For KP-ABE, the notion of *selective security* requires the adversary to commit to a target set of attributes for the challenge ciphertext that will be attacked at the start of the security game. The earliest constructions of ABE using bilinear groups were proven secure in this model [17,34]. The notion of *semi-adaptive security* [18] requires the adversary to commit to a target set of attributes, but allows the adversary to see the public parameters first. These notions are obviously not realistic attack scenarios, so a KP-ABE scheme would ideally satisfy the notion of *adaptive security* (full security), where the challenge attribute set can be chosen adaptively (in response to public parameters and any amount of secret keys received). The first construction of ABE achieving adaptive security appeared in [21], employing the dual system encryption methodology [33] in its security reduction.

Another way to make proving security of ABE schemes easier is to reduce security to parameterized assumptions like *q-type assumptions*, where the size of

© Springer Nature Switzerland AG 2019
K. Lee (Ed.): ICISC 2018, LNCS 11396, pp. 23–36, 2019.
https://doi.org/10.1007/978-3-030-12146-4_2

the elements included in the assumption's challenge grows with some property of the adversary. q-type assumptions were used in the ABE constructions of [24,34] to prove security. However, the security of dynamic assumptions like q-type assumptions is not well-understood, and the assumptions are often closely related to the scheme in which they are used. For example, the assumption may include a number of group elements that scales with the number of queries made by the adversary in the security proof. Further, it is known that many q-type assumptions become stronger as q grows [11], so we would ideally like to reduce security of ABE constructions to better understood assumptions of a static size, like the Decisional Linear Assumption (DLIN) or the Symmetric External Diffie-Hellman Assumption (SXDH).

A natural class of access policies one would like to be able to support in an ABE construction is that of general Boolean formulas. Unfortunately, it has proven tremendously difficult to construct efficient ABE for general Boolean formulas with adaptive security under static assumptions. All constructions except for [32] suffer from a "one-use restriction". That is, they only natively support read-once Boolean formulas, or formulas where attributes are used at most once in inputs. One way to extend such constructions to support formulas that use attributes more than once (say, k times) is to use k copies of new "meta-attributes" that stand for each use of the original attribute, and are handled as a group [21]. The downside of this approach is that it destroys the compactness of the construction – for KP-ABE, the size of the ciphertexts no longer depends on just the attribute set of the ciphertext, but also on the complexity of the formulas that the scheme supports (namely, the ciphertexts grow linearly with the maximum number of attribute uses in any formula supported). Ciphertexts associated with n' attributes in a scheme like [21] where policies can reuse attributes at most k times are of size $O(n' \cdot k)$.

Takashima presented the first KP-ABE scheme (proven adaptively secure from static assumptions) with ciphertexts that do not grow directly with the number of attribute uses [32], but unfortunately, the construction still has a dependence on the set of allowed policies. Specifically, ciphertexts are of size $O(n + r)$, where n is the maximum size of any supported attribute set and r is the maximum number of columns in any policy matrix supported (this is the policy dependency). For (fan-in 2) Boolean formulas, standard techniques [22] to translate the formula into a policy matrix result in r being equal to the number of AND gates in the formula. Additionally, this dependence on n, the maximum size of any supported attribute set rather than only the attribute set of the relevant ciphetext is undesirable, since one can imagine the size of each ciphertext's associated attribute set varying wildly from the worst-case maximum-sized set supported by the system. In fact, it is unclear whether $O(n + r)$-size ciphertexts are ever an asymptotic improvement over the $O(n' \cdot k)$-size ciphertexts of all other known ABE schemes proved adaptively secure under static assumptions.

reference	$|sk|$	$|ct|$	assumption
[21]	$O(|f|)$	$O(n' \cdot k)$	DLIN
[28]	$O(|f|)$	$O(n' \cdot k)$	DLIN
[9]	$O(|f|)$	$O(n' \cdot k)$	k-LIN
[10]	$O(|f|)$	$O(n' \cdot k)$	SXDH
[32]	$O(|f|)$	$O(n + r)$	DLIN
Ours	$O(|f|)$	$O(n' + r)$	SXDH

Fig. 1. Summary of several KP-ABE schemes proven adaptively secure under static assumptions for monotone span programs. Here, n' is the number of attributes associated to the ciphertext, n is the maximum size of any supported attribute set, r is the maximum number of columns in any policy matrix, and k is the maximum number of attribute reuses in any policy (except in the name for the "k-LIN" assumption, which is unrelated and an unfortunate overloading).

1.1   Our Result

In this work, we describe a KP-ABE construction that mitigates one of the two undesirable dependencies of [32], featuring ciphertexts of size $O(n' + r)$ instead of $O(n+r)$ (while remaining adaptively secure from a static assumption: the Symmetric Diffie-Hellman Assumption (SXDH) and allowing the reuse of attributes in its monotone span program policies). This significant improvement allows us to rigorously argue that there exist classes of access policies for which our construction enjoys an asymptotic improvement over the state of the art. We note that our construction is for the small-universe setting, where attributes come from a polynomial (in the security parameter) sized universe that is fixed upon setup, whereas the construction of [32] supports an attribute universe that may be exponentially large. This allows us to focus on the techniques required to asymptotically improve the ciphertext size. Our scheme is likely translatable to accommodate a large attribute universe without sacrificing asymptotic efficiency, but we leave this for future work.

Our construction avoids a dependence on k, the multiplicity of attribute-reuse in supported policies, but retains the dependence on r, the number of columns in supported policy matrices. We view reducing this last dependence to achieve truly compact adaptively secure ABE from a static assumption as an interesting open problem.

1.2   Comparing Perfomance

Figure 1 contains a comparison of several KP-ABE schemes proven adaptively secure under static assumptions for monotone span programs.

An obvious question in comparing our construction to the state-of-the-art is: how does r compare to k? Is $n' + r$ ever better than $n' \cdot k$? It is easy to come up with individual formulas where this is the case, but it's not obvious that such a formula can't always be "compressed" to an equivalent formula that has less attribute-reuse. In general, circuit/formula minimization questions like this are difficult to answer.

Fortunately, we can make a simple counting argument to show that indeed there are classes of functions which cannot be expressed using Boolean formulas with much smaller maximum attribute reuse than the maximum number of AND gates within the class. To see this, consider some subset of x attributes in the attribute universe. There are 2^{2^x} Boolean functions on these attributes, and we can express each function as a DNF in the naive way as a formula that uses at most $O(2^x)$ AND gates. So, for this class of functions, $r = O(2^x)$.

However, counting the number of different Boolean formulas that could attempt to realize these functions using a maximum k reuses of any attribute shows that at least $k = \Omega(2^x)$ attribute-reuses are required to realize all of the functions in this class. In this case, we see our construction enjoys a multiplicative to additive improvement (from $n' \cdot \Omega(2^x)$ to $n' + O(2^x)$).

1.3   Technical Details

Our construction can be seen as combining the best of both worlds between the construction of [32], which is the first to not directly depend on the number of attribute-reuses (while adaptively secure from a static assumption), and the lineage of [10,17,19,21], which enjoys ciphertexts that are independent of the size of the attribute universe (they depend only on the number of attributes actually associated with the ciphertext).

Specifically, all of these schemes are based on linear secret sharing and are built using bilinear groups. Given a matrix M representing a monotone span program, linear secret shares of α are constructed by choosing randomness r_i, then computing $M \cdot (\alpha, r_2, ..., r_m)$ to obtain a vector of shares λ. The constructions of [10,17,19,21] embed these shares into their constructions' secret keys, where they are hidden by attribute-randomness that can only be removed using corresponding elements from a ciphertext. See Fig. 2 for an example. A crucial step of the dual-system proof [33] of adaptive security occurs when secret shares in the dual "semi-functional" space of a secret key are changed from sharing 0 to sharing a random element α' (in [24], this is the change from "nominal semi-functional" to "temporary semi-functional"). This is the step of the proof that uses the fact that the keys requested by an adversary are not allowed to decrypt the challenge ciphertext, to argue that there exists different randomness r_i' where a sharing of 0 using the r_i randomness looks identically distributed to a sharing of random α' using the r_i' randomness, as long as the only shares seen are not allowed to reconstruct the secret. Crucially, the alternative randomness r_i' is not defined until the challenge ciphertext is requested (as the challenge ciphertext defines which shares in the key are allowed to be seen). The constructions in the [21] lineage therefore require that the change in the secret shares in their keys be information theoretic (so they can be implicitly changed upon challenge ciphertext creation). This turns out to be the root of the one-time attribute use restriction (reusing attributes prevents this information-theoretic argument from working).

$$\{g^{\lambda_j + a_{\rho(j)} y_j}, \quad g^{y_j}\}_{j \in M}$$

Fig. 2. Example secret key

[32] employs a technique of delayed share construction to get around this problem. Specifically, the construction does not construct a secret sharing $\{\lambda_j\}$ which is embedded in the secret key, but instead keeps the components that generate λ_j (vectors M_j and $(\alpha, r_2, ..., r_m)$) separate *until decryption*. The M_j portion is embedded in the key and the randomness $(\alpha, r_2, ..., r_m)$ is stored *in the ciphertext*. Decryption computes the dot product of these two components to implicitly construct λ_j that function in the same way as before. The advantage of this approach is that the randomness used in the secret shares is not needed until the challenge ciphertext is requested, so computational assumptions can be used to side-step the one-time attribute use restriction that comes with information-theoretic changes.

Like [21,32] also masks secret key components, making them only available to ciphertexts associated with the appropriate attributes. However [32] does this via a somewhat blunt tool: namely, its secret keys contain a vector y which can encode orthogonality relationships with any subset of the attributes associated with a ciphertext and whose length is as large as the maximum attribute set supported by the system.

In contrast, the "share encapsulation" in [21] demonstrated in Fig. 2 can be thought of as using a vector of dimension 2 to perform the same job. $(a_{\rho(j)} y_j, y_j)$ is being used to hide the share λ_j and share retrieval will be allowed only give a ciphertext with an "orthogonal" vector: $(s, -sa_{\rho(j)})$. Our construction can be seen as essentially replacing [32]'s vector y with constant-dimensioned vectors like this, resulting in a ciphertext dependent only on the number of attributes associated with it, just like all previous schemes. Essentially, an information-theoretic "encapsulation" argument supported by the vector y for all shares is replaced with a computational one using a vectors of a constant size for each attribute. Doing so makes the dual-system hybrid more delicate, as it requires careful management of rerandomization across the now greatly reduced dimensions.

1.4   Related Work

Additional work on ABE in the bilinear setting includes various constructions of KP-ABE and CP-ABE schemes (e.g. [5,16,18,30]), schemes supporting multiple authorities (e.g. [7,8,22,29]), and schemes supporting large attribute universes (e.g. [1–3,6,10,15,19,23,28,31]).

The construction of [14] supports circuit access policies rather than monotone span programs or Boolean formulas, which makes it more expressive than any known bilinear scheme. It was proven selectively secure under the standard LWE assumption. The construction of [6] later extended this to semi-adaptive security for circuit access policies from LWE. Proving full adaptive security for a ABE scheme supporting circuits from LWE or an assumption on bilinear maps is an interesting open problem.

Circuit policies are supported by the construction in [12] based on multilinear maps. This scheme is proven selectively secure, under a particular computational hardness assumption for multilinear groups. The multilinear scheme in [13] achieves adaptive security, relying on computational hardness assumptions in multilinear groups.

2   Preliminaries

We will write $a \leftarrow \mathbb{Z}_p$ to denote choosing a uniformly at random from set $\mathbb{Z}_p$ and will abuse notation to use $j \in M$ as a subscript to denote each index j of the rows M_j of matrix M.

2.1   Prime Order Bilinear Groups

We construct our system in prime order asymmetric bilinear groups. We let $\mathcal{G}$ denote a group generator - an algorithm which takes a security parameter λ as input and outputs (p, G, H, G_T, e), where p is a prime, G, H and G_T are cyclic groups of order p, and $e : G \times H \rightarrow G_T$ is a map with the following properties:

1. (Bilinear) $\forall g \in G, h \in H, a, b \in \mathbb{Z}_p, e(g^a, h^b) = e(g, h)^{ab}$
2. (Non-degenerate) $\exists g \in G, h \in H$ such that $e(g, h)$ has order p in G_T.

We refer to G, H as the *source groups* and G_T as the *target group*. We assume that the group operations in G, H and G_T and the map e are computable in polynomial time with respect to λ, and the group descriptions of G, H and G_T include a generator of each group.

2.2   Dual Pairing Vector Spaces

We will employ the concept of dual pairing vector spaces from [26, 27], where we'll denote choosing random dual orthogonal bases as: $(\mathbb{B}, \mathbb{B}^*) \leftarrow Dual(\mathbb{Z}_p^n)$. Such bases are collections of linearly independent vectors chosen at random up to orthogonality constraints ($\boldsymbol{b}_i \cdot \boldsymbol{b}_i^* = 1, \boldsymbol{b}_i \cdot \boldsymbol{b}_j^* = 0$ for $i \neq j$). For example, one can implement $Dual(\mathbb{Z}_p^n$ by choosing a random invertible matrix B, setting $\mathbb{B} := B$ which then defines $\mathbb{B}^*$ as $\mathbb{B}^* := (B^{-1})^T$. Note that the dual basis generation procedure satisfies the property that, if R is an invertible matrix, then $(\mathbb{B}, \mathbb{B}^*)$ and $(R \cdot \mathbb{B}, (R^{-1})^T \cdot \mathbb{B}^*)$ are distributed identically when $(\mathbb{B}, \mathbb{B}^*) \leftarrow Dual(\mathbb{Z}_p^n)$. We will use this fact in our security proof to introduce new randomness into free dimensions of the construction as well as to embed computational assumptions. Finally, we will write $g^{\boldsymbol{v}}$ to denote the vector of group elements $(g^{v_1}, ..., g^{v_n})$, and will use the notation: $(x_1, ..., x_n)_{\mathbb{B}}$ to denote $g^{x_1 \boldsymbol{b}_1} \cdot ... \cdot g^{x_n \boldsymbol{b}_n}$.

2.3   Complexity Assumptions

The security of our system will be reduced to the Symmetric External Diffie-Hellman assumption (SXDH). We use the notation $x \leftarrow S$ to express that element x is chosen uniformly at random from the finite set S.

Symmetric External Diffie-Hellman Assumption (SXDH). The SXDH problem in G is stated as follows: given an asymmetric bilinear group (G, H) of prime order p with respective generators g, h, and given g^a, g^b and $T = g^{ab+r^*} \in G$ where $a, b \leftarrow \mathbb{Z}_p$ and either $r^* = 0$ or $r \leftarrow \mathbb{Z}_p$, output "yes" if r is a random element of $\mathbb{Z}_p$ and "no" otherwise. The SXDH problem in H is stated symmetrically, swapping the role of G and H.

Definition 1. *SXDH Assumption in (G, H): no polynomial time algorithm can achieve non-negligible advantage in deciding the SXDH problem in G or the SXDH problem in H.*

2.4   Background for ABE

We now give required background material on Linear Secret Sharing Schemes, the formal definition of a KP-ABE scheme, and the security definition we will use.

Monotone Span Programs/Linear Secret Sharing Schemes. Our construction uses linear secret-sharing schemes (LSSS) to realize monotone span program access structures [25]. We use the following definition (adapted from [4]). In the context of ABE, attributes will play the role of parties and will be represented as indexes $i \in [|\mathcal{U}|]$ for a fixed universe $\mathcal{U}$.

Definition 2. *(Linear Secret-Sharing Schemes (LSSS)) A secret sharing scheme Π over a set of attributes is called* linear *(over $\mathbb{Z}_p$) if*

1. *The shares belonging to all attributes form a vector over $\mathbb{Z}_p$.*
2. *There exists an $\ell \times n$ matrix Λ called the share-generating matrix for Π. The matrix Λ has ℓ rows and n columns. For all $j = 1, \ldots, \ell$, the jth row of Λ is labeled by an attribute $i = \rho(j)$ (ρ is a mapping that maintains the relationship between matrix rows and attributes). When we consider the column vector $v = (s, r_2, \ldots, r_n)$, where $s \in \mathbb{Z}_p$ is the secret to be shared and $r_2, \ldots, r_n \in \mathbb{Z}_p$ are randomly chosen, then Λv is the vector of ℓ shares of the secret s according to Π. The share $(\Lambda v)_j = \lambda_j$ belongs to attribute $i = \rho(j)$.*

We note the *linear reconstruction* property: we suppose that Π is an LSSS. We let S denote an authorized set. Then there is a subset $S^* \subseteq S$ such that the vector $(1, 0, \ldots, 0)$ is in the span of rows of Λ indexed by S^*, and there exist constants $\{\omega_i \in \mathbb{Z}_p\}_{i \in S^*}$ such that, for any valid shares $\{\lambda_i\}$ of a secret s according to Π, we have: $\displaystyle\sum_{i \in S^*} \omega_i \lambda_i = s$. These constants $\{\omega_i\}$ can be found in time polynomial in the size of the share-generating matrix Λ [4]. For unauthorized sets, no such S^*, $\{\omega_i\}$ exist.

For any set S of unauthorized shares, since the vector $(1, 0, ..., 0)$ is not in the span of rows indexed by S, then there is some vector w that is orthogonal to all of the rows of Λ indexed by S but is not orthogonal to $(1, 0, ..., 0)$. By scaling this vector, we can maintain these orthogonality relationships and force the first coordinate w_1 to be 1. Our proof of security will use the existence of this vector.

KP-ABE Definition. A key-policy attribute-based encryption system consists of four algorithms: Setup, Encrypt, KeyGen, and Decrypt.

$Setup(\lambda, \mathcal{U}) \rightarrow (PP, MSK)$. The setup algorithm takes in the security parameter λ and the attribute universe description $\mathcal{U}$. It outputs the public parameters PP and a master secret key MSK.

$Encrypt(PP, m, S) \rightarrow CT$. The encryption algorithm takes in the public parameters PP, the message m, and a set of attributes S. It will output a ciphertext CT. We assume that S is implicitly included in CT.

$KeyGen(MSK, PP, \mathbb{A}) \rightarrow SK$. The key generation algorithm takes in the master secret key MSK, the public parameters PP, and an access structure $\mathbb{A}$ over the universe of attributes. It outputs a private key SK which can be used to decrypt ciphertexts encrypted under a set of attributes which satisfies $\mathbb{A}$. We assume that $\mathbb{A}$ is implicitly included in SK.

$Decrypt(PP, CT, SK) \rightarrow m$. The decryption algorithm takes in the public parameters PP, a ciphertext CT encrypted under a set of attributes S, and a private key SK for an access structure $\mathbb{A}$. If the set of attributes of the ciphertext satisfies the access structure of the private key, it outputs the message m.

Adaptive Security for KP-ABE Systems. We define adaptive security for KP-ABE Systems in terms of the following game:

Setup. The challenger runs the Setup algorithm and gives the public parameters to the attacker.

Phase 1. The attacker queries the challenger for private keys corresponding to access structures.

Challenge. The attacker declares two equal length messages M_0, M_1 and a set of attributes $A \subseteq \mathcal{U}$ where $\mathcal{U}$ is the attribute universe such that A does not satisfy the access structure of any of the keys requested in Phase 1. The challenger flips a random coin $\beta \in \{0, 1\}$, encrypts M_β under S to yield ciphertext CT_β and gives CT_β to the attacker.

Phase 2. The attacker queries the challenger for private keys corresponding to access structures that are not satisfied by S.

Guess. The attacker outputs a guess β'.

Definition 3. *The advantage of an attacker $\mathcal{A}$ in this game is defined as* $Adv_{\mathcal{A}}^{KP-ABE}(\lambda) = \Pr[\beta = \beta'] - \frac{1}{2}$.

Definition 4. *A key-policy attribute based encryption scheme is adaptively secure if no polynomial time algorithm can achieve a non-negligible advantage in the above security game.*

3   Construction

Setup$(\lambda, \mathcal{U}) \rightarrow PP, MSK$. The setup algorithm chooses an asymmetric bilinear group $\mathcal{G}(\lambda) \rightarrow (p, G, H, G_T, e)$. It then chooses random generators $g \in G, h \in H$. For $i \in [k]$ where $k = |\mathcal{U}|$ it chooses values $a_i \leftarrow \mathbb{Z}_p$. It then generates random dual orthonormal sets:

$$(\mathbb{D}, \mathbb{D}^*) \leftarrow Dual(\mathbb{Z}_p^6)$$
$$(\mathbb{B}, \mathbb{B}^*) \leftarrow Dual(\mathbb{Z}_p^{3(r+1)})$$
$$(\mathbb{A}_i, \mathbb{A}_i^*) \leftarrow Dual(\mathbb{Z}_p^3) \text{ for } i \in [k]$$

The public parameters PP are:

$$e(g, h)$$
$$(e_1)_{\mathbb{D}^*}, (e_2)_{\mathbb{D}^*}$$
$$\{(e_i)_{\mathbb{B}^*}\}_{i \in [r+1]}$$
$$\{(a_i, 0, 0)_{\mathbb{A}_i^*}\}_{i \in [k]}$$

The MSK is:

$$(e_1)_{\mathbb{D}}, (e_2)_{\mathbb{D}}$$
$$\{(e_i)_{\mathbb{B}}\}_{i \in r+1}$$
$$\{(1, 0, 0)_{\mathbb{A}_i}\}_{i \in [k]}$$

Such a construction is equipped to create keys for access policies which include attributes $i \in \mathcal{U}$.

Encrypt$(m, S, PP) \rightarrow CT$. The encryption algorithm draws $\alpha, \Delta, s, z_i \leftarrow \mathbb{Z}_p$ (for $i \in [r]$) and forms the ciphertext as:

$$CT_S = (C_0, C_1, C_2, \{C_{3,i}\}_{i \in S})$$

where

$$C_0 := m \cdot e(g, h)^\alpha$$
$$C_1 := (\alpha, -\Delta, 0^2, 0^2)_{\mathbb{D}^*}$$
$$C_2 := (\Delta, z_2, ..., z_r, s, 0^{r+1}, 0^{r+1})_{\mathbb{B}^*}$$
$$C_{3,i} := (sa_i, 0, 0)_{\mathbb{A}_i^*}$$

(This implicitly includes S)

KeyGen$(MSK, M, PP) \rightarrow SK$. The key generation algorithm takes in the public parameters, master secret key, and LSSS access matrix M. It chooses a random exponent $x \leftarrow \mathbb{Z}_p$. For each row j (associated with attribute $\rho(j)$) in the policy matrix M, it chooses exponent $y_j \leftarrow \mathbb{Z}_p$ and outputs the secret key:

$$SK_M = (K_1, \{K_{2,j}, K_{3,j}\}_{j \in M})$$

where:

$$K_1 := (1, x, \mathbf{0}^2, \mathbf{0}^2)_{\mathbb{D}}$$
$$K_{2,j} := (\text{———} x M_j \text{———}, a_{\rho(j)} y_j, \mathbf{0}^{r+1}, \mathbf{0}^{r+1})_{\mathbb{B}}$$
$$K_{3,j} := (-y_j, 0, 0)_{\mathbb{A}_{\rho(j)}}$$

Decrypt$(CT_S, SK_M, PP) \rightarrow m$. Given ciphertext $CT_S = (C_0, C_1, C_2, \{C_{3,i}\}_{i \in S})$ and secret key $SK_M = (K_1, \{K_{2,j}, K_{3,j}\}_{j \in M})$, if S satisfies M, then there is a set S^* of policy row indices such that $j \in S^* \implies \rho(j) \in S$ and there exist efficiently computable constants ω_j such that $\sum_{j \in S^*} \omega_j M_j \cdot \mathbf{z} = \Delta$ (recall Sect. 2.4).
The decryption algorithm computes these ω_j and then computes:

$$B = \prod_{j \in S^*} e(C_2, K_{2,j})^{\omega_j} \cdot e(C_{3,\rho(j)}, K_{3,j})^{\omega_j}$$

$$D = e(C_1, K_1)$$

and finally, computes and outputs:

$$\frac{C_0}{B \cdot D} = m$$

4   Correctness

This scheme satisfies correctness since:

$$B = \prod_{j \in S^*} e(C_2, K_{2,j})^{\omega_j} \cdot e(C_{3,\rho(j)}, K_{3,j})^{\omega_j}$$

$$= \prod_{j \in S^*} e \left(\begin{matrix} (\Delta, z_2, ..., z_r, & s, & \mathbf{0}^{r+1}, \mathbf{0}^{r+1})_{\mathbb{B}^*}, \\ (\text{———} x M_j \text{———}, & a_{\rho(j)} y_j, \mathbf{0}^{r+1}, \mathbf{0}^{r+1})_{\mathbb{B}} \end{matrix} \right)^{\omega_j} \cdot e \left(\begin{matrix} (sa_{\rho(j)}, 0, 0)_{\mathbb{A}_{\rho(j)}^*}, \\ (-y_j, 0, 0)_{\mathbb{A}_{\rho(j)}} \end{matrix} \right)^{\omega_j}$$

$$= \prod_{j \in S^*} e(g, h)^{x \omega_j \lambda_j + s \omega_j a_{\rho(j)} y_j} \cdot e(g, h)^{-s \omega_j a_{\rho(j)} y_j}$$

$$= e(g, h)^{x \sum_{j \in S^*} \omega_j \lambda_j}$$

$$= e(g, h)^{x \Delta}$$

$$D = e(C_1, K_1)$$

$$= e\left(\begin{array}{c} (\alpha, -\Delta, \mathbf{0}^2, \mathbf{0}^2)_{\mathbb{D}^*} \\ (1, \quad x, \ \mathbf{0}^2, \mathbf{0}^2)_{\mathbb{D}} \end{array}\right)$$

$$= e(g, h)^{\alpha - x\Delta}$$

and finally:

$$\frac{C_0}{B \cdot D} = \frac{m \cdot e(g, h)^{\alpha}}{e(g, h)^{x\Delta} \cdot e(g, h)^{\alpha - x\Delta}}$$

$$= m$$

5   Proof of Security

Our proof of security will consist of a hybrid sequence of games where the keys and challenge ciphertext are constructed according to various types. At a high level, the proof follows a typical dual-system hybrid structure, where the challenge ciphertext is first made "semi-functional," then the hybrid continues over the secret keys requested, transforming each key into a "semifunctional" variant which is useless to the attacker relative to the challenge (semifunctional) ciphertext.

There are two parts to the key hybrid: one that makes semifunctional keys which were requested before the challenge ciphertext and another that makes semifunctional keys which were requested after the challenge ciphertext. The high level reason for this difference is that for keys requested after the challenge ciphertext, the challenge attribute set is already known. This makes it easy to follow a standard selective security argument to make each key semifunctional. The harder part of the hybrid deals with making keys requested before the challenge ciphertext (and the challenge attribute set) is known. This is where we use the delayed randomness contained within our ciphertext as well as the fact that we allow the semifunctional ciphertext distributions to depend on the current key of the hybrid. This bifurcated approach to handling secret keys in a dual-system proof was first employed in [24] and later refined by [2,3].

A key step in our proof (and of [32]) is a lemma where each policy matrix row is isolated in turn against the ciphertext's w alternative randomness component and their dot product's distribution is used to argue that the row can be multiplied by an uncorrelated x^*. In [32], this argument takes advantage of the inefficient y vector, but for us, we need to delicately thread just enough randomness through the single attribute element a_i hiding each row to accomplish the same feat.

Theorem 1. *Under the SXDH assumption, our KP-ABE construction is adaptively secure against any polynomial time adversary $\mathcal{A}$.*

We give the proof of Theorem 1 in the full version of this paper [20].

34   L. Kowalczyk et al.

Acknowledgements. This work was supported in part by The Leona M. & Harry B. Helmsley Charitable Trust; NSF grant CCF-1423306; and the Defense Advanced Research Project Agency (DARPA) and Army Research Office (ARO) under Contract W911NF-15-C-0236. The first author is additionally supported in part by an NSF Graduate Research Fellowship DGE-16-44869. Any opinions, findings and conclusions or recommendations expressed are those of the authors and do not necessarily reflect the views of the the Defense Advanced Research Projects Agency, Army Research Office, the National Science Foundation, or the U.S. Government.

References

1. Agrawal, S., Chase, M.: FAME: fast attribute-based message encryption. In: CCS (2017)
2. Attrapadung, N.: Dual system encryption via doubly selective security: framework, fully secure functional encryption for regular languages, and more. In: Nguyen, P.Q., Oswald, E. (eds.) EUROCRYPT 2014. LNCS, vol. 8441, pp. 557–577. Springer, Heidelberg (2014). https://doi.org/10.1007/978-3-642-55220-5_31
3. Attrapadung, N.: Dual system encryption framework in prime-order groups via computational pair encodings. In: Cheon, J.H., Takagi, T. (eds.) ASIACRYPT 2016. LNCS, vol. 10032, pp. 591–623. Springer, Heidelberg (2016). https://doi.org/10.1007/978-3-662-53890-6_20
4. Beimel, A.: Secure schemes for secret sharing and key distribution. Ph.D. thesis, Israel Institute of Technology, Technion, Haifa, Israel (1996)
5. Bethencourt, J., Sahai, A., Waters, B.: Ciphertext-policy attribute-based encryption. In: Proceedings of the IEEE Symposium on Security and Privacy, pp. 321–334 (2007)
6. Brakerski, Z., Vaikuntanathan, V.: Circuit-ABE from LWE: unbounded attributes and semi-adaptive security. In: Robshaw, M., Katz, J. (eds.) CRYPTO 2016. LNCS, vol. 9816, pp. 363–384. Springer, Heidelberg (2016). https://doi.org/10.1007/978-3-662-53015-3_13
7. Chase, M.: Multi-authority attribute based encryption. In: Vadhan, S.P. (ed.) TCC 2007. LNCS, vol. 4392, pp. 515–534. Springer, Heidelberg (2007). https://doi.org/10.1007/978-3-540-70936-7_28
8. Chase, M., Chow, S.S.M.: Improving privacy and security in multi-authority attribute-based encryption. In: Proceedings of the 2009 ACM Conference on Computer and Communications Security, pp. 121–130 (2009)
9. Chen, J., Gay, R., Wee, H.: Improved dual system ABE in prime-order groups via predicate encodings. In: Oswald, E., Fischlin, M. (eds.) EUROCRYPT 2015. LNCS, vol. 9057, pp. 595–624. Springer, Heidelberg (2015). https://doi.org/10.1007/978-3-662-46803-6_20
10. Chen, J., Gong, J., Kowalczyk, L., Wee, H.: Unbounded ABE via bilinear entropy expansion, revisited. In: Nielsen, J.B., Rijmen, V. (eds.) EUROCRYPT 2018. LNCS, vol. 10820, pp. 503–534. Springer, Cham (2018). https://doi.org/10.1007/978-3-319-78381-9_19
11. Cheon, J.H.: Security analysis of the strong Diffie-Hellman problem. In: Vaudenay, S. (ed.) EUROCRYPT 2006. LNCS, vol. 4004, pp. 1–11. Springer, Heidelberg (2006). https://doi.org/10.1007/11761679_1

12. Garg, S., Gentry, C., Halevi, S., Sahai, A., Waters, B.: Attribute-based encryption for circuits from multilinear maps. In: Canetti, R., Garay, J.A. (eds.) CRYPTO 2013. LNCS, vol. 8043, pp. 479–499. Springer, Heidelberg (2013). https://doi.org/10.1007/978-3-642-40084-1_27

13. Garg, S., Gentry, C., Halevi, S., Zhandry, M.: Fully secure attribute based encryption from multilinear maps. IACR Cryptology ePrint Archive 2014, 622 (2014). http://eprint.iacr.org/2014/622

14. Gorbunov, S., Vaikuntanathan, V., Wee, H.: Attribute-based encryption for circuits. In: STOC, pp. 545–554 (2013)

15. Goyal, R., Koppula, V., Waters, B.: Semi-adaptive security and bundling functionalities made generic and easy. In: Hirt, M., Smith, A. (eds.) TCC 2016. LNCS, vol. 9986, pp. 361–388. Springer, Heidelberg (2016b). https://doi.org/10.1007/978-3-662-53644-5_14

16. Goyal, V., Jain, A., Pandey, O., Sahai, A.: Bounded ciphertext policy attribute based encryption. In: Aceto, L., Damgård, I., Goldberg, L.A., Halldórsson, M.M., Ingólfsdóttir, A., Walukiewicz, I. (eds.) ICALP 2008. LNCS, vol. 5126, pp. 579–591. Springer, Heidelberg (2008). https://doi.org/10.1007/978-3-540-70583-3_47

17. Goyal, V., Pandey, O., Sahai, A., Waters, B.: Attribute based encryption for fine-grained access control of encrypted data. In: ACM Conference on Computer and Communications Security, pp. 89–98 (2006)

18. Chen, J., Wee, H.: Semi-adaptive attribute-based encryption and improved delegation for boolean formula. In: Abdalla, M., De Prisco, R. (eds.) SCN 2014. LNCS, vol. 8642, pp. 277–297. Springer, Cham (2014). https://doi.org/10.1007/978-3-319-10879-7_16

19. Kowalczyk, L., Lewko, A.B.: Bilinear entropy expansion from the decisional linear assumption. In: Gennaro, R., Robshaw, M. (eds.) CRYPTO 2015. LNCS, vol. 9216, pp. 524–541. Springer, Heidelberg (2015). https://doi.org/10.1007/978-3-662-48000-7_26

20. Kowalczyk, L., Liu, J., Malkin, T., Meiyappan, K.: Mitigating the one-use restriction in attribute-based encryption. IACR Cryptology ePrint Archive 2018, 645 (2018). https://eprint.iacr.org/2018/645.pdf

21. Lewko, A., Okamoto, T., Sahai, A., Takashima, K., Waters, B.: Fully secure functional encryption: attribute-based encryption and (hierarchical) inner product encryption. In: Gilbert, H. (ed.) EUROCRYPT 2010. LNCS, vol. 6110, pp. 62–91. Springer, Heidelberg (2010). https://doi.org/10.1007/978-3-642-13190-5_4

22. Lewko, A., Waters, B.: Decentralizing attribute-based encryption. In: Paterson, K.G. (ed.) EUROCRYPT 2011. LNCS, vol. 6632, pp. 568–588. Springer, Heidelberg (2011). https://doi.org/10.1007/978-3-642-20465-4_31

23. Lewko, A., Waters, B.: Unbounded HIBE and attribute-based encryption. In: Paterson, K.G. (ed.) EUROCRYPT 2011. LNCS, vol. 6632, pp. 547–567. Springer, Heidelberg (2011). https://doi.org/10.1007/978-3-642-20465-4_30

24. Lewko, A., Waters, B.: New proof methods for attribute-based encryption: achieving full security through selective techniques. In: Safavi-Naini, R., Canetti, R. (eds.) CRYPTO 2012. LNCS, vol. 7417, pp. 180–198. Springer, Heidelberg (2012). https://doi.org/10.1007/978-3-642-32009-5_12

25. Karchmer, M., Wigderson, A.: On span programs. In: CCC, pp. 102–111 (1993)

26. Okamoto, T., Takashima, K.: Homomorphic encryption and signatures from vector decomposition. In: Galbraith, S.D., Paterson, K.G. (eds.) Pairing 2008. LNCS, vol. 5209, pp. 57–74. Springer, Heidelberg (2008). https://doi.org/10.1007/978-3-540-85538-5_4

27. Okamoto, T., Takashima, K.: Hierarchical predicate encryption for inner-products. In: Matsui, M. (ed.) ASIACRYPT 2009. LNCS, vol. 5912, pp. 214–231. Springer, Heidelberg (2009). https://doi.org/10.1007/978-3-642-10366-7_13

28. Okamoto, T., Takashima, K.: Fully secure unbounded inner-product and attribute-based encryption. In: Wang, X., Sako, K. (eds.) ASIACRYPT 2012. LNCS, vol. 7658, pp. 349–366. Springer, Heidelberg (2012). https://doi.org/10.1007/978-3-642-34961-4_22

29. Okamoto, T., Takashima, K.: Decentralized attribute-based signatures. In: Kurosawa, K., Hanaoka, G. (eds.) PKC 2013. LNCS, vol. 7778, pp. 125–142. Springer, Heidelberg (2013). https://doi.org/10.1007/978-3-642-36362-7_9

30. Ostrovksy, R., Sahai, A., Waters, B.: Attribute based encryption with non-monotonic access structures. In: ACM Conference on Computer and Communications Security, pp. 195–203 (2007)

31. Rouselakis, Y., Waters, B.: Practical constructions and new proof methods for large universe attribute-based encryption. In: 2013 ACM Conference on Computer and Communications Security, pp. 463–474 (2013)

32. Takashima, K.: New proof techniques for DLIN-based adaptively secure attribute-based encryption. In: Pieprzyk, J., Suriadi, S. (eds.) ACISP 2017. LNCS, vol. 10342, pp. 85–105. Springer, Cham (2017). https://doi.org/10.1007/978-3-319-60055-0_5

33. Waters, B.: Dual system encryption: realizing fully secure IBE and HIBE under simple assumptions. In: Halevi, S. (ed.) CRYPTO 2009. LNCS, vol. 5677, pp. 619–636. Springer, Heidelberg (2009). https://doi.org/10.1007/978-3-642-03356-8_36

34. Waters, B.: Ciphertext-policy attribute-based encryption: an expressive, efficient, and provably secure realization. In: Catalano, D., Fazio, N., Gennaro, R., Nicolosi, A. (eds.) PKC 2011. LNCS, vol. 6571, pp. 53–70. Springer, Heidelberg (2011). https://doi.org/10.1007/978-3-642-19379-8_4

Attacking Noisy Secret CRT-RSA
Exponents in Binary Method

Kento Oonishi$^{(\boxtimes)}$ and Noboru Kunihiro$^{(\boxtimes)}$

The University of Tokyo, Tokyo, Japan
kento_oonishi@mist.i.u-tokyo.ac.jp, kunihiro@k.u-tokyo.ac.jp

Abstract. In this paper, we perform a security evaluation on the RSA encryption scheme with the Chinese remainder theorem (CRT-RSA), against side-channel attacks. We discuss the methods for recovering the CRT-RSA secret keys by observing physical information. In the CRT-RSA scheme, we calculate the exponentiations by repeated squaring and multiplication operations during decryption. The square-and-multiply sequences of the exponentiation can be obtained by side-channel attacks. However, errors occur in the square and multiply sequences because of physical-information observation errors, due to which the secret keys cannot be recovered by using Bernstein et al.'s method, even if window size $w = 1$ in sliding window exponentiation. In this paper, we propose an algorithm for correcting the errors in the square-and-multiply sequences, and for obtaining the correct secret keys, when the square-and-multiply sequences are generated at $w = 1$, namely, the binary method. Moreover, we theoretically prove that the expected time complexity of our algorithm is in polynomial time, when the error rate is less than 5.8%.

Keywords: CRT-RSA encryption scheme · Exponentiation ·
Error correction · Side-channel attacks

1 Introduction

1.1 Background

RSA encryption scheme [25] is an extensively used public-key cryptosystem. It is composed of public keys (N, e), and a secret key d. The public key e, is a sufficiently small parameter, such as $2^{16} + 1 = 65537$, which is used in many systems; whereas, N is the product of two distinct $n/2$ bit primes, p and q. The value-pair $(e, d) \in \mathbb{Z}^*_{(p-1)(q-1)} \times \mathbb{Z}^*_{(p-1)(q-1)}$, satisfies $ed \equiv 1 \mod (p-1)(q-1)$. In this paper, we consider the PKCS#1 standard [26], RSA with the Chinese remainder theorem (CRT-RSA). In the CRT-RSA scheme, the public keys are the same as those of the standard RSA scheme, while secret keys are (p, q, d, d_p, d_q, q_p). Parameters $d_p \in \mathbb{Z}^*_{p-1}, d_q \in \mathbb{Z}^*_{q-1}$, and $q_p \in \mathbb{Z}_p$ are defined as $d_p := d \mod p - 1, d_q := d \mod q - 1$, and $q_p := q^{-1} \mod p$, respectively. These additional secret keys enable faster decryption using the CRT.

© Springer Nature Switzerland AG 2019
K. Lee (Ed.): ICISC 2018, LNCS 11396, pp. 37–54, 2019.
https://doi.org/10.1007/978-3-030-12146-4_3

The security of the RSA scheme is based on the difficulty of the factorization problem. However, even if the factorization problem is difficult, the implemented RSA is not always secure against physical attacks, such as cold boot attacks [7] and side-channel attacks [14]. As the secret keys can be leaked by physical attacks, the RSA scheme must be secure against such attacks.

In previous research, security analysis of the RSA scheme against physical attacks, such as cold boot and side-channel attacks, were conducted. In cold boot attacks, attackers observe the DRAM data remanence and read the secret-key bits. In side-channel attacks, they observe physical data based on secret keys, such as the power consumption [11,15], implementation time [14], electromagnetic [5], sound [6], and cache access [2,12,23,30,31], during RSA decryption. We review these physical attacks on the RSA scheme.

Cold Boot Attacks on the RSA Scheme. Before cold boot attacks were introduced, there were several studies on the consecutive partial leakage of the RSA secret-key bits. The first key recovery method was proposed by Rivest and Shamir [24]. They proved that the RSA scheme can be broken, when 2/3 of the most or least significant bits of p are known. Further, Coppersmith [4] proved that the RSA scheme can be broken, when half of the most significant bits of p are known. In the Coppersmith method, attackers recover the RSA secret keys by solving equations on the small unknown values of the secret keys using LLL reduction [19]. Recent studies on key recovery from known consecutive bits are based on the Coppersmith method [29].

After the emergence of cold boot attacks, the recovery of the RSA secret keys, based on cold boot attacks, were extensively researched. In cold boot attacks, the RSA secret keys are given as bits with non-consecutive erasure and error; therefore, the setting of the given consecutive bits in previous research is an unnatural assumption. Hence, several models for the partial RSA secret keys have been proposed, and a key recovery algorithm was constructed, based on Heninger and Shacham's work [9] in CRYPTO 2009, in which RSA secret keys with erasure bits were considered, and the search for the RSA secret keys was thorough a binary tree. Based on this, studies on the various settings of noisy bits were conducted. For example, Henecka et al. [8] considered bits with random errors, and Kunihiro et al. [17] considered bits with both erasure and error. Similarly, RSA-secret-key recovery has been considered for various settings in discrete [22] and analog data [16,18]. The above mentioned methods recover the RSA secret keys from the least significant bits. In addition to these methods, methods for recovering the RSA secret keys from the most significant bits by changing the method of constructing the tree have been proposed [13,20,27,28].

Side-Channel Attacks on the RSA Scheme. Since the proposal of the timing attack [14] in 1996, there have been many side-channel attacks on the RSA scheme. These attacks monitor the exponentiations during decryption and extract the RSA secret keys as square-and-multiply sequences. Obtaining these sequences is easier than obtaining the bits because cold boot attacks require the device itself of a regular user, whereas side-channel attacks do not.

Exponentiation is implemented efficiently using the binary, fixed window, and sliding window methods. Recent studies focus on attacking the fixed window [5, 11,31] and sliding window methods [2,5,11,12]. In particular, Bernstein et al. [2] have proposed a method for recovering CRT-RSA secret keys from square-and-multiply sequences without knowing the multiplier.

1.2 Motivation

Previous studies on side-channel attacks on the RSA scheme have mainly focused on observing the physical data precisely; they assumed that there were no errors in the square-and-multiply sequences. Bernstein et al. attacked the exponentiation using the sliding window method. They demonstrated that, if the square-and-multiply sequences can be obtained correctly, the CRT-RSA secret keys can be recovered, when the window size is $w \leq 4$.

However, there are errors in the square-and-multiply sequences obtained from physical data [1,2]. Bernstein et al. [2], in particular, report that there is an average of 14 errors in the square-and-multiply sequences of 1024-bit CRT-RSA decryption, under the sliding window method at $w = 4$. This corresponds to a 1.1% error rate in the square-and-multiply sequences. If there are errors in these sequences, Bernstein's method fails even if $w = 1$. In order to correct the errors in the square-and-multiply sequences, they proposed a majority vote by observing the physical data repeatedly, and declared that their method succeeds when the physical data is observed 20 times. However, there is no guarantee that attackers can observe the physical data repeatedly because decryption depends on the regular user. A similar problem arises in other methods, such as the differential power analysis [15].

In view of the above, we focus on correcting the errors in the square-and-multiply sequences, based on one observation alone. In cold boot attacks, various bits models are studied. In these researches, if key recovery succeeds, the correct CRT-RSA secret keys can be obtained. Therefore, the errors in the bits can be corrected. In side-channel attacks, the errors in the square-and-multiply sequences are considered in [12]; they proposed a key recovery algorithm, using square-and-multiply sequences with error. However, this study is ambiguous regarding the construction of the algorithm. Moreover, it does not reveal the error that can be recovered by their algorithm.

Thus, we require a key recovery algorithm using square-and-multiply sequences with small errors, extracted from one physical data alone. In addition, the error, in the square-and-multiply sequences, that can be recovered must be revealed.

1.3 Our Contribution

In this paper, we present a key recovery algorithm using the square-and-multiply sequences on exponentiations, c^{d_p} and c^{d_q}, with small errors at $w = 1$, namely, the binary method. In order to recover the CRT-RSA secret keys from square-and-multiply sequences with error, we focus on the similarity between our attacking situation and that of Henecka et al.'s [8], whose algorithm recovers CRT-RSA secret key bits with error using a binary tree. We focus on the relationship between the bits and the square-and-multiply sequences, and propose a key recovery algorithm based on the square-and-multiply sequences.

In addition, we analyze our proposed algorithm theoretically. We prove that we can recover with an error rate less than 5.8% in the square-and-multiply sequences in the expected polynomial time, if square-and-multiply sequences with error are provided from one physical data, by observing c^{d_p} and c^{d_q}. Our algorithm works well on the errors in the square-and-multiply sequences, as demonstrated experimentally in [2]; the average error rate is 1.1%.

2   Preliminaries

In this section, we introduce the encryption and decryption of the CRT-RSA scheme and the binary method. In addition, we present the leakage model of a square-and-multiply sequence, and define the problem dealt with in this study. Finally, we introduce the notations used in our proposed algorithm, and the analysis of our algorithm.

2.1   Encryption and Decryption of the CRT-RSA Scheme

First, we introduce the encryption and decryption of the standard RSA scheme. In standard RSA scheme, the public keys are (N, e) and the secret key is d. The encryption of a message m, is performed as $C = m^e \mod N$, and the decryption as $m = C^d \mod N$.

Next, we present the encryption and decryption of the CRT-RSA scheme. In the CRT-RSA scheme, the public keys are (N, e) and the secret keys are (p, q, d, d_p, d_q, q_p). The mathematical relationship between the public and secret keys can be expressed using $k, k_p, k_q \in \mathbb{Z}$ as follows:

$$N = pq,$$
$$ed = 1 + k(p-1)(q-1),$$
$$ed_p = 1 + k_p(p-1),$$
$$ed_q = 1 + k_q(q-1).$$

The encryption is the same as that of the standard RSA scheme. The decryption algorithm is given by Algorithm 1.

While the standard RSA decryption calculates under modulus N, the CRT-RSA decryption calculates under moduli p and q. Therefore, the exponentiation deals with half bits compared to the standard RSA scheme; decryption in the CRT-RSA scheme is approximately four times faster than that in the standard RSA scheme.

Algorithm 1. CRT-RSA Decryption Algorithm [26]

Input: Ciphertext c, and secret keys (p, q, d, d_p, d_q, q_p).
Output: Message m
Compute $m_1 = c^{d_p} \mod p, m_2 = c^{d_q} \mod q$
Compute $h = (m_1 - m_2)q_p \mod p$
Compute $m = m_2 + qh$
return m

2.2 Exponentiation (Binary Method)

The binary method is performed as Algorithm 2. With this Algorithm 2, we read d from the most-significant-bit side. Exponentiation includes two operations: squaring and multiplication. When the bit is zero, we perform squaring; when the bit is one, we perform squaring followed by multiplication. In this paper, **S** denotes squaring and **M** denotes multiplication. We write the implementation record using **S** and **M**. Using Algorithm 2, when we convert c^d as a square-and-multiply sequence; we convert a zero bit of d into **S**, and a one bit of d into **SM**. For example, when we calculate c^5, the binary representation of 5 is $5 = 101_2$; therefore, the square-and-multiply sequence is **SMSSM**. In a square-and-multiply sequence, the first operation is always **S**, and the **M**s are not consecutive. Moreover, the number of **S**s is the same as the number of bits of d, and the number of **M**s is the same as the number of one bits in d.

Algorithm 2. Exponentiation (binary method) [21]

Input: c and $d = (d_t \ldots d_0)_2$
Output: c^d
Set $x = 1$
for $i = t$ **down to** 0
   Compute $x = x^2$ (Squaring)
   if $d_i = 1$
      Compute $x = cx$ (Multiplication)
end for
return x

2.3 Leakage Model

In CRT-RSA decryption, we execute modular exponentiations, c^{d_p} and c^{d_q}. From these operation, we obtain the square-and-multiply sequence by physical attacks. If these sequences can be obtained correctly, we can obtain the secret keys easily because the bits and the square-and-multiply sequence have a one-to-one correspondence. However, we do not always obtain the square-and-multiply sequence correctly.

In this paper, we consider that attackers obtain a square-and-multiply sequence, in which each operation, **S** (resp. **M**), is incorrectly judged as another operation, **M** (resp. **S**), with probability $\delta > 0$, independently. When $\delta = 0$, the above attacking situation is the same as that in Bernstein et al. However, in our model, the attacker knows the public keys and the square-and-multiply sequence of c^{d_p} and c^{d_q} with error. We propose an algorithm that can solve our problem, and analyze the upper bound of the error rate δ, that can be recovered in the expected polynomial time.

2.4   Notations

We use notations similar to those in [9]. First, we introduce the notations used in our algorithm. We rewrite each bit of integer $x \geq 0$, as $x = x_{n-1}x_{n-2}\ldots x_0$ and $x[i] = x_i(0 \leq i \leq n-1)$. Next, we define $\tau(x) = \max_{m \in \mathbb{Z}} 2^m | x$. Moreover, we define Slice(i) as $(p[i], q[i], d_p[i + \tau(k_p)], d_q[i + \tau(k_q)])$. Slice($i$) denotes the corresponding bits of secret keys p, q, d_p, d_q.

Next, we introduce the notations used in the analysis of our algorithm. In the analysis, we define the entropy function $H(x)$, as $H(x) = -x \log x - (1 - x) \log(1-x)$. The base of log is two. In addition, we define the inverse function of the entropy function. We define $y = H^{-1}(x)(0 < x \leq 1)$ as y such that $H(y) = x$ and $0 < y \leq 1/2$. In addition, we define α_t and β_t as following.

$$\alpha_t = \frac{1}{2} - \frac{\log(2t+1) + 1}{2t}, \quad \beta_t = H^{-1}(\alpha_t).$$

3   Previous CRT-RSA Secret Key Recovery Algorithm [8]

In this section, we review Henecka et al.'s key recovery algorithm. Their method is based on the RSA-key-candidate tree [9]. We review the construction of the binary tree and Henecka et al.'s work.

First, we review the key candidate tree proposed in [9]. When the public keys (N, e), and parameters (k_p, k_q), are given, we can construct the key candidate tree. The tree depth corresponds to the number of recovered bits of p and q.

In order to construct the tree, we calculate the root of tree, initially, as Slice(0) = $(1, 1, d_p[\tau(k_p)], d_q[\tau(k_q)])$, and the lower bits of d_p, d_q, namely, $d_p[i]$ $(0 \leq i \leq \tau(k_p) - 1)$ and $d_q[i]$ $(0 \leq i \leq \tau(k_q) - 1)$. We can calculate these using known data, such as the public keys (N, e), and parameters (k_p, k_q).

From the root, we can calculate the candidates for the CRT-RSA secret keys. If bits below Slice($i-1$) are given in each variable, we can calculate Slice(i) as follows: We define $p' = \sum_{j=0}^{i-1} p[j]2^j, q' = \sum_{j=0}^{i-1} q[j]2^j, d_p' = \sum_{j=0}^{i+\tau(k_p)-1} d_p[j]2^j, d_q' = \sum_{j=0}^{i+\tau(k_q)-1} d_q[j]2^j$. We calculate these values using known information. Then, we solve

$$p[i] + q[i] \equiv (N - p'q')\,[i] \quad \text{mod } 2, \tag{1}$$

$$d_p[i + \tau(k_p)] + p[i] \equiv (k_p\,(p' - 1) + 1 - ed'_p)\,[i + \tau(k_p)] \quad \text{mod } 2, \tag{2}$$

$$d_q[i + \tau(k_q)] + q[i] \equiv (k_q\,(q' - 1) + 1 - ed'_q)\,[i + \tau(k_q)] \quad \text{mod } 2. \tag{3}$$

We can calculate the right-hand sides of (1)–(3) from known information. These simultaneous equations have two solutions; therefore, if the bits below Slice$(i-1)$ are given in each variable, we can calculate two candidates for Slice(i). Thereby, we can calculate the candidates for the CRT-RSA secret keys through the binary tree.

By repeating this operation, until we calculate Slice$(n/2-1)$, we can calculate the candidates for the CRT-RSA secret keys; the CRT-RSA secret keys always have $2^{n/2}$ candidates. However, finding the correct secret keys consumes tremendous time. Therefore, we adopt the branch and bound algorithm to searching for the secret key candidates. Previous studies prune leaves that do not resemble the observed data, in the binary tree.

Next, we review Henecka et al.'s work. Attackers are given the public keys (N, e), and the secret keys bits with errors. Each bit is judged incorrectly as another bit with probability $\delta > 0$, independently. Their algorithm calculates t slices using (1)–(3), and compares the Hamming distance between the t new bits and observed data, pruning leaves, whose Hamming distance is greater than C. Using this algorithm, when we are given two secret keys d_p and d_q, we can recover with an error rate of 11% in the expected polynomial time.

4   Proposed CRT-RSA Secret Key Recovering Algorithm

4.1   Proposed Algorithm

First, we calculate the candidates for (k_p, k_q), as described in [9]. The number of candidates for (k_p, k_q) is $2(e - 1)$; therefore, we construct $2(e - 1)$ trees.

Next, we calculate the key candidate in each tree, using the branch and bound strategy. We calculate t new unknown slices using simultaneous Eqs. (1)–(3). We set parameter t, to obtain the expected time complexity of our algorithm's polynomial time (see Sect. 4.2). After calculating the t new unknown slices, we convert t bits of d_p and d_q into square-and-multiply sequences, 0 to **S** and 1 to **SM**.

We then prune the leaves that do not match the observed sequence. We further calculate the disagreement rate between the calculated and observed sequences. We pay attention to the t new bits of d_p and d_q, and when the disagreement rate is strictly more than Y, we prune the leaf. We set parameter Y, to obtain the expected time complexity of our algorithm's polynomial time (see Sect. 4.2).

An example of pruning is illustrated in Fig. 1. We compare candidates and observed data from the rightmost operation. For candidate 1, we consider 10

	d_p	d_q	Disagreement Rate	Result
Observed Data	SMSSSM	MSSMSS		
Candidate 1	SMSS	SMSMSM	5/10=0.5	Prune
Candidate 2	SSSM	SSMSM	1/9=0.111	Remain
Candidate 3	SSS	SSSM	3/7=0.429	Prune

Fig. 1. Example of pruning, when $Y = 0.15$

operations: four in d_p and six in d_q. For d_p, we compare **SMSS** in candidate 1 and **SSSM** in the observed data; therefore, the number of disagreements is two. Similarly, for d_q, we compare **SMSMSM** of candidate 1 and **MSSMSS** in the observed data; therefore, the number of disagreements is three. The number of disagreements between candidate 1 and the observed data is five; thus, the disagreement rate is $5/10 = 0.5$. Therefore, candidate 1 is pruned because $0.5 > 0.15$. Candidates 2 and 3 are similar.

We repeat this step, until Slice($n/2 - 1$) are calculated, and search for the leaf that satisfies $N = pq$. The proposed algorithm is given by Algorithm 3.

Algorithm 3. Proposed CRT-RSA Secret Key Recovery Algorithm

Input: Public keys (N, e), square-and-multiply sequences of c^{d_p}, c^{d_q} with error, number of expansions t, threshold Y
Output: Secret keys (p, q, d_p, d_q)
Each k_p, k_q
  Calculate the root of the tree.
  for $i = 1$ **to** $\lceil (n/2 - 1)/t \rceil$
    Compute Slice$((i - 1)t + 1), \cdots,$ Slice(it).
    Transform the new bits of d_p, d_q into square-and-multiply sequences.
    Prune the leaves, whose disagreement rates are strictly more than Y.
  end for
  Search for the leaf that satisfies $N = pq$.

4.2   Analysis of the Proposed Algorithm

In this subsection, we analyze Algorithm 3. We introduce a heuristic assumption [9] where, if the values of k_p and k_q are correct, the slice calculated from the incorrect leaf is random. Moreover, we assume that, when the values of k_p and k_q are incorrect, the slice calculated is random.

The goal of our analysis is to prove the following Theorem 1:

Theorem 1. Let $\varepsilon > 0$, $t = \lceil \ln(n)/(4\varepsilon^2) \rceil$. Moreover, let Y satisfy $Y \leq \beta_t/(2 - \beta_t)$. Then, when $\delta < Y - \varepsilon$, the expected time complexity of Algorithm 3 is $O\left(n^{2+\frac{\ln 2}{2\varepsilon^2}}\right)$, and the success rate of Algorithm 3 is at least $1 - \left(\frac{2\varepsilon^2}{\ln n} + \frac{1}{n}\right)$.

Remark 1. When $n \to \infty$, β_t converges to $H^{-1}(1/2) = 0.11$. Therefore, Theorem 1 implies that $0.11/(2 - 0.11) = 0.058$ is the upper bound of the error rate that can be recovered in expected polynomial time. Moreover, the success rate converges to 1 in $n \to \infty$.

Analysis Strategy. Before proving Theorem 1, we review the analysis in [8], wherein the parameters satisfy the condition that the expected number of new leaves generated from incorrect partial keys is $1/2$, after calculating t slices and pruning. This guarantees that the bound of the expected number of leaves after pruning is constant. Then, because the expected number of all leaves in Henecka et al.'s algorithm is polynomial in n, the expected time complexity of their algorithm is polynomial in n.

We evaluate the number of remaining leaves, when we calculate t slices and prune the candidates for parameter Y. When t slices are calculated, 2^t leaves are generated from one leaf. As heuristic assumption in [9], the new $2t$ bits of d_p, d_q calculated from the incorrect partial keys are independent of the 2^{2t} elements. Thus, if these elements are reduced to less than 2^{t-1} by pruning, the expected number of new leaves generated from the incorrect partial keys is less than $1/2$.

In [8], the analytical result is obtained directly from the Hoeffding inequality [10]. However, in square-and-multiply sequences, there are restrictions between each operation. Moreover, same-length key bits have different lengths in a square-and-multiply sequence. These restrictions render analysis hard, when square-and-multiply sequences are given. In order to address this problem, we analyze each length of a square-and-multiply sequence.

When we calculate t slices, we convert t bits of d_p and d_q into square-and-multiply sequences. We deal these $2t$ bits together. In these bits, the number of **M** in the square-and-multiply sequence is equal to the number of "one" bits. When we define the number of one bits in $2t$ bits as $t_{\mathbf{M}}$ ($0 \leq t_{\mathbf{M}} \leq 2t$), the length of the sequence is $2t + t_{\mathbf{M}}$. If there are less than $2^{t-1}/(2t + 1)$ leaves in each $t_{\mathbf{M}}$ after pruning, the number of leaves is reduced to 2^{t-1} because $\sum_{t_{\mathbf{M}}=0}^{2t} 2^{t-1}/(2t + 1) = 2^{t-1}$. We now analyze the condition, Y, where less than $2^{t-1}/(2t + 1)$ leaves remain in each $t_{\mathbf{M}}$ after pruning. Y indicates the condition for the error rate that can be corrected in the expected polynomial time. We focus on the analysis of Y and the details of the proof of Theorem 1 with respect to the setting of t, the time complexity, and the success rate, shown in full version.

Analysis of the Upper Bound of Y. We now prove the following Lemma 1 that shows the condition, Y, where less than $2^{t-1}/(2t + 1)$ leaves remain in each $t_{\mathbf{M}}$, after pruning.

Lemma 1. When $0 \leq t_{\mathbf{M}} \leq 2t\beta_t$ or $2t(1 - \beta_t) \leq t_{\mathbf{M}} \leq 2t$, the number of leaves is always less than $2^{t-1}/(2t+1)$. When $2t\beta_t < t_{\mathbf{M}} < 2t(1 - \beta_t)$, less than $2^{t-1}/(2t+1)$ leaves remain after pruning, under $Y \leq 2t\beta_t/(2t + t_{\mathbf{M}})$.

In order to prove Lemma 1, we use the following the Lemma 2 proved in Appendix, and the Lemma 3 from [3].

Lemma 2. Let $L \in \mathbb{N}$, t_M be an integer that satisfies $0 \leq t_\mathrm{M} \leq \lfloor L/2 \rfloor$, and C be an integer that satisfies $0 \leq C \leq L$. Then, the number of sequences with length L, including t_M Ms, generated from the bits, and whose number of disagreements is less than C compared to the observed data, is less than

$$
\begin{cases}
\dbinom{L - t_\mathrm{M}}{C} & \text{if } 0 \leq C \leq \min(t_\mathrm{M}, L - 2t_\mathrm{M}) \\
\dbinom{L - t_\mathrm{M}}{t_\mathrm{M}} & \text{if } \min(t_\mathrm{M}, L - 2t_\mathrm{M}) \leq C \leq L.
\end{cases}
$$

Lemma 3. [3] Let a, b be nonnegative integers. When $b \leq a$, it holds that $\dbinom{a}{b} \leq 2^{aH(b/a)}$.

Remark 2. We explain the relationship between Lemma 1 and the error rate that can be recovered. We now consider the average and the worst cases, when $t \to \infty$ that β_t converges to $H^{-1}(1/2) = 0.11$. First, we consider the average case. In practice, as there are random 0 and 1 values in the $2t$ bits, $t_\mathrm{M} = t$ with a high probability. Therefore, $Y \leq 2/3 \times 0.11 = 0.073$, and we can correct a 7.3% error rate in the square-and-multiply sequences practically. From [8], we can correct an 11% error rate in the bits. This difference occurs because one bit corresponds to 1.5 operations on an average, and the information in a character in an operation is $2/3$ times as much as that in a bit. We next consider the worst case. In Lemma 1, the strictest upper bound of Y is $\beta_t / (2 - \beta_t)$. Therefore, $0.11/(2 - 0.11) = 0.058$ is the worst case of the error rate that can be recovered in the expected polynomial time. We consider the latter worst case in Theorem 1.

Proof. We consider a sequence with length, $L = 2t + t_\mathrm{M}$, that is generated from $2t$ bits and includes t_M Ms. When $L = 2t + t_\mathrm{M}$, the condition, where the leaf is not pruned is $C = (2t + t_\mathrm{M}) Y$. Therefore, from Lemma 2, the number of leaves after pruning is less than

$$
\begin{cases}
\dbinom{2t}{(2t + t_\mathrm{M}) Y} & \text{if } 0 \leq (2t + t_\mathrm{M}) Y \leq \min(t_\mathrm{M}, 2t - t_\mathrm{M}) \\
\dbinom{2t}{t_\mathrm{M}} & \text{if } \min(t_\mathrm{M}, 2t - t_\mathrm{M}) \leq (2t + t_\mathrm{M}) Y \leq 2t + t_\mathrm{M}.
\end{cases}
$$

Thus, from Lemma 3, the number of leaves after pruning is less than

$$
\begin{cases}
2^{2tH((2t + t_\mathrm{M})Y/2t)} & \text{if } 0 \leq (2t + t_\mathrm{M}) Y \leq \min(t_\mathrm{M}, 2t - t_\mathrm{M}) \\
2^{2tH(t_\mathrm{M}/2t)} & \text{if } \min(t_\mathrm{M}, 2t - t_\mathrm{M}) \leq (2t + t_\mathrm{M}) Y \leq 2t + t_\mathrm{M}.
\end{cases}
\tag{4}
$$

We must prove that, when we set Y as Lemma 1, the number of (4) is less than $2^{t-1}/2t + 1$. From the definiton of α_t, $2t\alpha_t = (t - 1) - \log(2t + 1)$. Therefore, $2^{2t\alpha_t} = 2^{t-1}/(2t + 1)$. Thus, we must show the following, under Y, in Lemma 1:

$$
\begin{cases}
H\left(\dfrac{(2t + t_\mathrm{M}) Y}{2t} \right) \leq \alpha_t & \text{if } 0 \leq (2t + t_\mathrm{M}) Y \leq \min(t_\mathrm{M}, 2t - t_\mathrm{M}) \\
H\left(\dfrac{t_\mathrm{M}}{2t} \right) \leq \alpha_t & \text{if } \min(t_\mathrm{M}, 2t - t_\mathrm{M}) \leq (2t + t_\mathrm{M}) Y \leq 2t + t_\mathrm{M}.
\end{cases}
\tag{5}
$$

Further, we show this is satisfied for all $0 \leq t_M \leq 2t$, under Y, in Lemma 1.

First, when $0 \leq t_M \leq 2t\beta_t$ or $2t\,(1 - \beta_t) \leq t_M \leq 2t$, then $0 \leq H(t_M/2t) \leq \alpha_t$. Thus, in $0 \leq (2t + t_M)\,Y \leq \min\,(t_M, 2t - t_M)$,

$$H\left(\frac{(2t + t_M)\,Y}{2t}\right) \leq H\left(\frac{t_M}{2t}\right) \leq \alpha_t.$$

Therefore, the first inequality in (5) is satisfied. The second inequality in (5) is also satisfied obviously.

Next, we consider in the case, where $2t\beta_t < t_M < 2t\,(1 - \beta_t)$. In Lemma 1, when $Y \leq 2t\beta_t/\,(2t + t_M)$, we insist that the number of leaves after pruning, is less than $2^{t-1}/(2t + 1)$. Now, we show that, when $Y \leq 2t\beta_t/\,(2t + t_M)$, (5) is satisfied.

When $Y \leq 2t\beta_t/\,(2t + t_M)$, for all $0 \leq x \leq Y$,

$$H\left(\frac{(2t + t_M)\,x}{2t}\right) \leq H\,(\beta_t) = \alpha_t.$$

However, if there are Y that satisfy $(2t + t_M)\,Y > \min\,(t_M, 2t - t_M)$, there are x in $[0, Y]$ such that

$$H\left(\frac{(2t + t_M)\,x}{2t}\right) > H\left(\frac{t_M}{2t}\right) > H\,(\beta_t) = \alpha_t.$$

Therefore, if there are Y that satisfy $(2t + t_M)\,Y > \min\,(t_M, 2t - t_M)$, contradiction occurs. Therefore, Y satisfies $(2t + t_M)\,Y \leq \min\,(t_M, 2t - t_M)$. Thus, we consider only the first inequality in (5), which is satisfied obviously. Therefore, we show (5) in all t_M, we prove Lemma 1.   □

5   Numerical Experiments with the Proposed Algorithm

We implemented our proposed Algorithm 3 in C++ using NTL library version 10.3.0. Our tests were run on an Intel Core i7, at 2.40 GHz with 16-GB memory. To render the proposed Algorithm 3 more efficient, we used following techniques. We implemented the proposed Algorithm 3 for the first depth search. We pruned the leaves containing $4tY$ errors, before calculating t slices completely. This is because the errors monotonically increase, during the calculation of the slices.

In the experiment, we tested the proposed Algorithm 3 on a 1024-bit CRT-RSA. We generated 100 random keys. In each key, we added noise in the square-and-multiply sequences as $\delta = 0.011$ observed in [2]. We then set parameters t, Y, in Algorithm 3. In this experiment, we set $t = 40, 60, 80, 100$ for $Y = 0.03$, $t = 40, 60$ for $Y = 0.04$. We executed our proposed Algorithm 3 for each square-and-multiply sequence with error. When the correct secret keys were output, the proposed Algorithm 3 was considered successful. We measured the success rate, average time for all the trials, and average time for successful trials for 100 pairs of square-and-multiply sequences with noise, when given the correct k_p and k_q. The experimental results are listed in Tables 1 and 2.

Table 1. Experimental data under $\delta = 0.011, Y = 0.03$ for correct k_p and k_q

	t			
	40	60	80	100
Average time for all the trials (ms)	22.2	237	1530	17973
Average time for successful trials (ms)	25.2	234	1537	17188
Success rate (%)	67	93	98	97

Table 2. Experimental data under $\delta = 0.011, Y = 0.04$ for correct k_p and k_q

	t	
	40	60
Average time for all the trials (ms)	124	2287
Average time for successful trials (ms)	130	2287
Success rate (%)	88	100

From Tables 1 and 2, it can be established that the secret keys were recovered with a high success rate. In Table 1, when we set $t = 40$, we recovered 67% of the secret keys in 25.2 ms; thus, we recovered 2/3 of the secret keys in 30 ms. Moreover, when we set a larger value of t, the running time was more and the success rate was higher. For example, when we set $t = 60$, we recovered 93% of the secret keys in 0.24 s. For $t = 80, 100$, almost all the secret keys were recovered; however, the success rate was not 100%.

Thus, we set the larger value of Y as $Y = 0.04$, for achieving a success rate of 100%. In Table 2, when we set $t = 40$, we recovered 88% of the secret keys in 0.13 s. When we set $t = 60$, we recovered all the secret keys in 2.3 s.

In addition to these experiments, we executed Algorithm 3 for $\delta = 0.011$, $t = 40, Y = 0.05$. In this experiment, we recovered 99% of the secret keys in 1.1 s. Thus, for $\delta = 0.011$, almost all the secret keys were recovered in approximately 1–2 s using our proposed Algorithm 3.

6   Conclusion

In this paper, we presented a key recovery algorithm, using the square-and-multiply sequences on exponentiations c^{d_p} and c^{d_q} with small errors, at $w = 1$, namely, the binary method.

We theoretically proved that we can correct square-and-multiply sequences with error rates less than 5.8% in the expected polynomial time, when c^{d_p} and c^{d_q} are observed only once. In addition, we experimentally demonstrated that our proposed Algorithm 3 recovers small errors of 1.1%, in 1–2 s under correct (k_p, k_q).

Acknowledgements. This research was partially supported by JST CREST Grant Number JPMJCR14D6, Japan and JSPS KAKENHI Grant Number 16H02780.

A   Appendix: Proof of Lemma 2

In this section, initially, we prepare the tools used for proving Lemma 2, after which we prove the Lemma 2.

A.1   Tools Used for Proving Lemma 2

In order to prove Lemma 2, we define certain sets and prove Lemma 4. First, we define the set of square-and-multiply sequences.

Definition 1. Let $L \in \mathbb{N}$. $A(L)$ is defined as the set of all square-and-multiply sequences with length L.

$A(L)$ includes elements that cannot be converted into bits. The elements in set $A(L)$, merely display $\mathbf{S}$ and $\mathbf{M}$. Therefore, the number of elements in set $A(L)$, is 2^L. In the following discussion, we refer to operations in sequence as the first operation, second operation, etc., from the left operation.

We then define the square-and-multiply sequences that can be converted into bits.

Definition 2. Let $L \in \mathbb{N}$ and $t_{\mathbf{M}}$ be non-negative integers. $T(L, \mathbf{S}, t_{\mathbf{M}})$ is defined as the set of square-and-multiply sequences that satisfy the following conditions: the length of the sequence is L, the first operation of the sequence is $\mathbf{S}$, the number of $\mathbf{M}$s is $t_{\mathbf{M}}$, and the $\mathbf{M}$s are not consecutive. Similarly, $T(L, \mathbf{M}, t_{\mathbf{M}})$ is defined as the set of square-and-multiply sequence that satisfies the following conditions: the length of the sequences is L, the first operation of the sequence is $\mathbf{M}$, the number of $\mathbf{M}$s is $t_{\mathbf{M}}$, and the $\mathbf{M}$s are not consecutive.

We now calculate the number of elements in $T(L, \mathbf{S}, t_{\mathbf{M}})$, used in our analysis. The element of $T(L, \mathbf{S}, t_{\mathbf{M}})$ that satisfies the first operation is $\mathbf{S}$, and the $\mathbf{M}$s are not consecutive. Moreover, the number of $\mathbf{S}$s is $L - t_{\mathbf{M}}$ and the number of $\mathbf{M}$s is $t_{\mathbf{M}}$. Thus, the elements of $T(L, \mathbf{S}, t_{\mathbf{M}})$ are generated from $L - t_{\mathbf{M}}$ bits, including $t_{\mathbf{M}}$ one bits. Therefore,

$$|T(L, \mathbf{S}, t_{\mathbf{M}})| = \begin{cases} \dbinom{L - t_{\mathbf{M}}}{t_{\mathbf{M}}} & 0 \leq t_{\mathbf{M}} \leq \lfloor L/2 \rfloor \\ 0 & \text{otherwise.} \end{cases}$$

Next, we consider a situation, where an observed square-and-multiply sequence is $\tilde{o}$. We define the set of square-and-multiply sequences at a certain distance from $\tilde{o}$.

Definition 3. Let $L \in \mathbb{N}$, $t_{\mathbf{M}}, C$ be non-negative integers, and $\tilde{o} \in A(L)$. Then, $B(L, \mathbf{S}, t_{\mathbf{M}}, C, \tilde{o})$ is defined as the set of square-and-multiply sequences in $T(L, \mathbf{S}, t_{\mathbf{M}})$ satisfying the condition that the number of disagreements with $\tilde{o}$, excluding the first operation, is less than C. In addition, $B(L, \mathbf{M}, t_{\mathbf{M}}, C, \tilde{o})$ is defined as the set of square-and-multiply sequences in $T(L, \mathbf{M}, t_{\mathbf{M}})$ satisfying the condition that the number of disagreements with $\tilde{o}$, excluding the first operation, is less than C.

By Definition 3, for all $\tilde{o} \in A(L)$,

$$|B(L, \mathbf{S}, t_\mathbf{M}, C - 1, \tilde{o})| \le |B(L, \mathbf{S}, t_\mathbf{M}, C, \tilde{o})|. \tag{6}$$

Finally, we define the upper bound of the number of elements in set B.

Definition 4. Let $L \in \mathbb{N}$, k, C be non-negative integers. Then, $\tilde{b}(L, \mathbf{S}, t_\mathbf{M}, C)$ is defined as $\tilde{b}(L, \mathbf{S}, t_\mathbf{M}, C) = \max_{\tilde{o} \in A(L)} |B(L, \mathbf{S}, t_\mathbf{M}, C, \tilde{o})|$.

From (6),

$$\tilde{b}(L, \mathbf{S}, t_\mathbf{M}, C - 1) \le \tilde{b}(L, \mathbf{S}, t_\mathbf{M}, C). \tag{7}$$

In order to prove Lemma 2, we present the upper bound of $\tilde{b}(L, \mathbf{S}, t_\mathbf{M}, C)$ in Lemma 4.

Lemma 4. Let $L \in \mathbb{N}$, $t_\mathbf{M}$ be an integer satisfying $0 \le t_\mathbf{M} \le \lfloor L/2 \rfloor$, and C be an integer satisfying $0 \le C \le L - 1$. Then,

$$\begin{cases} \tilde{b}(L, \mathbf{S}, t_\mathbf{M}, C) \le \dbinom{L - t_\mathbf{M}}{C} & \text{if } 0 \le C \le \min(t_\mathbf{M}, L - 2t_\mathbf{M}) \\ \tilde{b}(L, \mathbf{S}, t_\mathbf{M}, C) \le \dbinom{L - t_\mathbf{M}}{t_\mathbf{M}} & \text{if } \min(t_\mathbf{M}, L - 2t_\mathbf{M}) \le C \le L - 1 \end{cases}$$

is satisfied.

Proof. We prove Lemma 4 by mathematical induction. The recurrence formula on $\tilde{b}$ is following. When $L \ge 3$,

$$\tilde{b}(L, \mathbf{S}, t_\mathbf{M}, C)$$
$$\le \max\Big(\tilde{b}(L - 1, \mathbf{S}, t_\mathbf{M}, C) + \tilde{b}(L - 2, \mathbf{S}, t_\mathbf{M} - 1, C - 1),$$
$$\tilde{b}(L - 1, \mathbf{S}, t_\mathbf{M}, C - 1) + \tilde{b}(L - 2, \mathbf{S}, t_\mathbf{M} - 1, C)\Big). \tag{8}$$

We now consider the case where $L = 1$ and $L = 2$. For $L = 1$ and $L = 2$, because $|T(1, \mathbf{S}, 0)| = 1$ and $|T(2, \mathbf{S}, 0)| = |T(2, \mathbf{S}, 1)| = 1$, $\tilde{b}(L, \mathbf{S}, t_\mathbf{M}, C)$ is always less than one. Therefore, when $L = 1$ and $L = 2$, Lemma 4 is true.

We now consider the case, where Lemma 4 is true, when $1 \le L \le m (m \in \{x \in \mathbb{N} | x \ge 2\})$. We prove that Lemma 4 is true, when $L = m + 1$. We consider the set $T(m + 1, \mathbf{S}, t_\mathbf{M})$. The condition $T(m + 1, \mathbf{S}, t_\mathbf{M}) \ne \emptyset$ is $0 \le t_\mathbf{M} \le \lfloor (m + 1)/2 \rfloor$, and the possible value of C is $0 \le C \le m$. For these $(t_\mathbf{M}, C)$, we prove

$$\begin{cases} \tilde{b}(m + 1, \mathbf{S}, t_\mathbf{M}, C) \le \dbinom{m + 1 - t_\mathbf{M}}{C} & \text{if } 0 \le C \le X \\ \tilde{b}(m + 1, \mathbf{S}, t_\mathbf{M}, C) \le \dbinom{m + 1 - t_\mathbf{M}}{t_\mathbf{M}} & \text{if } X \le C \le m \end{cases}$$

under $X = \min(t_\mathbf{M}, m + 1 - 2t_\mathbf{M})$.

First, we consider the case, where $C = 0$. $C = 0$ indicates that no mismatch exists between the sequence and the observed data; thus, the number of the sequences that satisfy $C = 0$ is no more than one. Thus, $\tilde{b}(m + 1, \mathbf{S}, t_\mathbf{M}, C) \le 1$.

We then consider the case, where $1 \leq C \leq t_M - 1$. From the assumption by mathematical induction, Lemma 4 is true, when $L = m, m - 1$. When $L = m$,

$$\begin{cases} \tilde{b}(m, \mathbf{S}, t_M, C) \leq \binom{m - t_M}{C} & \text{if } 0 \leq C \leq \min(t_M, m - 2t_M) \\ \tilde{b}(m, \mathbf{S}, t_M, C) \leq \binom{m - t_M}{t_M} & \text{if } \min(t_M, m - 2t_M) \leq C \leq m - 1 \end{cases}$$

is satisfied. Thus, $\binom{m - t_M}{t_M} \leq \binom{m - t_M}{C}$ under $\min(t_M, m - 2t_M) \leq C \leq t_M$. Therefore, $\tilde{b}(m, \mathbf{S}, t_M, C) \leq \binom{m - t_M}{C}$ under $0 \leq C \leq t_M$. Similarly, when $L = m - 1$,

$$\begin{cases} \tilde{b}(m - 1, \mathbf{S}, t_M, C) \leq \binom{m - 1 - t_M}{C} & \text{if } 0 \leq C \leq \min(t_M, m - 1 - 2t_M) \\ \tilde{b}(m - 1, \mathbf{S}, t_M, C) \leq \binom{m - 1 - t_M}{t_M} & \text{if } \min(t_M, m - 1 - 2t_M) \leq C \leq m - 2 \end{cases}$$

is satisfied. Thus, $\binom{m - 1 - t_M}{t_M} \leq \binom{m - 1 - t_M}{C}$ under $\min(t_M, m - 1 - 2t_M) \leq C \leq t_M$. Therefore, $\tilde{b}(m - 1, \mathbf{S}, t_M, C) \leq \binom{m - 1 - t_M}{C}$ under $0 \leq C \leq t_M$.

Thus, under $1 \leq C \leq t_M - 1$,

$$\tilde{b}(m, \mathbf{S}, t_M, C) + \tilde{b}(m - 1, \mathbf{S}, t_M - 1, C - 1)$$
$$\leq \binom{m - t_M}{C} + \binom{(m - 1) - (t_M - 1)}{C - 1} = \binom{m + 1 - t_M}{C}$$

and

$$\tilde{b}(m, \mathbf{S}, t_M, C - 1) + \tilde{b}(m - 1, \mathbf{S}, t_M - 1, C)$$
$$\leq \binom{m - t_M}{C - 1} + \binom{(m - 1) - (t_M - 1)}{C} = \binom{m + 1 - t_M}{C}.$$

Therefore, from (8), under $1 \leq C \leq t_M - 1$,

$$\tilde{b}(m + 1, \mathbf{S}, t_M, C) \leq \binom{m + 1 - t_M}{C}.$$

However, $|T(m + 1, \mathbf{S}, t_M)| = \binom{m + 1 - t_M}{t_M}$. Thus, under $1 \leq C \leq t_M - 1$,

$$\tilde{b}(m + 1, \mathbf{S}, t_M, C) \leq \min\left(\binom{m + 1 - t_M}{C}, \binom{m + 1 - t_M}{t_M}\right).$$

Therefore, under $X = \min(t_{\mathbf{M}}, m + 1 - 2t_{\mathbf{M}})$,

$$
\begin{cases}
\tilde{b}(m+1, \mathbf{S}, t_{\mathbf{M}}, C) \leq \dbinom{m+1-t_{\mathbf{M}}}{C} & \text{if } 1 \leq C \leq X \\[3mm]
\tilde{b}(m+1, \mathbf{S}, t_{\mathbf{M}}, C) \leq \dbinom{m+1-t_{\mathbf{M}}}{t_{\mathbf{M}}} & \text{if } X \leq C \leq t_{\mathbf{M}} - 1
\end{cases}
$$

is satisfied.

Finally, we consider the case, where $t_{\mathbf{M}} \leq C \leq m$. Because $|T(m+1, \mathbf{S}, t_{\mathbf{M}})| = \dbinom{m+1-t_{\mathbf{M}}}{t_{\mathbf{M}}}$, $\tilde{b}(m+1, \mathbf{S}, t_{\mathbf{M}}, C) \leq \dbinom{m+1-t_{\mathbf{M}}}{t_{\mathbf{M}}}$ under $t_{\mathbf{M}} \leq C \leq m$.

In conclusion,

$$
\begin{cases}
\tilde{b}(m+1, \mathbf{S}, t_{\mathbf{M}}, C) \leq \dbinom{m+1-t_{\mathbf{M}}}{C} & \text{if } 0 \leq C \leq X \\[3mm]
\tilde{b}(m+1, \mathbf{S}, t_{\mathbf{M}}, C) \leq \dbinom{m+1-t_{\mathbf{M}}}{t_{\mathbf{M}}} & \text{if } X \leq C \leq m
\end{cases}
$$

under $X = \min(t_{\mathbf{M}}, m + 1 - 2t_{\mathbf{M}})$. Therefore, when $L = m + 1$, Lemma 4 is true. In conclusion, Lemma 4 is proved. □

A.2   Proof of Lemma 2

In Lemma 4, the upper bound of $\tilde{b}(L, \mathbf{S}, t_{\mathbf{M}}, C)$ is calculated. This function does not consider the first operation of the sequence. We now consider the first operation of the sequence. Let $\tilde{o}$ be a partially observed sequence, and let its length be L. We calculate the upper bound of the number of elements $\tilde{u}(L, t_{\mathbf{M}}, C, \tilde{o})$ in $T(L, \mathbf{S}, t_{\mathbf{M}})$, such that the number of disagreements with $\tilde{o}$ is less than C.

First, we consider the case, where $0 \leq C \leq L - 1$. From Lemma 4, if the first operation of $\tilde{o}$ is $\mathbf{S}$, then $\tilde{u}(L, t_{\mathbf{M}}, C, \tilde{o}) \leq \tilde{b}(L, \mathbf{S}, t_{\mathbf{M}}, C)$, and if the first operation of $\tilde{o}$ is $\mathbf{M}$, then $\tilde{u}(L, t_{\mathbf{M}}, C, \tilde{o}) \leq \tilde{b}(L, \mathbf{S}, t_{\mathbf{M}}, C - 1)$. Because of (7), $\tilde{u}(L, t_{\mathbf{M}}, C, \tilde{o}) \leq \tilde{b}(L, \mathbf{S}, t_{\mathbf{M}}, C)$. From Lemma 4, $\tilde{u}(L, t_{\mathbf{M}}, C, \tilde{o})$ is less than

$$
\begin{cases}
\dbinom{L-t_{\mathbf{M}}}{C} & \text{if } 0 \leq C \leq \min(t_{\mathbf{M}}, L - 2t_{\mathbf{M}}) \\[3mm]
\dbinom{L-t_{\mathbf{M}}}{t_{\mathbf{M}}} & \text{if } \min(t_{\mathbf{M}}, L - 2t_{\mathbf{M}}) \leq C \leq L - 1
\end{cases}
$$

We next consider the case, where $C = L$. Because of $|T(L, \mathbf{S}, t_{\mathbf{M}})| = \dbinom{L-t_{\mathbf{M}}}{t_{\mathbf{M}}}$, $\tilde{u}(L, t_{\mathbf{M}}, C, \tilde{o})$ is less than $\dbinom{L-t_{\mathbf{M}}}{t_{\mathbf{M}}}$. Therefore, the number of sequences, whose length is L, including $t_{\mathbf{M}}$ **M**s, generated from the bits, and whose number of disagreements is less than C, compared to the observed data is less than

$$
\begin{cases}
\dbinom{L-t_{\mathbf{M}}}{C} & \text{if } 0 \leq C \leq \min(t_{\mathbf{M}}, L - 2t_{\mathbf{M}}) \\[3mm]
\dbinom{L-t_{\mathbf{M}}}{t_{\mathbf{M}}} & \text{if } \min(t_{\mathbf{M}}, L - 2t_{\mathbf{M}}) \leq C \leq L
\end{cases}
$$

□

References

1. Bauer, S.: Attacking exponent blinding in RSA without CRT. In: Schindler, W., Huss, S.A. (eds.) COSADE 2012. LNCS, vol. 7275, pp. 82–88. Springer, Heidelberg (2012). https://doi.org/10.1007/978-3-642-29912-4_7

2. Bernstein, D.J., et al.: Sliding right into disaster: left-to-right sliding windows leak. In: Fischer, W., Homma, N. (eds.) CHES 2017. LNCS, vol. 10529, pp. 555–576. Springer, Cham (2017). https://doi.org/10.1007/978-3-319-66787-4_27

3. Bruen, A.A., Forcinito, M.A.: Cryptography, Information Theory, and Error-Correction: A Handbook for the 21st Century. Wiley, Hoboken (2005)

4. Coppersmith, D.: Small solutions to polynomial equations, and low exponent RSA vulnerabilities. J. Crypt. **10**, 233–260 (1997). https://doi.org/10.1007/s001459900030

5. Genkin, D., Pachmanov, L., Pipman, I., Tromer, E.: Stealing keys from PCs using a radio: cheap electromagnetic attacks on windowed exponentiation. In: Güneysu, T., Handschuh, H. (eds.) CHES 2015. LNCS, vol. 9293, pp. 207–228. Springer, Heidelberg (2015). https://doi.org/10.1007/978-3-662-48324-4_11

6. Genkin, D., Shamir, A., Tromer, E.: RSA key extraction via low-bandwidth acoustic cryptanalysis. In: Garay, J.A., Gennaro, R. (eds.) CRYPTO 2014. LNCS, vol. 8616, pp. 444–461. Springer, Heidelberg (2014). https://doi.org/10.1007/978-3-662-44371-2_25

7. Halderman, J.A., et al.: Lest we remember: cold-boot attacks on encryption keys. Commun. ACM. **52**, 91–98 (2009). https://doi.org/10.1145/1506409.1506429

8. Henecka, W., May, A., Meurer, A.: Correcting errors in RSA private keys. In: Rabin, T. (ed.) CRYPTO 2010. LNCS, vol. 6223, pp. 351–369. Springer, Heidelberg (2010). https://doi.org/10.1007/978-3-642-14623-7_19

9. Heninger, N., Shacham, H.: Reconstructing RSA private keys from random key bits. In: Halevi, S. (ed.) CRYPTO 2009. LNCS, vol. 5677, pp. 1–17. Springer, Heidelberg (2009). https://doi.org/10.1007/978-3-642-03356-8_1

10. Hoeffding, W.: Probability inequalities for sums of bounded random variables. J. Am. Stat. Assoc. **58**, 13–30 (1963). https://doi.org/10.1080/01621459.1963.10500830

11. Homma, N., Miyamoto, A., Aoki, T., Satoh, A., Shamir, A.: Comparative power analysis of modular exponentiation algorithms. IEEE Trans. Comput. **59**, 795–807 (2010). https://doi.org/10.1109/TC.2009.176

12. İnci, M.S., Gulmezoglu, B., Irazoqui, G., Eisenbarth, T., Sunar, B.: Cache attacks enable bulk key recovery on the cloud. In: Gierlichs, B., Poschmann, A.Y. (eds.) CHES 2016. LNCS, vol. 9813, pp. 368–388. Springer, Heidelberg (2016). https://doi.org/10.1007/978-3-662-53140-2_18

13. Jagnere, P., Sanket, S., Chauhan, A., Jaiswal, R.: Better algorithms for MSB-side RSA reconstruction. In: WCC 2015 (2015)

14. Kocher, P.C.: Timing attacks on implementations of Diffie-Hellman, RSA, DSS, and other systems. In: Koblitz, N. (ed.) CRYPTO 1996. LNCS, vol. 1109, pp. 104–113. Springer, Heidelberg (1996). https://doi.org/10.1007/3-540-68697-5_9

15. Kocher, P., Jaffe, J., Jun, B.: Differential power analysis. In: Wiener, M. (ed.) CRYPTO 1999. LNCS, vol. 1666, pp. 388–397. Springer, Heidelberg (1999). https://doi.org/10.1007/3-540-48405-1_25

16. Kunihiro, N., Honda, J.: RSA meets DPA: recovering RSA secret keys from noisy analog data. In: Batina, L., Robshaw, M. (eds.) CHES 2014. LNCS, vol. 8731, pp. 261–278. Springer, Heidelberg (2014). https://doi.org/10.1007/978-3-662-44709-3_15

17. Kunihiro, N., Shinohara, N., Izu, T.: Recovering RSA secret keys from noisy key bits with erasures and errors. IEICE Trans. Fundam. **E97-A**, 1273–1284 (2014). https://doi.org/10.1587/transfun.E97.A.1273
18. Kunihiro, N., Takahashi, Y.: Improved key recovery algorithms from noisy RSA secret keys with analog noise. In: Handschuh, H. (ed.) CT-RSA 2017. LNCS, vol. 10159, pp. 328–343. Springer, Cham (2017). https://doi.org/10.1007/978-3-319-52153-4_19
19. Lenstra, A.K., Lenstra, H.W., Lovász, L.: Factoring polynomials with rational coefficients. Math. Ann. **261**, 515–534 (1982). https://doi.org/10.1007/BF01457454
20. Maitra, S., Sarkar, S., Sen Gupta, S.: Factoring RSA modulus using prime reconstruction from random known bits. In: Bernstein, D.J., Lange, T. (eds.) AFRICACRYPT 2010. LNCS, vol. 6055, pp. 82–99. Springer, Heidelberg (2010). https://doi.org/10.1007/978-3-642-12678-9_6
21. Menezes, A.J., van Oorschot, P.C., Vanstone, S.A.: Handbook of Applied Cryptography. CRC Press, Boca Raton (1996)
22. Paterson, K.G., Polychroniadou, A., Sibborn, D.L.: A coding-theoretic approach to recovering noisy RSA keys. In: Wang, X., Sako, K. (eds.) ASIACRYPT 2012. LNCS, vol. 7658, pp. 386–403. Springer, Heidelberg (2012). https://doi.org/10.1007/978-3-642-34961-4_24
23. Percival, C.: Cache Missing for Fun and Profit (2005). http://www.daemonology.net/papers/htt.pdf
24. Rivest, R.L., Shamir, A.: Efficient factoring based on partial information. In: Pichler, F. (ed.) EUROCRYPT 1985. LNCS, vol. 219, pp. 31–34. Springer, Heidelberg (1986). https://doi.org/10.1007/3-540-39805-8_3
25. Rivest, R.L., Shamir, A., Adleman, L.: A method for obtaining digital signatures and public-key cryptosystems. Commun. ACM. **21**, 120–126 (1978). https://doi.org/10.1145/359340.359342
26. RSA Laboratories: PKCS#1 v2.2: RSA Cryptography Standard (2012). https://www.emc.com/collateral/white-papers/h11300-pkcs-1v2-2-rsa-cryptography-standard-wp.pdf
27. Sarkar, S., Sen Gupta, S., Maitra, S.: Reconstruction and error correction of RSA secret parameters from the MSB side. In: WCC 2011 (2011)
28. Sarkar, S., Sen Gupta, S., Maitra, S.: Error correction of partially exposed RSA private keys from MSB side. In: Bagchi, A., Ray, I. (eds.) ICISS 2013. LNCS, vol. 8303, pp. 345–359. Springer, Heidelberg (2013). https://doi.org/10.1007/978-3-642-45204-8_26
29. Takayasu, A., Kunihiro, N.: A tool kit for partial key exposure attacks on RSA. In: Handschuh, H. (ed.) CT-RSA 2017. LNCS, vol. 10159, pp. 58–73. Springer, Cham (2017). https://doi.org/10.1007/978-3-319-52153-4_4
30. Yarom, Y., Falkner, K.: FLUSH+RELOAD: a high resolution, low noise, L3 cache side-channel attack. In: USENIX 2014, pp. 719–732 (2014)
31. Yarom, Y., Genkin, D., Heninger, N.: CacheBleed: a timing attack on OpenSSL constant time RSA. In: Gierlichs, B., Poschmann, A.Y. (eds.) CHES 2016. LNCS, vol. 9813, pp. 346–367. Springer, Heidelberg (2016). https://doi.org/10.1007/978-3-662-53140-2_17

Compact Implementation of Modular Multiplication for Special Modulus on MSP430X

Hwajeong Seo[1]([⊠]), Kyuhwang An[1], Hyeokdong Kwon[1], and Zhi Hu[2]

[1] Hansung University, Seoul, Republic of Korea
hwajeong84@gmail.com, tigerk9212@gmail.com, hdgwon@naver.com
[2] Central South University, Changsha, China
huzhi_math@csu.edu.cn

Abstract. For the pre/post-quantum Public Key Cryptography (PKC), such as Elliptic Curve Cryptography (ECC) and Supersingular Isogeny Diffie–Hellman key exchange (SIDH), modular multiplication is the most expensive operation among basic arithmetic of these cryptographic schemes. For this reason, the execution timing of such cryptographic schemes in an implementation level, which may highly determine the service availability for the low-end microprocessors (e.g., 8-bit AVR and 16-bit MSP430X), is mainly relied on the efficiency of modular multiplication on the target processors.

In this paper, we present new optimal modular multiplication techniques based on interleaved Montgomery multiplication on 16-bit MSP430X microprocessors, where the multiplication part is performed in a hardware multiplier and the reduction part is performed in a basic Arithmetic Logic Unit (ALU) with optimal modular multiplication routine, respectively. This approach is effective for special modulus of NIST curves, SM2 curves, and SIDH. In order to demonstrate the superiority of proposed Montgomery multiplication, we applied the proposed method to the NIST P–256 curve, of which the implementation improves the previous modular multiplication and squaring operations by 39% and 37.1% on 16-bit MSP430X microprocessors, respectively. Moreover, secure countermeasures against timing attack and simple power analysis is also applied to the scalar multiplication of NIST P–256, which achieves the 9,285,578 clock cycles and only requires 0.575 s (@16 MHz). The proposed Montgomery multiplication has broad applications to other cryptographic schemes and microprocessors.

Keywords: Montgomery multiplication · Public Key Cryptography · MSP430X · Software implementation

1 Introduction

Internet of Things (IoT) technology has been actively studied in academic and industry fields due to its useful applications, ranging from home automation,

© Springer Nature Switzerland AG 2019
K. Lee (Ed.): ICISC 2018, LNCS 11396, pp. 55–66, 2019.
https://doi.org/10.1007/978-3-030-12146-4_4

surveillance system, and health–care services. Unlike traditional service models, the IoT applications are able to provide highly customized services for each user by recognizing the customer's needs or preferences through actively collected data from remotely deployed IoT devices. However, the low-end IoT sensors are usually placed in the public space (building, road, and street), which are easily accessible and manipulated by any legitimate or malicious users. If the adversaries illegally capture the installed IoT devices and perform the sophisticated reverse engineering or any effective hacking measures, the secret information can be easily leaked.

In order to prevent the potential threats, the information of the IoT devices should be securely encrypted through the cryptography algorithm, namely Public Key Cryptography (PKC). However, the PKC requires the complicated computations and the low-end IoT devices have very limited resources, in terms of storage, energy, and computation power. In order to meet the sufficient service availabilities, the careful optimization techniques of implementations should be considered. The PKC instantiations such Elliptic Curve Cryptography (ECC) in pre-quantum case or Supersingular Isogeny Diffie–Hellman key exchange (SIDH) in post-quantum case highly rely on the efficient implementation of modular multiplication, which is the most expensive operations in finite field arithmetic. For this reason, the execution timing of modular multiplication determines the service availability for the low-end microprocessors (e.g., 8-bit AVR and 16-bit MSP430X embedded processors).

In this paper, we present new optimal modular multiplication techniques based on interleaved Montgomery multiplication on 16-bit MSP430X microprocessors, which are effective for special modulus of NIST curves, SM2 curves, and SIDH. In the proposed interleaved Montgomery multiplication, the multiplication part is performed in a hardware multiplier, while the reduction part is performed in a basic Arithmetic Logic Unit (ALU) with optimal routine. Specially, we applied the proposed method to the NIST P–256 curve, of which the implementation improves the previous modular multiplication and squaring operations by 39% and 37.1% for 16-bit MSP430X microprocessors, respectively. Moreover, secure countermeasures against timing attack and simple power analysis are applied to the scalar multiplication on NIST P–256 curve, which achieves the 9,285,578 clock cycles and only requires 0.575 s (@16 MHz). Our implementations imply that the proposed Montgomery multiplication would have broad applications to more cryptographic schemes (e.g., SM2 and SIDH) and microprocessors (e.g., 8-bit AVR).

The rest of this paper is organized as follows. In Sect. 2, we explore the previous works of Montgomery multiplication and target MSP430X processor. In Sect. 3, we present implementations of Montgomery multiplication and NIST P-256 on the MSP430X processor. In Sect. 4, we evaluate the proposed implementations on the target embedded processors. Finally, we conclude the paper in Sect. 5.

Algorithm 1. Calculation of the Montgomery reduction

Require: An odd m-bit modulus M, Montgomery radix $R = 2^m$, an operand T where
$T = A \cdot B$ or $T = A \cdot A$ in the range $[0, 2M - 1]$, and pre-computed constant
$M' = -M^{-1} \bmod R$
Ensure: Montgomery product $Z = \text{MonRed}(T, R) = T \cdot R^{-1} \bmod M$
1: $Q \leftarrow T \cdot M' \bmod R$
2: $Z \leftarrow (T + Q \cdot M)/R$
3: **if** $Z \geq M$ **then** $Z \leftarrow Z - M$ **end if**
4: **return** Z

2 Preliminaries and Related Works

2.1 Montgomery Multiplication

The modular reduction in School-book approach requires an expensive division operation, which is a high overheads on the low-end devices. Such expensive division operation can be transformed to the relatively cheap multiplication operation through Montgomery reduction, of which the detailed description is given in Algorithm 1.

The Montgomery reduction is proceeded as: given the intermediate result of multiplication $T = A \cdot B$ or $T = A \cdot A$ (where A and B are operands), T is multiplied by the inverse of modulus (M') and then the results are reduced by R and stored into Q. Afterward, the equation $((T + Q \times M)/R)$ is performed. Finally, the calculation of the Montgomery multiplication may require a final subtraction of the modulus (M) to get a reduced result in the range of $[0, M)$. Recently, Gueron and Krasnov presented the implementation of Montgomery multiplication friendly modulus [6]. When the modulus has a special pattern (0xFFFFFFFF in hexadecimal), this can be performed in addition and subtraction operations rather than multiplication. The approach is widely used in recent ECC and SIDH implementations and shows the highest performance [2,4,8–10].

2.2 Target Processors

The MSP430 family of microcontrollers are widely used in IoT fields, such as small satellite applications [12]. The most popular IoT platform is TelosB and TmoteSky. The MSP430 microcontrollers have 16-bit instruction sets and 12 general-purpose registers. The specifications of clock frequency and ROM/RAM varies for each model. The MSP430 supports a number of instruction sets, including addition, subtraction, and basic operations. The detailed basic arithmetic is given in Table 1.

In particular, the integer multiplication is carried out with a memory–mapped hardware multiplier. The cost of multiplication is the cost of writing the operands and reading the result to/from a multiplier's memory address in the MSP430 embedded processors. The operands can be accessed by four different addressing modes, including register direct, indexed, register indirect, and indirect with auto-increment.

Table 1. Instruction set summary for MSP

asm	Operands	Description	Operation	#Clock
ADD	Rr, Rd	Add without carry	Rd ← Rd+Rr	1
ADDC	Rr, Rd	Add with carry	Rd ← Rd+Rr+C	1
SUB	Rr, Rd	Sub without borrow	Rd ← Rd-Rr	1
SUBC	Rr, Rd	Sub with borrow	Rd ← Rd-Rr-B	1
MOV	Rr, Rd	Move	Rd ← Rr	1
CLR	Rd	Clear	Rd ← 0	1

Recently, advanced MSP430X microcontrollers have been introduced. The MSP430X supports 20-bit addressing pointers and a new 32-bit hardware multiplier. This sophisticated 32-bit hardware multiplier significantly improves the performance of traditional MSP430 implementation based on 16-bit hardware multiplier. The hardware multiplier supports both 32-bit multiplication and 32-bit Multiplication & ACcumulation (MAC) modes. In order to select the multiplication modes, the 32-bit operands should be written into specific memory addresses (multiplication: MPY32L, MPY32H, MAC: MAC32L, MAC32H) by two 16-bit. Particularly, the MAC mode efficiently accumulates the intermediate results into the result memory (RES0, RES1, RES2, RES3) and sets the carry bit into the carry memory (SUMEXT). The multiplier is triggered by writing the 32-bit operands into the operand memory (OP2L, OP2H). Afterward, the 65-bit results are accessible through result and carry memory addresses (RES0, RES1, RES2, RES3, SUMEXT).

Many previous works used the product-scanning multiplication over MSP430X hardware multiplier since the MAC mode efficiently accumulates the intermediate results in a column-wise fashion with small number of memory accesses [5,14]. In this work, we also adopted the product-scanning method for multiplication, but we used a basic ALU for reduction over the MSP430X microprocessors for special modulus.

3   Proposed Montgomery Multiplication

In this section, we explore the efficient implementation of Montgomery multiplication for special modulus. The target modulus consists of special patterns (0x00000000, 0x00000001, and 0xFFFFFFFF in hexadecimal), which can be performed in simple addition and subtraction operations rather than complicated multiplication. Though we target the NIST P–256, the proposed method can be applied to the other cryptographic algorithms, such as SM2 and SIDH.

3.1   Constant Modular Addition/Subtraction for Special Modulus

Finite field addition (resp. subtraction) operation requires the final subtraction (resp. addition) with target modulus after addition (resp. subtraction) to fit the intermediate results in the range of target field. When the data format is **unsigned**, the reduced result should not generate the overflow bits. If we perform the conditional final subtraction or addition operation, the execution timing or power consumption becomes varied depending on the conditional statements. Since the program routines are highly correlated with secret values, the adversary may get the secret information from conditional execution of final subtraction for reduction [18].

In order to avoid the conditional statements, the constant-time reduction is introduced by Liu et al. in [11], which utilizes the conditional reduction (i.e. a multi-precision subtraction) of field arithmetic with the mask. After executing the first part of modular addition (i.e. $A + B$), it first generates the 2's complement of carry, and it can be the value ($mask$). When the carry bit is set, the $mask$ is always set to 0xFF. Otherwise, the value is set to zero (0x00). The masked modulo is then subtracted without the comparison. In [19], the optimized reduction technique for special modulus is introduced. For the NIST P-256 curve, the modulus $p_{256} = 2^{256} - 2^{224} + 2^{192} + 2^{96} - 1$ can be rewritten as

0xFFFFFFFF00000001000000000000000000000000FFFFFFFFFFFFFFFFFFFFFFFF

in hexadecimal[1], which consists of only three special patterns as 0xFFFF, 0x0000, and 0x0001, in 16-bit wise hexadecimal way. Among them two patterns (i.e. 0xFFFF and 0x0001) are only masked and utilized them for reduction since 0x0000 pattern does not require the masked reduction. These features are highly utilized in MSP430X microprocessors. The pattern (0x0001) is obtained from carry and the remaining pattern is obtained through one subtraction with ZERO register and CARRY register (i.e. 0x0000 - 0x0001 = 0xFFFF). The details are given in Algorithm 2.

Algorithm 2. Masked subtraction for NIST P-256 on MSP430X

Input: carry register (CARRY), temporal
    register (MASK)
Output: result pointer (RESULT)
1: CLR MASK
2: SUB CARRY, MASK

3: SUB MASK, 2*0(RESULT)
4: SUBC MASK, 2*1(RESULT)
5: SUBC MASK, 2*2(RESULT)
6: SUBC MASK, 2*3(RESULT)
7: SUBC MASK, 2*4(RESULT)
8: SUBC MASK, 2*5(RESULT)

9: SBC 2*6(RESULT)
10: SBC 2*7(RESULT)
11: SBC 2*8(RESULT)
12: SBC 2*9(RESULT)
13: SBC 2*10(RESULT)
14: SBC 2*11(RESULT)

15: SUBC CARRY, 2*12(RESULT)
16: SBC 2*13(RESULT)

17: SUBC MASK, 2*14(RESULT)
18: SUBC MASK, 2*15(RESULT)

As above demonstration, the MASK register is firstly set to zero, and then subtracted by CARRY register. When the CARRY register is set to 1, the MASK

---

[1] SM2 curve also has similar special patterns of modulus (0xFFFFFFFFEFFFFFFF FFFFFFFFFFFFFFFFFFFFFFFF00000000FFFFFFFFFFFFFFFF).

register is always set to 0xFFFF (in hexadecimal). Otherwise, both CARRY and MASK registers are set to 0. By using an efficient memory based operation of MSP430X processor, the masked values are directly subtracted from the intermediate results (i.e. RESULT). For the case of modular subtraction, the borrow bit is used for MASK register and the least significant bit of MASK register is extracted to CARRY register through AND instruction with value (0x0001).

3.2 Interleaved Montgomery Multiplication/Squaring for Special Modulus

Generic n-word Montgomery multiplication requires (n^2+n) multiplication. The Montgomery multiplication consists of multiplication and reduction parts. Both parts can be implemented in interleaved or separated way. On one hand, the advantage of separated version combines any multiplication and reduction methods without difficulties. On the other hand, the interleaved version optimizes the number of memory access for intermediate results. In this paper, the interleaved version is adopted since the hardware multiplier of MSP430X is very efficient to handle the accumulation of intermediate results. In Fig. 1, the comparison of procedures for interleaved Montgomery multiplication in hardware utilization are described.

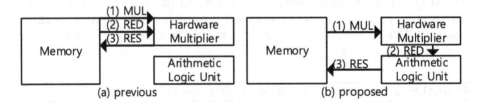

Fig. 1. Procedures for Montgomery multiplication in hardware utilization, MUL: multiplication, RED: reduction, RES: result

Note that previous methods only utilized the hardware multiplier for both multiplication and reduction. This approach is efficient for original Montgomery multiplication. However, the proposed method performs the multiplication in the hardware multiplier and the reduction in the basic arithmetic logic unit. This approach shows better performance than previous work when it comes to special modulus.

Register and Memory Utilization. Since the performance is highly relied on the number of memory accesses, the optimized register utilization is very important for high-speed implementations. MSP430X microprocessor equips only 12 general purpose registers, among which five, two, one, and three registers are assigned for intermediate results, temporal storage, memory address of intermediate results in hardware multiplier, and memory address of operands as well as results, respectively. Every operand of multiplication is directly assigned to hardware multiplier and the 96-bit wise intermediate results are cached in the five 16-bit registers, which is used for efficient reduction based on the basic arithmetic. Montgomery multiplication needs to keep Q operands to perform the reduction, which are dynamically loaded/stored from/to the STACK.

Modular Reduction. Our modular multiplication combined both hardware-aided multiplication and basic Arithmetic Logic Unit (ALU) based modular reduction. At first we follow the product–scanning multiplication (i.e. column-wise multiplication) routines, which can be implemented with Multiplication-and–ACcumulation mode of hardware multiplier. Afterward the intermediate results are loaded to some 16-bit registers and then reduced. Different from previous Montgomery reduction which utilizes the product–scanning based multiplication, in our reduction we exploited the properties of special modulus and thus replaced the expensive multiplication into addition/subtraction operations.

Algorithm 3. Montgomery Multiplication in second column for NIST P-256 on MSP430X

Input: operand pointers (APTR and BPTR), memory address of carry bit in hardware multiplier (SPTR), temporal registers (TO and T1)

Output: stack pointer R1, intermediate results (C0, C1, C2, C3, CARRY)

...

```
 1: MOV @APTR+, &MPY32L
 2: MOV @APTR+, &MPY32H
 3: MOV @BPTR+, &OP2L
 4: MOV @BPTR+, &OP2H

 5: MOV @APTR+, &MAC32L
 6: MOV @APTR+, &MAC32H
 7: SUB #2*4, BPTR
 8: MOV @BPTR+, &OP2L
 9: MOV @BPTR+, &OP2H

10: ADD @RL+, C0
11: ADDC @RL+, C1
12: ADDC @RL+, C2
13: ADDC @RL+, C3
14: ADDC @SPTR, CARRY
15: SUB #2*4, RL
```

```
16: MOV @R1+, TO
17: MOV @R1+, T1
18: ADD TO, C2
19: ADDC T1, C3
20: ADC CARRY

21: SUB TO, C0
22: SUBC T1, C1
23: SBC C2
24: SBC C3
25: SBC CARRY

26: SUB #2*2, R1
27: MOV C0, 2*2(R1)
28: MOV C1, 2*3(R1)

29: ADD C2, C0
30: ADDC C3, C1
31: CLR C2
32: CLR C3
33: ADDC CARRY, C2
34: CLR CARRY
...
```

For example, the modulus for NIST P-256 curve consists of three patterns in hexadecimal way, which includes 0x00000000, 0x00000001, and 0xFFFFFFFF.

Since the 0x00000000 pattern does not require any computations, the routine is optimized away. The 0x00000001 pattern only requires five 16-bit wise addition, and the operands are directly loaded from memory and added to the memory. The 0xFFFFFFFF pattern requires three 16-bit wise addition and five 16-bit wise subtraction operations, where both operations requires identical 32-bit operands. We firstly load the 32-bit operands to two 16-bit registers (temporal storages) and used the operands twice for 32-bit addition and 32-bit subtraction, respectively. When the 0xFFFFFFFF pattern appears before operand generation, five 16-bit wise subtraction operations are optimized away because the least significant double-word is always set to zero.

The detailed descriptions of Montgomery multiplication in second column for NIST P-256 on MSP430X are given in Algorithm 3. It can be viewed that from Step 1 to 15, two partial products are obtained in the product-scanning way, while from Step 16 to 34, Montgomery reduction with 0xFFFFFFFFFFFFFFFF is performed in simple addition and subtraction.

Final Reduction. The last step of Montgomery multiplication may require the final subtraction to get reduced results. We adopted the masked subtraction described in Algorithm 2.

Modular Squaring. The squaring operation is also frequently called in the cryptographic implementations. For the straight-forward squaring implementation, we can directly use the multiplication for squaring by setting both operands to identical values. However, the multiplication routine does not ensure the highest performance for squaring operation since some memory accesses/partial products can be optimized by loading/performing once rather than twice. The detailed descriptions are given in Algorithm 4. Note that from Step 1 to 6, the partial product is obtained. When the part of operand for partial product is identical, we only need to assign it rather than full operands.

Algorithm 4. Partial products for squaring operations on MSP430X

Input: operand pointers (APTR and BPTR), memory address of carry bit in hardware multiplier (SPTR)
Output: intermediate results (CARRY)
...
1: MOV @APTR+, &MAC32L
2: MOV @APTR+, &MAC32H
3: SUB #2*4, BPTR
4: MOV @BPTR+, &OP2L

5: MOV @BPTR+, &OP2H
6: ADD @SPTR, CARRY

7: SUB #2*2, BPTR
8: MOV @BPTR+, &OP2L
9: MOV @BPTR+, &OP2H
10: ADD @SPTR, CARRY
...

3.3   Implementation of NIST P–256 on MSP430X Microprocessors

The first implementation of ECC on MSP430X belongs to Gouvêa et al. [5], where they utilized the new 32-bit hardware multiplier instructions of MSP430X. Particularly, the new 32-bit hardware multiplier enhances the previous 16-bit hardware multiplier based prime field multiplication by about 45%. The combination of optimized algorithms and hardware shows that ECC at the security level of 128-bit is feasible for the MSP430X. Seo et al. intensively studied on multi-precision multiplication and squaring operations on MSP430 processors [15–17], where they optimized the register usages by caching the operands and memory access through incremental addressing mode.

In LatinCrypt'14, Hinterwälder et al. suggested Curve25519 for MSP430 microcontrollers [7], in which they avoided conditional jumps and loads to prevent timing attacks. Moreover, they provided a comprehensive evaluation of different implementations of the modular multiplication, based on which the Curve25519 implementations on MSP430X having 16-bit and 32-bit hardware multipliers achieved 9.1M and 6.5M cycles, respectively. Düll et al. in [3] optimized the X25519 key-exchange protocol for MSP430X 16-bit microcontrollers, and their implementations for MSP430X takes 5,301,792 cycles (32-bit multiplier) and 7,933,296 cycles (16-bit multiplier) for the computation of Diffie–Hellman key exchange. The computation is performed in less than a second if clocked at 16 MHz for a security level of 128 bits. Recently, Seo in [14] presented size optimized implementation of Curve25519, where he utilized hardware multiplier and accelerated the performance through the optimized multiplication routines in product-scanning way.

In this work, we targeted the special modulo of NIST P–256, and implemented desired cryptographic primitives. The NIST P–256 elliptic curve is given by

$$E/\mathbb{F}_{p_{256}} : y^2 = x^3 - 3 \cdot x + b, \ p_{256} = 2^{256} - 2^{224} + 2^{192} + 2^{96} - 1,$$

and other details can be referred to the FIPS 186-2 standard [1]. For finite field arithmetic, we mainly follow the proposed techniques described in Sect. 3 to do the modular addition/subtraction and modular multiplication/squaring operations. Moreover, we adopted the constant-time finite field inversion of NIST P–256, which is performed by powering $p_{256} - 2$. Such inversion can be computed at a cost of 255S + 13M by following Algorithm 2 in [19]. For elliptic curve group arithmetic, we utilized the Montgomery ladder using co-Z Jacobian arithmetic with X and Y coordinates only, which ensures the fast and regular Montgomery ladder algorithm for scalar multiplication [13]. Since the regular Montgomery ladder algorithm does not require conditional statements, the implementation is always constant timing, and thus secure against the simple power analysis and timing attacks.

4   Evaluation

We implemented the NIST P–256 by using the proposed method on 16-bit MSP430X microprocessors (i.e. MSP430F5529) and evaluated the performance of implementations in execution time (clock cycles).

In Table 2, the detailed descriptions of performance evaluation for finite field operations are given. Note that addition and subtraction operations are much cheaper than multiplication and squaring operations (i.e., 8.x faster). It is also natural that the squaring operation is faster (by 4.6%) than multiplication through dedicated squaring routine in this paper. What's more, the inversion is implemented based on Fermat's little theorem, which is a regular fashion and ensures constant timing.

Table 2. Performance evaluation (execution timing in clock cycles) of finite field addition, subtraction, multiplication, squaring and inversion operations for NIST P–256 on 16-bit MSP430X microprocessors.

ADD	SUB	MUL	SQR	INV
227	228	2,019	1,926	522,040

We also give the comparison results of NIST P–256 with previous work as Table 3. For the most performance-critical operations, our proposed modular multiplication and squaring operations improve the performance of those in [5] by 39% and 37.1%, respectively. Such performance enhancements are achieved through optimized memory access, register utilization, and efficient modular reduction techniques. Moreover, this performance improvement directly influences the performance of scalar multiplication.

Table 3. Comparison of NIST P–256 implementations on 16-bit MSP430X microprocessors

Method	MUL	SQR	Scalar MUL	Cache Attack	Timing Attack
Gouvêa et al. [5]	3,315	3,064	5,321,776	–	–
This work	2,019	1,926	9,122,988	√	√

Though previous implementation of scalar multiplication requires 5,321,776 clock cycles [5], which is faster than ours. This is mainly because their implementation utilized the NAF method for scalar multiplication, which requires pre-computed Look-Up Table (LUT) to accelerate the performance. However, the frequent LUT access increases cache hit rates and may cause cache attack. It should be noted that in [5] the point addition and doubling chain is not a regular fashion, which would be vulnerable to timing attack and leak the secret information.

In order to avoid the potential side channel attacks, we also implemented the scalar multiplication on NIST P–256 in regular fashion as the Montgomery ladder algorithm. Thus constant timing finite field arithmetic and regular elliptic curve group arithmetic result in constant timing scalar multiplication implementation. Even though we sacrifice the performance, the implementation is much secure than previous works.

5 Conclusion

In this paper, we present new optimal modular multiplication techniques for special modulus based on interleaved Montgomery multiplication on 16-bit MSP430X microprocessors. The multiplication part of Montgomery multiplication is performed in the hardware multiplier, while the reduction operation is performed in the basic Arithmetic Logic Unit (ALU) with an optimal routine. Furthermore, the final subtraction is efficiently handled through masked subtraction for the target embedded processors.

The proposed implementation improves the previous modular multiplication and squaring operations for NIST P–256 curve by 39% and 37.1% for 16-bit MSP430X microprocessors, respectively. Based on the improved Montgomery multiplication, the scalar multiplication of NIST P–256 is efficiently constructed. The implementation utilized the Co-Z representation and security countermeasures against timing attack and simple power analysis. The proposed implementation of scalar multiplication achieves 9,122,988 clock cycles and requires only 0.575 s (@16 MHz).

We hope that such techniques for modular multiplication with special modulus on MSP430X microprocessor would improve the performance (as well as implementation security) of cryptographic primitives, which are thus applicable for more cryptographic schemes (such as SM2/NIST ECC and SIDH) and more platforms (such as 8-bit AVR).

Acknowledgement. This work was partly supported by the National Research Foundation of Korea (NRF) grant funded by the Korea government (MSIT) (No. NRF-2017R1C1B5075742) and the MSIT(Ministry of Science and ICT), Korea, under the ITRC (Information Technology Research Center) support program (2014-1-00743) supervised by the IITP (Institute for Information & communications Technology Promotion). The work of Zhi Hu is partially supported by the Natural Science Foundation of China (Grant No. 61602526).

References

1. FIPS 186-2: Digital signature standard (DSS). Federal Information Processing Standards Publication 186–2, National Institute of Standards and Technology (2000)
2. Adalier, M.: Efficient and secure elliptic curve cryptography implementation of Curve P-256. In: Workshop on Elliptic Curve Cryptography Standards (2015)

3. Düll, M., et al.: High-speed Curve25519 on 8-bit, 16-bit, and 32-bit microcontrollers. Des. Codes Crypt. **77**(2–3), 493–514 (2015)
4. Faz-Hernández, A., López, J., Ochoa-Jiménez, E., Rodríguez-Henríquez, F.: A faster software implementation of the supersingular isogeny Diffie-Hellman key exchange protocol. IEEE Trans. Comput. **67**(11) (2017)
5. Gouvêa, C.P., Oliveira, L.B., López, J.: Efficient software implementation of public-key cryptography on sensor networks using the MSP430X microcontroller. J. Cryptogr. Eng. **2**(1), 19–29 (2012)
6. Gueron, S., Krasnov, V.: Fast prime field elliptic-curve cryptography with 256-bit primes. J. Cryptogr. Eng. **5**(2), 141–151 (2015)
7. Hinterwälder, G., Moradi, A., Hutter, M., Schwabe, P., Paar, C.: Full-size high-security ECC implementation on MSP430 microcontrollers. In: Aranha, D.F., Menezes, A. (eds.) LATINCRYPT 2014. LNCS, vol. 8895, pp. 31–47. Springer, Cham (2015). https://doi.org/10.1007/978-3-319-16295-9_2
8. Jalali, A., Azarderakhsh, R., Kermani, M.M., Jao, D.: Supersingular isogeny Diffie-Hellman key exchange on 64-bit ARM. IEEE Trans. Dependable Secure Comput. (2017)
9. Koziel, B., Jalali, A., Azarderakhsh, R., Jao, D., Mozaffari-Kermani, M.: NEON-SIDH: efficient implementation of supersingular isogeny Diffie-Hellman key exchange protocol on ARM. In: Foresti, S., Persiano, G. (eds.) CANS 2016. LNCS, vol. 10052, pp. 88–103. Springer, Cham (2016). https://doi.org/10.1007/978-3-319-48965-0_6
10. Liu, Z., Seo, H., Castiglione, A., Choo, K.K.R., Kim, H.: Memory-efficient implementation of elliptic curve cryptography for the Internet-of-Things. IEEE Trans. Dependable Secure Comput. (2018)
11. Liu, Z., Seo, H., Großschädl, J., Kim, H.: Efficient implementation of NIST-compliant elliptic curve cryptography for 8-bit AVR-based sensor nodes. IEEE Trans. Inf. Forensics Secur. **11**(7), 1385–1397 (2016)
12. Peters, D., Raskovic, D., Thorsen, D.: An energy efficient parallel embedded system for small satellite applications. ISAST Trans. Comput. Intell. Syst. **1**(2), 8–16 (2009)
13. Rivain, M.: Fast and regular algorithms for scalar multiplication over elliptic curves. IACR Cryptology Eprint Archive (2011)
14. Seo, H.: Compact software implementation of public-key cryptography on MSP430X. ACM Trans. Embed. Comput. Syst. (TECS) **17**(3), 66 (2018)
15. Seo, H., Kim, H.: Multi-precision squaring on MSP and ARM processors. In: 2014 International Conference on Information and Communication Technology Convergence, ICTC, pp. 356–361. IEEE (2014)
16. Seo, H., Lee, Y., Kim, H., Park, T., Kim, H.: Binary and prime field multiplication for public key cryptography on embedded microprocessors. Secur. Commun. Netw. **7**(4), 774–787 (2014)
17. Seo, H., Shim, K.A., Kim, H.: Performance enhancement of TinyECC based on multiplication optimizations. Secur. Commun. Netw. **6**(2), 151–160 (2013)
18. Walter, C.D., Thompson, S.: Distinguishing exponent digits by observing modular subtractions. In: Naccache, D. (ed.) CT-RSA 2001. LNCS, vol. 2020, pp. 192–207. Springer, Heidelberg (2001). https://doi.org/10.1007/3-540-45353-9_15
19. Zhou, L., Su, C., Hu, Z., Lee, S., Seo, H.: Lightweight implementations of NIST P-256 and SM2 ECC on 8-bit resource-constraint embedded device. ACM Trans. Embed. Comput. Syst. (TECS) (2018)

Homomorphic Encryption

Multi-identity IBFHE and Multi-attribute ABFHE in the Standard Model

Xuecheng Ma[1,2] and Dongdai Lin[1,2(✉)]

[1] State Key Laboratory of Information Security,
Institute of Information Engineering, Chinese Academy of Sciences,
Beijing 100093, China
{maxuecheng,ddlin}@iie.ac.cn
[2] School of Cyber Security, University of Chinese Academy of Sciences,
Beijing 100049, China

Abstract. The notion of multi-identity IBFHE is an extension of identity based fully homomorphic (IBFHE) encryption. In 2015, Clear and McGoldrick (CRYPTO 2015) proposed a multi-identity IBFHE scheme that is selectively secure in the random oracle model under the hardness of Learning with Errors (LWE). At TCC 2016, Brakerski et al. presented multi-target ABFHE in the random oracle where the evaluator should know the target policy. In this paper, we present a multi-identity IBFHE scheme and a multi-attribute ABFHE scheme in the standard model. Our schemes can support evaluating circuits of unbounded depth but with one limitation: there is a bound N on the number of ciphertexts under the same identity or attribute involved in the computation. The bound N could be thought of as a bound on the number of independent senders. Our schemes allow N to be exponentially large so we do not think it is a limitation in practice. Our construction combines *fully* multi-key FHE and leveled *single-identity* IBFHE or *single-attribute* ABFHE, both of which have been realized from LWE, and therefore we can instantiate our construction that is secure under LWE. Moreover, our multi-attribute ABFHE is non-target where the public evaluator do not need to know the policy.

Keywords: Multi-identity · Multi-attribute ·
Homomorphic encryption · Standard model

1 Introduction

Identity Based Encryption (IBE) is proposed in 1984 by Shamir [Sha84] which is a generalization of public key encryption where the public key of a user can be arbitrary string such as an email address, IP address or staff number, depending on the application. The first realizations of IBE are given by [SOK00, BF01] using groups equipped with bilinear maps. Subsequently, realizations from

© Springer Nature Switzerland AG 2019
K. Lee (Ed.): ICISC 2018, LNCS 11396, pp. 69–84, 2019.
https://doi.org/10.1007/978-3-030-12146-4_5

bilinear maps [BB04a, BB04b, Wat05, Wat09], from quadratic residues modulo composite [Coc01, BGH07], from lattices [GPV08, CHKP10, ABB10] and from the computational Diffie-Hellman assumption [DG17] have been proposed.

Attribute-based encryption (ABE)[1] [SW05, GPSW06] is a generalization of IBE that allows to implement access control. A (master) public key mpk is used for encryption, and users are associated to secret keys sk_f corresponding to policy functions $f : \{0, 1\}^\ell \to \{0, 1\}$. The encryption of a message μ is labeled with a public attribute $x \in \{0, 1\}^\ell$, and can be decrypted using sk_f if and only if $f(x) = 0$. The security guarantee of ABE is collusion resistance: a coalition of users learns nothing about the plaintext message μ if none of their individual keys are authorized to decrypt the ciphertext. Goyal, Pandey, Sahai and Waters [GPSW06] constructed ABE for log-depth circuits using bilinear maps. Gorbunov, Vaikuntanathan and Wee [GVW13] presented the first ABE scheme where the policies can be arbitrary (a-priori bounded) polynomial circuits from LWE. Boneh et al. [BGG+14] showed an ABE scheme improving the size of secret key.

Fully homomorphic encryption (FHE) is first presented in 1987 by Rivest, Adleman and Dertouzos [RAD78]. Then Gentry [Gen09a, Gen09b] proposed the first construction in a breakthrough work in 2009. Since then, there are some follow-up works [BV11, BGV12, Bra12, GSW13, AP14] for improving efficiency and security. In 2013, Gentry, Sahai and Waters [GSW13] proposed a FHE scheme without an evaluation key which makes it enable to compile IBE with some properties into identity based (leveled) fully homomorphic encryption (IBFHE). Their compiler can also be applied to ABE that yields an attribute based (leveled) fully homomorphic encryption. Clear and McGoldrick [CM14] make IBFHE and ABFHE bootstrappable by using programm obfuscation.

López-Alt, Tromer and Vaikuntanthan [LTV12] considered an extension of homomorphic encryption into the multi-key setting, where it is possible to compute on encrypted messages even if they were not encrypted using the same key. In multi-key FHE (MKFHE), a public evaluator takes ciphertexts encrypted under different keys, and evaluates arbitrary (maybe with bounded depth) functions on them. The resulting ciphertext can then be decrypted using the collection of keys of all parties involved in the computation. Note that the security of the encryption scheme compels that all secret keys need to be used for decryption. [LTV12] constructed an on-the-fly multiparty computation (MPC) protocol by applying multi-key FHE. The next step forward was by Clear and McGoldrick [CM15] who proposed a multi-key FHE in the standard model and a multi-identity IBFHE in the random oracle. Note that the compiler in [GSW13] can only yield a *single-identity* IBFHE. As a stepping stone, they were able to construct a multi-key FHE scheme based on the hardness of the learning with errors (LWE) problem [Reg05, Reg09], which is related to the hardness of certain short vector problems (such as GapSVP, SIVP) in worst case lattices.

---

[1] There are other variants such as ciphertext-policy ABE, but we focus on key-policy ABE here.

The first multi-identity IBFHE proposed by [CM15] is a leveled multi-identity IBFHE in the random oracle. [CM16] constructed a *fully* multi-identity IBFHE scheme that supports unbounded-depth circuits with bounded inputs by combining a MKFHE and multi-identity IBFHE. A natural question is that *Can we construct a multi-identity IBFHE supporting unbounded-depth circuits in the standard model on the standard assumption?*

Brakerski et al. [BCTW16] proposed a leveled multi-target ABFHE in the random model where the evaluator should know the target policy. Hiromasa and Kawai [HK17] extended leveled multi-target ABFHE to dynamic homomorphic evaluation. But these multi-target ABFHE schemes are all in the random oracle model. *Can we construct a multi-attribute ABFHE supporting unbounded-depth circuits in the standard model on the standard assumption?*

1.1 Our Contributions

We propose constructions of multi-identity IBFHE and multi-attribute ABFHE in the standard model. Our schemes support unbounded evaluation circuit depth but with one limitation: the number of ciphertexts joining the function computation under the same identity or attribute is bounded but can be exponential. Our construction combines a *fully* MKFHE and *single-identity* IBFHE or *single-attribute* ABFHE. There are instantiations of *single-identity* IBFHE, *fully* MKFHE [CM15, BP16, PS16, CZW17] and *single-attribute* ABFHE [GSW13] from LWE, so our construction can be instantiated from LWE. In order to construct a multi-identity IBFHE or multi-attribute ABFHE, Clear and McGoldrick [CM16] combine a fully MKFHE and leveled multi-identity IBFHE or multi-attribute ABFHE. So their proposal is not in the standard model without a multi-identity IBFHE or multi-attribute ABFHE in the standard model. In fact, their purpose is to make the multi-identity IBFHE support unbounded evaluation circuit depth while our goal is to construct a multi-identity IBFHE or multi-attribute ABFHE that supports *unbounded* evaluation circuit depth in the *standard model*. Brakerski et al. [BCTW16] proposed a multi-target ABFHE in the random oracle where the policy should be known to the evaluator while ours is *non-target* in the *standard model*.

1.2 Our Construction

Our construction is combining a MKFHE and a *single-identity* IBFHE or *single-attribute* ABFHE. The constructions of multi-identity IBFHE and multi-attribute ABFHE are similar, so we will show our high level idea of the construction of multi-identity IBFHE, the detail will be presented in Sect. 4. Our construction is similar to [CM16]. Both of our construction is using MKFHE to evaluate the circuit and then decrypts the evaluated MKFHE ciphertext by evaluation with the IBFHE encryptions of corresponding secret keys which makes the final resulting ciphertext compact. The difference here is that [CM16] decrypts the evaluated MKFHE ciphertext completely which needs a multi-identity IBFHE while we just partially decrypts it which makes the final resulting

IBFHE ciphertexts compact. It is the point that *single-identity* IBFHE works here.

We present the overview of our construction as follows: The setup algorithm generates params of MKFHE and (mpk, msk) of IBFHE by running their setup algorithms respectively. The extract algorithm is the same as that of *single-identity* IBFHE. When encrypts a plaintext μ_i, the sender generates a pair of key (pk_i, sk_i) of MKFHE, and encrypts sk_i using encryption algorithm of IBFHE and then encrypts the plaintext using pk_i. When evaluates a function f, run the evaluation algorithm of MKFHE and obtain the MKFHE encryption of $f(\mu_1, ..., \mu_\ell)$, then partially decrypt the evaluated MKFHE ciphertext with IBFHE ciphertexts of the collection of secret keys corresponding the same identity. The resulting ciphertext is compact because the number of compact IBFHE ciphertexts in the final resulting ciphertexts are independent on the number of input ciphertexts and the size of the evaluation function. If the ciphertext is "fresh", we just obtain the secret key of MKFHE by decrypting IBFHE ciphertext and then decrypt the MKFHE ciphertext. If the ciphertext is evaluated, we can obtain the partial decryption by decrypting the IBFHE ciphertexts and finish the remaining decryption procedure of MKFHE.

1.3   Other Related Work

Clear and McGoldrick [CM15] extended the scheme of [GSW13] to the multi-identity setting and obtain a multi-identity IBFHE scheme that is selectively secure in the random oracle model under the hardness of Learning with Errors (LWE). Their scheme was simplified by Mukherjee and Wichs [MW16] who used multi-key FHE to construct a 2-round MPC protocols in the CRS model. Recently, Peikert and Shiehian [PS16] put forth a notion of *multi-hop* MKFHE, in which the result ciphertexts of homomorphic evaluations can be used in further homomorphic computations involving additional parties. Chen et al. [CZW17] then presented a compact *multi-hop* MKFHE which is based on Brakerski-Gentry-Vaikuntanathan (BGV) FHE scheme. Brakerski and Perlman [BP16] presented a similar notion called *fully dynamic* MKFHE that supports an unbounded number of homomorphic operations for an unbounded number of parties. Canetti et al. [CRRV17] show that CPA secure multi-identity IBFHE can be used to construct CCA1 secure homomorphic encryption.

2   Preliminaries

Let ℓ_q denote $\lfloor \log q \rfloor + 1$ and $\hat{m} = m \cdot \ell_q$. Let $a \in \mathbb{Z}_q^m$ be a vector of some dimension m over $\mathbb{Z}_q$ and $A \in \mathbb{Z}_q^{n \times m}$ be a matrix. $A[i]$ means the i-th row of A. We can see a vector as a matrix where $n = 1$. BitDecomp(a): We define an algorithm BitDecomp that takes as input a vector $a \in \mathbb{Z}_q^m$ and outputs an $\hat{m}$-dimensional vector $(a_{1,0}, ..., a_{1,\ell_q-1}, ..., a_{m,0}, ..., a_{m,\ell_q-1})$ where $a_{i,j}$ is the j-th bit in a_i's binary representation (ordered from least significant to most significant).

Binary(A): It takes a matrix $A \in \mathbb{Z}_q^{n \times m}$ and outputs a $(n \cdot \hat{m})$-dimensional vector $(\text{BitDecomp}(A[1]), ..., \text{BitDecomp}(A[n]))$.

Definition 1 ([BHHO08]). *A public key encryption scheme PKE is said to be* **weakly circular secure** *if it is secure even against an adversary who gets encryptions of the bits of the secret key.*

2.1 Multi-identity IBFHE

A Multi-Identity IBFHE scheme is defined with respect to a message space $\mathcal{M}$, an identity space $\mathcal{I}$, a class of circuits $\mathbb{C} \subset \mathcal{M}^* \rightarrow \mathcal{M}$ and ciphertext space C. A Multi-identity IBFHE scheme is a tuple of *ppt* algorithms (Setup, KeyGen, Encrypt, Decrypt, Eval) defined as follows:

- Setup(1^λ): On input (in unary) a security parameter λ, generate public parameters MPK and a master secret key MSK. Output (MPK, MSK).
- KeyGen(MSK, id): On input master secret key MSK and an identity id: derive and output a secret key $\mathsf{sk}_{\mathsf{id}}$ for identity id.
- Encrypt(MPK, id, μ): On input public parameters MPK, an identity id, and a message $\mu \in \mathcal{M}$, output a ciphertext $c \in C$ that encrypts μ under identity id.
- Decrypt($\mathsf{sk}_{\mathsf{id}_1}, ..., \mathsf{sk}_{\mathsf{id}_n}, c$): On input n secret keys $\mathsf{sk}_{\mathsf{id}_1}, ..., \mathsf{sk}_{\mathsf{id}_n}$ for (resp.) identities $\mathsf{id}_1, ..., \mathsf{id}_n$ and a ciphertext $c \in C$, output $\mu \in \mathcal{M}$ if c is a valid encryption under identities $\mathsf{id}_1, ..., \mathsf{id}_n$; output a failure symbol $\perp$ otherwise.
- Eval(MPK, C, $c_1, ..., c_\ell$): On input public parameters MPK, a circuit $\mathsf{C} \in \mathbb{C}$ and ciphertexts $c_1, ..., c_\ell \in C$, output an evaluated ciphertext $\hat{c} \in C$.

For all choices of Setup(1^λ) $\rightarrow$ (MPK, MSK),$\mathsf{id}_1, ..., \mathsf{id}_n, j_1, ..., j_\ell \in [n]$, $c_i =$ Encrypt(MPK, id_{j_i}, μ_i) $(\mu_i \in \mathcal{M})$,C : $\mathcal{M}^\ell \rightarrow \mathcal{M}$,$\hat{c} =$ Eval(MPK, C,$c_1, ..., c_\ell$)

- **Correctness.**

$$\mathsf{Decrypt}(\mathsf{sk}_{\mathsf{id}_1}, ..., \mathsf{sk}_{\mathsf{id}_n}, \hat{c}) = \mathsf{C}(\mu_0, ... \mu_\ell)$$

- **Compactness.**

$$|\hat{c}| \leq \mathsf{poly}(\lambda, n)$$

where n is the number of distinctive identities.
- **Security.** The security of multi-identity IBFHE is the same with the security of IBE.

3 Building Blocks from Previous Works

3.1 Fully Multi-key FHE

A homomorphic encryption scheme is multi-key if it can evaluate circuits on ciphertexts encrypted under different public keys. It is called leveled MKFHE if its setup algorithm needs to take a supported evaluation circuit depth as a input. Any leveled MKFHE [CM15, PS16, CZW17] with additional weakly circular security assumption can be converted into a *fully* MKFHE scheme. To decrypt an evaluated ciphertext, the decryption algorithm uses the secret keys of all parties involved in the computation. In fact, we need the MKFHE with threshold decryption property. We will define a generalized threshold decryption property called *subset threshold decryption* and show that we can realize it by modifying existing threshold decryption multi-key FHE.

A multi-key homomorphic encryption scheme MKFHE = (MKFHE.Setup, MKFHE.Keygen, MKFHE.Encrypt, MKFHE.Decrypt, MKFHE.Eval) is a 5-tuple of *ppt* algorithms as follows:

- **Setup** params ← MKFHE.Setup(1^λ): Takes the security parameter as input and outputs the public parametrization params of the system.
- **Key generation** (pk, sk) ← MKFHE.Keygen(params): Outputs a public encryption key pk and a secret decryption key sk.
- **Encryption** c ← MKFHE.Encrypt(pk, μ): Using the public key pk, encrypts a single bit message $\mu \in \{0, 1\}$ into a ciphertext c.
- **Decryption** μ ← MKFHE.Decrypt($(sk_1, ..., sk_{\hat{N}}), c$): Using the sequence of secret keys $(sk_1, ..., sk_{\hat{N}})$, decrypts a ciphertext c to recover the message $\mu \in \{0, 1\}$.
- **Evaluation** $\hat{c}$ ← MKFHE.Eval(C, $(c_1, ..., c_\ell)$, $(pk_1, ..., pk_{\hat{N}})$): Using the sequence of public keys $(pk_1, ..., pk_{\hat{N}})$, applies a circuit C : $\{0, 1\}^\ell \to \{0, 1\}$ to $(c_1, ..., c_\ell)$, where each ciphertext c_j is evaluated under a sequence of public keys $V_j \subset \{pk_1, ..., pk_{\hat{N}}\}$ (we assume that V_j is implicit in c_j). Upon termination, outputs a ciphertext $\hat{c}$.

Remark 1. *In multi-key GSW scheme, there is a* Expand *algorithm which takes a ciphertext c_j under pk_j and V_j where $V_j \subset \{pk_1, ..., pk_{\hat{N}}\}$ and $pk_j \in V_j$ as inputs and outputs a expanded ciphertext $\hat{c}_j$ which is the encryption of the same plaintext encrypted by c_j under all of public keys in V_j.*

Definition 2 (fully multi-key FHE). *A scheme* MKFHE *is fully multi-key FHE, if the following holds. Let $\hat{N} = \hat{N}_\lambda$ be any polynomial in the security parameter, C $= C_\lambda$ be a sequence of circuits. For all* params ← MKFHE.Setup(1^λ), (pk_i, sk_i) ← MKFHE.Keygen(params)$(i \in [\hat{N}])$, $\mu_j \in \{0, 1\}_{j \in [\ell]}$. MKFHE.Decrypt($c_j, sk_{i_j}$) $= \mu_j$ *where* $\{sk_{i_j} \in \{sk_1, ..., sk_{\hat{N}}\}\}_{j \in [\ell]}$, $\hat{c}$ ← MKFHE.Eval(C, $(c_1, ..., c_\ell)$, $(pk_1, ..., pk_{\hat{N}})$).

– *Correctness.*

$$C(\mu_0, ..., \mu_\ell) = \text{MKFHE.Decrypt}(\hat{c})$$

– *Compactness.*

$$|\hat{c}| \leq \text{poly}(\lambda, \hat{N})$$

where $\hat{N}$ is the number of distinctive public keys whose corresponding ciphertexts joining the computation.

Semantic Security. The definition of IND-CPA security for MKFHE is the same as that for standard public-key encryption. It works for the multi-key setting because if any adversary $\mathcal{A}$ who can distinguish expanded (possibly evaluated) ciphertexts of two equal-length plaintext can be used to distinguish two equal-length plaintext encryptions of public-key encryption. There exists a simulator $\mathcal{B}$ that can break IND-CPA security of PKE[2] with the help of $\mathcal{A}$. The challenger generates $(pk_1, sk_1) \leftarrow$ MKFHE.KeyGen(params) (we suppose the params here is common information). $\mathcal{B}$ receives pk_1 from the challenger and sends it to $\mathcal{A}$. $\mathcal{A}$ generates $\hat{N} - 1$ pairs of keys $\{(pk_i, sk_i) \leftarrow$ MKFHE.KeyGen(params)$\}_{i \in \{2, ..., \hat{N}\}}$ and sends $pk_2, ...pk_{\hat{N}}$ and two equal-length messages (μ_0, μ_1) to $\mathcal{B}$. $\mathcal{B}$ forwards (μ_0, μ_1) to the challenger and obtains the challenge ciphertext c from it. $\mathcal{B}$ can expand c into a ciphertext $\hat{c}$ under $pk_1, ..., pk_{\hat{N}}$ and sends it to $\mathcal{A}$. $\mathcal{B}$ just forwards $\mathcal{A}$'s guess. If $\mathcal{A}$ can guess right with probability $\frac{1}{2} + \xi$, then the advantage of $\mathcal{B}$ breaks IND-CPA of PKE is ξ. The reason we define the security of multi-identity IBFHE and multi-attribute ABFHE as the security of IBE and ABE respectively is similar.

We now define a multi-key FHE which supports a one-round generalized threshold distributed decryption protocol called subset threshold decryption. Such a protocol consists of two components: (1) given an expanded ciphertext (possibly evaluated) c each subset can compute a partial decryption using its corresponding secret keys, (2) there is a way to combine the partial decryptions computed by each subset to recover the plaintext. It is easy to know that threshold decryption is just a special case that there is only one element in each subset.

Definition 3. *A* Subset Threshold multi-key FHE scheme *is a MKFHE scheme with two additional algorithms* MFHE.SubsetDec, MFHE.CombineDec *described as follows:*

- $h_i \leftarrow$ MKFHE.SubsetDec$(c, (pk_1, ..., pk_{\hat{N}}), I_{i_1}, ...I_{i_{|\mathcal{T}_i|}}, sk_{I_{i_1}}, ..., sk_{I_{i_{|\mathcal{T}_i|}}})$: *On input an expanded ciphertext (possibly evaluated) under a sequence of $\hat{N}$ public keys and corresponding secret keys $sk_{i_1}, ..., sk_{i_{|\mathcal{T}_i|}}$ of the i-th index subset $\mathcal{T}_i$ and outputs a partial decryption h_i. Here $\mathcal{T}_i = \{I_{i_1}, ..., I_{i_{t_i}}\}$ where $I_{i_j} \in [\hat{N}]$, $t_i = |\mathcal{T}_i|$.*
- $\mu \leftarrow$ MKFHE.CombineDec$(h_1, ..., h_n)$: *On input n partial decryption outputs the plaintext μ.*

Along with the properties of multi-key FHE we require the scheme to satisfy the correctness and security.

Correctness. *Let* params $\leftarrow$ MKFHE.Setup(1^λ). *For any sequences of $\hat{N}$ correctly generated key pairs $\{(pk_i, sk_i) \leftarrow$ MKFHE.Keygen(params)$\}_{i \in [\hat{N}]}$ and any ℓ-tuple of messages $(\mu_1, ..., \mu_\ell)$. For set of indices $\mathcal{T} = \{1, ..., \hat{N}\}$ and any n subsets of $\mathcal{T}$ $\mathcal{T}_1, ...\mathcal{T}_n$, where $\mathcal{T}_i \cap \mathcal{T}_j = \emptyset$ $(i \neq j)$ and $\mathcal{T}_1 \cup ... \cup \mathcal{T}_n = \mathcal{T}$. We denote $\mathcal{T}_i$*

---

[2] The PKE is not a general PKE here. Its setup, encryption, decryption algorithms are the same as the MKFHE scheme.

as $\{I_{i_1}, ..., I_{i_{t_i}}\}$. Let $R : [\ell] \rightarrow [\hat{N}]$ denote a function from indices of plaintexts to indices of public keys and $\{c_k \leftarrow \mathsf{Encrypt}(pk_{R(k)}, \mu_k)\}_{k \in [\ell]}$ be encryptions of the messages μ_k under the $R(k)$-th public key. Let C be any (boolean) circuit and let $\hat{c} := Eval(\mathsf{C}, (c_1, ..., c_\ell)$ be the evaluated ciphertext. The below equation should hold with probability 1.

$$\mathsf{MKFHE.CombineDec}(\mathsf{h}_1, ..., \mathsf{h}_n) = \mathsf{C}(\mu_1, ..., \mu_\ell)$$

$\{\mathsf{h}_i \leftarrow \mathsf{MKFHE.SubsetDec}(c, pk_1, ..., pk_{\hat{N}}, I_{i_1}, ..., I_{i_{|\mathcal{T}_i|}}, sk_{I_{i_1}}, ..., sk_{I_{i_{|\mathcal{T}_i|}}})\}_{i \in [n]}$ are partial decryptions and $\{\mathcal{T}_i = \{I_{i_1}, ..., I_{i_{t_i}}\}\}_{i \in [n]}$ in above equation.

Security. *The semantic security of* MKFHE *with subset threshold decryption should hold. It is trivial because the* IND-CPA *security does not dependent on decryption algorithm.*

We will show that we can easily convert the threshold decryption of multi-key GSW into our subset threshold decryption. In fact, threshold decryption defined in [MW16] also has two similar algorithms[3] PartDec and FinDec where Part-Dec takes the evaluated ciphertext c, all parties' public keys $(pk_1, ..., pk_{\hat{N}})$ and one party's secret key sk_i and outputs the partial decryption p_i, and FinDec takes all partial decryptions $(p_1, ..., p_{\hat{N}})$ as inputs and outputs the plaintext μ. We observe that the FinDec algorithm of GSW-type scheme is $\sum_{i=1}^{\hat{N}} p_i$. So we can instantiate our SubsetDec and CombineDec algorithms as follows: $\mathsf{SubsetDec}(c, pk_1, ..., pk_{\hat{N}}, I_{i_1}, ..., I_{i_{|\mathcal{T}_i|}}, sk_{I_{i_1}}, ..., sk_{I_{i_{|\mathcal{T}_i|}}})$:

$$\{p_{i_j} \leftarrow \mathsf{PartDec}(c, pk_1, ..., pk_{\hat{N}}, I_{i_j}, sk_{I_{i_j}})\}_{j \in [|\mathcal{T}_i|]}, \mathsf{h}_i = \sum_{j=1}^{|\mathcal{T}_i|} p_{i_j}$$

$\mathsf{CombineDec}(\mathsf{h}_1, ..., \mathsf{h}_n) : \sum_{i=1}^n \mathsf{h}_i$

We refer to $\overline{\mathsf{SubsetDec}}[\mathcal{T}_i]$ as the circuit that SubsetDec algorithm takes $\mathcal{T}_i$ as the indices components of inputs.

3.2 Leveled IBFHE

A leveled IBFHE scheme is defined with respect to a message space $\mathcal{M}$, an identity space $\mathcal{I}$, a class of circuits $\mathbb{C} \subset \mathcal{M}^* \rightarrow \mathcal{M}$ and ciphertext space C. An IBFHE scheme is a tuple of *ppt* algorithms (Setup, KeyGen, Encrypt, Decrypt, Eval) defined as follows:

- Setup($1^\lambda, L$): On input (in unary) a security parameter λ and the bounded evaluation circuit depth L supported, generate public parameters MPK and a master secret key MSK. Output (MPK, MSK).

---

[3] More details of the two algorithms can be found in [MW16].

- KeyGen(MSK, id): On input master secret key MSK and an identity id: derive and output a secret key sk_{id} for identity id.
- Encrypt(MPK, id, μ): On input public parameters MPK, an identity id, and a message $\mu \in \mathcal{M}$, output a ciphertext $c \in C$ that encrypts μ under identity id.
- Decrypt(sk_{id}, c): On input secret key sk_{id} for (resp.) identity id and a ciphertext $c \in \mathcal{C}$, output $\mu \in \mathcal{M}$ if c is a valid encryption under identities id; output a failure symbol $\perp$ otherwise.
- Eval(MPK, C, id, $c_1, ..., c_\ell$): On input public parameters MPK, a circuit $C \in \mathbb{C}$ and ciphertexts $c_1, ..., c_\ell \in C$ under id, output an evaluated ciphertext $\hat{c} \in C$ under id.

For all choices of Setup($1^\lambda, L$) $\rightarrow$ (MPK, MSK), $c_i =$ Encrypt(MPK, id, μ_i) ($\mu_i \in \mathcal{M}$), C : $\mathcal{M}^* \rightarrow \mathcal{M}$ whose depth is less than L, $\hat{c} =$ Eval(MPK, C, $c_1, ..., c_\ell$)

– **Correctness.**

$$\text{Decrypt}(sk_{id}, \hat{c}) = C(\mu_0, ...\mu_\ell)$$

– **Compactness.**

$$|\hat{c}| \leq \text{poly}(\lambda, L)$$

4 Multi-identity IBFHE

4.1 Construction

We combine a multi-key FHE and *single-identity* IBFHE to construct our multi-identity IBFHE scheme. Setup algorithm outputs public parameters and master secret key of IBFHE and params of MKFHE by running their setup algorithms respectively. When encrypt a plaintext $\mu \in \{0, 1\}$, the sender generates $(pk, sk) \leftarrow$ MKFHE.KeyGen(params), then encrypts sk under id and μ under pk. The evaluator evaluates the circuit on MKFHE ciphertexts and obtain an evaluated ciphertext $\hat{c}$. Then it evaluates with the leveled IBFHE scheme the partial decryption circuit $\overline{\text{SubsetDec}[\mathcal{T}_j]}$ for all $j \in [n]$ where $\mathcal{T}_j$ is the set of indices of corresponding public keys for id_j. The number of (compact) evaluated IBFHE ciphertext is independent on the number of senders which makes the whole resulting ciphertext compact. Receivers can obtain partial decryption of the evaluated MKFHE ciphertext by decrypting the IBFHE ciphertext under its identity. Then they can jointly decrypt the evaluated MKFHE ciphertext. Our construction is *fully* multi-identity IBFHE with additional weakly circular security where we do not need to take circuit depth as input in the Setup algorithm. In order to compute a function in our construction we will assign every plaintexts, every pair of (public and secret) keys of MKFHE and identities of IBFHE indices. Let the pair of keys and plaintext share the same index because each public key of MKFHE only encrypts one plaintext. For example, if we use pk to encrypts μ_i, we denote pk as pk_i and sk as sk_i. We can use lexicographic order of identities as their indices. Suppose there are ℓ plaintexts and n different identities involved in the computation, we can define a function $\hat{R} : [\ell] \to [n]$ where $\hat{R}(i) = j$ if pk_i is generated in the encryption process for id_j. Set the preimages of j as $\mathcal{T}_j = \{I_{(\mathsf{id}_j,1)}, ..., I_{(\mathsf{id}_j,t_j)}\}$ where $I_{(\mathsf{id}_j,1)}, ..., I_{(\mathsf{id}_j,t_j)}$ are indices of the ciphertexts for the same identity id_j.

- Setup($1^\lambda, N$): Take the security parameter and the bound of number of ciphertexts under the same identity that the system can tolerate. Compute params $\leftarrow$ MKFHE.Setup(1^λ), (MPK′, MSK′) $\leftarrow$ IBFHE.Setup($1^\lambda, L$), where $L = \tau(N, \lambda)$ is the depth of the decryption circuit of MKFHE for parameters[4] λ and N. Output (MPK, MSK) = ((MPK′, params), MSK′).
- KeyGen(MSK, id): This algorithm is the same as IBFHE. Just output $\mathsf{sk}_{\mathsf{id}} = $ IBFHE.KeyGen(MSK, id).
- Encrypt(MPK, id, $\mu \in \{0,1\}$): Run $(pk, sk) \leftarrow$ MKFHE.KeyGen(params). Compute $c' \leftarrow$ MKFHE.Encrypt(pk, μ), $\phi \leftarrow$ IBFHE.Encrypt(MPK′, id, sk). Output $c = (\mathsf{type} := 0, \mathsf{enc} := (c', \phi, \mathsf{id}, pk))$.
- Eval(MPK, C, $c_1, ..., c_\ell$): The ciphertexts are assumed to be "fresh" ciphertexts generated with the encryption algorithm. In other words, their type components are all 0. Otherwise the evaluator outputs $\bot$. Parse c_i as $(\mathsf{type} := 0, \mathsf{enc} := (c'_i, \phi_i, \mathsf{id}_{\hat{R}(i)}, pk_i)$. Firstly, evaluate the circuit on MKFHE ciphertexts. Compute $\hat{c} = $ MKFHE.Eval(C, $(c'_1, ..., c'_\ell), (pk_1, ..., pk_\ell))$. For all $j \in [n]$, proceed as following two steps. Step 1: encrypt the evaluated MKFHE ciphertext. Let $\hat{c}_{\mathsf{bin}} = $ Binary($\hat{c}$). Compute $\{\bar{c}_i \leftarrow $ IBFHE.Encrypt(MPK′, $\mathsf{id}_j, \hat{c}_{\mathsf{bin}}[i])\}_{i \in [|\hat{c}_{\mathsf{bin}}|]}$ the IBFHE encryption of every bit of evaluated ciphertext $\hat{c}$. Step 2: evaluate partial decryption circuit $\overline{\mathsf{SubsetDec}}[\mathcal{T}_j]$ on $(\{\phi_{I_{(\mathsf{id}_j,k)}}\}_{k \in |\mathcal{T}_j|}, \{\bar{c}_i\}_{i \in [|\hat{c}_{\mathsf{bin}}|]})$ and obtain the IBFHE encryption c_{id_j} under id_j of the partial decryption of $\hat{c}$ where $c_{\mathsf{id}_j} = $ IBFHE.Eval(MPK′, $\overline{\mathsf{SubsetDec}}[\mathcal{T}_j], \mathsf{id}_j, \phi_{I_{(\mathsf{id}_j,1)}}, ..., \phi_{I_{(\mathsf{id}_j,|\mathcal{T}_j|)}}, \bar{c}_1, ...\bar{c}_{|\hat{c}_{\mathsf{bin}}|})$. Finally, outputs $c = (\mathsf{type} := 1, (c_{\mathsf{id}_1}, ..., c_{\mathsf{id}_n}))$
- Decrypt($\mathsf{sk}_{\mathsf{id}_1}, ..., \mathsf{sk}_{\mathsf{id}_n}, c$): If c is a "fresh" ciphertext where $\mathsf{type} = 0$, we parse enc as $(c', \phi, \mathsf{id}, pk)$ and computes $sk = $ IBFHE.Decrypt($\mathsf{sk}_{\mathsf{id}}, \phi$). Computes $\mu = $ MKFHE.Decrypt(c', sk) and outputs μ if $sk \neq \bot$. If c is an evaluated ciphertext (i.e. $\mathsf{type} = 1$), parse c as $(c_{\mathsf{id}_1}, ..., c_{\mathsf{id}_n})$, compute $\mathsf{h}_i = $ IBFHE.Decrypt($\mathsf{sk}_{\mathsf{id}_i}, c_{\mathsf{id}_i}$) and outputs $\mu = $ MKFHE.CombineDec($\mathsf{h}_1, ..., \mathsf{h}_n$). Otherwise output $\bot$.

Remark 2. *We can instantiate our MKFHE with the scheme of GSW-MKFHE [MW16, BP16, PS16] where its decryption circuit depth is $O(log(N \cdot \lambda))$. We set N to be a large value which dominates λ, so its decryption circuit depth is roughly $O(logN)$. For example, suppose we set N as 2^{64}, we need a leveled IBFHE that can evaluate 64-depth circuits.*

---

[4] In fact, if there exists a "pure" IBFHE, we don't need take N as input that makes our construction be a "pure" multi-key IBFHE. Unfortunately, [CM16] can only yield almost "pure" scheme which does not work here.

4.2   Main Results

Theorem 1. *Let N be a positive integer. Let λ be the security parameter. Let n be any polynomial in λ. Suppose there exists an* IND-CPA *secure subset threshold decryption* MKFHE *scheme that evaluates circuits of depth d, and its subset threshold decryption circuit depth is $\tau(N, \lambda)$. Suppose that there exists an* IBFHE *scheme that can compactly evaluate circuits depth of τ. Then there exists a multi-identity* IBFHE *scheme supporting n identities that can compactly evaluate all d-depth boolean circuits in $\{0, 1\}^* \to \{0, 1\}$ with a limitation that the number of ciphertexts under the same identity is no more than N.*

Correctness

The construction is correct if MKFHE and leveled IBFHE are both correct. The decryption correctness of fresh ciphertext is guaranteed by the decryption correctness of MKFHE and IBFHE. If we set the parameters of IBFHE to support evaluation circuits depth larger than the depth of the decryption algorithm of MKFHE scheme, combining the evaluation and subset threshold decryption correctness of MKFHE and decryption correctness of IBFHE, the evaluated ciphertext can be decrypted correctly. So the correctness of our construction is guaranteed.

Compactness

If ciphertexts of IBFHE are compact, our construction is likewise compact. If ciphertexts of IBFHE is compact, $|c_{\mathsf{id}_j}| \leq \mathsf{poly}(\lambda, L)$, where L is a polynomial[5] in λ and larger than the depth of the decryption circuit of MKFHE. The evaluated ciphertext c is $n \cdot |h_i|$ *compact* IBFHE ciphertexts where $|h_i|$ is independent on the evaluated function and N. So we can conclude that $|c| \leq \mathsf{poly}(\lambda, n)$.

Security

Theorem 2. *Suppose that* MKFHE *is* IND-CPA *secure and single-identity* IBFHE *is* IND-X-CPA *secure, our construction is* IND-X-CPA *secure multi-identity* IBFHE *where* $\mathsf{X} \in \{\mathsf{Selective}, \mathsf{Adaptive}\}$.

Proof. We will prove the security by hybrid argument as follows.

Hybrid $\mathcal{H}_0$: This is identical to the IND-X-CPA game of multi-identity IBFHE.

Hybrid $\mathcal{H}_1$: Let id^* be the challenge identity the adversary sends. There is only one difference in the challenge ciphertext with $\mathcal{H}_0$. The challenger replaces the encryption of the secret key sk of the MKFHE (i.e. ϕ component of the challenge ciphertext) with $\phi \leftarrow \mathsf{IBFHE.Encrypt}(\mathsf{MPK}', \mathsf{id}^*, 0^{|sk|})$, where $0^{|sk|}$ is zeros whose length is the same as sk.

$\mathcal{H}_0$ and $\mathcal{H}_1$ is indistinguishable. In fact, if any ppt adversary $\mathcal{A}$ can distinguish them there exists a simulator $\mathcal{B}$ that can use $\mathcal{A}$ to break the IND-X-CPA of IBFHE. In the challenge phase, when $\mathcal{A}$ chooses a challenge identity id^*, $\mathcal{B}$ generates a pair of key for MKFHE i.e. it computes $\mathsf{params} \leftarrow \mathsf{MKFHE.Setup}(1^\lambda)$ and

---

[5] We see N as a constant here.

$(pk, sk) \leftarrow$ MKFHE.KeyGen(params). Then $\mathcal{B}$ sends id^* and $(m_0 = sk, m_1 = 0^{|sk|})$ to its challenger. $\mathcal{B}$ obtains the challenge ciphertext from the challenger and set it as the ϕ component of its own challenge ciphertext c^* and then computes the remaining components of c^* via the encryption algorithm. $\mathcal{B}$ sends $\mathcal{A}$'s guess to its challenger. If ϕ is the encryption of sk, the view of $\mathcal{A}$ is identical to $\mathcal{H}_0$. If ϕ is the encryption of $0^{|sk|}$, the view of $\mathcal{A}$ is identical to $\mathcal{H}_1$. So the advantage of $\mathcal{B}$ breaks IND-X-CPA of IBFHE is equal to the advantage of $\mathcal{A}$ distinguishing $\mathcal{H}_0$ and $\mathcal{H}_1$. It is concluded that $\mathcal{H}_0$ and $\mathcal{H}_1$ are indistinguishable.

Hybrid $\mathcal{H}_2$: This is same as $\mathcal{H}_1$ except that the challenger dose not encrypt μ_0 or μ_1 sent by the adversary $\mathcal{A}$ in the challenge phase. It encrypts 0 instead. If $\mathcal{A}$ can distinguish $\mathcal{H}_1$ and $\mathcal{H}_2$ with a non-negligible advantage there exists a simulator $\mathcal{B}$ can break the IND-CPA security of MKFHE. $\mathcal{B}$ sends the public key pk obtained from its challenger to $\mathcal{A}$. $\mathcal{A}$ chooses two plaintext $\mu_0 \in \{0, 1\}$ and $\mu_1 \in \{0, 1\}$ as the challenge plaintext pair to $\mathcal{B}$. $\mathcal{B}$ randomly choose a bit b and sends $(\mu_b, 0)$ to its challenger. $\mathcal{B}$ obtains a ciphertext c' from its challenger and set it as the MKFHE component of its challenge ciphertext c^* answered to $\mathcal{A}$. $\mathcal{B}$ computes the remaining components of c^*. $\mathcal{B}$ outputs 0 if $\mathcal{A}$'s guess is $\mathcal{H}_1$, and 1 otherwise. If c' encrypts μ_b, the view of $\mathcal{A}$ is identical to $\mathcal{H}_1$. If c' encrypts 0, the view of $\mathcal{A}$ is identical to $\mathcal{H}_2$. So $\mathcal{H}_1$ is indistinguishable from $\mathcal{H}_2$ if MKFHE is IND-CPA secure. In $\mathcal{H}_2$, the advantage of the adversary is zero because there are no information about the bit the challenger chooses in the challenge ciphertext.

Optimization

In fact, we can choose an integer ω in the setup stage and encrypt ω bits under one public key of MKFHE. We need an additional hybrid argument of multiple encryptions of MKFHE in the proof of security.

5   Multi-attribute ABFHE

In this section, we will show the construction of multi-attribute ABFHE. The construction and proof are similar to those of multi-identity IBFHE. We give the proof in full version. Let $\mathcal{X}$ denotes attribute space and $\mathcal{F}$ denotes policy space. Suppose there are ℓ plaintexts and n different attributes involved in the computation, we can define a function $R' : [\ell] \rightarrow [n]$ where $R'(i) = j$ if pk_i is generated in the encryption process for x_j. Set the preimages of j as $\mathcal{T}_j = \{I_{(x_j,1)}, ..., I_{(x_j,t_j)}\}$ where $I_{(x_j,1)}, ..., I_{(x_j,t_j)}$ are indices of the ciphertexts for the same identity x_j.

- Setup($1^\lambda, N$): Take the security parameter and the bound of number of ciphertexts for the same attribute that the system can tolerate. Compute params $\leftarrow$ MKFHE.Setup(1^λ), (MPK′, MSK′) $\leftarrow$ ABFHE.Setup($1^\lambda, L$), where L is the depth of the decryption circuit of MKFHE for parameters, λ and N. Output (MPK, MSK) = ((MPK′, params), MSK′).
- KeyGen(MSK, $f \in \mathcal{F}$): This algorithm is the same as ABFHE. Just output sk$_f$ = IBFHE.KeyGen(MSK, f).

- Encrypt(MPK, $x \in \mathcal{X}, \mu \in \{0,1\}$): $(pk, sk) \leftarrow$ MKFHE.KeyGen(params). Compute $c' \leftarrow$ MKFHE.Enc(pk, μ), $\phi \leftarrow$ Encrypt(MPK$'$, x, sk). Output $c =$ (type $:= 0$, enc $:= (c', \phi, x, pk)$).
- Eval(MPK, C, $c_1, ..., c_\ell$): Firstly, the ciphertexts are assumed to be "fresh" ciphertexts generated with the encryption algorithm. In other words, their type components are all 0. Otherwise the evaluator outputs $\perp$. Parse c_i as (type $:= 0$, enc $:= (c'_i, \phi_i, x_{R'(i)}, pk_i)$. Compute $\hat{c} =$ MKFHE.Eval(C, $(c'_1, ..., c'_\ell), (pk_1, ..., pk_\ell))$. Let $\hat{c}_{\mathsf{bin}} =$ Binary$(\hat{c})$. Compute $\{\bar{c}_i \leftarrow$ ABFHE. Encrypt(MPK$'$, $x_j, \hat{c}_{\mathsf{bin}}[i])\}_{i \in [|\hat{c}_{\mathsf{bin}}|]}$ the ABFHE encryption of every bit of evaluated ciphertext $\hat{c}$. Then evaluate partial decryption circuit $\overline{\mathsf{SubsetDec}}[\mathcal{T}_j]$ on ciphertexts $(\{\phi_{I_{x_j,k}}\}_{k \in |\mathcal{T}_j|}, \{\bar{c}_i\}_{i \in |c_{\mathsf{bin}}|})$ and obtain ABFHE encryption c_{x_j} under x_j of partial decryption of $\hat{c}$ where $c_{x_j} =$ ABFHE.Eval(MPK$'$, $\overline{\mathsf{SubsetDec}}[\mathcal{T}_j], x_j, \phi_{I_{(x_j,1)}}, ..., \phi_{I_{(x_j,|\mathcal{T}_j|)}}, \bar{c}_1, ... \bar{c}_{|\hat{c}_{\mathsf{bin}}|})$ for all $j \in [n]$. Outputs $c =$ (type $:= 1, (c_{x_1}, ..., c_{x_n})$)
- Decrypt(sk$_{\mathsf{f}_1}, ..., sk_{\mathsf{f}_n}, c$): For simplicity[6], we suppose $f_i(x_i) = 0$ here. If c is a "fresh" ciphertext where type $= 0$, we parse enc as (c', ϕ, x, pk) and compute $sk =$ ABFHE.Decrypt(sk_{f_i}, ϕ) where $x = x_i$. Compute $\mu =$ MKFHE.Decrypt(c', sk) and output μ if $sk \neq \perp$. If c is an evaluated ciphertext (i.e. type $= 1$), parse c as $(c_{x_1}, ..., c_{x_n})$, compute h$_i =$ ABFHE.Decrypt(sk_{f_i}, c_{x_i}) and outputs $\mu =$ MKFHE.CombineDec(h$_1, ..., h_n$). Otherwise output $\perp$.

References

[ABB10] Agrawal, S., Boneh, D., Boyen, X.: Efficient lattice (H)IBE in the standard model. In: Gilbert, H. (ed.) EUROCRYPT 2010. LNCS, vol. 6110, pp. 553–572. Springer, Heidelberg (2010). https://doi.org/10.1007/978-3-642-13190-5_28

[AP14] Alperin-Sheriff, J., Peikert, C.: Faster bootstrapping with polynomial error. In: Garay, J.A., Gennaro, R. (eds.) CRYPTO 2014. LNCS, vol. 8616, pp. 297–314. Springer, Heidelberg (2014). https://doi.org/10.1007/978-3-662-44371-2_17

[BB04a] Boneh, D., Boyen, X.: Efficient selective-ID secure identity-based encryption without random oracles. In: Cachin, C., Camenisch, J.L. (eds.) EUROCRYPT 2004. LNCS, vol. 3027, pp. 223–238. Springer, Heidelberg (2004). https://doi.org/10.1007/978-3-540-24676-3_14

[BB04b] Boneh, D., Boyen, X.: Secure identity based encryption without random oracles. In: Franklin, M. (ed.) CRYPTO 2004. LNCS, vol. 3152, pp. 443–459. Springer, Heidelberg (2004). https://doi.org/10.1007/978-3-540-28628-8_27

[BCTW16] Brakerski, Z., Cash, D., Tsabary, R., Wee, H.: Targeted homomorphic attribute-based encryption. In: Hirt, M., Smith, A. (eds.) TCC 2016. LNCS, vol. 9986, pp. 330–360. Springer, Heidelberg (2016). https://doi.org/10.1007/978-3-662-53644-5_13

---

[6] It also works if sk_{f_i} can decrypt ciphertexts under many different attributes.

[BF01] Boneh, D., Franklin, M.: Identity-based encryption from the Weil pairing. In: Kilian, J. (ed.) CRYPTO 2001. LNCS, vol. 2139, pp. 213–229. Springer, Heidelberg (2001). https://doi.org/10.1007/3-540-44647-8_13

[BGG+14] Boneh, D., et al.: Fully key-homomorphic encryption, arithmetic circuit ABE and compact garbled circuits. In: Nguyen, P.Q., Oswald, E. (eds.) EUROCRYPT 2014. LNCS, vol. 8441, pp. 533–556. Springer, Heidelberg (2014). https://doi.org/10.1007/978-3-642-55220-5_30

[BGH07] Boneh, D., Gentry, C., Hamburg, M.: Space-efficient identity based encryption without pairings. IACR Cryptology ePrint Archive, vol. 2007, no. 177 (2007)

[BGV12] Brakerski, Z., Gentry, C., Vaikuntanathan, V.: (Leveled) fully homomorphic encryption without bootstrapping. In: Innovations in Theoretical Computer Science 2012, Cambridge, MA, USA, 8–10 January 2012, pp. 309–325 (2012)

[BHHO08] Boneh, D., Halevi, S., Hamburg, M., Ostrovsky, R.: Circular-secure encryption from decision Diffie-Hellman. In: Wagner, D. (ed.) CRYPTO 2008. LNCS, vol. 5157, pp. 108–125. Springer, Heidelberg (2008). https://doi.org/10.1007/978-3-540-85174-5_7

[BP16] Brakerski, Z., Perlman, R.: Lattice-based fully dynamic multi-key FHE with short ciphertexts. In: Robshaw, M., Katz, J. (eds.) CRYPTO 2016. LNCS, vol. 9814, pp. 190–213. Springer, Heidelberg (2016). https://doi.org/10.1007/978-3-662-53018-4_8

[Bra12] Brakerski, Z.: Fully homomorphic encryption without modulus switching from classical GapSVP. In: Safavi-Naini, R., Canetti, R. (eds.) CRYPTO 2012. LNCS, vol. 7417, pp. 868–886. Springer, Heidelberg (2012). https://doi.org/10.1007/978-3-642-32009-5_50

[BV11] Brakerski, Z., Vaikuntanathan, V.: Efficient fully homomorphic encryption from (standard) LWE. In: IEEE 52nd Annual Symposium on Foundations of Computer Science, FOCS 2011, Palm Springs, CA, USA, 22–25 October 2011, pp. 97–106 (2011)

[CHKP10] Cash, D., Hofheinz, D., Kiltz, E., Peikert, C.: Bonsai trees, or how to delegate a lattice basis. In: Gilbert, H. (ed.) EUROCRYPT 2010. LNCS, vol. 6110, pp. 523–552. Springer, Heidelberg (2010). https://doi.org/10.1007/978-3-642-13190-5_27

[CM14] Clear, M., McGoldrick, C.: Bootstrappable identity-based fully homomorphic encryption. In: Gritzalis, D., Kiayias, A., Askoxylakis, I. (eds.) CANS 2014. LNCS, vol. 8813, pp. 1–19. Springer, Cham (2014). https://doi.org/10.1007/978-3-319-12280-9_1

[CM15] Clear, M., McGoldrick, C.: Multi-identity and multi-key leveled FHE from learning with errors. In: Gennaro, R., Robshaw, M. (eds.) CRYPTO 2015. LNCS, vol. 9216, pp. 630–656. Springer, Heidelberg (2015). https://doi.org/10.1007/978-3-662-48000-7_31

[CM16] Clear, M., McGoldrick, C.: Attribute-based fully homomorphic encryption with a bounded number of inputs. In: Pointcheval, D., Nitaj, A., Rachidi, T. (eds.) AFRICACRYPT 2016. LNCS, vol. 9646, pp. 307–324. Springer, Cham (2016). https://doi.org/10.1007/978-3-319-31517-1_16

[Coc01] Cocks, C.: An identity based encryption scheme based on quadratic residues. In: Honary, B. (ed.) Cryptography and Coding 2001. LNCS, vol. 2260, pp. 360–363. Springer, Heidelberg (2001). https://doi.org/10.1007/3-540-45325-3_32

[CRRV17] Canetti, R., Raghuraman, S., Richelson, S., Vaikuntanathan, V.: Chosen-ciphertext secure fully homomorphic encryption. In: Fehr, S. (ed.) PKC 2017. LNCS, vol. 10175, pp. 213–240. Springer, Heidelberg (2017). https://doi.org/10.1007/978-3-662-54388-7_8

[CZW17] Chen, L., Zhang, Z., Wang, X.: Batched multi-hop multi-key FHE from ring-LWE with compact ciphertext extension. In: Kalai, Y., Reyzin, L. (eds.) TCC 2017. LNCS, vol. 10678, pp. 597–627. Springer, Cham (2017). https://doi.org/10.1007/978-3-319-70503-3_20

[DG17] Döttling, N., Garg, S.: Identity-based encryption from the Diffie-Hellman assumption. In: Katz, J., Shacham, H. (eds.) CRYPTO 2017. LNCS, vol. 10401, pp. 537–569. Springer, Cham (2017). https://doi.org/10.1007/978-3-319-63688-7_18

[Gen09a] Gentry, C.: A fully homomorphic encryption scheme. Ph.D. thesis, Stanford University (2009). crypto.stanford.edu/craig

[Gen09b] Gentry, C.: Fully homomorphic encryption using ideal lattices. In: Proceedings of the 41st Annual ACM Symposium on Theory of Computing, STOC 2009, Bethesda, MD, USA, 31 May–2 June 2009, pp. 169–178 (2009)

[GPSW06] Goyal, V., Pandey, O., Sahai, A., Waters, B.: Attribute-based encryption for fine-grained access control of encrypted data. In: Proceedings of the 13th ACM Conference on Computer and Communications Security, CCS 2006, Alexandria, VA, USA, 30 October–3 November 2006, pp. 89–98 (2006)

[GPV08] Gentry, C., Peikert, C., Vaikuntanathan, V.: Trapdoors for hard lattices and new cryptographic constructions. In: Proceedings of the 40th Annual ACM Symposium on Theory of Computing, Victoria, British Columbia, Canada, 17–20 May 2008, pp. 197–206 (2008)

[GSW13] Gentry, C., Sahai, A., Waters, B.: Homomorphic encryption from learning with errors: conceptually-simpler, asymptotically-faster, attribute-based. In: Canetti, R., Garay, J.A. (eds.) CRYPTO 2013. LNCS, vol. 8042, pp. 75–92. Springer, Heidelberg (2013). https://doi.org/10.1007/978-3-642-40041-4_5

[GVW13] Gorbunov, S., Vaikuntanathan, V., Wee, H.: Attribute-based encryption for circuits. In: Symposium on Theory of Computing Conference, STOC 2013, Palo Alto, CA, USA, 1–4 June 2013, pp. 545–554 (2013)

[HK17] Hiromasa, R., Kawai, Y.: Fully dynamic multi target homomorphic attribute-based encryption. IACR Cryptology ePrint Archive, vol. 2017, no. 373 (2017)

[LTV12] López-Alt, A., Tromer, E., Vaikuntanathan, V.: On-the-fly multiparty computation on the cloud via multikey fully homomorphic encryption. In: Proceedings of the 44th Symposium on Theory of Computing Conference, STOC 2012, New York, NY, USA, 19–22 May 2012, pp. 1219–1234 (2012)

[MW16] Mukherjee, P., Wichs, D.: Two round multiparty computation via multi-key FHE. In: Fischlin, M., Coron, J.-S. (eds.) EUROCRYPT 2016. LNCS, vol. 9666, pp. 735–763. Springer, Heidelberg (2016). https://doi.org/10.1007/978-3-662-49896-5_26

[PS16] Peikert, C., Shiehian, S.: Multi-key FHE from LWE, revisited. In: Hirt, M., Smith, A. (eds.) TCC 2016. LNCS, vol. 9986, pp. 217–238. Springer, Heidelberg (2016). https://doi.org/10.1007/978-3-662-53644-5_9

[RAD78] Rivest, R.L., Adleman, L., Dertouzos, M.L.: On data banks and privacy homomorphisms. Found. Sec. Comput. 4, 169–179 (1978)

[Reg05] Regev, O.: On lattices, learning with errors, random linear codes, and cryptography. In: Proceedings of the 37th Annual ACM Symposium on Theory of Computing, Baltimore, MD, USA, 22–24 May 2005, pp. 84–93 (2005)

[Reg09] Regev, O.: On lattices, learning with errors, random linear codes, and cryptography. J. ACM **56**(6), 34:1–34:40 (2009)

[Sha84] Shamir, A.: Identity-based cryptosystems and signature schemes. In: Blakley, G.R., Chaum, D. (eds.) CRYPTO 1984. LNCS, vol. 196, pp. 47–53. Springer, Heidelberg (1985). https://doi.org/10.1007/3-540-39568-7_5

[SOK00] Sakai, R., Ohgishi, K., Kasahara, M.: Cryptosystem based on pairings, 01 2000

[SW05] Sahai, A., Waters, B.: Fuzzy identity-based encryption. In: Cramer, R. (ed.) EUROCRYPT 2005. LNCS, vol. 3494, pp. 457–473. Springer, Heidelberg (2005). https://doi.org/10.1007/11426639_27

[Wat05] Waters, B.: Efficient identity-based encryption without random oracles. In: Cramer, R. (ed.) EUROCRYPT 2005. LNCS, vol. 3494, pp. 114–127. Springer, Heidelberg (2005). https://doi.org/10.1007/11426639_7

[Wat09] Waters, B.: Dual system encryption: realizing fully secure IBE and HIBE under simple assumptions. In: Halevi, S. (ed.) CRYPTO 2009. LNCS, vol. 5677, pp. 619–636. Springer, Heidelberg (2009). https://doi.org/10.1007/978-3-642-03356-8_36

Approximate Homomorphic Encryption over the Conjugate-Invariant Ring

Duhyeong Kim[1] and Yongsoo Song[2]([⊠])

[1] Department of Mathematical Sciences and RIM, Seoul National University,
Seoul, South Korea
doodoo1204@snu.ac.kr
[2] University of California, San Diego, USA
yongsoosong@ucsd.edu

Abstract. The Ring Learning with Errors (RLWE) problem over a cyclotomic ring has been the most widely used hardness assumption for the construction of practical homomorphic encryption schemes. However, this restricted choice of a base ring may cause a waste in terms of plaintext space usage. For example, an approximate homomorphic encryption scheme of Cheon et al. (ASIACRYPT 2017) is able to store a complex number in each of the plaintext slots since its canonical embedding of a cyclotomic field has a complex image. The imaginary part of a plaintext is not underutilized at all when the computation is performed over the real numbers, which is required in most of the real-world applications such as machine learning.

In this paper, we are proposing a new homomorphic encryption scheme which supports arithmetic over the real numbers. Our scheme is based on RLWE over a subring of a cyclotomic ring called conjugate-invariant ring. We show that this problem is no easier than a standard lattice problem over ideal lattices by the reduction of Peikert et al. (STOC 2017). Our scheme allows real numbers to be packed in a ciphertext without any waste of a plaintext space and consequently we can encrypt twice as many plaintext slots as the previous scheme while maintaining the same security level, storage, and computational costs.

Keywords: Ring Learning with Errors · Homomorphic encryption · Real number arithmetic

1 Introduction

Learning with Errors (LWE) is a computational problem which asks to distinguish a system of linear equations with small errors from a uniformly random one. After Regev [35] firstly introduced the LWE problem, it has been one of the standard assumptions for the construction of cryptographic primitives due to its security and versatility. Lyubashevsky, Peikert, and Regev [32] proposed a variant of LWE called the Ring Learning with Errors (RLWE) problem. They showed

© Springer Nature Switzerland AG 2019
K. Lee (Ed.): ICISC 2018, LNCS 11396, pp. 85–102, 2019.
https://doi.org/10.1007/978-3-030-12146-4_6

that the (decisional) RLWE problem over a cyclotomic ring can be reduced from the Shortest Independent Vectors Problem (SIVP) over ideal lattices.

Homomorphic Encryption (HE) is a cryptographic scheme which enables arithmetic operations on encrypted data without decryption. This technology is a promising solution which can prevent leakage of sensitive personal information such as financial, medical and genomic data. A number of HE schemes [5, 7, 8, 13, 15, 16, 18, 19, 21, 23, 24] have been suggested following Gentry's blueprint [22]. Currently, most of the practical HE schemes [13, 15, 21, 23] rely their security on the hardness of RLWE over a cyclotomic ring. For years, the choice of base ring was restricted because nothing was known about the hardness of (decisional) RLWE over non-cyclotomic rings.

Cheon et al. [13] proposed a HE scheme (HEAAN) that supports the arithmetic of approximate numbers. In addition to homomorphic addition and multiplication, the HEAAN scheme can compute the rounding operation (extraction of the most significant bits) efficiently, which has traditionally been considered a challenging subject on HE system. Because of this functionality, HEAAN has showed a remarkable performance in many of the applications [6, 14, 17, 28–30], requiring computations of real numbers.

Motivation. The HEAAN scheme exploits a variant of the (complex) canonical embedding over a cyclotomic field to pack a number of plaintext values in a single ciphertext. Hence, each of the plaintext slots could store a complex number. We point out that this complex encoding method has some problems in terms of efficiency and precision. Since most of the real-world applications (e.g. machine learning) require computations over purely real numbers, the imaginary part of a plaintext of HEAAN is underutilized. It can be viewed as a waste of a plaintext space. In addition, homomorphic operations of HEAAN, such as multiplication and rounding, generate additional complex errors which can reduce the computational accuracy.

Peikert et al. [34] recently showed that the RLWE problem over the ring of integers of an arbitrary number field is no easier than SIVP over ideal lattices of the same number field. So we aimed to find a new number field and construct a HE scheme over its ring of integers, which utilizes a fully packed plaintext space over real numbers to overcome the existing problem.

Our Contribution. We consider the maximal real subfield of a cyclotomic field as a base number field and define the RLWE problem over its ring of integers which is called the *conjugate-invariant* ring. We first show that the conjugate-invariant ring is the set of real numbers in the ring of integers of a cyclotomic field and adapt the reduction of [34] to guarantee the hardness of RLWE problem over the conjugate-invariant ring.

Based on this problem, we construct a new HE scheme that supports approximate arithmetic of real numbers. Our scheme can store a real number in each of the plaintext slots since the image of conjugate-invariant ring with respect to the canonical embedding belongs to the set of real vectors. We also propose a specialized Fast Fourier Transformation (FFT) algorithm over the residue ring of conjugate-invariant ring to minimize the complexity of arithmetic operations.

As a result, our HE scheme can encrypt *twice* as many plaintext slots as the original HEAAN scheme while maintaining the same security level and computational costs, i.e., the amortized complexity per slot is reduced by half.

Technical Details. Let m be a power-of-two integer, $n = \phi(m) = m/2$ and $\Phi_m(X) = X^n + 1$. Let $\zeta = \exp(2\pi i/m)$ be an m-th primitive root of unity and let $F = \mathbb{Q}(\xi)$ be the maximal real subfield of the cyclotomic field $K = \mathbb{Q}(\zeta)$ for $\xi = \zeta + \zeta^{-1}$. Then the ring of integers of $F = \mathbb{Q}(\xi)$ is $R = \mathbb{Z}[\xi]$, and we call this ring the conjugate-invariant ring. By adapting the reduction in [34], we can show that RLWE over the ring R is no easier than SIVP over ideal lattices in K. This hardness proof reasonably motivates us to exploit R as a base ring for the construction of a HE scheme. We also give a cryptanalysis of RLWE over the conjugate-invariant ring $R = \{a(X) \in \mathbb{Z}[X]/(X^n + 1) : a(X) = a(X^{-1})\}$ to study the concrete security level. We consider all known attacks on RLWE and conclude that this problem requires the same attack complexity as the ordinary $(n/2)$-dimensional LWE problem.

The plaintext encoding technique of HEAAN utilizes the canonical embedding map for the packing of plaintexts in a single ciphertext. Similarly, we consider the canonical embedding map $\tau : F \to \mathbb{C}^{n/2}$ of the number field F. Since ξ and its conjugate elements are real, the image of F with respect to its canonical embedding actually lies in $\mathbb{R}^{n/2}$. Therefore, we can successfully define a ring homomorphism from F into the vector of purely real numbers, and make the use of plaintext encoding/decoding algorithms between R and $\mathbb{R}^{n/2}$ based on this canonical embedding.

We construct a new HE scheme whose security relies on the hardness of RLWE over R. We first propose a vector representation for the elements F, which is efficient for the rounding operation into R and the modulo operation of the residue ring $R_q = R/qR$. Then, we describe a HE scheme over the real numbers, which provides approximate arithmetic operations and an approximate rounding operation.

We also explain how to represent the elements of R_q and perform the arithmetic operations between them. We present a specialized Fast Fourier Transform (FFT) algorithm for an efficient Number Theoretic Transform (NTT) on the residue ring R_q and fast multiplication between ring elements. This optimization technique constructs a simply computable ring isomorphism from R_q to $\mathbb{Z}_q[X]/(X^{n/2} - 1)$, so the ordinary NTT conversion on $\mathbb{Z}_q[X]/(X^{n/2} - 1)$ can be applied to R_q whose dimension is one quarter of that of a naive method.

In conclusion, our approximate HE scheme over R can encrypt $(n/2)$ plaintext slots in a single ciphertext, twice as many plaintext slots compared to $(n/4)$ of the ordinary HEAAN scheme over $\mathbb{Z}_q[X]/(X^{n/2} + 1)$, while keeping the same concrete security level, storage, and computational costs.

Related Works. Arita and Handa [3] proposed a HE scheme based on RLWE over the decomposition ring, which is a subring of cyclotomic ring. Their subring technique is applied to HElib [26]: they consider the plaintext space as $\mathbb{Z}_p \oplus \cdots \oplus \mathbb{Z}_p$, which is a subring of the plaintext space $GF(p^d) \oplus \cdots \oplus GF(p^d)$ of HElib for some integers p and d, where $GF(p^d)$ denotes the Galois field of the cardinality

p^d. They claimed that RLWE over the decomposition ring is at least as hard as its search version. However, there is no known reduction from lattice problems over ideal lattices to the search version, since the decomposition ring is not known to be a ring of integers of some number field so far. In contrary, RLWE over the conjugate-invariant ring which we desired in this paper has a reduction from SIVP over ideal lattices.

Road-Map. In Sect. 2, we present notations of our paper and some backgrounds for RLWE. In Sect. 3, we define RLWE over the conjugate-invariant ring and discuss about its hardness. In Sect. 4, we present our new approximate HE scheme constructed over the conjugate-invariant ring, describe encoding/decoding algorithms for real numbers, and propose a specialized FFT algorithm for the desired ring. In last section, we give a summary on our approximate HE scheme compared to original HEAAN.

2   Background

2.1   Notation

All logarithms are base 2 unless otherwise indicated. For an integer $m \geq 2$, $\mathbb{Z}_m := \mathbb{Z}/m\mathbb{Z}$, and $\mathbb{Z}_m^\times$ is the multiplicative group of units in $\mathbb{Z}_m$. For a ring R, its residue ring R/qR modular an integer q is denoted by R_q. For a real number r, $\lfloor r \rceil$ denotes the nearest integer to r, rounding upwards in case of a tie. For a vector $\boldsymbol{u}$ of (complex) numbers, $\|\boldsymbol{u}\|_2$ (resp. $\|\boldsymbol{u}\|_\infty$) denotes the ℓ_2-norm (resp. ℓ_∞-norm) of $\boldsymbol{u}$. For an element a of a number field K, $\|a\|_2^{\mathsf{can}}$ (resp. $\|a\|_\infty^{\mathsf{can}}$) denotes the ℓ_2-norm (resp. ℓ_∞-norm) of the image vector of a via the canonical embedding map. For vectors $\boldsymbol{a}$ and $\boldsymbol{b}$ of the same dimension, $\boldsymbol{a} \odot \boldsymbol{b}$ denotes the component-wise multiplication of $\boldsymbol{a}$ and $\boldsymbol{b}$. We denote by $\phi(\cdot)$ the Euler's totient function and $\Phi_m(X)$ the m-th cyclotomic polynomial. For a complex number $z \in \mathbb{C}$, $\bar{z}$ denotes the complex conjugation of z.

2.2   Number Fields and Ideal Lattices

For any number field K, there exists an element ζ of K such that $K = \mathbb{Q}(\zeta)$. Hence K is isomorphic to $\mathbb{Q}[X]/(f(X))$ for the minimal polynomial $f(X)$ of ζ over $\mathbb{Q}$. The degree n of $f(X)$ equals to the extension degree $[K : \mathbb{Q}]$. There are exactly n injective ring homomorphisms $\{\sigma_j\}_{1 \leq j \leq n}$ from K to $\mathbb{C}$. The canonical embedding is defined as the n-tuple of these embeddings as follows:

$$\sigma : K \to \mathbb{C}^n$$
$$a \mapsto (\sigma_j(a))_{1 \leq j \leq n}.$$

Let s_1 be the number of real embeddings of K, then $n = s_1 + 2s_2$ for some non-negative integer s_2. Without loss of generality, let $\sigma_1, \ldots, \sigma_{s_1}$ be real embeddings of K. Then the image of σ lies in the space $H := \{(x_1, \ldots, x_n) \in \mathbb{C}^n : x_{s_1+s_2+j} = \overline{x_{s_1+j}}, 1 \leq j \leq s_2\}$. Let $\{\boldsymbol{e}_j\}_{1 \leq j \leq n}$ be a canonical basis of $\mathbb{C}^n$. Let $\boldsymbol{h}_j = \boldsymbol{e}_j$ for $1 \leq$

$j \leq s_1$, $\boldsymbol{h}_{s_1+j} = (\boldsymbol{e}_{s_1+j} + \boldsymbol{e}_{s_1+s_2+j})/\sqrt{2}$ and $\boldsymbol{h}_{s_1+s_2+j} = (\boldsymbol{e}_{s_1+j} - \boldsymbol{e}_{s_1+s_2+j})/\sqrt{-2}$ for $1 \leq j \leq s_2$. Then, $\{\boldsymbol{h}_j\}_{1 \leq j \leq n}$ forms an orthogonal $\mathbb{R}$-basis of H.

An element of K is called an algebraic integer if its minimal polynomial over $\mathbb{Q}$ has integral coefficients. The set of all algebraic integers, denoted by $\mathcal{O}_K$, is called the ring of integers of K. A fractional ideal I of K is $\mathcal{O}_K$-submodule of K such that there exists a non-zero element $r \in \mathcal{O}_K$ which satisfies $rI \subseteq \mathcal{O}_K$. If $I \subseteq \mathcal{O}_K$, then we call I an (integral) ideal. The image $\sigma(I)$ of a fractional ideal I via the canonical embedding forms a lattice in $\mathbb{C}^n$, and we call it an ideal lattice generated by I. The dual of I in K is a fractional ideal in K defined as $I^\vee := \{a \in K : \mathrm{Tr}(aI) \subseteq \mathbb{Z}\}$.

For $1 \leq k \leq n$, the k-th successive minima of the lattice $\mathcal{L}$, denoted by $\lambda_i(\mathcal{L})$, is the minimum value of $r > 0$ such that $\mathcal{L}$ has k linearly independent vectors of length at most r. If $\mathcal{L}$ is an ideal lattice $\sigma(I)$ for a fractional ideal $I \in K$, we simply denote by $\lambda_k(I)$. The SIVP over ideal lattices in K is defined as follow.

Definition 1 *(SIVP over ideal lattices). For a number field K of degree n and an approximation factor $\gamma \geq 1$, the K-SIVP$_\gamma$ problem is: given a fractional ideal I of K, output n linearly independently vectors in the ideal lattice $\sigma(I)$ of length at most $\gamma \cdot \lambda_n(I)$.*

2.3 Ring Learning with Errors

For positive integers n and q, let R be the ring of integers of a number field K, $R_q = R/qR$ and $K_{\mathbb{R}} = K \otimes_{\mathbb{Q}} \mathbb{R}$. Let χ_{key} and χ_{err} be distributions over $R^\vee$ and $K_{\mathbb{R}}$, respectively. For $s \in R_q^\vee$, $A_{q,\chi_{err}}^{\text{R-LWE}}(s)$ is a distribution which draws $a \leftarrow R_q$ and $e \leftarrow \chi_{err}$, and output the pair $(a, a \cdot s + e)$ in $R_q \times K_{\mathbb{R}}/qR^\vee$. The (decisional) RLWE problem is defined as follows.

Definition 2 (Ring Learning with Errors). *Let n, q be positive integers, and χ_{key} (resp. χ_{err}) be a distribution over $R_q^\vee$ (resp. $K_{\mathbb{R}}$). The RLWE problem, denoted by R-LWE$_{q,\chi_{err}}(\chi_{key})$, is to distinguish between the uniform distribution over $R_q \times K_{\mathbb{R}}/qR^\vee$ and $A_{q,\chi_{err}}^{\text{R-LWE}}(s)$ where $s \leftarrow \chi_{key}$.*

Since $K_{\mathbb{R}}$ is isomorphic to the vector space H, a distribution over H can be identified as a distribution over $K_{\mathbb{R}}$. If χ_{err} is a (spherical) Gaussian distribution $D_{\alpha q}$ over H with respect to the basis $\{\boldsymbol{h}_i\}_{1 \leq i \leq n}$ and χ_{key} is the uniform distribution over $R_q^\vee$, we simply denote by R-LWE$_{q,\alpha}$.

Lyubashevsky et al. [32] proposed a polynomial-time quantum reduction from lattice problems over ideal lattices to the RLWE problem, which holds only for the cyclotomic fields with some special conditions on the modulus q. Peikert et al. [34] gave a new reduction from the same problem which can be applied to an arbitrary number field and modulus.

Theorem 1 *([34, Corollary 7.3]). Let n, q be positive integers, $0 < \alpha < 1$ be a real number such that $\alpha q = \omega(1)$, K be an arbitrary number field of degree n and $R = \mathcal{O}_K$. Then there exists a polynomial-time quantum reduction from K-SIVP$_\gamma$ to R-LWE$_{q,\alpha}$ given ℓ samples for $\gamma = \max\{\omega(\sqrt{n \log n}/\alpha) \cdot (n\ell/\log(n\ell))^{1/4}, \sqrt{2n}\}$.*

Recently, it was shown by Rosca et al. [36] that the non-dual RLWE problem, i.e., RLWE with the distribution of the secret over R_q rather than $R_q^\vee$, is at least as hard as the original RLWE problem. In addition, the rounding technique of Peikert [33] allows us to sample errors from a discrete Gaussian distribution rather than a continuous Gaussian distribution. With these settings, an RLWE sample lies in $R_q \times R_q$ rather than $R_q \times K_{\mathbb{R}}/qR^\vee$.

3   RLWE over the Conjugate-Invariant Ring

The cyclotomic rings have been the most commonly used as base rings for RLWE for two main reasons. The ring of integers of the m-th cyclotomic field is isomorphic to $\mathbb{Z}[X]/(\Phi_m(X))$, and its structure was particularly well suitable in the construction of cryptographic schemes with the perspective of efficiency and some functionalities. In addition, there have been no known reduction to the RLWE over a non-cyclotomic ring for years until Peikert et al. [34] proposed a reduction from SIVP over ideal lattices to (decisional) RLWE for arbitrary number fields recently.

In this section, we introduce a new number field which has not been exploited in the lattice-based cryptography so far, and compute the ring of integers of the number field. Then we study on the hardness of RLWE problem over a new ring in two ways: we give a reduction from a standard lattice problem and study the concrete security level by considering all known attacks.

Let $m \geq 2$ be an integer and $n = \phi(m)$ for Euler's totient function $\phi(\cdot)$. For the m-th primitive root of unity $\zeta = \exp(2\pi i/m)$, the m-th cyclotomic field is defined by $K = \mathbb{Q}(\zeta)$. Let σ_{-1} be the element of $\mathrm{Gal}(K/\mathbb{Q})$ defined by $\sigma_{-1} : \zeta \mapsto \zeta^{-1}$, and $G = \{id, \sigma_{-1}\}$ be the cyclic subgroup of $\mathrm{Gal}(K/\mathbb{Q})$ generated by σ_{-1}. We denote by $F = K^G$ the G-invariant subfield of K which is defined as $F = \{a \in K : \tau(a) = a, \forall \tau \in G\}$. We first remark that $F = \mathbb{Q}(\xi)$ for $\xi = \zeta + \zeta^{-1}$. It is clear that $\mathbb{Q}(\xi) \subseteq F \subseteq \mathbb{Q}(\zeta)$ and $[\mathbb{Q}(\zeta) : F] = |G| = 2$. Since ζ is a root of $X^2 - \xi \cdot X + 1 \in \mathbb{Q}(\xi)[X]$, the inequality $[\mathbb{Q}(\zeta) : \mathbb{Q}(\xi)] \leq 2$ holds and it implies $F = \mathbb{Q}(\xi)$. In particular, we are interested in the set of integer coefficient elements in $\mathbb{Q}(\xi)$ with respect to the $\mathbb{Q}$-basis $\{1, \xi, \xi^2, \ldots, \xi^{\frac{n}{2}-1}\}$. We will call this set $\mathbb{Z}[\xi]$ as the conjugate-invariant ring.

3.1   Reduction from SIVP

Some well-known reductions [32,34] from standard problems over ideal lattices to RLWE requires a condition that the base ring exploited in RLWE should be a ring of integers of a number field. Therefore, it is crucial to study the ring of integers of a number field to define and show the hardness of RLWE problem.

We consider the subfield $F = \mathbb{Q}(\xi)$ of $K = \mathbb{Q}(\zeta)$ as a base number field, and compute its ring of integers $R := \mathcal{O}_F$ in this section. Fortunately, the structure of a cyclotomic field derives a quite simple and nice result on the conjugate-invariant ring as follows.

Lemma 1. $\mathbb{Z}[\xi]$ *is the ring of integers of* $F = \mathbb{Q}(\xi)$ *(Fig. 1).*

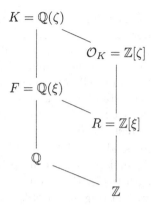

Fig. 1. Number fields and their rings of the integers

Proof. It is clear that $\mathbb{Z}[\xi] \subseteq \mathcal{O}_F$. Since $\mathcal{O}_F \subseteq \mathcal{O}_K = \mathbb{Z}[\zeta]$, every element $a \in \mathcal{O}_F$ is uniquely expressed as $a = \sum_{-\frac{n}{2} \leq j < \frac{n}{2}} a_j \cdot \zeta^j$ for some integers $a_{-\frac{n}{2}}, \ldots, a_{\frac{n}{2}-1}$. From the definition of F, we obtain $\sigma_{-1}(a) = a$, i.e., $\sum_{j=-\frac{n}{2}}^{\frac{n}{2}-1} a_j \zeta^j = \sum_{j=-\frac{n}{2}+1}^{\frac{n}{2}} a_{-j} \zeta^j$ which implies $a_j = a_{-j}$ for $0 \leq i < \frac{n}{2}$ and $a_{-\frac{n}{2}} = 0$. Then, $a = a_0 + \sum_{j=1}^{\frac{n}{2}-1} a_i(\zeta^j + \zeta^{-j}) \in \mathbb{Z}[\xi]$, since $\zeta^j + \zeta^{-j} \in \mathbb{Z}[\xi]$ for $1 \leq j < \frac{n}{2}$. Therefore, $\mathcal{O}_F \subseteq \mathbb{Z}[\xi]$, which directly implies $\mathbb{Z}[\xi] = \mathcal{O}_F$. □

It is derived from Lemma 1 that the RLWE problem over $R = \mathbb{Z}[\xi]$, simply denoted by $R\text{-LWE}_{q,\alpha}$, is at least as hard as F-SIVP from Theorem 1. We can naturally identify R with the ring of polynomials $\mathbb{Z}[Y]/(g(Y))$ for the minimal polynomial $g(Y) \in \mathbb{Z}[Y]$ of ξ over $\mathbb{Q}$ via mapping $a(Y) \mapsto a(\xi)$. However, it is more convenient to consider R as the subring

$$R = \{a(X) \in \mathbb{Z}[X]/(\Phi_m(X)) : a(X) = a(X^{-1})\}$$

of $\mathcal{O}_K = \mathbb{Z}[X]/(\Phi_m(X))$. Note that the condition $a(X) = a(X^{-1})$ corresponds to the conjugation-invariant property. We will follow this subring perspective in the rest of paper.

3.2   Cryptanalysis

In this section, we discuss the attack complexity of RLWE over the conjugate-invariant ring. In general, the RLWE problem does not guarantee the same security level as LWE with the same parameter. For example, there have been several attempts to attack the RLWE (or Poly-LWE) problem over a ring $\mathbb{Z}[X]/(f(X))$ by exploiting its ring structure [9,10,20]. One common limitation of these attacks is that $f(X)$ should have a root modulo q satisfying some strong conditions.

The RLWE assumption can be viewed as a specific case of LWE ($A, \boldsymbol{b} = A\boldsymbol{s} + \boldsymbol{e}$) where the random matrix A has a special algebraic structure. In the case of RLWE over a power-of-two cyclotomic ring, an RLWE sample can be understood as a variant of n-dimensional LWE instance where A is a random anti-circulant matrix. However, there has been no known attack achieving a lower complexity by exploiting this property. As a result, the current best known attacks are standard lattice attacks on the ordinary LWE problem such as dual attack and primal attack, which are well described in [1].

Now we explain how to understand an R-LWE instance as an LWE instance with a special structure. Let m be a power-of-two integer so that $n = m/2$ and $\Phi_m(X) = X^n + 1$. An element of $R = \{a(X) \in \mathbb{Z}[X]/(X^n + 1) : a(X) = a(X^{-1})\}$ can be uniquely expressed as $a(X) = a_0 + \sum_{j=1}^{\frac{n}{2}-1} a_j \cdot (X^j + X^{-j})$ for some integers $a_0, \ldots, a_{\frac{n}{2}-1}$. Therefore, $a(X)$ can be identified with the vector $\boldsymbol{a} = (a_0, a_1, \ldots, a_{\frac{n}{2}-1})$ of length $(n/2)$. Based on this identification, an RLWE sample over the conjugate-invariant ring $(a(X), b(X) = a(X) \cdot s(X) + e(X)) \in R_q^2$ with secret $s(X)$ can be transformed to $(A, \boldsymbol{b} = A\boldsymbol{s} + \boldsymbol{e}) \in \mathbb{Z}_q^{\frac{n}{2} \times \frac{n}{2}} \times \mathbb{Z}_q^{\frac{n}{2}}$ where A is a square matrix of size $(n/2)$ whose (i,j)-th component is given by

$$
A_{ij} = \begin{cases} a_{|i-j|} & j = 0, \text{ or } i+j = \frac{n}{2} \\ a_{|i-j|} + a_{i+j} & j > 0, \text{ and } i+j < \frac{n}{2} \\ a_{|i-j|} - a_{n-(i+j)} & j > 0, \text{ and } i+j > \frac{n}{2} \end{cases}
$$

for $0 \le i, j < n/2$. This transformation shows that R-LWE can be viewed as a variant of the $(n/2)$-dimensional LWE problem where the random matrix A has this special form. We consider all known attacks on RLWE and claim that they do not achieve a lower complexity than the standard lattice attacks on LWE, i.e., currently there is no special attack on R-LWE which exploits the ring structure of R corresponding to this special structural distribution of A, similar to the case of RLWE over a power-of-two cyclotomic ring. Therefore, we conclude that the current best attacks on R-LWE$_{q,\alpha}$ are the standard lattice attacks, which require the same attack complexity as the lattice attacks on the $(n/2)$-dimensional LWE problem.

4 Approximate Homomorphic Encryption over the Real Numbers

The HEAAN scheme of Cheon et al. [12,13] is the first HE system which supports an efficient rounding operation for approximate arithmetic. It allows us to encrypt a number of complex numbers in a single ciphertext and perform an approximate arithmetic between encrypted vectors in a SIMD manner. However, there remained one significant problem about the plaintext space.

Most of the real-world applications require computations over the purely real numbers, but the original HEAAN scheme could encrypt a complex number in each of plaintext slots. The previous researches [29,30] used the set of real numbers as a subring of complex numbers, but this approach cannot be a fundamental solution for the following reason. Every algorithm of the original HEAAN scheme, such as homomorphic arithmetic and rounding operation,

$$K = \mathbb{Q}(\zeta) \xrightarrow{\;\simeq\;} \mathbb{Q}[X]/(X^n + 1)$$

$$\uparrow \qquad \circlearrowleft \qquad \uparrow Y \mapsto X + X^{-1}$$

$$F = \mathbb{Q}(\xi) \xrightarrow{\;\simeq\;} \mathbb{Q}[Y]/(g(Y)) \xrightarrow{\;\tau\;} \mathbb{R}^{n/2}$$

Fig. 2. Polynomial representation of number fields and canonical embedding

adds a small complex error to the plaintext vector. The imaginary part of an encrypted plaintext can gradually increase as the computation progressed, and finally the desired result (real part) can no longer be recovered after its imaginary part becomes larger than the ciphertext modulus. Consequently, every circuit in previous applications had a limited depth to bound the size of imaginary parts during its evaluation.

In this section, we describe a HE scheme which is optimized in the approximate computation over the *real* numbers compared to the original HEAAN scheme with complex plaintext slots. The security of our scheme relies on the RLWE assumption over the ring $R = \mathbb{Z}[\xi]$ introduced in the previous section. For simplicity, the integer m will be chosen as a power of two so that $n = m/2$ and $\Phi_m(X) = X^n + 1$.

4.1 Canonical Embedding and Packing Technique

In this subsection, we describe the canonical embedding map of the conjugate-invariant field and explain how to represent its elements. As mentioned in the previous section, the conjugate-invariant field $F = \mathbb{Q}(\xi)$ can be identified with the polynomial ring $\mathbb{Q}[Y]/(g(Y))$ for the minimal polynomial $g(Y) \in \mathbb{Z}[Y]$ of ξ over $\mathbb{Q}$. Note that $g(Y)$ is a polynomial of degree $(n/2)$ satisfying $g(X + X^{-1}) = X^{\frac{n}{2}} + X^{-\frac{n}{2}}$. Let $\xi_j = \zeta^{4j+1} + \zeta^{-(4j+1)}$ for $0 \le j < n/2$. Then $\{\xi_0, \ldots, \xi_{\frac{n}{2}-1}\}$ forms the set of distinct roots of $g(Y)$ since $X^n + 1 = (X - \zeta)(X - \zeta^3) \ldots (X - \zeta^{m-1}) = \prod_{j=0}^{\frac{n}{2}-1}(X^2 - \xi_j \cdot X + 1)$. Therefore, we have a commute diagram (Fig. 2) for a polynomial representation of number fields by identifying $Y \mapsto X + X^{-1}$.

Let us denote by τ the canonical embedding of $F = \mathbb{Q}[Y]/(g(Y))$ into $\mathbb{C}^{n/2}$. It sends an element $a(Y)$ to the vector of its evaluations $\tau(a) = (a(\xi_j))_{0 \le j < \frac{n}{2}}$ at the roots of $g(Y)$. Since all roots of $g(Y)$ are real, F is a totally real number field and the image of τ is a subring of $\mathbb{R}^{n/2}$. The canonical embedding norm of an element of a number field is defined by the norm of its canonical embedding. For example, we write $\|a\|_{\infty}^{\mathsf{can}} := \|\tau(a)\|_{\infty}$ and $\|a\|_2^{\mathsf{can}} := \|\tau(a)\|_2$ for $a \in F$.

The packing technique of HE system allows us to encrypt a multiple number of messages in a single ciphertext and supports the parallel computation in a SIMD manner. It has been one of the most important techniques for the performance improvements of HE schemes in terms of expansion rate and amortized computational cost. The packing method of approximate HE scheme [13] is based on the canonical embedding over the complex numbers.

We present a new packing method over the real numbers, by modifying the previous solution over the complex plane. The core idea is to restrict the domain of canonical embedding τ to the ring of integers $R = \mathbb{Z}[Y]/(g(Y))$. In other words, the decoding algorithm transforms an element $a(Y)$ of R into the vector $\tau(a) = (a(\xi_j))_{0 \leq j < n/2}$ of dimension $(n/2)$. This vector is real as noted above. Conversely, the encoding map takes a real vector $x = (x_j)_{0 \leq j < n/2} \in \mathbb{R}^{n/2}$ as an input. It first computes the rounding $x' = \lfloor x \rceil_{\tau(R)} \in \mathbb{R}^{n/2}$, which is an element of $\tau(R)$ with a small rounding error $\|x - x'\|_2^{\mathsf{can}}$. The output is obtained by computing the inverse of x' which is an integral polynomial in $R = \mathbb{Z}[Y]/(g(Y))$. Our packing method is explicitly described as follows.

- Ecd(x). For given $x = (x_j)_{0 \leq j < n/2} \in \mathbb{R}^{n/2}$, discretize x into $\tau(R)$. Output the corresponding polynomial $\mathsf{m}(Y) = \tau^{-1}\left(\lfloor x \rceil_{\tau(R)}\right) \in R$.
- Dcd(m). For given $\mathsf{m} \in R$, output the vector $x = (\mathsf{m}(\xi_j))_{0 \leq j < n/2} \in \mathbb{R}^{n/2}$.

The Ecd algorithm can be viewed as an approximate inverse of the decoding function with a small rounding error. One can multiply a scale factor to an input vector before the rounding operation to reduce the relative size of rounding error and preserve the precision of plaintexts.

As a toy example, let $n = m/2 = 4$. Then $\zeta_8 = \exp(\pi i/4) = (1 + i)/\sqrt{2}$ is an m-th primitive root of unity, and we have $\{\xi_0, \xi_1\} = \{\sqrt{2}, -\sqrt{2}\}$. For a real vector $x = (1.1, 2.3)$, its encoding polynomial with the scaling factor $\Delta = 64$ is obtained by $\mathsf{m}(Y) = \tau^{-1}\left(\lfloor \Delta \cdot x \rceil_{\tau(R)}\right) = 109 - 27Y$. Conversely, the decoded vector of $109 - 27Y$ is computed by $\Delta^{-1} \cdot \mathsf{Dcd}(\mathsf{m}) = \frac{1}{64}(109 - 27\sqrt{2}, 109 + 27\sqrt{2}) \approx (1.1065, 2.2997)$, which is a good approximation of the original vector x.

4.2   Scheme Description

This subsection gives a explicit description of our HE scheme over the real numbers. Our scheme is very similar to the original HEAAN scheme, but it exploits a different ring structure $R = \mathbb{Z}[\xi]$. We first propose a method to represent the elements of the conjugate-invariant field F.

The number field F can be identified with $\mathbb{Q}^{n/2}$ as a $\mathbb{Q}$-module. For example, an arbitrary element of $F = \mathbb{Q}[Y]/(g(Y))$ can be uniquely expressed as the sum $\sum_{j=0}^{\frac{n}{2}-1} a_j \cdot Y^j$ for some $a_j \in \mathbb{Q}$, which corresponds to the isomorphism $a \mapsto (a_0, \ldots, a_{\frac{n}{2}-1})$ between two modules. However, this representation is not the best choice for the construction of HE system. One major reason is that the image $\{\tau(1), \tau(Y), \ldots, \tau(Y^{\frac{n}{2}-1})\}$ of the basis $\{1, Y, \ldots, Y^{\frac{n}{2}-1}\}$ does not form an orthogonal set in the space $\mathbb{R}^{n/2}$.

The conjugate-invariant field $F = \mathbb{Q}[Y]/(g(Y))$ can be understood as a subfield of $K = \mathbb{Q}[X]/(X^n + 1)$ by identifying $Y = X + X^{-1}$ as noted in the previous subsection. Every element $a(X)$ of $F \leq K$ can be uniquely expressed as a Laurent polynomial $a(X) = a_0 + \sum_{i=1}^{\frac{n}{2}-1} a_i(X^i + X^{-i})$ of degree and order strictly less then $(n/2)$ for some $a_0, \ldots, a_{\frac{n}{2}-1} \in \mathbb{Q}$. In the following, an arbitrary element $a(X)$ of F will be identified with its vector of coefficients $(a_0, \ldots, a_{\frac{n}{2}-1}) \in \mathbb{Q}^{n/2}$. Note that the set $\{1, X + X^{-1}, \ldots, X^{\frac{n}{2}-1} + X^{1-\frac{n}{2}}\}$ is a basis of F (resp. R) as a module over $\mathbb{Q}$ (resp. $\mathbb{Z}$). In addition, the image of this basis with respect to the canonical embedding map τ forms an orthogonal basis in $\mathbb{R}^{n/2}$.

This orthogonal property allows us to use an efficient rounding operation on F as well as a modulo operation over R. We define the rounding operation $\lfloor \cdot \rceil : F \to R$ by sending each of coefficients $a_i \in \mathbb{Q}$ to the closest integer $\lfloor a_i \rceil \in \mathbb{Z}$. Note that $\lfloor a \rceil$ is an element of R which minimizes the rounding error $\|a - \lfloor a \rceil\|_2^{\mathsf{can}}$ with respect to the ℓ_2 canonical embedding norm. Similar to the rounding operation, the modulo q operation is simply defined by the coefficient-wise modular reduction, i.e., $[a]_q$ is the element of $a + qR$ which minimizes the size $\|[a]_q\|_2^{\mathsf{can}}$.

- $\mathsf{Setup}(p, 1^\lambda, L)$.

  – The base integer p, the number of levels L and the security parameter λ are given as input. Set moduli $q_1, q_2, \ldots, q_L$, which are usually chosen as $q_i = p^i$.
  – Choose integers m and P, and small distributions χ_{key}, χ_{enc}, and χ_{err} over the ring R.
  – Return the parameter set $\mathsf{params} \leftarrow (m, P, \chi_{key}, \chi_{enc}, \chi_{err})$.

The setup step should generate a HE parameter set that achieves λ-bit of security level against the best known attacks on RLWE. A security proof will be given at the end of this subsection.

- $\mathsf{KeyGen}(\mathsf{params})$.

  – Sample $s \leftarrow \chi_{key}$. Set the secret key as $\mathsf{sk} \leftarrow (1, s)$.
  – Sample $a \leftarrow U(R_{q_L})$ and $e \leftarrow \chi_{err}$. Set the public key as $\mathsf{pk} \leftarrow (b, a) \in R_{q_L}^2$ where $b \leftarrow -as + e \pmod{q_L}$.
- $\mathsf{KSGen}(s_1, s_2)$. For $s_1, s_2 \in R$, sample $a' \leftarrow U(R_{P \cdot q_L})$ and $e' \leftarrow \chi_{err}$. Output the switching key as $\mathsf{swk} \leftarrow (b', a') \in R_{P \cdot q_L}^2$ where $b' \leftarrow -a's_2 + e' + P \cdot s_1 \pmod{P \cdot q_L}$.
  – Set the evaluation key as $\mathsf{evk} \leftarrow \mathsf{KSGen}(s^2, s)$.
- $\mathsf{Enc}_{\mathsf{pk}}(\mathsf{m})$. For $\mathsf{m} \in R$, sample $v \leftarrow \chi_{enc}$ and $e_0, e_1 \leftarrow \chi_{err}$. Output $v \cdot \mathsf{pk} + (\mathsf{m} + e_0, e_1) \pmod{q_L}$.
- $\mathsf{Dec}_{\mathsf{sk}}(\mathsf{ct})$. For $\mathsf{ct} = (c_0, c_1) \in R_{q_\ell}^2$, output $\mathsf{m}' = c_0 + c_1 \cdot s \pmod{q_\ell}$.

The decryption algorithm can be simply written by $\mathsf{m}' \leftarrow [\langle \mathsf{ct}, \mathsf{sk} \rangle]_{q_\ell}$. The encryption procedure returns a level L ciphertext ct which satisfies $[\langle \mathsf{ct}, \mathsf{sk} \rangle]_{q_L} \approx \mathsf{m}$, i.e., we can only recover an approximate value of m from its encryption. We use the canonical embedding norm to measure the size of polynomials in R.

- $\mathsf{Add}(\mathsf{ct}, \mathsf{ct}')$. For $\mathsf{ct}, \mathsf{ct}' \in R_{q_\ell}^2$, output $\mathsf{ct}_{add} \leftarrow \mathsf{ct} + \mathsf{ct}' \pmod{q_\ell}$.
- $\mathsf{Mult}_{\mathsf{evk}}(\mathsf{ct}, \mathsf{ct}')$. For $\mathsf{ct} = (c_0, c_1)$, $\mathsf{ct}' = (c_0', c_1') \in R_{q_\ell}^2$, let $(d_0, d_1, d_2) = (c_0 c_0', c_0 c_1' + c_1 c_0', c_1 c_1') \pmod{q_\ell}$. Output $\mathsf{ct}_{mult} \leftarrow (d_0, d_1) + \lfloor P^{-1} \cdot d_2 \cdot \mathsf{evk} \rceil \pmod{q_\ell}$.

- $RS_{\ell \to \ell'}(ct)$. For a ciphertext $ct \in R^2_{q_\ell}$ at level ℓ, output $ct' \leftarrow \lfloor (q_{\ell'}/q_\ell) \cdot ct \rceil$ (mod $q_{\ell'}$). We will omit the subscript ($\ell \to \ell'$) when $\ell' = \ell - 1$.

The algorithms Add and Mult$_{evk}$ perform the arithmetic operations over encrypted plaintexts. The *rescaling* procedure $RS_{\ell \to \ell'}(\cdot)$ transforms a level ℓ encryption of m into an encryption of $(q_{\ell'}/q_\ell) \cdot m$ of level ℓ' securely. We refer the refer to the full version of this paper for the correctness proof and noise estimation.[1]

Security. We claim that our HE scheme is IND-CPA secure under the hardness of RLWE problems over the ring R. It can be shown by considering the following three distributions:

$$\mathcal{D}_1 = \{(pk, ct) : pk \leftarrow KeyGen(params), ct \leftarrow Enc_{pk}(0)\},$$
$$\mathcal{D}_2 = \{(pk, ct) : pk \leftarrow U(\mathcal{R}^2_q), ct \leftarrow Enc_{pk}(0)\},$$
$$\mathcal{D}_3 = \{(pk, ct) : pk \leftarrow U(\mathcal{R}^2_q), ct \leftarrow U(\mathcal{R}^2_q)\}.$$

First, the distributions $\mathcal{D}_1$ and $\mathcal{D}_2$ are computationally indistinguishable under the assumption of $R\text{-LWE}_{q_L, \chi_{err}}(\chi_{key})$ since the key generation step samples s from χ_{key} and generates an RLWE sample pk of parameter (q_L, χ_{err}). The second and third distributions are computationally indistinguishable as long as $R\text{-LWE}_{q_L, \chi_{err}}(\chi_{enc})$ since a sample from $\mathcal{D}_2$ forms two independent RLWE samples of parameter (q_L, χ_{err}) with a secret $v \leftarrow \chi_{enc}$. Finally, the evaluation key $evk \leftarrow KSGen(s^2, s)$ can be viewed as an encryption of s^2 encrypted by the secret s. The distribution of evk can be indistinguishable from the uniform distribution on $\mathcal{R}^2_{P \cdot q_L}$ under the assumption of circular security when the $R\text{-LWE}_{P \cdot q_L, \chi_{err}}(\chi_{key})$ problem is hard.

4.3   Implications of the Conjugate-Invariant Ring

This section compares our approximate HE scheme over the real numbers with the original HEAAN scheme from a variety of perspectives. We claim that our scheme can have twice as many plaintext slots as HEAAN while guaranteeing the same security level and performance. Furthermore, the utilization of the conjugate-invariant ring fundamentally blocks the complex explosion problem of HEAAN which possibly effect on the most significant bits of real messages.

Representation of Ring Elements. Our HE scheme is constructed over the residue ring $R_q = \{a(X) \in \mathbb{Z}_q[X]/(X^n + 1) : a(X) = a(X^{-1})\}$ for an integer q. We introduce two methods to represent the ring elements of R_q with different pros and cons.

Basically we use the coefficient representation $(a_0, \ldots, a_{n-1}) \in \mathbb{Z}_q^{n/2}$ of $a(X) \in R_q$ as described in the previous subsection. The coefficient representation is useful to perform the non-arithmetic operations such as the rounding operation in rescaling procedure. However, we have to consider the following representation for an efficient multiplication between polynomials in R_q.

---

[1] https://eprint.iacr.org/2018/952.

Suppose that q is an integer such that there exists an m-th primitive root ω_m of unity in $\mathbb{Z}_q$. Note that $\omega_n := \omega_m^2$ (resp. $\omega_{\frac{n}{2}} := \omega_m^4$) is an n-th (resp. $(n/2)$-th) primitive root of unity in $\mathbb{Z}_q$. The map $\mathbb{Z}_q[X]/(X^n+1) \to \mathbb{Z}_q^n$, $a \mapsto (a(\omega_m), a(\omega_m^3), \ldots, a(\omega_m^{m-1}))$ is a ring isomorphism since the m-th cyclotomic polynomial is expressed as a product $X^n+1 = (X-\omega_m)(X-\omega_m^3)\ldots(X-\omega_m^{2n-1})$ modulo q. We point out that an element $a \in \mathbb{Z}_q[X]/(X^n+1)$ is contained in the subring R_q if and only if $a(\omega_m^j) = a(\omega_m^{2n-j})$ for all $j = 1, 3, \ldots, n-1$. Therefore, the map $a \mapsto \hat{a} = (a(\omega_m), a(\omega_m^5), \ldots, a(\omega_m^{m-3}))$ is an ring isomorphism from R_q to $\mathbb{Z}_q^{n/2}$ satisfying $\widehat{a \cdot b} = \hat{a} \odot \hat{b}$ for any $a, b \in R_q$, where $\odot$ denotes the Hadamard (component-wise) multiplication between vectors. It enables us to perform an arithmetic operation of R_q in $O(n)$ modulo q operations, but the rescaling procedure cannot be done under this representation.

Complexity of Ring Operations. The conversion between two representations $a \mapsto \hat{a}$ is one of the most important parts to improve the efficiency of the HE system on R_q. It can be viewed as a linear transformation on $\mathbb{Z}_q^{n/2}$ by identifying the elements of R_q with their coefficient vectors.

The NTT is a discrete Fourier transform over a finite field. Specifically, the NTT over the finite field $\mathbb{Z}_q$ with an m-th primitive root ω_m of unity modulo q, denoted by $\mathtt{NTT}_m(\cdot)$, converts a polynomial in $\mathbb{Z}_q[X]/(X^m-1)$ into a vector in $\mathbb{Z}_q^m$ by $a \mapsto (a(\omega_m^j))_{0 \le j < m}$. The NTT is a ring isomorphism between $\mathbb{Z}_q[X]/(X^m-1)$ and $\mathbb{Z}_q^m$, and its inverse is denoted by $\mathtt{INTT}_m(\cdot)$. The NTT conversion can be understood as a linear map from $\mathbb{Z}_q^n$ to $\mathbb{Z}_q^n$ whose matrix representation is the $m \times m$ Vandermonde matrix generated by $\{1, \omega_m, \ldots, \omega_m^{m-1}\}$. The FFT algorithm can compute $\mathtt{NTT}_m(\cdot)$ in $O(m \cdot \log m)$ operations in $\mathbb{Z}_q$.

There have been suggested several methods to modify the NTT conversion to perform some operations used in cryptographic schemes. For example, Alkim et al. [2] and Longa-Naehrig [31] exploit a variant of NTT to make an efficient conversion between distinct representations of a ring element in $\mathbb{Z}_q[X]/(X^n+1)$. In the following, we propose a specialized FFT algorithm to perform the linear transformation $a \mapsto \hat{a}$ on R_q efficiently.

The main idea is to express the linear transformation $a \mapsto \hat{a}$ by a composition of $(n/2)$-dimensional NTT conversion and a few simple arithmetic operations. To be precise, the equality

$$a(\omega_m^{4j+1}) = a(\omega_m \cdot \omega_{\frac{n}{2}}^j) = a_0 + \sum_{i=1}^{\frac{n}{2}-1} a_i \left(\omega_m^i \cdot \omega_{\frac{n}{2}}^{ij} + \omega_m^{-i} \cdot \omega_{\frac{n}{2}}^{-ij} \right)$$

$$= a_0 + \sum_{i=1}^{\frac{n}{2}-1} a_i \cdot \omega_m^i \cdot \omega_{\frac{n}{2}}^{ij} + \sum_{i=1}^{\frac{n}{2}-1} a_{\frac{n}{2}-i} \cdot \omega_m^{-(\frac{n}{2}-i)} \cdot \omega_{\frac{n}{2}}^{ij}$$

$$= a_0 + \sum_{i=1}^{\frac{n}{2}-1} \left(a_i \cdot \omega_m^i + a_{\frac{n}{2}-i} \cdot \omega_m^{-(\frac{n}{2}-i)} \right) \omega_{\frac{n}{2}}^{ij} = \tilde{a}(\omega_{\frac{n}{2}}^j)$$

holds for any $0 \leq j < \frac{n}{2}$ where

$$\tilde{a}(X) = a_0 + \left(a_1 \cdot \omega_m + a_{\frac{n}{2}-1} \cdot \omega_m^{1-\frac{n}{2}}\right) X + \cdots + \left(a_{\frac{n}{2}-1} \cdot \omega_m^{\frac{n}{2}-1} + a_1 \cdot \omega_m^{-1}\right) X^{\frac{n}{2}-1}.$$

Therefore, the linear transformation $a \mapsto \hat{a}$ can be written by the composition of $\mathrm{NTT}_{n/2}$ and a simple arithmetic operation

$$\left(a_0, \ldots, a_{\frac{n}{2}-1}\right) \mapsto \left(a_0, a_1 \cdot \omega_m + a_{\frac{n}{2}-1} \cdot \omega_m^{1-\frac{n}{2}}, \ldots, a_{\frac{n}{2}-1} \cdot \omega_m^{\frac{n}{2}-1} + a_1 \cdot \omega_m^{-1}\right),$$

and we can compute its inverse by

$$a = \left(\tilde{a}_0, 2^{-1} \cdot (\tilde{a}_1 \cdot \omega_m^{-1} + \tilde{a}_{\frac{n}{2}-1} \cdot \omega_m), \ldots, 2^{-1} \cdot (\tilde{a}_{\frac{n}{2}-1} \cdot \omega_m^{1-\frac{n}{2}} + \tilde{a}_1 \cdot \omega_m^{\frac{n}{2}-1})\right)$$

for $\tilde{a} = (\tilde{a}_0, \ldots, \tilde{a}_{\frac{n}{2}-1}) \leftarrow \mathrm{INTT}_{n/2}(\hat{a})$.

Now let us consider the multiplication of polynomials in the conjugate-invariant ring R. For given polynomials $a, b \in R_q$ with coefficient representation, we compute their product $c = a \cdot b$ by computing $\hat{c} = \widehat{a \cdot b} = \hat{a} \odot \hat{b}$ and recovering c from $\hat{c}$. It consists of three Hadamard multiplications on $\mathbb{Z}_q^{n/2}$, two $\mathrm{NTT}_{n/2}$ conversions, and a single $\mathrm{INTT}_{n/2}$. Since the Hadamard multiplication takes only $O(n)$, the complexity of a multiplication over the special ring R_q can be estimated by three NTT conversions of dimension $(n/2)$, while a multiplication over the ring $\mathbb{Z}_q[X]/(X^n + 1)$ includes three NTT conversions of dimension n. As a result, the computational cost of an arithmetic operation on R_q is almost half that of the m-th cyclotomic ring.

4.4   Application to Fixed-Point Operation

The HEAAN scheme is able to evaluate a circuit approximately, and specifically our variant is optimized in an arithmetic over the real numbers. We explain how to use our scheme to perform the fixed-point operation with a finite precision.

As described in Sect. 4.1, a real-valued vector can be identified with a polynomial in the conjugate-invariant ring R via the canonical embedding τ. For the use of our scheme in fixed-point operation, the base p in scheme description will be chosen as a scaling factor. So an arbitrary real vector $\boldsymbol{x} \in \mathbb{R}^{n/2}$ is encoded to a polynomial $\mathsf{m} \in R$ such that $\mathsf{m} \approx p \cdot \tau^{-1}(\boldsymbol{x})$ with a small rounding error. An encryption procedure induces an additional error so that an encryption of m is a pair $\mathsf{ct} = (c_0, c_1) \in R_{q_L}^2$ satisfying $[c_0 + c_1 \cdot s]_{q_L} = \mathsf{m} + e \approx p \cdot \tau^{-1}(\boldsymbol{x})$ for some small error e. The precision of an encrypted plaintext is decided by a scaling factor p and the size of errors, i.e., we can use a larger scaling factor to keep more significant bits.

Let ct_i be an encryption of $\mathsf{m}_i \approx p \cdot \tau^{-1}(\boldsymbol{x}_i)$ for $i = 1, 2$. Then their homomorphic multiplication returns a ciphertext ct_{mult} encrypting

$$\mathsf{m}_1 \cdot \mathsf{m}_2 \approx p^2 \cdot \tau^{-1}(\boldsymbol{x}_1) \cdot \tau^{-1}(\boldsymbol{x}_2) = p^2 \cdot \tau^{-1}(\boldsymbol{x}_1 \odot \boldsymbol{x}_2)$$

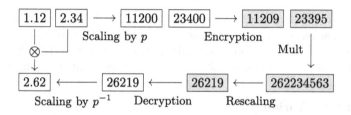

Fig. 3. An example of fixed-point operation

which is an encoding of the slot-wise product $x_1 \odot x_2$ with scaling factor p^2. Then, we can use the rescaling procedure $\mathsf{RS}(\cdot)$ to obtain an encryption of $p \cdot \tau^{-1}(x_1 \odot x_2)$ and recover the initial scaling factor p. In Fig. 3, we describe an example of fixed-point multiplication between 1.12 and 2.34 with scaling factor $p = 10^4$. Numbers in gray boxes represent the encrypted values in plaintext slots.

The scaling factor stays the same and the rescaling procedure reduces a ciphertext level by one. Therefore, for the evaluation of a circuit with depth L, the bitsize of largest ciphertext modulus should be $O(L \cdot \log p)$ which grows linearly on the depth and bit precision of plaintext, compared to the exponential growth based on the HE schemes for exact computations without rounding operation [8,21].

5 Discussions

5.1 Comparison with HEAAN

The security of our scheme relies on the hardness of R-LWE problem. From the cryptanalysis on RLWE over the conjugate-invariant ring in Sect. 3.2, our approximate HE scheme over $R = \{a(X) \in \mathbb{Z}[X]/(X^{2n} + 1) : a(X) = a(X^{-1})\}$ has (approximately) the same security level as the original HEAAN over $\mathbb{Z}[X]/(X^n + 1)$ for a power-of-two integer n, while the other parameters are set equal. In this setting, the maximum number of plaintexts packed in a single ciphertext in our scheme is n, while that of HEAAN is $(n/2)$. This implies our approximate HE scheme supports twice more parallel computations than HEAAN in a SIMD manner (Table 1).

Since it requires $n \log q$ bits to express an element of the form $a_0 + \sum_{i=1}^{n-1} a_i(X^i + X^{-i}) \in R_q$, both schemes essentially have the same key size and

Table 1. Comparison of our scheme and HEAAN

Approximate HE	Ours $(2n, q)$	HEAAN (n, q)
Number of plaintext slots	n	$n/2$
NTT dimension	n	n
Bit size of ciphertexts	$2n \log q$	$2n \log q$

ciphertext size. Furthermore, both schemes exploit the NTT of dimension n for a ring multiplication, so they have almost same arithmetic complexity. As a result, our scheme over the dimension $2n$ actually performs as well as HEAAN over the dimension n while carrying a definite advantage in the number of plaintext slots.

5.2 Full RNS Variant

Many of ring-based HE schemes such as BGV [8,23] and BFV [7,21] require polynomial arithmetic over a huge modulus. Recent implementations of HE schemes [27,37] exploit the Residue Number System (RNS) for the performance improvements. In particular, there have been suggested some variants of BFV [4,25] which can be implemented without high-precision arithmetic.

In both the original HEAAN and our scheme, ciphertext moduli are chosen to be powers of a base because the scaling factor of a rescaling procedure is equal to the ratio of two consecutive ciphertext moduli. Unfortunately, this restriction makes it difficult to apply the existing RNS techniques to HEAAN.

Cheon et al. [11] recently proposed a method to fully eliminate the high-precision arithmetic of HEAAN based on the *approximate base*. We leave it to the reader to check that this idea can be directly applied to our scheme.

Acknowledgement. Duhyeong Kim was supported in part by Research Foundation of Korea (NRF) Grant funded by the Korean Government (Global Ph.D. Fellowship Program) under Grant 2016H1A2A1906584, and in part by NRF Grant funded by the Korean Government (MSIT) under Grant 2017R1A5A1015626.

References

1. Albrecht, M., et al.: Homomorphic encryption security standard. Technical report, Cambridge MA, March 2018. HomomorphicEncryption.org
2. Alkim, E., Ducas, L., Pöppelmann, T., Schwabe, P.: Post-quantum key exchange—a new hope. In: Proceedings of the 25th USENIX Security Symposium, pp. 327–343. USENIX Association (2016)
3. Arita, S., Handa, S.: Subring homomorphic encryption. In: Kim, H., Kim, D.-C. (eds.) ICISC 2017. LNCS, vol. 10779, pp. 112–136. Springer, Cham (2018). https://doi.org/10.1007/978-3-319-78556-1_7
4. Bajard, J.-C., Eynard, J., Hasan, M.A., Zucca, V.: A full RNS variant of FV like somewhat homomorphic encryption schemes. In: Avanzi, R., Heys, H. (eds.) SAC 2016. LNCS, vol. 10532, pp. 423–442. Springer, Cham (2017). https://doi.org/10.1007/978-3-319-69453-5_23
5. Bos, J.W., Lauter, K., Loftus, J., Naehrig, M.: Improved security for a ring-based fully homomorphic encryption scheme. In: Stam, M. (ed.) IMACC 2013. LNCS, vol. 8308, pp. 45–64. Springer, Heidelberg (2013). https://doi.org/10.1007/978-3-642-45239-0_4
6. Boura, C., Gama, N., Georgieva, M.: Chimera: a unified framework for B/FV, TFHE and HEAAN fully homomorphic encryption and predictions for deep learning. Cryptology ePrint Archive, Report 2018/758 (2018). https://eprint.iacr.org/2018/758

7. Brakerski, Z.: Fully homomorphic encryption without modulus switching from classical GapSVP. In: Safavi-Naini, R., Canetti, R. (eds.) CRYPTO 2012. LNCS, vol. 7417, pp. 868–886. Springer, Heidelberg (2012). https://doi.org/10.1007/978-3-642-32009-5_50

8. Brakerski, Z., Gentry, C., Vaikuntanathan, V.: (Leveled) fully homomorphic encryption without bootstrapping. In: Proceedings of ITCS, pp. 309–325. ACM (2012)

9. Castryck, W., Iliashenko, I., Vercauteren, F.: Provably weak instances of ring-LWE revisited. In: Fischlin, M., Coron, J.-S. (eds.) EUROCRYPT 2016. LNCS, vol. 9665, pp. 147–167. Springer, Heidelberg (2016). https://doi.org/10.1007/978-3-662-49890-3_6

10. Chen, H., Lauter, K., Stange, K.E.: Attacks on the search RLWE problem with small errors. SIAM J. Appl. Algebr. Geom. 1(1), 665–682 (2017)

11. Cheon, J.H., Han, K., Kim, A., Kim, M., Song, Y.: A full RNS variant of approximate homomorphic encryption. Cryptology ePrint Archive, Report 2018/931 (2018). https://eprint.iacr.org/2018/931

12. Cheon, J.H., Han, K., Kim, A., Kim, M., Song, Y.: Bootstrapping for approximate homomorphic encryption. In: Nielsen, J.B., Rijmen, V. (eds.) EUROCRYPT 2018. LNCS, vol. 10820, pp. 360–384. Springer, Cham (2018). https://doi.org/10.1007/978-3-319-78381-9_14

13. Cheon, J.H., Kim, A., Kim, M., Song, Y.: Homomorphic encryption for arithmetic of approximate numbers. In: Takagi, T., Peyrin, T. (eds.) ASIACRYPT 2017. LNCS, vol. 10624, pp. 409–437. Springer, Cham (2017). https://doi.org/10.1007/978-3-319-70694-8_15

14. Cheon, J.H., Kim, D., Kim, Y., Song, Y.: Ensemble method for privacy-preserving logistic regression based on homomorphic encryption. IEEE Access 6, 46938–46948 (2018)

15. Chillotti, I., Gama, N., Georgieva, M., Izabachène, M.: Faster fully homomorphic encryption: bootstrapping in less than 0.1 seconds. In: Cheon, J.H., Takagi, T. (eds.) ASIACRYPT 2016. LNCS, vol. 10031, pp. 3–33. Springer, Heidelberg (2016). https://doi.org/10.1007/978-3-662-53887-6_1

16. Costache, A., Smart, N.P.: Which ring based somewhat homomorphic encryption scheme is best? In: Sako, K. (ed.) CT-RSA 2016. LNCS, vol. 9610, pp. 325–340. Springer, Cham (2016). https://doi.org/10.1007/978-3-319-29485-8_19

17. Dathathri, R., et al.: CHET: compiler and runtime for homomorphic evaluation of tensor programs. arXiv preprint arXiv:1810.00845 (2018)

18. van Dijk, M., Gentry, C., Halevi, S., Vaikuntanathan, V.: Fully homomorphic encryption over the integers. In: Gilbert, H. (ed.) EUROCRYPT 2010. LNCS, vol. 6110, pp. 24–43. Springer, Heidelberg (2010). https://doi.org/10.1007/978-3-642-13190-5_2

19. Ducas, L., Micciancio, D.: FHEW: bootstrapping homomorphic encryption in less than a second. In: Oswald, E., Fischlin, M. (eds.) EUROCRYPT 2015. LNCS, vol. 9056, pp. 617–640. Springer, Heidelberg (2015). https://doi.org/10.1007/978-3-662-46800-5_24

20. Elias, Y., Lauter, K.E., Ozman, E., Stange, K.E.: Provably weak instances of ring-LWE. In: Gennaro, R., Robshaw, M. (eds.) CRYPTO 2015. LNCS, vol. 9215, pp. 63–92. Springer, Heidelberg (2015). https://doi.org/10.1007/978-3-662-47989-6_4

21. Fan, J., Vercauteren, F.: Somewhat practical fully homomorphic encryption. IACR Cryptology ePrint Archive 2012, 144 (2012)

22. Gentry, C.: Fully homomorphic encryption using ideal lattices. In: Proceedings of the Forty-First Annual ACM Symposium on Theory of Computing, STOC 2009, pp. 169–178. ACM (2009)
23. Gentry, C., Halevi, S., Smart, N.P.: Homomorphic evaluation of the AES circuit. In: Safavi-Naini, R., Canetti, R. (eds.) CRYPTO 2012. LNCS, vol. 7417, pp. 850–867. Springer, Heidelberg (2012). https://doi.org/10.1007/978-3-642-32009-5_49
24. Gentry, C., Sahai, A., Waters, B.: Homomorphic encryption from learning with errors: conceptually-simpler, asymptotically-faster, attribute-based. In: Canetti, R., Garay, J.A. (eds.) CRYPTO 2013. LNCS, vol. 8042, pp. 75–92. Springer, Heidelberg (2013). https://doi.org/10.1007/978-3-642-40041-4_5
25. Halevi, S., Polyakov, Y., Shoup, V.: An improved RNS variant of the BFV homomorphic encryption scheme. Cryptology ePrint Archive, Report 2018/117 (2018). https://eprint.iacr.org/2018/117
26. Halevi, S., Shoup, V.: Design and implementation of a homomorphic-encryption library. IBM Research (Manuscript) (2013)
27. Halevi, S., Shoup, V.: An implementation of homomorphic encryption (2014). https://github.com/shaih/HElib/
28. Jiang, X., Kim, M., Lauter, K., Song, Y.: Secure outsourced matrix computation and application to neural networks. In: Proceedings of the 2018 ACM SIGSAC Conference on Computer and Communications Security, pp. 1209–1222. ACM (2018)
29. Kim, A., Song, Y., Kim, M., Lee, K., Cheon, J.H.: Logistic regression model training based on the approximate homomorphic encryption. BMC Med. Genomics 11(4), 83 (2018)
30. Kim, M., Song, Y., Wang, S., Xia, Y., Jiang, X.: Secure logistic regression based on homomorphic encryption: design and evaluation. JMIR Med. Inform. 6(2) (2018)
31. Longa, P., Naehrig, M.: Speeding up the number theoretic transform for faster ideal lattice-based cryptography. In: Foresti, S., Persiano, G. (eds.) CANS 2016. LNCS, vol. 10052, pp. 124–139. Springer, Cham (2016). https://doi.org/10.1007/978-3-319-48965-0_8
32. Lyubashevsky, V., Peikert, C., Regev, O.: On ideal lattices and learning with errors over rings. In: Gilbert, H. (ed.) EUROCRYPT 2010. LNCS, vol. 6110, pp. 1–23. Springer, Heidelberg (2010). https://doi.org/10.1007/978-3-642-13190-5_1
33. Peikert, C.: An efficient and parallel Gaussian sampler for lattices. In: Rabin, T. (ed.) CRYPTO 2010. LNCS, vol. 6223, pp. 80–97. Springer, Heidelberg (2010). https://doi.org/10.1007/978-3-642-14623-7_5
34. Peikert, C., Regev, O., Stephens-Davidowitz, N.: Pseudorandomness of ring-LWE for any ring and modulus. In: Proceedings of the 49th Annual ACM SIGACT Symposium on Theory of Computing, pp. 461–473. ACM (2017)
35. Regev, O.: On lattices, learning with errors, random linear codes, and cryptography. In: Proceedings of the Thirty-Seventh Annual ACM Symposium on Theory of Computing, STOC 2005, pp. 84–93. ACM (2005)
36. Rosca, M., Stehlé, D., Wallet, A.: On the ring-LWE and polynomial-LWE problems. In: Nielsen, J.B., Rijmen, V. (eds.) EUROCRYPT 2018. LNCS, vol. 10820, pp. 146–173. Springer, Cham (2018). https://doi.org/10.1007/978-3-319-78381-9_6
37. Simple Encrypted Arithmetic Library (release 3.0.0). Microsoft Research, Redmond, WA, October 2018. http://sealcrypto.org

Excalibur Key-Generation Protocols for DAG Hierarchic Decryption

Louis Goubin[1], Geraldine Monsalve[2,3], Juan Reutter[2,3],
and Francisco Vial-Prado[2,3(✉)]

[1] Laboratoire de Mathématiques de Versailles, UVSQ, Université Paris-Saclay,
Versailles, France
[2] DCC, Pontificia Universidad Católica de Chile, Santiago, Chile
`fovial@uc.cl`
[3] IMFD Chile, Santiago, Chile
`https://www.imfd.cl`

Abstract. Public-key cryptograpy applications often require structuring decryption rights according to some hierarchy. This is typically addressed with re-encryption procedures or relying on trusted parties, in order to avoid secret-key transfers and leakages. Using a novel approach, Goubin and Vial-Prado (2016) take advantage of the Multikey FHE-NTRU encryption scheme to establish decryption rights at key-generation time, thus preventing leakage of all secrets involved (even by powerful key-holders). Their algorithms are intended for two parties, and can be reused to form chains of users with inherited decryption rights. In this article, we provide new protocols for generating *Excalibur* keys under any DAG-like hierarchy, and present formal proofs of security against semi-honest adversaries. Our protocols are compatible with the homomorphic properties of FHE-NTRU, and the base case of our security proofs may be regarded as a more formal, simulation-based proof of said work.

1 Introduction

In some public-key cryptography applications, parties own decryption rights over ciphertexts according to hierarchic structures. For instance, in a mail redirection scenario it may be required that Alice is able to decrypt all of Bob's ciphertexts, and not conversely. If Bob simply transfers his secret key to Alice, she may leak or sell Bob's secret, causing a lot more damage than leaking Bob's plaintexts only. Overcoming this, proxy re-encryption and hierarchical identity-based encryption schemes rely on trusted parties to generate master secret keys or involve public re-encryption procedures. Using a different approach, authors in [4] proposed two-party computation protocols that securely perform a key generation procedure of the celebrated NTRU-based Multikey-FHE [8]. As a result of this, Alice receives a key pair $(\mathsf{sk}_A, \mathsf{pk}_A)$ such that sk_A can decrypt all of Bob's ciphertexts, and no information about Bob's secret key can be deduced from this key pair, the execution of the protocol or any public values. Moreover, Alice's and Bob's

© Springer Nature Switzerland AG 2019
K. Lee (Ed.): ICISC 2018, LNCS 11396, pp. 103–120, 2019.
https://doi.org/10.1007/978-3-030-12146-4_7

secrets are tied together, thus effectively avoiding leakage by Alice (assuming she is not willing to reveal her own secret key sk_A). The newly-created *Excalibur* key pair behaves as a regular key of the system, even allowing multikey homomorphic operations when using sufficiently large parameters such as the ones suggested in [8]. In addition, these keys can be used as inputs to generate more powerful Excalibur keys, allowing decryption inheritance for a bounded chain of users. In addition, this whole procedure can be regarded as an *automatic* N-hop proxy re-encryption scheme, addressing a re-encryption paradigm in fully homomorphic encryption scenarios, as pointed out in [4].

In this article we extend these key-generation protocols and provide multiparty computation protocols in the advised cyclotomic polynomial ring of Stehlé and Steinfeld [9] that can securely generate FHE-NTRU keys for some types of DAG-like hierarchies, in such a way that the key of a particular node n in this hierarchy can decrypt all messages encrypted for nodes below it, while not having access to secrets (other than own private keys). In order to do this, we address the case where Bob above is replaced by a set of participants.

As typical MPC applications, our protocols require the composition of several routines. In all generality, this poses additional security restrictions, as a composition of two secure protocols is not necessarily secure (some examples are given in [3]). This has been the subject of extensive research, and we highlight the Universal Composability (UC) framework proposed by Cannetti [2] in which any two UC-secure protocols may be arbitrarily composed ensuring the inheritance of security. In our case, only non-concurrent composition is performed, and as pointed out in [3], security is proven directly with simulation-based proofs.

Our Contributions. This article provides new protocols that extend those from [4], addressing the case where $k > 2$ parties jointly generate an NTRU-FHE key pair with additional decryption rights, and preventing leakage of secret keys from the receiving party. To this end, we propose secure multi-party computation protocols in cyclotomic polynomial rings between parties $P_1, \ldots, P_k$ that generate a key-pair for party P_1 with decryption rights over all parties involved, and such that no information about other secret-keys can be leaked from the execution of the protocol, inputs and outputs, and public values, even if some parties collude.

We provide security analysis of these protocols in the semi-honest but colluded setting, where parties follow the protocols and sample from the correct distributions but may cooperate with each other to deduce secrets. The base case of our security analysis may be regarded as a more formal, simulation-based proof of the protocols in [4]. Achieving security in malicious adversarial settings (where parties may deviate or sabotage the protocol) is a challenging problem for $k > 2$ parties, which we leave for future work.

In order to generate an Excalibur key pair that inherits decryption of 3 other keys with 128 bits of security in mind, parties using our protocols need to perform around 2^{24} 1-out-of-2 OT protocols and 2^{24} multiplications in $\mathbb{Z}_q[x]/(x^n+1)$ with secure NTRU parameters (this can be performed in range of minutes on a regular laptop, as per our simulations and [1]). We are confident that there is much to optimize in these procedures, opening another interesting angle for future work.

Overview of the Article. We begin in Sect. 2 by revisiting necessary concepts to construct our protocols. In Sects. 3 and 4, we define the notions of security we want to achieve, and state the corresponding underlying assumptions. Our protocols are presented in Sect. 5, with proofs and security analysis described in Sect. 6.

2 Preliminaries

Let q be a large prime. For an integer $x \in \mathbb{Z}$, $(x \mod q)$ represents the modular reduction of x into the set $\{-\lfloor q/2 \rfloor, \ldots \lfloor q/2 \rfloor\}$, which we denote by $\mathbb{Z}_q$. The indicator function of a set S is any function that outputs 1 if the preimage is in S and 0 otherwise. For a distribution χ that samples from some set S, $e \leftarrow \chi$ means that e is sampled from S with the distribution χ, and $e \xleftarrow{\$} S$ means that e is sampled from S according to the uniform distribution. We denote tuples of elements with bold letters. For n a power of 2, let $R_q \overset{\text{def}}{=} \mathbb{Z}_q[X]/(X^n + 1)$ be the cyclotomic ring of polynomials modulo $X^n + 1$ and coefficients in $\mathbb{Z}_q$.

Invertible Bounded Gaussian Distributions in the Quotient Ring. We first recall that the ring $R_q = \mathbb{Z}_q[X]/(X^n + 1)$ is not a unique factorization domain as $X^n + 1$ is generally not irreducible in $\mathbb{Z}_q$, and units of this ring with small coefficients are used as NTRU secret keys. For instance, if $q = 1 \mod 2n$, $X^n + 1$ splits into n linear factors in $\mathbb{Z}_q$, ensuring a large key space. See [4,9] for more details. In the following definition, for a real number $r > 0$, let Γ_r be the Gaussian distribution on $\mathbb{R}^n$ centered about 0 and with standard-deviation r.

Definition 1 (Bounded Discrete Gaussian distribution over R_q). *For a real number $0 < B \ll q$, let $\mathcal{G}_B$ be the B-bounded discrete Gaussian distribution over R_q, that is, the distribution that samples polynomials from R_q as follows:*

1. *Sample a vector $\boldsymbol{x} \leftarrow \Gamma_B$, and restart if $\|\boldsymbol{x}\|_\infty > B$.*
2. *Output the polynomial $p \in R_q$ whose coefficients vector is $\lfloor \boldsymbol{x} \rceil$.*

Let $\mathcal{G}_B^\times$ be the distribution that samples from $\mathcal{G}_B$ until the output is invertible.

FHE-NTRU Encryption and the Multikey Property. The Multikey FHE scheme presented in [8] uses a modified version of NTRU (N^{th}-truncated) encryption scheme, which we present here for the sake of completeness.

    Parameters: Let n be a power of 2, q be a large prime such that $q = 1 \mod 2n$ and $0 < B \ll q$. Recall that $R_q = \mathbb{Z}_q[x]/(x^n + 1)$.

    Key Generation: Sample a polynomial $f \leftarrow \mathcal{G}_B$ and set $\mathsf{sk} \leftarrow 2f + 1$ until sk is invertible. Sample $g \leftarrow \mathcal{G}_B^\times$ and define $\mathsf{pk} \leftarrow 2g \cdot \mathsf{sk}^{-1}$.

Encryption: For a message $m \in \{0, 1\}$ and public-key pk, sample $s, e \leftarrow \mathcal{G}_B$, and output $c \leftarrow m + 2e + s \cdot \text{pk} \mod q$.

Decryption: For ciphertext c and secret-key sk, output $m = c \cdot \text{sk} \mod 2$.

The linearity of the decryption equation allowed authors of [8] to construct the first Multikey FHE scheme, where the result of homomorphic operations involving ciphertexts related to different entities can be jointly decrypted by these parties. As noted in [4], this linearity also implies that a secret key with small extra multiplicative factors (such as other secret keys) is able to correctly decrypt, i.e. a polynomial multiplication in R_q of a small number of secret keys acts as a regular key and inherits the decryption rights of all its factors. For two parties in [4], the R_q-product of secret keys is performed in a secure MPC fashion in order to attain the desired property.

3   Security Definitions

We present the usual secure MPC definitions that capture the security of our key-generation protocols. We distinguish two adversarial settings: the *semi-honest* case in which players cooperate with the execution of the protocol and sample from the correct distributions, but may collude and try to learn secrets from their shared views; and the *malicious* case, in which adversaries are not guaranteed to follow the protocol. In this paper, we only consider semi-honest adversaries, leaving the malicious case for future work.

3.1   Simulation-Based MPC Security Against Semi-honest Adversaries

Fix a set of $P = \{P_1, \ldots, P_k\}$ of parties. Following e.g. [7], our notion of security is based on the idea of emulating *functionalities*, which are k-ary functions $f : (\{0, 1\}^*)^k \rightarrow (\{0, 1\}^*)^k$. To describe a functionality f we usually write $f = (f_1, \ldots, f_k)$, where each f_i is a random k-ary function that outputs a string.

As usual, the idea is to show that a protocol computing a functionality f is secure if all possible information that can be computed by a collusion of some parties can be simulated by means of the combined input and output of these parties when executing the protocol.

Let f be a functionality, and π a protocol for computing f. The *view* of the i-th party when executing π on input $\mathbf{x} = \{x_1, \ldots, x_k\}$ and security parameter λ, denoted as $\text{view}_i^\pi(\mathbf{x}, \lambda)$, is a tuple $(x_i, r_i, m_1^i, \ldots, m_j^i)$, where r^i is the content of the internal random tape of the i-th party, and $m_1^i, \ldots, m_j^i$ represents the messages sent and received with other parties during the execution of π. For a set $S \subseteq P$ of parties, we set $\text{view}_S^\pi(\mathbf{x}, \lambda)$ as the concatenation of each tuple $\text{view}_i^\pi(\mathbf{x}, \lambda)$. We also write f_S as the tuple formed of each f_i, for $i \in S$, and $\mathbf{x}_S$ as the tuple formed of each x_i, $i \in S$.

The output of the i-th party when executing π on input $\mathbf{x} = \{x_1, \ldots, x_k\}$ and security parameter λ is denoted as $\text{output}^\pi(\mathbf{x}, \lambda)$.

Definition 2. *Let P be a set of k parties, and $f = (f_1, \ldots, f_k)$ a functionality. We say that π securely computes f in the presence of semi-honest adversaries if for every set $S \subsetneq P$ of colluded parties there is a PPT algorithm $\mathcal{I}_S$ such that*

$$(\mathcal{I}_S(1^\lambda, \boldsymbol{x}_S, f_S(\boldsymbol{x}, \lambda)), f(\boldsymbol{x}, \lambda)) \overset{s}{\approx} (view_S^\pi(\boldsymbol{x}, \lambda), output^\pi(\boldsymbol{x}, \lambda))$$

We now give a proposition that allows us to prove security for protocols that involve executing other protocols as non-concurrent sub-routines. Let $\pi_1, \ldots, \pi_\ell$ be protocols computing functionalities $\phi_1, \ldots, \phi_\ell$, and let $\rho^{\pi_1, \ldots, \pi_\ell}$ be a protocol computing a functionality g that makes use of $\pi_1, \ldots, \pi_\ell$ in a non-concurrent fashion, so that π_i is called only after π_{i-1} returns, and additionally, $\rho^{\pi_1, \ldots, \pi_\ell}$ pauses when executing each π_i. Denote by $\rho^{\pi_1 \to \phi_1, \ldots, \pi_\ell \to \phi_\ell}$ the protocol where instead of calling to each π_i, an oracle computes the functionality f_i. We have the following.

Proposition 1. *If every π_i securely computes ϕ_i, and $\rho^{\pi_1 \to \phi_1, \ldots, \pi_\ell \to \phi_\ell}$ securely computes g, then $\rho^{\pi_1, \ldots, \pi_\ell}$ securely computes g.*

Please refer to [7] for more on simulation techniques. Note that the restriction that sub-protocols are invoked non-concurrently is key for stating this result in such a simplified way, instead of using the machinery proposed in [2]. We do highlight that the restriction of Canetti's framework into our scenario yields security requirements equivalent to that of Definition 2 (see [3], Sect. 5.2).

4   Hardness Assumptions

The security of our protocols against semi-honest adversaries is based on two well-known assumptions (RLWE and DSPR), and the difficulty of new factorization problems in R_q that extend those from [4].

Definition 3 (Decisional Small Polynomial Ratio assumption, from [9]). *For some parameters q, n, B, it is computationally hard to distinguish between the following two distributions over R_q: (1) A polynomial $\mathsf{pk} = 2g(2f + 1)^{-1} \in R_q$ where $f, g \leftarrow \mathcal{G}_B^\times$, and (2) a uniformly random polynomial $u \overset{\$}{\leftarrow} R_q$.*

Definition 4 (Gaussian Product Distribution). *Let ξ_B^l be the distribution that samples polynomials $p_i \leftarrow \mathcal{G}_B^\times$ for $i = 1, \ldots, l$ and outputs $\prod_{i=1}^l p_i$.*

Definition 5 (Special factors problem). *Let $\alpha \leftarrow \xi_B^l$ and $\beta \leftarrow \xi_B^m$. The Special Factors Problem is to output α, β with the knowledge of $c = \alpha \cdot \beta$ and access to the indicator function of $\{\alpha, \beta\}$.*

In other words, the task is to find the correct factorization of c. Recall that R_q is not a UFD, so for any unit $u \in R_q$ there is a posible factorization $c = u \cdot (u^{-1}c)$. In order to find α, β, the solver must query the indicator function. In our construction, secret keys play the role of the individual factors of c, and the indicator function consists in encrypt-decrypt key-guessing routines. When $l = m = 1$, this is the small factors problem from [4].

Definition 6 (Special GCD problem). *Let $\alpha, \beta \leftarrow \mathcal{G}_B^{\times}$ and $y \leftarrow \xi_B^l$. Given $u = \alpha \cdot y$ and $v = \alpha \cdot \beta$ and access to the indicator function of $\{\alpha, \beta, y\}$, output α, β and y.*

Definition 7 (Special Factors Assumption). *For some set of parameters, it is computationally hard to solve the Special Factors or the Special GCD problems.*

As noted in [4], Special GCD reduces to a version of DSPR, and the SF problem may be expressed as a quadratic system of equations in $\mathbb{Z}_q$ in the underdetermined setting, which is considered secure [10]. Moreover, as in [4] we put it as a conjecture that the additional cyclic structure provided by R_q does not help an attacker to solve this system.

5　MPC Key Generation Protocols

Our protocols assume a set of participants $P = \{P_1, \ldots, P_k\}$, and the objective is to create a key pair $(\mathsf{sk}_1, \mathsf{pk}_1)$ for participant P_1, based on the set $(\mathsf{sk}_i, \mathsf{pk}_i)$ of all other participants. As we have mentioned, the pair $(\mathsf{sk}_1, \mathsf{pk}_1)$ can decrypt any message encrypted with the public key of any other participant.

We use a basic protocol SP_m for computing a certain scalar product, that works as follows. Party A holds m bits $\boldsymbol{b} = (b_1, \ldots, b_m)$ and party B holds two vectors $\boldsymbol{r}^{(0)} = (r_1^{(0)}, \ldots, r_m^{(0)})$ and $\boldsymbol{r}^{(1)} = (r_1^{(1)}, \ldots, r_m^{(1)})$. At the end of the protocol, A learns $\sum_{i=1}^m r_i^{(b_i)}$ and B learns nothing. Note that when $m = 1$ this is simply a $\binom{2}{1}$-OT (1-out-of-2 oblivious transfer). The construction of this protocol is straightforward, and based on [6]. It is outlined in Appendix A.

5.1　Secure MPC Protocols for Multiplication in R_q

The building blocks of our key-generating scheme are two protocols, that we name *k-Multiplication Protocol* and *k-Shared Multiplication Protocol*. Both of these protocols share the goal of performing a multiparty multiplication of elements in R_q, but differ in the final output learned by the participants.

Our *k-Multiplication Protocol* is a nontrivial extension of algorithm 2-MP, from [4]. We need this algorithm for defining our protocols, so we recall it bellow.

k-Multiplication Protocol (k-MP). We use this protocol to multiply k elements in our ring. Every participant P_ℓ begins with a secret element x_ℓ given as input, as well as a uniformly random polynomial r_ℓ. Upon finishing, participant P_1 learns $\prod_{\ell=1}^k x_\ell + \sum_{\ell=2}^k r_\ell$, and the rest of the participants learn nothing.

Algorithm 2 contains the detail of this protocol, and Algorithm 1 provides the base case. The idea is to use (k-1)-MP to perform a secure multiplication of all but participant's P_1 ring elements (see step 8). In turn, the multiplication for participant P_1 is carefully masked with additive uniform noise in order to avoid input leaking. In the end (Step 12), P_1 performs SP_m with each other participant with the goal of cancelling noise.

Algorithm 1. Two-party R_q multiplication 2-MP, from [4].

Require: Player P_1 holds $x_1 \in R_q$ and P_2 holds a pair $(x_2, r_2) \subset R_q^2$. Let $m \in \mathbb{N}$ be such that it is unfeasible to compute 2^m additions in R_q.

Ensure: Player P_1 learns $x_1 \cdot x_2 + r_2$.

1: **procedure** k-MP
2:      Player P_1 generates m polynomials $(x_{1i} \overset{\$}{\leftarrow} R_q)_{i=1}^m$, such that $\sum_{i=1}^m x_{1i} = x_1$.
3:      Player P_2 samples m polynomials $(r_{2i} \overset{\$}{\leftarrow} R_q)_{i=1}^m$ such that $\sum_{i=1}^m r_{2i} = r_2$.
4:      **for** $i = 1, \ldots, m$ **do**
5:          Player P_1 generates a random bit $b \overset{\$}{\leftarrow} \{0,1\}$, polynomials $(v_0, v_1) \overset{\$}{\leftarrow} R_q^2$ and sets $v_b = x_{1i}$.
6:          Player P_1 sends (v_0, v_1) to P_2.
7:          Player P_2 computes $(e_0, e_1) = (v_0 \cdot x_2 + r_{2i}, v_1 \cdot x_2 + r_{2i})$.
8:          With a $\binom{2}{1}$-OT protocol, player P_1 extracts e_b from P_2.
9:      Let $\hat{e}_1, \ldots, \hat{e}_m$ be the polynomials extracted by P_1 in each of the m steps. Player P_1 computes $\sum_{i=1}^m \hat{e}_i = x_1 \cdot x_2 + r_2$.

Algorithm 2. Multiparty multiplication of k elements in R_q

Require: A number of players $k \geq 3$. Player P_1 holds $x_1 \in R_q$ and each other player P_ℓ holds a pair $(x_\ell, r_\ell) \subset R_q^2$. Let $m \in \mathbb{N}$ be such that it is unfeasible to compute 2^m additions in R_q.

Ensure: Player P_1 learns $\prod_{\ell=1}^k x_\ell + \sum_{\ell=2}^k r_\ell$.

1: **procedure** k-MP
2:      Player P_1 generates m polynomials $(x_{1i} \overset{\$}{\leftarrow} R_q)_{i=1}^m$, such that $\sum_{i=1}^m x_{1i} = x_1$.
3:      Each player P_ℓ in $P \backslash \{\hat{P}_1\}$ samples $(r_{\ell i} \overset{\$}{\leftarrow} R_q)_{i=1}^m$ such that $\sum_{i=1}^m r_{\ell i} = r_\ell$, and $2m$ polynomials $(\hat{r}_{\ell i}^b \overset{\$}{\leftarrow} R_q)_{i=1}^m$ for $b = 0, 1$. Let $s_{\ell i}^b = r_{\ell i} + \hat{r}_{\ell i}^b$.
4:      **for** $i = 1, \ldots, m$ **do**
5:          Player P_1 generates a random bit $b \overset{\$}{\leftarrow} \{0,1\}$, and polynomials $(v_0, v_1) \overset{\$}{\leftarrow} R_q^2$ such that $v_b = x_{1i}$.
6:          Player P_1 sends (v_0, v_1) to P_2.
7:          **for** $j = 0, 1$ **do**
8:              Players $P_2, \ldots, P_k$ perform $[k$-$1]$-MP$(v_j \cdot x_2, (x_3, s_{3i}^j), \ldots, (x_k, s_{ki}^j))$. P_2 learns $v_j \cdot \prod_{\ell=2}^k x_\ell + \sum_{\ell=3}^k s_{\ell i}^j$.
9:              Player P_2 adds s_{2i}^j to this output, obtaining $e_j = v_j \cdot \prod_{\ell=2}^k x_\ell + \sum_{\ell=2}^k s_{\ell i}^j$
10:          With a $\binom{2}{1}$-OT protocol, player P_1 extracts e_b from P_2. Note that $e_b = x_{1i} \cdot \prod_{\ell=2}^k x_\ell + \sum_{\ell=2}^k s_{\ell i}^b$.
11:      Let $\hat{e}_i$ be the polynomials extracted in each of these m steps, and b_i the random bits. P_1 computes $\theta := \sum_{i=1}^m \hat{e}_i = \prod_{\ell=1}^k x_\ell + \sum_{\ell=2}^k r_\ell + \sum_{\ell=2}^k \left(\sum_{i=1}^m \hat{r}_{\ell i}^{b_i} \right)$.
12:      **for** $\ell = 2, \ldots, k$ **do**
13:          P_1 extracts $\hat{s}_\ell = \sum_{i=1}^m \hat{r}_{\ell i}^{b_i}$ from P_ℓ with SP$(\boldsymbol{b}, (\boldsymbol{r}_\ell^0, \boldsymbol{r}_\ell^1))$, where $\boldsymbol{b} = (b_1, \ldots, b_m)$ and $\boldsymbol{r}_\ell^j = (\hat{r}_{\ell 1}^j, \ldots, \hat{r}_{\ell m}^j)$.
14:      Finally, P_1 computes $\theta - \sum_{\ell=2}^k \hat{s}_\ell = \prod_{\ell=1}^k x_\ell + \sum_{\ell=2}^k r_\ell$.

---

k-Shared Multiplication Protocol(k-sMP). In this protocol every participant starts with a pair of additive shares (x_i, y_i) of elements $x, y \in R_q$, and in the end learns an additive share π_i of the product $\pi = x \cdot y$, i.e., $\sum \pi_i = (\sum x_i) \cdot (\sum y_i)$.

The details of this protocol are shown in Algorithm 3. Players perform $k(k-1)$ pair-wise multiplications of shares using 2-MP (steps 3–5). The random noise added by 2-MP serves us to mask the value of the correct shares, and it is then cancelled out when adding up all polynomials (step 6).

5.2 Excalibur Key Generation Protocols

In out key-generating protocols, players $P_2, \ldots, P_k$ start with their secret keys β_i, and all players sample a random polynomial s_i from $\mathcal{G}_B$. These polynomials act as additive shares of P_1's secret, called α (thus P_1 does not know α either). Upon finishing, participant P_1 learns the secret key $\mathsf{sk}_1 = \alpha \prod_{i=2}^{k} \beta_i$, as well as its public key pk_1. On the other hand, all other participants only learn pk_1. As advised in [4], parties generate the public key first, and P_1 commits to it.

Public Key Generation ($\mathsf{Exc}_{\mathsf{pk}}$). Protocol $\mathsf{Exc}_{\mathsf{pk}}$ is used to generate the public key for participant P_1. Every participant P_i apart from P_1 holds a key pair $(\mathsf{sk}_i, \mathsf{pk}_i) = (\beta_i, 2h_i\beta_i^{-1})$. Player P_2 plays a special role computing some products. Upon finishing, a public key pk_1 is broadcast to everyone. This public key is a polynomial of the form $2g(\alpha \prod_{i=2}^{k} \beta_i)^{-1}$, for additively shared elements $\alpha = 2(\sum_{i=1}^{k} s_i) + 1$ and $g = \sum_{i=1}^{k} g_i$.

The protocol is shown as Algorithm 4. It begins with participants sampling a gaussian share g_i, and random elements r_i, t_{ij} used to additively mask polynomials, as in protocol 3. Once the joint secret $\prod_{i=2}^{k} \beta_i$ is shared, P_2 has the task of inverting it in the ring, multiplying by α^{-1}, g and broadcasting. To avoid P_2 extracting or using these secrets, they are separated into multiplicative factors that do not leak secrets (or, more precisely, such that extracting secrets from them needs to solve SF or Special GCD problems).

Secret Key Generation ($\mathsf{Exc}_{\mathsf{sk}}$). Protocol $\mathsf{Exc}_{\mathsf{sk}}$ is used to generate the secret key sk_1 for participant P_1, given secret keys $\beta_2, \ldots, \beta_k$ of the other participants. This protocol needs the same additive share of α of the $\mathsf{Exc}_{\mathsf{pk}}$ protocol (hence the need of semi-honest players). Upon finishing, P_1 receives the secret key $\mathsf{sk}_1 = \alpha \prod_{i=2}^{k} \beta_i$. The protocol is shown as Algorithm 5, and again it uses our multiplication protocols together with carefully selected random noise (Fig. 1).

Algorithm 3. Multiparty shared multiplication of k elements in R_q

---

Require: Each participant P_i holds a pair (x_i, y_i) of elements from R_q.
Ensure: Each $P_i \in P$ learns an element π_i, such that $\sum_{j=1}^{k} \pi_j = (\sum_{j=1}^{k} x_j) \cdot (\sum_{j=i}^{k} y_j)$.
 1: **procedure** k-sMP
 2:     Each P_i samples $R_i = \{r_{ij} \overset{R}{\longleftarrow} R_q \mid j = [1,k] \wedge i \neq j\}$.
 3:     **for** $i = 1, \ldots, k$ **do**
 4:         **for** $j = 1, \ldots, k, \, j \neq i$ **do**
 5:             P_i, P_j perform 2-MP$(x_i, (y_j, r_{ji}))$. Thus P_i learns $u_{ij} = x_i \cdot y_j + r_{ji}$
 6:     Each participant P_i computes $\pi_i = x_i y_i + \sum_{j=1, j \neq i}^{k} u_{ij} - \sum_{r \in R_i} r$.

Algorithm 4. Excalibur Public Key Generation

Require: Participant P_1 holds an element $s_1 \leftarrow \mathcal{G}_B$ and each other participant holds
$\beta_i = \mathsf{sk}_i$ and $s_i \leftarrow \mathcal{G}_B$. Let $\alpha = 2(\sum_{i=1}^{k} s_i) + 1$.

Ensure: A public key $\mathsf{pk}_1 = 2g(\alpha \prod_{i=2}^{k} \beta_i)^{-1}$ for P_1.

1: **procedure** $\mathsf{Exc}_{\mathsf{pk}}$

2:       Each $P_i \in P$ samples $g_i \leftarrow \mathcal{G}_B$, $r_i \overset{\$}{\leftarrow} R_q$ and $t_{ij} \overset{\$}{\leftarrow} R_q$, for $j = 1, ..., k$.
       Let $r = \sum_{i=1}^{k} r_i$ and $g = \sum_{i=1}^{k} g_i$.

3:       All participants perform (k)-MP$(r_1, (\beta_2, t_{21}), \ldots, (\beta_k, t_{k1}))$. Thus,
       P_1 learns $r_1' = r_1 \cdot \prod_{i=2}^{k} \beta_i + \sum_{i=2}^{k} t_{i1}$.

4:       **for** $i = 2, \ldots, k$ **do**

5:          P_i and the rest of participants in $P \setminus \{P_1, P_i\}$ perform (k-1)-MP. P_i gives
          $r_i \beta_i$ as input, and each other player $P_j \in P \setminus \{P_1, P_i\}$ gives (β_j, t_{ji}).
          P_i learns $u_i = r_i \cdot \prod_{j=2}^{k} \beta_k + \sum_{j=2, j \neq i}^{k} t_{ji}$ and computes $r_i' = u_i - \sum_{j=1, j \neq i}^{k} t_{ij}$.

6:       With g_i, r_i and s_i, r_i', all players perform Shared k-MP twice to obtain shares
       of $w = g \cdot r$ and $z = \alpha \cdot r' = \alpha \prod_{i=2}^{k} \beta_i \cdot r$.
       Each participant reveal their shares to P_2, thus P_2 learns z, w.

7:       P_2 checks: if z is not invertible in R_q, restart the protocol.

8:       P_2 computes $2w(z\beta_2)^{-1} = 2g(\alpha \prod_{j=2}^{k} \beta_j)^{-1}$ and publishes it as pk_1.

Algorithm 5. Excalibur Secret Key Generation

Require: Let $\alpha = 2(\sum_{i=1}^{k} s_i) + 1$ be the same additive share as in protocol 4: Participant P_1 holds $s_1 \in R_q$ and each other participant holds a pair $(\beta_i, s_i) \in R_q^2$.

Ensure: A secret-key $\mathsf{sk}_1 = \alpha \prod_{i=2}^{k} \beta_i$ for P_1.

1: **procedure** $\mathsf{Exc}_{\mathsf{sk}}$

2:       Each participant P_i in $P \setminus \{P_1\}$ samples $r_{ij} \overset{R}{\leftarrow} R_q$, with $j = 2, \ldots, k$ and $j \neq i$.
       Let $r_i = \sum_{j=2, j \neq i}^{k} r_{ij}$.

3:       **for** $i = 2, \ldots, k$ **do**

4:          P_i and the rest of players from $P \setminus \{P_1, P_i\}$ perform (k-1)-MP,
          with P_i holding $2s_i \beta_i$ and each other P_ℓ holding $(\beta_\ell, r_{\ell i})$.
          P_i learns $u_i = 2s_i \prod_{j=2}^{k} \beta_j + \sum_{j=2, j \neq i}^{k} r_{ji}$ and computes
          $R_i = u_i - \sum_{j=2, j \neq i}^{k} r_{ij}$.

5:       All participants perform k-MP$(2s_1 + 1, (\beta_2, R_2), \ldots, (\beta_k, R_k))$, and P_1 obtains
       $\mathsf{sk}_1 := ((2s_1 + 1) \prod_{i=1}^{k} \beta_i) + \sum_{i=2}^{k} (2s_i \prod_{j=2}^{k} \beta_j) = \alpha \prod_{i=2}^{k} \beta_i \in R_q$

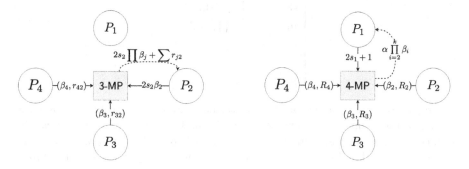

Fig. 1. Steps 4 and 5 from Algorithm 5 for 4 players.

6   Security Analysis

In this section we inspect the security of the proposed scheme against semi-honest adversaries. Throughout this section, parties $\{P_1, \ldots, P_k\}$ participate in the protocol, player P_1 receives the powerful key at the end, and P_2 has the special role of inverting a polynomial in protocol $\mathsf{Exc_{pk}}$.

6.1   Extracting Keys After the Protocol

Recall that P_1 is provided an Excalibur key pair of the form $(\mathsf{sk}_1, \mathsf{pk}_1) = (\alpha \cdot \prod_{i=2}^{k} \mathsf{sk}_i, 2g \cdot \mathsf{sk}_1^{-1}) \in R_q \times R_q$, and assume that some set of colluded parties P' try and deduce secrets. Note that extracting α is a successful attack, as sk_1/α can be leaked as a valid NTRU key decrypting messages intended to all parties excepting party P_1. Also, extracting a product of secret keys is also an attack even if individual keys are unknown, because of the multikey property.

Proposition 2. *Let $P' \subsetneq \{P_1, \ldots, P_k\}$ be a set of colluded parties. The problem of extracting α, g, r, r' or any secret key sk_j of a party $P_j \notin P'$ from public values, views of the protocol and secret keys of parties in P' reduces to instances of $\mathcal{G}_B^\times$-GCD or Special Factors problems. The same holds for the problem of extracting a product of secret keys of honest parties.*

Recall that the output of the proposed protocol is a key-pair of the form

$$(\mathsf{sk}_1, \mathsf{pk}_1) = (\alpha \cdot \prod_{i=2}^{k} \mathsf{sk}_i, 2g \cdot \mathsf{sk}_1^{-1}) \in R_q \times R_q,$$

where for $i = 2, \ldots, k$, $(\mathsf{sk}_i, \mathsf{pk}_i) \leftarrow \mathsf{NTRU - Keygen}()$ and $\alpha = 2(f_1 + \cdots + f_k) + 1$ for polynomials f_j sampled from $\mathcal{G}_B^\times$. The ring elements available to the uncolluded adversary are given by the output secret key, and public keys. Let $p_i = \mathsf{pk}^{-1} \in R_q$ for $i \in \{1, \ldots, k\}$. Note that $p_i = g_i \cdot \mathsf{sk}_i$ for some $g_i \in R_q$. The task of the adversary that receives sk_1 is to extract any element of the set $\{\alpha, \mathsf{sk}_2, \ldots, \mathsf{sk}_k\}$ from the view

$$\begin{cases} \mathsf{sk}_1 = \alpha \cdot \prod_{i=2}^{k} \mathsf{sk}_i, \\ p_1 = g_1 \cdot \alpha \cdot \prod_{i=2}^{k} \mathsf{sk}_i, \\ p_2 = g_2 \cdot \mathsf{sk}_1, \\ \vdots \\ p_k = g_k \cdot \mathsf{sk}_k. \end{cases}$$

We proceed by a series of claims. Without loss of generality, we assume that the attacker intends to extract an individual secret key, since keys can be grouped together in the view and equations are equivalent, but with different size parameters. For instance, an attacker extracting $\mathsf{sk}_2 \cdot \mathsf{sk}_3$ can reformulate the instance defining $\mathsf{sk}' = \mathsf{sk}_2 \cdot \mathsf{sk}_3, p' = p_2 \cdot p_3$ and extract sk' from a wider distribution.

Claim. Extracting α from sk_1 is an instance of the special factors problem.

Proof. Let $\beta = \prod_{i=2}^{k} \mathsf{sk}_i$. The task is to extract α from $\alpha \cdot \beta$. □

Claim. Extracting sk_i for $i \in \{2, \ldots, k\}$ from sk_1 is an instance of the special factors problem.

Proof. Let $\gamma = \alpha \cdot \prod_{j=2, j \neq i}^{k} \mathsf{sk}_j$. The task is to extract sk_i from $\mathsf{sk}_i \cdot \gamma$. □

Claim. Extracting sk_i for $i \in \{2, \ldots, k\}$ from the whole view is an instance of $\mathcal{G}_B^{\times}$-GCD problem, for some bound B.

Proof. Write $\mathsf{sk}_1 = \delta \cdot \mathsf{sk}_i$ for some $\delta \in R_q$ and consider $p_i = g_i \cdot \mathsf{sk}_i$. There are no other equations depending on sk_i or g_i, therefore solving for sk_i is exactly solving $\mathcal{G}_B^{\times}$-GCD. □

Claim. Extracting secret keys from the whole view and information from collusion with other parties are $\mathcal{G}_B^{\times}$-GCD or special factors problems.

Proof. If the attacker learns sk_i for $i \in \{2, \ldots, k\}$ by collusion, then defining $\mathsf{sk}'_* = \mathsf{sk}_*/\mathsf{sk}_i, p'_1 = p_1/p_i$ reduces to an equivalent instance of the problem of extracting another secret key. In other words, the view is now

$$
\begin{cases}
\mathsf{sk}_i \\
\mathsf{sk}'_* = \alpha \cdot \prod_{j=2, j \neq i}^{k} \mathsf{sk}_j, \qquad (*) \\
p_* = g'_* \cdot \alpha \cdot \prod_{j=2, j \neq i}^{k} \mathsf{sk}_j, \\
p_2 = g_1 \cdot \mathsf{sk}_1, \\
\vdots \\
p_k = g_k \cdot \mathsf{sk}_k,
\end{cases}
$$

and the only equations involving another secret-key sk_l for $l \neq i$ are $(*)$ and $p_l = g_l \cdot \mathsf{sk}_l$, defining an instance of $\mathcal{G}_B^{\times}$-GCD. The same holds for a larger set of colluded parties. □

Claim. Extracting α, g, r or any β_i from z, w and all public values is an instance of the special factors problem.

Proof. The task is to extract α, g, r from $w = g \cdot r$ and $z = \alpha \cdot r'$ with $z' = r \prod_{i=2}^{k} \beta_i$. □

6.2 Extracting Secrets During the Protocols

We address here the security of all our algorithms against semi-honest adversaries, during and after the execution.

Definition 8. *Consider the following functionalities. All variables are in R_q:*

$\mathcal{F}^{\mathsf{k\text{-}MP}} : (x_1, (x_2, r_2), \ldots, (x_k, r_k)) \mapsto ((\prod_{i=1}^{k} x_i + \sum_{i=2}^{k} r_i), \text{-}, \text{-}, \ldots, \text{-}),$

$\mathcal{F}^{\mathsf{k\text{-}sMP}} : ((x_1, y_2), \ldots, (x_k, y_k)) \mapsto (\pi_1, \ldots, \pi_k) \text{ where } \sum_i \pi_i = (\sum_i x_i) \cdot (\sum_i y_i);$

$\mathcal{F}^{\mathsf{Exc}_{\mathsf{pk}}} : (s_1, (\beta_2, s_2), \ldots, (\beta_k, s_k)) \mapsto (\mathsf{pk}_1, \mathsf{pk}_1, \cdots, \mathsf{pk}_1);$

$\mathcal{F}^{\mathsf{Exc}_{\mathsf{sk}}} : (s_1, (\beta_2, s_2), \ldots, (\beta_k, s_k)) \mapsto (\alpha \prod_{j=2}^{k} \beta_j, \text{-}, \cdots, \text{-}).$

Proposition 3. *For* $\rho \in \{\mathsf{k\text{-}MP}, \mathsf{k\text{-}sMP}, \mathsf{Exc_{pk}}, \mathsf{Exc_{sk}}\}$, ρ *securely computes* $\mathcal{F}^\rho$.

The proof of this result relies heavily on Proposition 1, as we need to show first that k-MP securely computes $\mathcal{F}^{\mathsf{k\text{-}MP}}$, then use this result to show that k-sMP securely computes $\mathcal{F}^{\mathsf{k\text{-}sMP}}$, and so forth. The proof of k-MP is also interesting because we need to perform an induction on k. In [4], authors discussed intuitive proofs of the base case of the above proposition, namely $k = 2$.

Proposition 4 (Simulation-based proof of [4], Sect. 7.1). The protocol 2-MP securely computes $\mathcal{F}^{\mathsf{2\text{-}MP}}$.

Proof. We first point out that the views of the protocol are semantically secure, that is, they do not leak any secrets from the protocol if our SF assumption holds. This is straightforward to see and is detailed in the proof of [4], Sect. 7.1.

As in Sect. 3.1 let $\mathbf{x} = \{x_1, (x_2, r_2)\}$ and $S \subsetneq P$ be a set of corrupted parties. Note that, as $k = 2$, we have $S \in \{\emptyset, \{P_1\}, \{P_2\}\}$. Now, according to Definition 2, for every possible set S we construct a PPT algorithm $\mathcal{I}_S$ such that

$$(\mathcal{I}_S(1^\lambda, \mathbf{x}_S, \mathcal{F}_S^{\mathsf{2\text{-}MP}}(\mathbf{x})), \mathcal{F}^{\mathsf{2\text{-}MP}}(\mathbf{x})) \stackrel{s}{\approx} (\mathrm{view}_S^{\mathsf{2\text{-}MP}}(\mathbf{x}, \lambda), \mathrm{output}^{\mathsf{2\text{-}MP}}(\mathbf{x}, \lambda)).$$

Case 1: $S = \{P_1\}$. The view of corrupt P_1 in 2-MP protocol is:

$$\mathrm{view}_S^{\mathsf{2\text{-}MP}}(\mathbf{x}, \lambda) = \begin{cases} x_1, x_{11}, x_{12}, \ldots, x_{1m}, \\ (b_1, v_0^1, v_1^1, \hat{e}_1), \cdots, (b_m, v_0^m, v_1^m, \hat{e}_m). \end{cases}$$

Recall that x_1 is P_1's input. The values x_{1i} are random polynomials such that they add up to x_1. The random bits b_i and the random polynomials v_j^i are such that $v_{b_i}^i$ is equal to x_{1i}. Finally, $\hat{e}_i$ is the output of the oblivious transfer functionality and $\sum_{i=1}^m \hat{e}_i = x_1 \cdot x_2 + r_2$.

Algorithm 6. Simulator for 2-MP corresponding to $S = \{P_1\}$

Require: 1^λ, x_1, $x_1 \cdot x_2 + r_2$
1: **procedure** $\mathcal{I}_{P_1}$
2:    Sample m random polynomials $(\tilde{x}_{1i} \stackrel{\$}{\leftarrow} R_q)_{i=1}^m$, such that $\sum_{i=1}^m \tilde{x}_{1i} = x_1$.
3:    **for** $i = 1 \ldots m$ **do**
4:        Sample $\tilde{b}_i \stackrel{\$}{\leftarrow} \{0,1\}$ and $(\tilde{v}_0^i, \tilde{v}_1^i) \stackrel{\$}{\leftarrow} R_q^2$. Set $\tilde{v}_{\tilde{b}_i} = x_{\mathcal{I}i}$.
5:    Sample m random polynomials $(\tilde{e}_i \stackrel{\$}{\leftarrow} R_q)_{i=1}^m$, such that $\sum_{i=1}^m \tilde{e}_i = x_1 \cdot x_2 + r_2$.
6:    Return x_1 together with all the values generated.

We define $\mathcal{I}_S$, a simulator of the view of P_1, in Algorithm 6. Its output is

$$\mathcal{I}_S(1^\lambda, \mathbf{x}) = \{ x_1, \tilde{x}_{11}, \tilde{x}_{12}, \ldots, \tilde{x}_{1m}, (\tilde{b}_1, \tilde{v}_0^1, \tilde{v}_1^1, \tilde{e}_1), \cdots, (\tilde{b}_m, \tilde{v}_0^m, \tilde{v}_1^m, \tilde{e}_m).$$

Recall that $\mathcal{F}^{\mathsf{2\text{-}MP}}(\mathbf{x})$ and $\mathrm{output}^{\mathsf{2\text{-}MP}}(\mathbf{x}, \lambda)$ are both equal to $x_1 \cdot x_2 + r_2$. Therefore, we only need to verify that $\mathcal{I}_S(1^\lambda, \mathbf{x}_S, \mathcal{F}_S^2(\mathbf{x})) \stackrel{s}{\approx} \mathrm{view}_S^{\mathsf{2\text{-}MP}}(\mathbf{x}, \lambda)$.

First, both views share x_1. The polynomials $x_{11}, x_{12}, \ldots, x_{1m}$ are uniformly generated by P_1 in 2-MP. On the other hand, $\tilde{x}_{11}, \tilde{x}_{12}, \ldots, \tilde{x}_{1m}$ are uniformly generated by $\mathcal{I}_S$. Also, we have that $\sum_{i=1}^{m} x_{1i} = \sum_{i=1}^{m} \tilde{x}_{1i} = x_1$, yielding that these sets of polynomials are indistinguishable.

In the same fashion, each b_i is a random bit and (v_0^i, v_1^i) are random polynomials in R_q chosen by P_1. On the other hand, $\tilde{b}_i$ is a random bit and $(\tilde{v}_0^i, \tilde{v}_0^i)$ are random polynomials in R_q generated by $\mathcal{I}_S$.

Finally $\tilde{e}_i$ is chosen at random, while $\hat{e}_i$ equals $x_{1i} \cdot x_2 + r_{2i}$. Note that this last value is indistinguishable from uniform because of the additive uniformly random polynomial r_{2i} selected by the honest player P_2. We conclude that $\text{view}_S^{\text{2-MP}}(\mathbf{x}, \lambda)$ and $\mathcal{I}_S(1^\lambda, \mathbf{x}_S, y_1)$ are statistically indistinguishable when $S = \{P_1\}$.

Case 2: $S = \{P_2\}$. The view of P_2 in 2-MP protocol is

$$\text{view}_S^{\text{2-MP}}(\mathbf{x}, \lambda) = \left\{ x_2, r_2, r_{21}, r_{22}, \ldots, r_{2m}, (v_0^1, v_1^1, e_0^1, e_1^1), \cdots, (v_0^m, v_1^m, e_0^m, e_1^m) \right\}$$

Algorithm 7. Simulator for 2-MP corresponding to $S = \{P_2\}$

Require: 1^λ, (x_2, r_2).
 1: **procedure** $\mathcal{I}_{P_2}$
 2:     Generate m random polynomials $(\tilde{r}_{2i} \xleftarrow{\$} R_q)_{i=1}^{m}$ such that $\sum_{i=1}^{m} \tilde{r}_{2i} = r_2$.
 3:     **for** $i = 1 \ldots m$ **do**
 4:         Generate random polynomials $(\tilde{v}_0^i, \tilde{v}_1^i) \xleftarrow{\$} R_q^2$ and compute
         $(\tilde{e}_0^i, \tilde{e}_1^i) = (\tilde{v}_0^i \cdot x_2 + r_{2i}, \tilde{v}_1^i \cdot x_2 + r_{2i})$.
 5:     Return (x_2, r_2) together with all the values generated.

We define $\mathcal{I}_S$ in Algorithm 7. Note that $\mathcal{F}_{P_2}^{\text{2-MP}}(x_1, (x_2, r_2))$ is empty. The output of $\mathcal{I}_S$ is:

$$\mathcal{I}_{\{P_2\}}(1^\lambda, \mathbf{x}_{P_2}) = \left\{ x_2, r_2, \tilde{r}_{21}, \tilde{r}_{22}, \ldots, \tilde{r}_{2m}, (\tilde{v}_0^1, \tilde{v}_1^1, \tilde{e}_0^1, \tilde{e}_1^1), \cdots, (\tilde{v}_0^m, \tilde{v}_1^m, \tilde{e}_0^m, \tilde{e}_1^m) \right\}$$

Analogously as before, is it clear that $\mathcal{I}_S(1^\lambda, \mathbf{x}_S) \overset{s}{\approx} \text{view}_S^{\text{2-MP}}(\mathbf{x}, \lambda)$.   □

The rest of the proof is similar to the above case, with an inductive step. We address the secure computation of k-MP here for the sake of completeness.

Proposition 5. *The protocol* k-MP *securely computes* $\mathcal{F}^{\text{k-MP}}$.

Remember that k-MP uses a functionality $\mathcal{F}^{SP_m}$ for the scalar product as in algorithm SP_m. We proceed with an inductive argument. First, assume that for all k' such that $2 \leq k' < k$, k'-MP securely computes $\mathcal{F}^{k'}$. The inductive step is to show that k-MP$^{(k\text{-}1)\text{-MP} \to \mathcal{F}^{k-1}, SP_m \to \mathcal{F}^{SP_m}}$ securely computes $\mathcal{F}^{\text{k-MP}}$, as we already established the base case in Proposition 4.

What follows are the views of parties P_1 (the key receiver), P_2, and P_ℓ for $\ell > 2$.

$$\text{view}_{P_1}^{\text{k-MP}}(\mathbf{x}, \lambda) = \begin{cases} x_1, \\ x_{11}, x_{12}, \ldots, x_{1m}, \\ (b_1, v_0^1, v_1^1, \hat{e}_1), \ldots, (b_m, v_0^m, v_1^m, \hat{e}_m), \\ \theta, \hat{s}_1, \ldots, \hat{s}_k \end{cases}$$

The elements x_{1i}, b_i, v_j^i and $\hat{e}_i$ are as in the proof of Proposition 4. On the other hand, the polynomial θ is the sum of $\hat{e}_i$ and $\hat{s}_\ell$ the sum of some random values $r_{\ell i}^j$ of player P_ℓ.

$$\text{view}_{P_2}^{\text{k-MP}}(\mathbf{x}, \lambda) = \begin{cases} x_2, r_2, \\ r_{21}, r_{22}, \ldots, r_{2m}, \\ (\hat{r}_{21}^0, \ldots, \hat{r}_{2m}^0), (\hat{r}_{21}^1, \ldots, \hat{r}_{2m}^1), \\ (v_0^1, v_1^1, e_0^1, e_1^1), \cdots, (v_0^m, v_1^m, e_0^m, e_1^m) \end{cases}$$

In P_2's view, the polynomials $r_{2i}, \hat{r}_{2i}^j$ are uniformly random values in R_q.

$$\text{view}_{P_\ell}^{\text{k-MP}}(\mathbf{x}, \lambda) = \begin{cases} x_\ell, r_\ell, \\ r_{\ell 1}, r_{\ell 2}, \ldots, r_{\ell m} \\ (\hat{r}_{\ell 1}^0, \ldots, \hat{r}_{\ell m}^0), (\hat{r}_{\ell 1}^1, \ldots, \hat{r}_{\ell m}^1) \end{cases}$$

Note that the view of P_ℓ is a subset of the view of P_2. The tuple (x_ℓ, r_ℓ) is the party's input, while $r_{\ell i}$ and $\hat{r}_{\ell i}^j$ are uniformly random polynomials.

For the construction of the algorithm $\mathcal{I}_S$, we consider the four cases: (i) $P_1, P_2 \in S$, (ii) $P_1 \in S, P_2 \notin S$, (iii) $P_1 \notin S, P_2 \in S$ and (iv) $\{P_1, P_2\} \cap S = \emptyset$. Proceeding as in the proof of Proposition 4, for each case we construct a simulator algorithm and then show indistinguishability between this simulator and the corresponding view. The complete proof for all claimed functionalities is available in the full version of this paper.   ∎

6.3   Parameters and Efficiency

The parameters n, q, B control the semantic (and multikey-homomorphic) security of the underlying NTRU encryption scheme, and the hardness of our new problems in R_q of Sect. 4. We consider them fixed and according to the suggested values in [4,8,9] for at least $\lambda = 128$ bits of security. The computational complexity of our key-generation protocol amounts to $O((2\lambda)^{k-1})$ instances of $\binom{2}{1}$-OT and $O((2\lambda)^{k-1})$ multiplications in R_q (see Appendix B for detailed computations). As a heuristic estimation, in order to securely generate an Excalibur key pair between 4 participants and with 128 bits of security (this is, create a key pair that inherits decryption of three parties), there is the need to perform approximatively 2^{24} OT's and 2^{24} products in R_q, which is feasible for secure n, q. With FFT or Karatsuba methods, polynomial multiplication can be carried out in time $\tilde{O}(n, q)$, and oblivious transfers can be efficiently performed using techniques as OT extensions. For instance, [1] reports computation of 700,000

$\binom{2}{1}$-OT per second over Wi-Fi, and [5] reduces an OT to three cryptographic hash computations. In a regular, commercially available laptop, 2^{24} products in R_q with $n = 512$ and $\log_2(q) \approx 256$ took us around fifteen minutes (in C++ with the bignum library GMP (https://gmplib.org/). Although there are relatively simple efficiency improvements to our protocols, on future work we will focus on attaining security against malicious adversaries before addressing efficiency concerns. We point out that, while our protocols may not be efficient enough for practical applications with a large number of parties, once key-generation procedures are finished, the resulting keys behave as regular NTRU keys without extra complexity other than coefficient size (which does not dramatically affect the efficiency of the NTRU scheme, and is analized in [9]).

7 Conclusion

Our paper extends the original Excalibur key-generation protocols for an arbitrary hierarchy of keys, and presents formal proofs for the security of these protocols. While we have defined our protocols with respect to a participant P_1 that aims to obtain a key that decrypts messages of participants $P_2, \ldots, P_k$, we can immediately extend these for any DAG-like hierarchy, as follows. Starting from the leaves, which already have their key pairs, first generate the keys of their parents. For a parent with k children, these keys are of the form $\alpha \prod \beta_i$, with $\alpha = 2(\sum s_i) + 1$ a sum of k elements sampled from $\mathcal{G}_B$, and each β_i the secret key of one of the leaves. In turn, these keys are used to generate the keys for nodes at higher levels, and so forth. Note that keys generated in this fashion are of the form $\alpha \prod \gamma_j$, where α is as above and γ is itself a product of secret keys of lower levels (which are either leaves or keys of the same form). Thus, secret keys for members of higher hierarchies are again products of elements distributing according to a gaussian distribution, so all of our security proofs can be extended for more complex hierarchies; we only need to update our hardness assumptions so that they hold with wider gaussian distributions, that is, bounded by $2kB + 1$ instead of B, where k is the outdegree of the hierarchy.

The problem of key-generation in the presence of malicious adversaries is an interesting direction for future work. In particular, we note that this case is not immediate form our results, as Definition 2 and Proposition 1 must be tightened when considering the malicious case, because tampering with intermediate values may affect the input of other protocols, even if they involve honest players only.

Acknowledgements. We would like to thank the anonymous reviewers for their comments. This work was supported by Instituto Milenio Fundamentos de los Datos, Vicuña Mackenna 4860, Santiago, Chile, and Fondecyt Chile (project number 1170866). The fourth author would like to thank Claudio Orlandi for his insight and for providing references on simulation-based proofs, and Martín Ugarte for his helpful comments.

A   Scalar Product Protocol SP_m

In our k-Multiplication protocol (Algorithm 2), parties rely on a multiparty scalar product protocol as a subroutine to cancel additive noise.

Definition 9. *For $m \in \mathbb{N}$, let SP_m be a two-party protocol performing the following. Party A has a sequence of bits ordered in a binary vector $\boldsymbol{b} = (b_1, \ldots, b_m)$. For each $i = 1, \ldots, m$, party B has a pair of polynomials $(p_i^{(0)}, p_i^{(1)})$ of R_q. In the end, party A learns $\gamma = p_1^{(b_1)} + p_2^{(b_2)} + \cdots + p_m^{(b_m)}$ and nothing more. Party B learns nothing.*

We refer to this functionality as a *scalar product*[1], since it computes

$$\gamma = \sum_{i=1}^{m} p_i^{(b_i)} = (p_1^0, \ldots, p_m^0) \cdot \boldsymbol{b}^c + (p_1^1, \ldots, p_m^1) \cdot \boldsymbol{b},$$

where $\boldsymbol{b}^c = (\bar{b}_1, \ldots, \bar{b}_m)$ is the binary complement of $\boldsymbol{b}$. The protocol is outlined in Algorithm 8 below.

Remark: This protocol can be restated as a $\binom{2^m}{1}$-OT protocol, as follows. For each $x \in \{0,1\}^m$, party B computes a mapping $x \mapsto \sum_{i=1}^{m} p_i^{(x[i])}$ where $x[i]$ is the i-th bit of x. Then, party A extracts the polynomial corresponding to $x' = \boldsymbol{b}$ with a $\binom{2^m}{1}$-OT protocol. We point out that this is highly inefficient, because B needs to compute $O(2^m)$ additions in R_q.

Algorithm 8. Scalar product protocol
___

Require: Alice holds $\boldsymbol{b} = (b_1, \ldots, b_m) \in \{0,1\}^m$. Bob holds $2m$ polynomials $((p_i^{(0)}, p_i^{(1)}) \in R_q^2)_{i=1}^m$. Let κ be such that it is unfeasible to compute 2^κ additions in R_q.
Ensure: Alice learns $(p_1^0, \ldots, p_m^0) \cdot \boldsymbol{b}^c + (p_1^1, \ldots, p_m^1) \cdot \boldsymbol{b}$, Bob learns nothing.
1: **procedure** SP_m
2:     Alice samples κ vectors $\boldsymbol{b}_1, \ldots, \boldsymbol{b}_\kappa \overset{\$}{\leftarrow} \mathbb{Z}^m$ such that $\boldsymbol{b}_1 + \cdots + \boldsymbol{b}_\kappa = \boldsymbol{b}$.
3:     **for** $i = 1 \ldots \kappa$ **do**
4:         Alice samples a bit σ and two vectors $\boldsymbol{a}_0, \boldsymbol{a}_1 \overset{\$}{\leftarrow} \{0,1\}^m$. She sets $\boldsymbol{a}_\sigma \leftarrow \boldsymbol{b}_i$.
5:         Alice sends the pair $(\boldsymbol{a}_1, \boldsymbol{a}_2)$ to Bob.
6:         Bob computes

$$d_0 = (p_1^0, \ldots, p_m^0) \cdot \boldsymbol{a}_0^c + (p_1^1, \ldots, p_m^1) \cdot \boldsymbol{a}_0$$
$$d_1 = (p_1^0, \ldots, p_m^0) \cdot \boldsymbol{a}_1^c + (p_1^1, \ldots, p_m^1) \cdot \boldsymbol{a}_1$$

7:         With a $\binom{2}{1}$-OT protocol, Alice extracts $\gamma_i := d_\sigma$ from Bob.
8:     Alice computes $\gamma = \sum_{i=1}^{\kappa} \gamma_i = (p_1^0, \ldots, p_m^0) \cdot \boldsymbol{b}^c + (p_1^1, \ldots, p_m^1) \cdot \boldsymbol{b}$.

B   Algorithmic Complexity

In this appendix we develop expressions for the computational complexity of our key generation protocols. In this section, let n, q, B be secure NTRU parameters,

---

[1] In the vector space R_q^m. Recall that $R_q \simeq \mathbb{F}_{q^n}$, the field of characteristic q^n.

m be such that it is unfeasible to compute 2^m additions in R_q, and k parties are involved in the key generation procedure.

As we show below, an Excalibur key pair $(\mathsf{sk}, \mathsf{pk})$ can be generated in $O((2m)^k)$ products in R_q and $O((2m)^{k-1})$ basic $\binom{2}{1}$-OT protocols. While this is certainly prohibitive for a large amount of parties and reasonable security, with fast polynomial multiplication and OT-extension techniques it is possible to generate a key pair with $k = 4$ and $m = 128$ in some minutes. Let us also mention that this key acts as other keys of the system, that is, after key generation is completed, no extra complexity is to be expected for encryption, decryption or homomorphic procedures (other than coefficient size, whose impact in complexity is analyzed in [8]).

Definition 10. *Let θ (resp. π) be the computational cost of performing a $\binom{2}{1}$-OT protocol (resp. performing a multiplication in R_q).*

Proposition 6. *The computational cost of performing k-MP is approximatively $(2m)^{k-1}\pi + (2m)^{k-1}\theta$. The computational cost of performing k-sMP is approximatively $mk(k-1)(2\pi + \theta)$.*

Proof. First, note that the computational cost of performing SP_m (with $\kappa = m$) is $m\theta$ (see Algorithm 8 from Appendix A and note that the scalar product is not expressed in terms of full R_q products), and the cost of performing 2-MP is $(2\pi + \theta)m$. Let u_k be the computational cost of performing k-MP. Given the description of the protocol in Algorithm 2, we have the following recurrence:

$$\begin{cases} u_k = 2mu_{k-1} + km\theta, \\ u_2 = (2\pi + \theta)m. \end{cases}$$

To see this, note that parties first perform $2m$ instances of $(\mathsf{k\text{-}1})$-MP, then m $\binom{2}{1}$-OT extractions, and finally $(k-1)$ scalar products SP_m. The solution to this equation for $k \geq 3$ is given by

$$u_k = (2m)^{k-2}u_2 + m\theta \sum_{i=3}^{k} i(2m)^{k-i},$$

and therefore the cost of k-MP is approximately $(2m)^{k-1}$ products in R_q and $(2m)^{k-1}$ $\binom{2}{1}$-OT protocols.

Let now v_k be the computational cost of performing k-sMP. Parties perform $k(k-1)$ instances of 2-MP (Algorithm 3), therefore we have

$$v_k = mk(k-1)(2\pi + \theta).$$

$\square$

Proposition 7. *The cost of performing both $\mathsf{Exc}_{\mathsf{pk}}$ and $\mathsf{Exc}_{\mathsf{sk}}$ between k parties is $O(u_k)$, that is, $O((2m)^{k-1})$ products in R_q and $O((2m)^{k-1})$ $\binom{2}{1}$-OT protocols.*

Proof. In $\mathsf{Exc_{pk}}$ (Algorithm 4), parties perform one k-MP and $(k-1)$ instances of (k-1)-MP. Also, in $\mathsf{Exc_{sk}}$ (Algorithm 5) parties perform $(k-1)$ instances of (k-1)-MP and one final k-MP. The leading term of computational cost in both cases is therefore $O((2m)^{k-1})$ products and $((2m)^{k-1})$ oblivious transfers.   □

Remark: With $m = 128$ bits of security against brute force additions in R_q, four parties need to compute around 2^{24} products in R_q and 2^{24} 1-out-of-2 oblivious transfer protocols.

References

1. Asharov, G., Lindell, Y., Schneider, T., Zohner, M.: More efficient oblivious transfer and extensions for faster secure computation, November 2013
2. Canetti, R.: Universally composable security: a new paradigm for cryptographic protocols. In: 2001 Proceedings of 42nd IEEE Symposium on Foundations of Computer Science, pp. 136–145. IEEE (2001)
3. Canetti, R.: Security and composition of cryptographic protocols: a tutorial (part I). SIGACT News **37**(3), 67–92 (2006)
4. Goubin, L., Vial Prado, F.J.: Blending FHE-NTRU keys – the excalibur property. In: Dunkelman, O., Sanadhya, S.K. (eds.) INDOCRYPT 2016. LNCS, vol. 10095, pp. 3–24. Springer, Cham (2016). https://doi.org/10.1007/978-3-319-49890-4_1
5. Ishai, Y., Kilian, J., Nissim, K., Petrank, E.: Extending oblivious transfers efficiently. In: Boneh, D. (ed.) CRYPTO 2003. LNCS, vol. 2729, pp. 145–161. Springer, Heidelberg (2003). https://doi.org/10.1007/978-3-540-45146-4_9
6. Li, S.-D., Dai, Y.-Q.: Secure two-party computational geometry. J. Comput. Sci. Technol. **20**(2), 258–263 (2005)
7. Lindell, Y.: How to simulate it – a tutorial on the simulation proof technique. Tutorials on the Foundations of Cryptography. ISC, pp. 277–346. Springer, Cham (2017). https://doi.org/10.1007/978-3-319-57048-8_6
8. López-Alt, A., Tromer, E., Vaikuntanathan, V.: On-the-fly multiparty computation on the cloud via multikey fully homomorphic encryption. In: Proceedings of the Forty-Fourth Annual ACM Symposium on Theory of Computing, STOC 2012, pp. 1219–1234. ACM, New York (2012)
9. Stehlé, D., Steinfeld, R.: Making NTRU as secure as worst-case problems over ideal lattices. In: Paterson, K.G. (ed.) EUROCRYPT 2011. LNCS, vol. 6632, pp. 27–47. Springer, Heidelberg (2011). https://doi.org/10.1007/978-3-642-20465-4_4
10. Thomae, E., Wolf, C.: Solving underdetermined systems of multivariate quadratic equations revisited. In: Fischlin, M., Buchmann, J., Manulis, M. (eds.) PKC 2012. LNCS, vol. 7293, pp. 156–171. Springer, Heidelberg (2012). https://doi.org/10.1007/978-3-642-30057-8_10

Secure Multiparty Computation

The Six-Card Trick: Secure Computation of Three-Input Equality

Kazumasa Shinagawa[1,2(✉)] and Takaaki Mizuki[3]

[1] Tokyo Institute of Technology, Meguro, Japan
shinagawakazumasa@gmail.com
[2] Institute of Advanced Industrial Science and Technology (AIST), Kōtō, Japan
[3] Tohoku University, Sendai, Japan

Abstract. Secure computation enables parties having secret inputs to compute some function of their inputs without revealing inputs beyond the output. It is known that secure computation can be done by using a deck of physical cards. The *five-card trick* proposed by den Boer in 1989 is the first card-based protocol, which computes the logical AND function of two inputs. In this paper, we design a new protocol for the three-input equality function using six cards, which we call the *six-card trick*.

1 Introduction

Suppose that during the two-candidate election, Alice, Bob, and Charlie wish to talk about the candidates only if they are supporting the same candidate. Unfortunately, they do not know each other's supporting candidate. If their supporting candidates do not coincide, they wish to hide their supporting candidates from each other in order not to break their friendship. Due to this secrecy condition, just revealing supporting candidates to others does not work. How can we solve this problem?

Secure computation, which is one of cryptographic techniques, serves a solution to such a situation in which parties wish to compare secret information without leaking it. Specifically, it enables a set of parties to compute some function of their inputs without leaking the inputs beyond the output. In the above situation, for the inputs $a, b, c \in \{0, 1\}$ indicating the candidates that Alice, Bob, and Charlie support, respectively, it is sufficient to securely compute a three-input function f such that $f(a, b, c) = 1$ if $a = b = c$ and 0 otherwise. When the output of the function is 1, they can start to talk about the election since in this case their supporting candidates are the same. When the output of the function is 0, they should talk about a topic which is not related to the election campaigns. In the latter case, thanks to the security of secure computation, their supporting candidates are hidden from each other. For example, Alice cannot obtain Bob's and Charlie's supporting candidates. The only thing obtained by Alice is that *at least one of Bob and Charlie supports the opposite*

K. Lee (Ed.): ICISC 2018, LNCS 11396, pp. 123–131, 2019.
https://doi.org/10.1007/978-3-030-12146-4_8

candidate from hers, but this is not problematic since it can be computable from the output of the function.

Secure computation is one of very active research areas in cryptography. In this paper, we focus on secure computation using a deck of cards, which is often called *card-based cryptography*, while most of secure computation is designed to be executed on computers. The reason why we choose card-based cryptography instead of secure computation running on computers, is that it is suitable to a situation in which *all parties gather together in the same place*. This is in contrast to secure computation running on computers which is designed to be executed among *parties whose locations are physically separated*. Card-based cryptography often provides a simple solution to such an everyday situation since the participants can watch each other to prevent malicious behavior. For example, in card-based cryptography, secret information is encoded by a sequence of face-down cards on a dinning table, instead of encryption as in the standard secure computation. This is sufficient to enforce honest behavior on parties since any malicious behavior can be detected by the parties. Moreover, due to its simplicity of model of card-based cryptography, it is easy to understand correctness and security even for non-experts, compared to secure computation running on computers. Therefore, this provides a solution such that all parties are convinced how their secret inputs are protected.

The first research on card-based cryptography is the *five-card trick* proposed by den Boer [2]. It provides a very simple solution to the case that two parties Alice and Bob having $a, b \in \{0, 1\}$, respectively, wish to securely compute the logical AND $a \wedge b$ (see Sect. 3). It has two nice properties. The first one is that it requires only a single shuffle, especially, a *random cut*. Since a random cut is accepted[1] as the most basic shuffle, this property means that it is the most efficient in terms of shuffle operation. We call a protocol that only requires a random cut a *single-cut protocol*. The second nice property is that at the end of an execution, all cards are face-up. We call such a protocol a *gabage-free protocol*.

Despite of these nice properties, as far as we know, no single-cut garbage-free protocols have been proposed except for the five-card trick. Are these properties special for the logical AND computation? In this paper, we show that the answer is NO by constructing a single-cut garbage-free protocol for the three-input equality function using six cards, which we call the *six-card trick*.

1.1 Related Works

Card-based protocol begins with the five-card trick proposed by den Boer [2] in 1989. Crépeau and Killian [1] showed that every function can be securely computed by applying a number of random cuts. A number of subsequent works [6,9,10,13,14,16] improved upon the protocols [1] in terms of numbers of cards and shuffles. While these works [1,6,9,13,14,16] aimed to achieve the feasibility

---

[1] A random cut is securely implemented by a Hindu cut [17], while most of other shuffles do not have a (direct) secure implementation. Koch and Walzer [5] showed that every uniform and closed shuffles are reduced to a number of random cuts.

and the efficiency of general secure computation, there is another line of research which focuses on a specific problem [2–4,7,8,11,12,15]. Our work focuses on a specific problem, specifically, the three-input equality function for Boolean inputs.

There are two types of card-based protocols: "committed format" protocols (e.g. [1,9]) and "non-committed format" protocols (e.g. [2,8]). The former produces a sequence of face-down cards that follows the input encoding. Thus, an output sequence of a committed format protocol can be used to be inputted to another protocol. In contrast, the latter outputs the result value directly. A committed format protocol implies a non-committed format protocol for the same function, by just turning over the output sequence. On the other hand, in general, non-committed format protocols can be more efficient in terms of the number of cards and shuffles, compared to committed format protocols. Our protocol, the six-card trick, is a non-committed format protocol for the three-input equality function.

1.2 Organization

In Sect. 2, we present basic definitions of card-based cryptography. In Sect. 3, we review a previous work called the five-card trick. In Sect. 4, we design our protocol for the three-input equality function, which we call the six-card protocol. In Sect. 5, we present some open problems. In Sect. 6, we conclude this paper.

2 Card-Based Cryptography

In this section, we present basic definitions of card-based cryptography. Although the first three subsections provide very common definitions of card-based cryptography, the last subsection defines new properties, which we call *single-cut* and *garbage-free*.

2.1 Basic Setting of Card-Based Protocols

Suppose that Alice, Bob, and Charlie having secret inputs $a, b, c \in \{0, 1\}$, respectively, wish to compute some function on their inputs without revealing the inputs beyond the output. (In the two-party case as in Sect. 3, Charlie is ignored.) They are in the same room together. There are a deck of cards and a flat space (e.g. dinner table) in the room. Cards will be arranged on the flat space. The deck of cards in our use contains two types of cards, ♣ and ♡, whose back sides are the same pattern ?. All cards having the same type (♣ or ♡) are indistinguishable. We say that ♣ and ♡ lying on the flat space are *face-up cards* and ? is a *face-down card*.

2.2   Commitment

We use the following encoding rule for Boolean inputs:

$$\boxed{\clubsuit}\boxed{\heartsuit} = 0, \quad \boxed{\heartsuit}\boxed{\clubsuit} = 1.$$

For a bit $x \in \{0, 1\}$, two face-down cards $\boxed{?}\boxed{?}$ having a value x according to the above encoding is called a *commitment* to x, denoted by

$$\underbrace{\boxed{?}\boxed{?}}_{x}.$$

2.3   Random Cut

A *random cut* is a random cyclic shift operation. Let X be a sequence of face-down cards $(x_0, x_1, \cdots, x_{n-1})$ as follows:

$$X = \underset{x_0}{\boxed{?}}\underset{x_1}{\boxed{?}} \cdots \underset{x_{n-1}}{\boxed{?}}.$$

For a sequence of cards X, a random cut generates a randomly shifted sequence $(x_i, x_{i+1}, \cdots, x_{i+n-1})$ for a uniformly distributed random integer i, $0 \le i \le n-1$, denoted by $\langle \cdot \rangle$, as follows:

$$\left\langle \boxed{?}\boxed{?}\boxed{?}\boxed{?}\boxed{?} \right\rangle \quad \rightarrow \quad \boxed{?}\boxed{?}\boxed{?}\boxed{?}\boxed{?}.$$

It must satisfy that the random number i is hidden from all parties. Ueda et al. [17] experimentally showed that a random cut is securely implementable by applying *Hindu cut*, which is a kind of shuffling operations widely used in card games.

2.4   Single-Cut and Garbage-Free Protocols

We say that a card-based protocol is *single-cut* if it requires only one random cut and does not require other shuffles. We say that a card-based protocol is *garbage-free* if at the end of the protocol, all cards are face-up.

In this paper, we focus on single-cut and garbage-free protocols of the following type:

1. Each party having two cards $\boxed{\clubsuit}\boxed{\heartsuit}$ privately places them in the face-down format according to his/her input. These pairs of cards (possibly with additional cards) are arranged according to a predetermined permutation.
2. Apply a random cut to the sequence of cards.
3. Turn over all of the cards. The output can be obtained from the pattern of the resulting sequence.

For ease of explanation, we define a term *cyclic set*. For a sequence of face-up cards, we define its cyclic set as a set of all its cyclic shifted sequences. For example, the cyclic set of ♡♣♣ is the set of three sequences:

$$\left\{ \boxed{♡}\,\boxed{♣}\,\boxed{♣}\,,\,\boxed{♣}\,\boxed{♡}\,\boxed{♣}\,,\,\boxed{♣}\,\boxed{♣}\,\boxed{♡} \right\}.$$

We say that a *single-cut garbage-free protocol computes a function* $f : \{0,1\}^n \to \{0,1\}$ if there exists two cyclic sets C_0 and C_1 such that for every input $x \in \{0,1\}^n$, a sequence of cards after the final step of the protocol is always contained in $C_{f(x)}$. For such a protocol, a proof of security is trivial since all cards are face-down except at the end of the protocol.

3 Five-Card Trick

Before presenting our construction of the six-card trick, we review the seminal work called the *five-card trick* proposed by den Boer [2]. This is a secure two-party protocol computing the logical AND function $a \wedge b$ for inputs $a, b \in \{0,1\}$. As the name suggests, it only requires five cards: ♣♣♡♡♡.

Now we are ready to explain the five-card trick. Suppose that Alice and Bob having $a, b \in \{0,1\}$, respectively, wish to securely compute the logical AND function. The five-card trick [2] proceeds as follows.

1. Alice privately arranges a commitment to negation $\bar{a}$ of bit a, and Bob privately arranges a commitment to b. These two commitments together with ♡ are arranged as follows:

$$\underbrace{\boxed{?}\boxed{?}}_{\bar{a}}\boxed{♡}\underbrace{\boxed{?}\boxed{?}}_{b} \;\; \rightarrow \;\; \underbrace{\boxed{?}\boxed{?}}_{\bar{a}}\boxed{?}\underbrace{\boxed{?}\boxed{?}}_{b}.$$

   It should be noted that the three middle cards would be ♡♡♡ only if $a = b = 1$. (See Table 1.)

2. A random cut is applied to the five cards as follows:

$$\left\langle \boxed{?}\boxed{?}\boxed{?}\boxed{?}\boxed{?} \right\rangle \;\; \rightarrow \;\; \boxed{?}\boxed{?}\boxed{?}\boxed{?}\boxed{?}.$$

3. Turn over all cards; then, we can consider cyclic sets of two sequences:

$$\boxed{♣}\boxed{♡}\boxed{♡}\boxed{♡}\boxed{♣} \;\text{ or }\; \boxed{♡}\boxed{♣}\boxed{♡}\boxed{♣}\boxed{♡}.$$

   The left cases, three (cyclically) consecutive ♡'s, imply $a \wedge b = 1$ and the right imply $a \wedge b = 0$.

The correctness of the five-card trick can be observed by Table 1 showing all possible sequences after Step 1. As described in Sect. 2.4, the security of the five-card trick is trivial because all cards are face-down except at the end of protocol.

Table 1. All possibilities of the sequence after Step 1 of the five-card trick.

(a, b)	Sequence
$(0, 0)$	♡♣♡♣♡
$(0, 1)$	♡♣♡♡♣
$(1, 0)$	♣♡♡♣♡
$(1, 1)$	♣♡♡♡♣

4 Six-Card Trick

In this section, we present the six-card trick.

The six-card trick is a protocol that securely computes the three-input equality function that takes three bits as inputs and outputs 1 if they are the same and 0 otherwise. As the name suggests, it only requires six cards: ♣♣♣♡♡♡. Suppose that Alice, Bob, and Charlie have $a, b, c \in \{0, 1\}$, respectively, and wish to securely compute the three-input equality function. The protocol proceeds as follows:

1. Alice, Bob, and Charlie privately arrange commitments to the inputs a, b, and c, respectively:

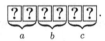

2. The six cards are arranged as follows:

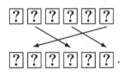

3. A random cut is applied to the six cards as follows:

$$\langle ??????\rangle \;\rightarrow\; ??????.$$

4. Turn over all cards; then, we can consider cyclic sets of two sequences:

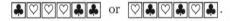

The output is 0 if it is the former case and 1 otherwise.

The correctness of the six-card trick can be easily observed by Tables 2 and 3 which show all possibilities of the sequence after Steps 1 and 2, respectively. As described in Sect. 2.4, the security of the six-card trick is trivial because all cards are face-down except at the end of protocol.

Table 2. All possibilities of the sequence after Step 1.

(a, b, c)	Sequence
$(0, 0, 0)$	♣♡♣♡♣♡
$(0, 0, 1)$	♣♡♣♡♡♣
$(0, 1, 0)$	♣♡♡♣♣♡
$(0, 1, 1)$	♣♡♡♣♡♣
$(1, 0, 0)$	♡♣♣♡♣♡
$(1, 0, 1)$	♡♣♣♡♡♣
$(1, 1, 0)$	♡♣♡♣♣♡
$(1, 1, 1)$	♡♣♡♣♡♣

Table 3. All possibilities of the sequence after Step 2.

(a, b, c)	Sequence
$(0, 0, 0)$	♣♡♣♡♣♡
$(0, 0, 1)$	♣♣♣♡♡♡
$(0, 1, 0)$	♣♡♡♡♣♣
$(0, 1, 1)$	♣♣♡♡♡♣
$(1, 0, 0)$	♡♡♣♣♣♡
$(1, 0, 1)$	♡♣♣♣♡♡
$(1, 1, 0)$	♡♡♡♣♣♣
$(1, 1, 1)$	♡♣♡♣♡♣

5 Open Problems

In Sect. 4, we obtained a six-card protocol for the three-input equality function. A natural question arises: Can we construct a $2n$-card protocol for the n-input equality function? Indeed, we obtain a four-card protocol for the two-input equality function. For inputs $a, b \in \{0, 1\}$, the protocol just applies a random cut to the following sequence:

$$\boxed{?}\boxed{?}\underbrace{}_{a}\boxed{?}\boxed{?}\underbrace{}_{b},$$

and outputs 1 if the resulting face-up sequence is ♣♡♣♡, and 0 otherwise[2]. Thus, the answer of the above question is YES for $n \in \{2, 3\}$.

Unfortunately, by using a computer, we found that there is no eight-card protocol for the four-input equality function. Thus, the answer of the above

---
[2] This technique was implicitly used in [1].

question is NO in general. Our conjecture is that the answer of the above question is NO for all $n \geq 4$. We left it as an open question.

Other open problems are as follows:

- Construct a single-cut and garbage-free protocol for the n-input equality function using more than $2n$ cards.
- Find single-cut and garbage-free protocols for other interesting functions.

6   Conclusion

In this paper, we designed the six-card trick, which is a single-cut garbage-free protocol for the three-input equality function. We leave several questions as open problems.

Acknowledgments. This work was supported in part by JSPS KAKENHI Grant Numbers 17J01169 and 17K00001. The authors would like to thank Osamu Watanabe for his valuable comments.

References

1. Crépeau, C., Kilian, J.: Discreet solitary games. In: Stinson, D.R. (ed.) CRYPTO 1993. LNCS, vol. 773, pp. 319–330. Springer, Heidelberg (1994). https://doi.org/10.1007/3-540-48329-2_27
2. den Boer, B.: More efficient match-making and satisfiability *the five card trick*. In: Quisquater, J.-J., Vandewalle, J. (eds.) EUROCRYPT 1989. LNCS, vol. 434, pp. 208–217. Springer, Heidelberg (1990). https://doi.org/10.1007/3-540-46885-4_23
3. Hashimoto, Y., Shinagawa, K., Nuida, K., Inamura, M., Hanaoka, G.: Secure grouping protocol using a deck of cards. In: Shikata, J. (ed.) ICITS 2017. LNCS, vol. 10681, pp. 135–152. Springer, Cham (2017). https://doi.org/10.1007/978-3-319-72089-0_8
4. Ishikawa, R., Chida, E., Mizuki, T.: Efficient card-based protocols for generating a hidden random permutation without fixed points. In: Calude, C.S., Dinneen, M.J. (eds.) UCNC 2015. LNCS, vol. 9252, pp. 215–226. Springer, Cham (2015). https://doi.org/10.1007/978-3-319-21819-9_16
5. Koch, A., Walzer, S.: Foundations for actively secure card-based cryptography. IACR Cryptology ePrint Archive, vol. 2017, p. 423 (2017)
6. Koch, A., Walzer, S., Härtel, K.: Card-based cryptographic protocols using a minimal number of cards. In: Iwata, T., Cheon, J.H. (eds.) ASIACRYPT 2015. LNCS, vol. 9452, pp. 783–807. Springer, Heidelberg (2015). https://doi.org/10.1007/978-3-662-48797-6_32
7. Mizuki, T., Asiedu, I.K., Sone, H.: Voting with a logarithmic number of cards. In: Mauri, G., Dennunzio, A., Manzoni, L., Porreca, A.E. (eds.) UCNC 2013. LNCS, vol. 7956, pp. 162–173. Springer, Heidelberg (2013). https://doi.org/10.1007/978-3-642-39074-6_16
8. Mizuki, T., Kumamoto, M., Sone, H.: The five-card trick can be done with four cards. In: Wang, X., Sako, K. (eds.) ASIACRYPT 2012. LNCS, vol. 7658, pp. 598–606. Springer, Heidelberg (2012). https://doi.org/10.1007/978-3-642-34961-4_36

9. Mizuki, T., Sone, H.: Six-card secure AND and four-card secure XOR. In: Deng, X., Hopcroft, J.E., Xue, J. (eds.) FAW 2009. LNCS, vol. 5598, pp. 358–369. Springer, Heidelberg (2009). https://doi.org/10.1007/978-3-642-02270-8_36

10. Mizuki, T., Uchiike, F., Sone, H.: Securely computing XOR with 10 cards. Australas. J. Comb. **36**, 279–293 (2006)

11. Nakai, T., Shirouchi, S., Iwamoto, M., Ohta, K.: Four cards are sufficient for a card-based three-input voting protocol utilizing private permutations. In: Shikata, J. (ed.) ICITS 2017. LNCS, vol. 10681, pp. 153–165. Springer, Cham (2017). https://doi.org/10.1007/978-3-319-72089-0_9

12. Nakai, T., Tokushige, Y., Misawa, Y., Iwamoto, M., Ohta, K.: Efficient card-based cryptographic protocols for millionaires' problem utilizing private permutations. In: Foresti, S., Persiano, G. (eds.) CANS 2016. LNCS, vol. 10052, pp. 500–517. Springer, Cham (2016). https://doi.org/10.1007/978-3-319-48965-0_30

13. Niemi, V., Renvall, A.: Secure multiparty computations without computers. Theor. Comput. Sci. **191**(1–2), 173–183 (1998)

14. Nishida, T., Hayashi, Y., Mizuki, T., Sone, H.: Card-based protocols for any Boolean function. In: Jain, R., Jain, S., Stephan, F. (eds.) TAMC 2015. LNCS, vol. 9076, pp. 110–121. Springer, Cham (2015). https://doi.org/10.1007/978-3-319-17142-5_11

15. Nishida, T., Mizuki, T., Sone, H.: Securely computing the three-input majority function with eight cards. In: Dediu, A.-H., Martín-Vide, C., Truthe, B., Vega-Rodríguez, M.A. (eds.) TPNC 2013. LNCS, vol. 8273, pp. 193–204. Springer, Heidelberg (2013). https://doi.org/10.1007/978-3-642-45008-2_16

16. Stiglic, A.: Computations with a deck of cards. Theor. Comput. Sci. **259**(1–2), 671–678 (2001)

17. Ueda, I., Nishimura, A., Hayashi, Y., Mizuki, T., Sone, H.: How to implement a random bisection cut. In: Martín-Vide, C., Mizuki, T., Vega-Rodríguez, M.A. (eds.) TPNC 2016. LNCS, vol. 10071, pp. 58–69. Springer, Cham (2016). https://doi.org/10.1007/978-3-319-49001-4_5

Unconditionally Secure Distributed Oblivious Polynomial Evaluation

Louis Cianciullo$^{(\boxtimes)}$ and Hossein Ghodosi

James Cook University, Townsville 4811, Australia
{louis.cianciullo,hossein.ghodosi}@jcu.edu.au

Abstract. Oblivious polynomial evaluation (OPE) was first introduced by Naor and Pinkas in 1999. An OPE protocol involves a receiver, R who holds a value, α and a sender, S with a private polynomial, $f(x)$. OPE allows R to compute $f(\alpha)$ without revealing either α or $f(x)$. Since its inception, OPE has been established as an important building block in many distributed applications.

In this article we investigate a method of achieving unconditionally secure distributed OPE (DOPE) in which the function of the sender is distributed amongst a set of n servers. Specifically, we introduce a model for DOPE based on the model for distributed oblivious transfer (DOT) described by Blundo et al. in 2002. We then describe a protocol that achieves the security defined by our model.

Our DOPE protocol is efficient and achieves a high level of security. Furthermore, our proposed protocol can also be used as a DOT protocol with little to no modification.

1 Introduction

Oblivious polynomial evaluation (OPE) was first introduced by Naor and Pinkas in 1999 [20]. An OPE protocol involves two parties, a receiver, R who holds a private value, α and a sender, S who holds a private polynomial, $f(x)$. Informally, an OPE protocol allows R to learn the evaluation of S's polynomial at his private value i.e. $f(\alpha)$, whilst keeping S from learning α and R from learning any more information about $f(x)$. A more formal definition, adapted from [7] is given below:

Definition 1 [7]. *An OPE protocol is composed of two parties, S who has a polynomial $f(x)$ over a finite field $\mathbb{F}$ and R who has an input value $\alpha \in \mathbb{F}$. Correctness is achieved if, at the end of the protocol, R learns $f(\alpha)$. Security is guaranteed if the following two conditions are met after the protocol has been executed:*

L. Cianciullo—This research is supported by an Australian Government Research Training Program (RTP) Scholarship.

K. Lee (Ed.): ICISC 2018, LNCS 11396, pp. 132–142, 2019.
https://doi.org/10.1007/978-3-030-12146-4_9

1. S cannot reduce his uncertainty of α, i.e. the probability of S computing α is $1/|\mathbb{F}|$.
2. R does not learn any information relating to $f(x)$, other than $f(\alpha)$.

OPE has been found to have a myriad of applications in such things as secure computation [12], oblivious neural learning [7], secure set intersection [15] and privacy preserving data mining [18]. As a result of this, an extensive amount of research has been conducted on this topic [7,13–15,17,19,24,25].

Within the literature, OPE protocols come in two flavours, (1) computationally (conditionally) secure protocols, which are secure against an adversary that is computationally bounded, and security is based on cryptographic assumptions; and (2) unconditionally (information theoretic) secure protocols, where the adversary is computationally unbounded. We limit the focus of this article to unconditionally secure OPE protocols.

To the best of the author's knowledge there exists only three unconditionally secure OPE protocols in the literature. The first unconditionally secure OPE was given by Chang and Lu [7]. To achieve information theoretic security they use a third party who takes an active role in the protocol execution. The second information theoretic secure OPE protocol was given by Hanaoka et al. in [14] (and was later expanded on in [24]). Their protocol also requires the use of a third party although, in their protocol the third party acts as an initialiser, in that he merely distributes some (unrelated, effectively random) information at the start of the protocol and then takes no further part in the protocol execution. The third OPE protocol that achieves information theoretic security was given by Li et al. [17]. Their protocol takes a different approach and instead utilises a set of servers to collectively implement the function of the sender. We denote such a scheme as a distributed oblivious polynomial evaluation (DOPE) protocol in order to differentiate this type of scheme from the other three-party protocols.

In the DOPE protocol of Li et al. [17] the sender initialises the protocol by distributing some information amongst a set of $n \geq 2$ servers. Following this, S takes no further part in the protocol. To compute his evaluation, R communicates with a subset composed of t amount of these servers where $t \leq n$ is known as the threshold. The sender's security is guaranteed against a coalition composed of $l - 1$ servers and R; whilst the receiver's privacy is guaranteed against a subset of $b - 1$ servers, where $b + l < t \leq n$. Li et al. also show how to improve this scheme allowing for the greater threshold of $b = l = t$ by introducing some publicly known information. However, we note that this increase in security comes at a cost. Namely, it increases the overall complexity of their protocol and it also allows both R and the servers to gain some extra information about $f(x)$. Since OPE protocols (and by extension DOPE protocols) are generally used as building blocks in larger multi-party protocols an OPE protocol that leaks information relating to $f(x)$ may result in security flaws in the overlying multi-party protocol.

As a result of this, an efficient DOPE protocol that does not leak any information and still achieves a high level of security is needed.

1.1 Our Contribution

In this paper we develop such a protocol by first describing a model of DOPE and then introducing an efficient DOPE protocol that achieves the security defined in our model. Specifically, our proposed protocol allows R to compute his evaluation by simply broadcasting some information and then receiving contact from t or more servers. The protocol achieves security for R against a coalition of $t - 1$ servers and security for S against a coalition composed of $t - 1$ servers and R and does not leak any information relating to either $f(x)$ or α.

To develop a model of DOPE we simply apply a slightly modified version of the already established and well studied security framework developed by Blundo et al. [4,5] for the purpose of distributed oblivious transfer (DOT) [1,8–10,21,22]. We then give the construction of a DOPE protocol that is secure under this model. An interesting property of our protocol is that it can also be utilised as a DOT protocol with little to no modification.

Our protocol achieves security equivalent to what Blundo et al. describe as a strong DOT protocol [4]. That is, our DOPE protocol is secure against a coalition composed of $t - 1$ servers and R even after R has received $f(\alpha)$.

2 Model

Similar to a DOT protocol a DOPE protocol consists of a sender, S, the receiver, R and n servers, $s_1, \cdots, s_n$. As per Definition 1 the sender has a polynomial, $f(x)$ of degree $k \geq 1$ over $\mathbb{F}$, whilst the receiver has a point $\alpha \in \mathbb{F}$, such that $|\mathbb{F}| = q$ where q is a prime number and $q > \max(k, n)$. We assume a standard model of communication present in many multi-party protocols [2] i.e. a synchronous broadcast connection exists between the servers and R, such that R can privately and simultaneously send the same message to all of the servers. Additionally, we assume each server has a secure channel that allows them to send private messages to R. DOPE consists of two phases:

1. **Initialisation:** S privately distributes some information relating to $f(x)$ to each of the n servers. Following this S takes no further part in the protocol.
2. **Evaluation:** R broadcasts some information to all of the servers. A set of t or more servers send a response to R who then uses this information to compute $f(\alpha)$.

In order to achieve both correctness and security a DOPE protocol must satisfy the following security conditions, originally given by Blundo et al. [4] and informally stated by Corniaux and Ghodosi [9] for the purpose of DOT:

1. **Correctness:** R is able to compute the requested evaluation after receiving information from t or more servers.
2. **Receiver's Privacy:** A coalition of $t - 1$ servers cannot compute any information relating to α.

3. **Sender's Privacy:** After the initialisation phase (but before the evaluation phase) a coalition composed of $t - 1$ servers and R cannot compute any information relating to $f(x)$.

4. **Sender's Privacy After Protocol Execution:** After communication between R and the servers has occurred and R has computed $f(\alpha)$, a coalition composed of $t-1$ servers and R cannot compute any information relating to $f(x)$; other than what the evaluation of R's chosen value (i.e. $f(\alpha)$) has already revealed.

In our model we assume that all participants follow the protocol exactly, i.e. they are semi-honest. A benefit of our model is that the degree of $f(x)$ (given as k) is not related to the threshold parameter, t. This allows for a flexible and easily changeable level of security. For instance, even if the degree of k is small S can ensure security against a large number of servers by assigning a high value to t.

In regards to the security conditions given by Blundo et al. it was shown that a DOT protocol that achieves all four security conditions could only be achieved in two rounds of communication between the servers and R or by allowing S to contact R during the initialisation phase. This also proves true for our DOPE protocol which is given in the next section. We note that, similar to Blundo's "Strong DOT Protocol" [4] our protocol assumes that S correctly distributes the information to the servers and does not try to initiate any further contact with R or the servers after the initialisation phase.

3 DOPE Protocol

In this section we describe our DOPE protocol and then evaluate the security of the protocol against the security conditions given in the previous section.

In our proposed protocol S utilises Shamir's secret sharing scheme to securely distribute his polynomial among the n servers. For completeness, we will firstly review Shamir's secret sharing scheme.

3.1 Shamir's Secret Sharing Scheme

In a threshold secret sharing scheme a special participant, known as the dealer, distributes shares of his secret value, s, amongst n participants, in such a way that any t of these participants can reconstruct s. Whilst $t - 1$ or fewer participants cannot compute any information relating to s. Secret sharing is a fundamental building block of many distributed protocols. The specific secret sharing scheme used in this article is Shamir's secret sharing scheme [23] which is briefly explained below.

Denote the n participants as $P_1, \cdots, P_n$, the dealer as $\mathcal{D}$ and let all calculations take place in the finite field $\mathbb{F}$ where $|\mathbb{F}| = q$ such that $q > n$ is a prime number. The scheme consists of two phases, the sharing phase and the reconstruction phase.

Sharing Phase

1. $\mathcal{D}$ constructs a random polynomial, $g(x)$, of degree at most $t - 1$, such that $g(0) = s$.
2. Each participant, P_i, is privately assigned the share $V_i = g(i)$.

Reconstruction Phase

1. A set of t or more participants perform Lagrange interpolation over their shares to compute $g(x)$.
2. The participants take $g(0)$ as the secret.

3.2   The Proposed DOPE Protocol

The underlying idea behind our protocol is similar to the protocol given by Li et al. [17], in that we have S utilise Shamir's secret sharing scheme to distribute shares of the coefficients of $f(x)$ to each server.

To achieve privacy for R we have S distribute some semi-random information along with the shares of the coefficients. Each server receives shares of this information whilst R receives the information in its entirety. Using the distributed information R can then easily distribute his value α among the servers, who then perform a computation and send the output back to R. Following this, R computes a polynomial of which the free term is his desired evaluation.

The actual method utilised to distribute shares of α was originally given in [11] as a means to securely introduce input values under a shared MAC key in multi-party computation. We specifically use it to allow the contacted servers to efficiently compute a share of α multiplied by a given coefficient of $f(x)$. The full OPE protocol is given in Fig. 1.

In Sect. 2 we stated the result of Blundo et al. [4] which proved that a strong DOT protocol can only be achieved in two rounds. The same is true for our DOPE protocol, we merely circumvented this limitation by allowing S to contact R in the initialisation phase. Specifically, in our protocol we have S directly send the values $r_1, \cdots, r_k$ to R in the initialisation phase. This is actually not strictly necessary, and to limit direct contact between S and R we could instead have S distribute shares of $r_1, \cdots, r_k$ to each server. At the start of the evaluation phase a set of t or more servers would then send R their shares of these values. This results in a two round protocol in which R only has to be present during the evaluation phase. This is, of course, the exact same approach taken by Blundo et al. [4] for their strong DOT protocol.

In fact, due to the similarity of the models our DOPE protocol can easily be converted to a strong $\binom{1}{m}$ DOT protocol. In a $\binom{1}{m}$ DOT protocol the receiver wishes to learn 1 of m secrets held by S. If we define S's secrets as $\omega_1, \cdots, \omega_m$ then we can achieve DOT by having S compute $f(x)$ so that $f(i) = \omega_i$ for $i = 1, \cdots, m$. In this case the degree of the polynomial is then $k = m - 1$. To learn the ith secret R sets $\alpha = i$ and then executes the rest of the protocol as before.

Input: S has the polynomial $f(x) = a_0 + a_1 x + \cdots + a_k x^k$ and R the value α.
Output: R receives $f(\alpha)$ and S gets nothing.

Initialisation

1. S creates a set of random values $r_1, \cdots, r_k$ and computes k values of the form $\gamma_i = r_i \cdot a_i$ for $i = 1, \cdots, k$.
2. For each coefficient, a_h ($h = 0, \cdots, k$), S computes a random polynomial, $A_h(x)$ of degree at most $t - 1$ such that $A_h(0) = a_h$. He does the same for each γ_i value, computing k polynomials of the form $\Gamma_i(x)$ with free term $\Gamma_i(0) = \gamma_i$.
3. Using Shamir's secret sharing scheme S distributes these values among the servers, giving server s_j ($j = 1, \cdots, n$) the following information:
   - k shares of the form $\gamma_{i_j} = \Gamma_i(j)$
   - $k + 1$ shares of the form $a_{h_j} = A_h(j)$
4. S privately sends to R the values $r_1, \cdots, r_k$ and then takes no further part in the protocol.

Evaluation

1. R broadcasts to all servers a set of k values of the form $\epsilon_i = \alpha^i - r_i$.
2. A set of t or more servers, denoted as $\mathcal{W}$ respond to R's broadcast values. Each server, $s_j \in \mathcal{W}$, computes and sends to R the share:

$$z_j = a_{0_j} + \sum_{i=1}^{k} (a_{i_j} \cdot \epsilon_i + \gamma_{i_j})$$

3. R performs Lagrange interpolation across each z_j value to compute the polynomial $Z(x)$ with free term $Z(0) = f(\alpha)$.

Fig. 1. The proposed DOPE protocol

Where our protocol differs from many DOT protocols [5] however, is that our proposed DOPE protocol allows the receiver to contact more than the threshold amount of t servers. In fact, in our protocol R actually contacts all n servers, and we require t or more servers to respond to R. The specific servers that do respond to R can be chosen in any arbitrary fashion, as long as there are t or more of them. This allows for a fairly robust protocol, in that the protocol can tolerate up to $n - t$ servers not responding to R.

3.3   Evaluation

In this section we evaluate the security of the proposed DOPE protocol by proving that it meets the conditions given in Sect. 2.

Correctness

Theorem 1. *If all participants follow the protocol correctly the receiver obtains* $f(\alpha)$ *by contacting* t *or more servers.*

Proof. At the end of the evaluation phase R will have received t or more (up to n) shares of the form:

$$z_j = a_{0_j} + \sum_{i=1}^{k}(a_{i_j} \cdot \epsilon_i + \gamma_{i_j})$$

Where the share z_j is from server s_j. Due to the homomorphic nature of Shamir's secret sharing scheme linear operations performed on shares also correspond to the secrets and polynomials these shares are computed from [3]. In other words the shares correspond to the polynomial:

$$Z(x) = A_0(x) + \sum_{i=1}^{k}(A_i(x) \cdot \epsilon_i + \Gamma_i(x))$$

The free term of each $A_i(x)$ is $A_i(0) = a_i$, similarly $\Gamma_i(0) = r_i \cdot a_i$, therefore:

$$Z(0) = a_0 + \sum_{i=1}^{k}(a_i \cdot \epsilon_i + r_i \cdot a_i)$$

Since $\epsilon_i = \alpha^i - r_i$ this becomes:

$$Z(0) = a_0 + \sum_{i=1}^{k}(a_i \cdot \alpha^i - a_i \cdot r_i + r_i \cdot a_i)$$

$$= a_0 + \sum_{i=1}^{k} a_i \cdot \alpha^i$$

$$= a_0 + a_1 \cdot \alpha + a_2 \cdot \alpha^2 + \cdots + a_k \cdot \alpha^k$$

$$= f(\alpha)$$

Receiver's Privacy

Theorem 2. *A coalition of* $t-1$ *servers cannot compute any information relating to* α.

Proof. Suppose that a set of $t - 1$ servers, who were all contacted by R, form a coalition. The goal of this coalition is to compute some information relating to α. Collectively the servers have a set of $t - 1$ shares relating to each coefficient of $f(x)$, (i.e. $a_0, \cdots, a_k$) as well as $t - 1$ shares relating to the product of each random value and a coefficient, i.e. $\gamma_i = a_i \cdot r_i$ for $i = 1, \cdots, k$. Additionally, the servers also have k values of the form $\epsilon_i = \alpha^i - r_i$ which gives the following system of equations:

$$\epsilon_1 = \alpha - r_1$$
$$\epsilon_2 = \alpha^2 - r_2$$
$$\vdots$$
$$\epsilon_k = \alpha^k - r_k$$

From the above system we can see that to compute α the coalition would first need to compute a given r_i value. However, due to the perfectly secure nature of Shamir's secret sharing scheme [6,23], $t - 1$ shares does not reveal any information relating to a given secret. As a result of this, the coalition of servers cannot compute any information relating to any of the coefficients of $f(x)$, the γ_i or the r_i values. Since each r_i value is chosen at random, and they cannot compute any information relating to these values the above system is composed of k independent equations and $k + 1$ unknowns (each r_i value in addition to α) which results in every possible value of α being equally likely.

Sender's Privacy

Theorem 3. *A coalition composed of $t - 1$ servers and R cannot compute any information relating to $f(x)$ during initialisation.*

Proof. At the end of the initialisation phase a coalition of $t - 1$ servers and R will have the following information:

1. The values $r_1, \cdots, r_k$.
2. $t - 1$ shares corresponding to each coefficient polynomial $(A_0(x), \cdots, A_k(x))$, which gives $(k + 1)(t - 1)$ shares.
3. $t - 1$ shares relating to the each of other set of polynomials $(\Gamma_1(x), \cdots, \Gamma_k(x))$, giving $k(t - 1)$ collective shares.

As per the proof of Theorem 1 it is impossible to compute any information about a given polynomial, of degree $t - 1$, with only $t - 1$ shares. However, the free term of each polynomial of the form $\Gamma_i(x)$ for $1 = 1 \cdots k$ is $\Gamma_i(0) = r_i a_i$ where r_i is a known quantity. The coalition can use this knowledge to compute a polynomial with free term a_i. This allows them to hold two polynomials with the free term a_i.

We note that even with this extra knowledge they cannot achieve anything as a_i is unknown to them and furthermore, holding two sets of $t - 1$ shares relating to two different polynomials with the same free term does not actually reveal any information [16,23].

Sender's Privacy After Protocol Execution

Theorem 4. *A coalition composed of $t-1$ servers and R cannot compute any information relating to $f(x)$ after the execution of the protocol, other than what the evaluation of R's chosen value, $f(\alpha)$, gives them.*

Proof. The proof of this is analogous to the previous proof with the addition of some extra information, namely the information given to R by the other servers who contacted him. For the sake of the proof we will assume the worst, i.e. that all n servers contact R. Without loss of generality and for the sake of convenience, assume that the coalition is composed of R and the first $t-1$ servers, $s_1, \cdots, s_{t-1}$. This coalition has the exact same information as before, this time however, they also have the added knowledge of the other $n-t$ server's responses to R. That is:

$$z_t = a_{0_t} + \sum_{i=1}^{k}(a_{i_t} \cdot \epsilon_i + \gamma_{i_t})$$

$$z_{t+1} = a_{0_{t+1}} + \sum_{i=1}^{k}(a_{i_{t+1}} \cdot \epsilon_i + \gamma_{i_{t+1}})$$

$$\vdots$$

$$z_n = a_{0_n} + \sum_{i=1}^{k}(a_{i_n} \cdot \epsilon_i + \gamma_{i_n})$$

If the coalition are able to compute any of the polynomials used to distribute the coefficients of the senders polynomial, $A_0(x), \cdots, A_k(x)$, or even the polynomials used to distribute the product of the random values and the coefficients, $\Gamma_1(x), \cdots, \Gamma_k(x)$, then they can easily compute the value of a given coefficient of $f(x)$. We must therefore prove that this is not possible.

First, let $h = 0, \cdots, k$ and let $i = 1, \cdots, k$ then any given server, s_j, contacted by R has $k+1$ shares of the form a_{h_j} corresponding to $A_h(x)$ and k shares of the form γ_{i_j} corresponding to $\Gamma_h(x)$. We can write these polynomials as:

$$A_h(x) = a_h + A_{h_1}x + A_{h_2}2x^2 + \cdots + A_{h_{t-1}}x^{t-1}$$
$$\Gamma_i(x) = r_i a_i + G_{i_1}x + G_{i_2}x^2 + \cdots + G_{i_{t-1}}x^{t-1}$$

Using this notation the response of each server, z_j for $j = 1, \cdots n$, can be written as:

$$z_j = \sum_{y=1}^{k}a_y\alpha^y + \sum_{h=0}^{k}\left(\epsilon_h\left(\sum_{v=1}^{t-1}A_{h_v}j^v\right)\right) + \sum_{i=1}^{k}\left(\sum_{v=1}^{t-1}G_{i_v}j^v\right)$$

Therefore, from n responses R obtains a system composed of n equations and $t(k+1) + k(t-1)$ unknowns, specifically, t unknowns from each of the $k+1$ polynomials of the form $A_h(x)$ and $t-1$ unknowns from each of the k amount of polynomials of the form $\Gamma_i(x)$.

However, we note that each z_j is composed of a linear combination of polynomials of degree $t - 1$. Therefore, the system that R constructs is only composed of, at most, t independent equations. We note that $t \geq 2$ and $k \geq 1$, meaning that the amount of unknowns will always be greater than the amount of independent equations. As a result of this, R and the coalition of $t - 1$ servers cannot compute anything from just the responses.

In fact, even with the direct shares of each of the $t - 1$ servers in the coalition they still cannot compute any information. This is because the equation used to describe a given share is not linearly independent to the equation used for a given z_j i.e. each z_j is simply a linear combination of a given participant's share and thus, is not a separate (independent) equation.

The net result for the coalition is a system composed of only t independent equations and $t(k + 1) + k(t - 1)$ unknowns, resulting in each value of a given coefficient of $f(x)$ being equally likely.

References

1. Beimel, A., Chee, Y.M., Wang, H., Zhang, L.F.: Communication-efficient distributed oblivious transfer. J. Comput. Syst. Sci. **78**(4), 1142–1157 (2012)
2. Ben-Or, M., Goldwasser, S., Wigderson, A.: Completeness theorems for non-cryptographic fault-tolerant distributed computation. In: Proceedings of the Twentieth Annual ACM Symposium on Theory of Computing, STOC 1988. ACM, New York (1988)
3. Benaloh, J.C.: Secret sharing homomorphisms: keeping shares of a secret secret (extended abstract). In: Odlyzko, A.M. (ed.) CRYPTO 1986. LNCS, vol. 263, pp. 251–260. Springer, Heidelberg (1987). https://doi.org/10.1007/3-540-47721-7_19
4. Blundo, C., D'Arco, P., De Santis, A., Stinson, D.: On unconditionally secure distributed oblivious transfer. J. Cryptol. **20**(3), 323–373 (2007)
5. Blundo, C., D'Arco, P., De Santis, A., Stinson, D.R.: New results on unconditionally secure distributed oblivious transfer. In: Nyberg, K., Heys, H. (eds.) SAC 2002. LNCS, vol. 2595, pp. 291–309. Springer, Heidelberg (2003). https://doi.org/10.1007/3-540-36492-7_19
6. Corniaux, C.L.F., Ghodosi, H.: An entropy-based demonstration of the security of Shamir's secret sharing scheme. In: 2014 International Conference on Information Science, Electronics and Electrical Engineering, vol. 1, pp. 46–48, April 2014
7. Chang, Y.-C., Lu, C.-J.: Oblivious polynomial evaluation and oblivious neural learning. In: Boyd, C. (ed.) ASIACRYPT 2001. LNCS, vol. 2248, pp. 369–384. Springer, Heidelberg (2001). https://doi.org/10.1007/3-540-45682-1_22
8. Cheong, K.Y., Koshiba, T., Nishiyama, S.: Strengthening the security of distributed oblivious transfer. In: Boyd, C., González Nieto, J. (eds.) ACISP 2009. LNCS, vol. 5594, pp. 377–388. Springer, Heidelberg (2009). https://doi.org/10.1007/978-3-642-02620-1_26
9. Corniaux, C.L.F., Ghodosi, H.: A verifiable distributed oblivious transfer protocol. In: Parampalli, U., Hawkes, P. (eds.) ACISP 2011. LNCS, vol. 6812, pp. 444–450. Springer, Heidelberg (2011). https://doi.org/10.1007/978-3-642-22497-3_33
10. Corniaux, C.L.F., Ghodosi, H.: An information-theoretically secure threshold distributed oblivious transfer protocol. In: Kwon, T., Lee, M.-K., Kwon, D. (eds.) ICISC 2012. LNCS, vol. 7839, pp. 184–201. Springer, Heidelberg (2013). https://doi.org/10.1007/978-3-642-37682-5_14

11. Damgård, I., Pastro, V., Smart, N., Zakarias, S.: Multiparty computation from somewhat homomorphic encryption. In: Safavi-Naini, R., Canetti, R. (eds.) CRYPTO 2012. LNCS, vol. 7417, pp. 643–662. Springer, Heidelberg (2012). https://doi.org/10.1007/978-3-642-32009-5_38

12. Döttling, N., Ghosh, S., Nielsen, J.B., Nilges, T., Trifiletti, R.: TinyOLE: Efficient actively secure two-party computation from oblivious linear function evaluation. In: Proceedings of the 2017 ACM SIGSAC Conference on Computer and Communications Security, CCS 2017, pp. 2263–2276. ACM, New York (2017)

13. Ghosh, S., Nielsen, J.B., Nilges, T.: Maliciously secure oblivious linear function evaluation with constant overhead. In: Takagi, T., Peyrin, T. (eds.) ASIACRYPT 2017. LNCS, vol. 10624, pp. 629–659. Springer, Cham (2017). https://doi.org/10.1007/978-3-319-70694-8_22

14. Hanaoka, G., Imai, H., Mueller-Quade, J., Nascimento, A.C.A., Otsuka, A., Winter, A.: Information theoretically secure oblivious polynomial evaluation: model, bounds, and constructions. In: Wang, H., Pieprzyk, J., Varadharajan, V. (eds.) ACISP 2004. LNCS, vol. 3108, pp. 62–73. Springer, Heidelberg (2004). https://doi.org/10.1007/978-3-540-27800-9_6

15. Hazay, C.: Oblivious polynomial evaluation and secure set-intersection from algebraic PRFs. In: Dodis, Y., Nielsen, J.B. (eds.) TCC 2015. LNCS, vol. 9015, pp. 90–120. Springer, Heidelberg (2015). https://doi.org/10.1007/978-3-662-46497-7_4

16. Ito, M., Saito, A., Nishizeki, T.: Secret sharing scheme realizing general access structure. Electron. Commun. Jpn. (Part III: Fundam. Electron. Sci.) 72(9), 56–64 (1989)

17. Li, H.D., Yang, X., Feng, D.G., Li, B.: Distributed oblivious function evaluation and its applications. J. Comput. Sci. Technol. 19(6), 942–947 (2004)

18. Lindell, Y., Pinkas, B.: Privacy preserving data mining. In: Bellare, M. (ed.) CRYPTO 2000. LNCS, vol. 1880, pp. 36–54. Springer, Heidelberg (2000). https://doi.org/10.1007/3-540-44598-6_3

19. Naor, M., Pinkas, B.: Oblivious polynomial evaluation. SIAM J. Comput. 35(5), 1254–1281 (2006)

20. Naor, M., Pinkas, B.: Oblivious transfer and polynomial evaluation. In: Proceedings of the Thirty-first Annual ACM Symposium on Theory of Computing, STOC 1999, pp. 245–254. ACM, New York (1999)

21. Naor, M., Pinkas, B.: Distributed oblivious transfer. In: Okamoto, T. (ed.) ASIACRYPT 2000. LNCS, vol. 1976, pp. 205–219. Springer, Heidelberg (2000). https://doi.org/10.1007/3-540-44448-3_16

22. Nikov, V., Nikova, S., Preneel, B., Vandewalle, J.: On unconditionally secure distributed oblivious transfer. In: Menezes, A., Sarkar, P. (eds.) INDOCRYPT 2002. LNCS, vol. 2551, pp. 395–408. Springer, Heidelberg (2002). https://doi.org/10.1007/3-540-36231-2_31

23. Shamir, A.: How to share a secret. Commun. ACM 22(11), 612–613 (1979)

24. Tonicelli, R., et al.: Information-theoretically secure oblivious polynomial evaluation in the commodity-based model. Int. J. Inf. Secur. 14(1), 73–84 (2015)

25. Zhu, H., Bao, F.: Augmented oblivious polynomial evaluation protocol and its applications. In: di Vimercati, S.C., Syverson, P., Gollmann, D. (eds.) ESORICS 2005. LNCS, vol. 3679, pp. 222–230. Springer, Heidelberg (2005). https://doi.org/10.1007/11555827_13

An Efficient Private Evaluation
of a Decision Graph

Hiroki Sudo[1,2] (ID), Koji Nuida[3] (ID), and Kana Shimizu[1(✉)] (ID)

[1] Waseda University, Tokyo, Japan
hsudo108@ruri.waseda.jp, shimizu.kana@waseda.jp
[2] AIST-Waseda University CBBD-OIL, Tokyo, Japan
[3] The University of Tokyo, Tokyo, Japan
nuida@mist.i.u-tokyo.ac.jp

Abstract. A decision graph is a well-studied classifier and has been used to solve many real-world problems. We assumed a typical scenario between two parties in this study, in which one holds a decision graph and the other wants to know the class label of his/her query without disclosing the graph and query to the other. We propose a novel protocol for this scenario that can obliviously evaluate a graph that is designed by an efficient data structure called the graph level order unary degree sequence (GLOUDS). The time and communication complexities of this protocol are linear to the number of nodes in the graph and do not include any exponential factors. The experiment results revealed that the actual runtime and communication size were well concordant with theoretical complexities. Our method can process a graph with approximately 500 nodes in only 11 s on a standard laptop computer. We also compared the runtime of our method with that of previous methods and confirmed that it was one order of magnitude faster than the previous methods.

Keywords: Decision graph · Homomorphic encryption · GLOUDS

1 Introduction

Classification is a central topic in machine learning (ML), which is aimed at training a classifier on a set of labeled samples so that the trained classifier can correctly assign one of the class labels to an input query, and has been successfully applied to various real-world problems such as credit scoring, drug discovery, and disease diagnostics [17, 20, 22]. One of the typical online services using ML is a classification service where a service provider has a trained classifier and a user can obtain classification results for his/her data. In fact, software platforms that easily achieve such scenario are already available [1–3], which enables service providers to publish the application programming interfaces (APIs) of the trained classifiers on the cloud server. Although both service providers and service users can benefit from such classification services, there are certain *privacy* concerns about the data. A natural scenario for an online classification service is

© Springer Nature Switzerland AG 2019
K. Lee (Ed.): ICISC 2018, LNCS 11396, pp. 143–160, 2019.
https://doi.org/10.1007/978-3-030-12146-4_10

for a user to send his/her query (an input to a classifier) to a server and the server to return the classification results based on the classifier. Suppose the online service involves disease diagnostics, where the input to the classifier includes the user's private information such as health records and genetic information. The server's classifier also includes data on donors' private information because the classifier was trained on private data. Various model-inversion attacks are possible [15,16] in this scenario; they can infer sensitive information being used for training by accessing the trained classifier. Therefore, it is necessary to conceal both the user's query and the server's classifier.

We focused on a decision graph (DG) as a classifier and tackled the problem of private evaluations of the decision graphs. A decision graph is an efficient data structure for the classification rules. It is also described as a decision diagram [7] in the logic synthesis literature and as a branching program (BP) [23] in computer science theory. Compared to complex models such as neural networks, the decision graph is easier to interpret and is therefore often preferred for problems like clinical diagnosis where the interpretation of decision-making is important. We assumed the underlying graph was a binary graph and defined a binary decision graph (BDG) as follows. BDG is a rooted directed acyclic graph (DAG) that consists of a set of nodes of in-degree ≥ 1 and the out-degree of two or zero. A node with the out-degree of zero is called a leaf and has a class label. Each internal node contains a split function that decides whether a query that reaches the node should visit a node connected to the right edge or the node connected to the left edge, depending on the corresponding attribute of the query. We assumed in our study that each split function computed whether or not the input was greater than a threshold t.

The problem setup for this study was as follows: one party (a server) has a BDG and the other party (a user) wants to obtain a classification result. The user's input is a private attribute vector, $x = (x_0, \ldots, x_{n-1})$. The length of the vector and the ordering of the attributes are common information between the server and the user. The user only knows common information and the height of the graph (maximum path length from a root node to a leaf). After computation, the user learns the classification result (a class label); he/she does not learn anything more than what he/she already knows.

1.1   Related Works

Many studies have addressed the problem of private evaluation of classifiers [13, 21,32]. Brickell et al. [6] and Barni et al. [4] respectively proposed methods which combine Yao's garbled circuit [36] and additively homomorphic encryption (AHE) for private evaluation of the BDG. We will present a detailed comparison of our method with those approaches in Sect. 4.2. Mohassel et al. [25] have also proposed a method of private evaluation of BDG; however, they assumed that a user knew all the outputs of all split functions of a server's internal nodes, which differs from our scenario.

Since the decision graph is regarded as a generalized form of a decision tree (DT), we also describe a series of studies for a private evaluation of DTs.

Bost et al. [5] proposed a secure decision tree evaluation protocol as part of their work. Their method evaluated a decision tree as a polynomial of Boolean variables using leveled fully homomorphic encryption. Although this method improved efficiency compared to other conventional methods, it still suffered from the problem of computation and communication costs. A recent work by Wu et al. [35] achieved more practical computational time and communication size. Their method was only based on AHE and performed efficiently for shallow trees; however, it did not perform well for the evaluation of deep trees because of its exponential time and communication complexity for the height of the trees. Cock et al. [10] proposed a protocol that achieved time complexity that was similar to Wu et al.'s algorithm and improved runtime by using arithmetic sharing to avoid heavy modular exponentiation. However, their protocol assumed a different problem setup where a trusted initializer participated in the protocol to generate multiplication triplets. The trusted initializer could be removed, but the additional costs of generating the multiplication triplets by the two parties was exponential to the height of the tree, which greatly deteriorated the runtime. Tai et al. [30] formulated a decision tree evaluation as a compact linear function to attain a protocol in which time complexity was only dependent on the number of internal nodes and was independent of the exponential of the tree height.

Protocols for DTs can theoretically be applied to private evaluations of BDG if the underlying graph is transformed into a tree. However, the number of nodes in the tree, that is equivalent to the graph, becomes very large. As we will discuss in Subsect. 2.4, two in-coming edges to an internal node cause a copy of all the subordinates of the node on transformation, which leads to the exponential growth in total tree size.

We also noted that a BDG achieved accurate predictions while it achieved lower model complexity than DT [19,26,27,29], and it even achieved considerably improved generalization [29]. A BDG with 3,000 nodes achieved the same accuracy as a DT with 22,000 nodes in the classification of a Kinect dataset in a study by Shotton et al. [29].

1.2   Our Contribution

The five main contributions of this paper are summarized below:

- We propose an efficient protocol for the oblivious evaluation of a BDG. More precisely, the protocol allows two parties, one holding a BDG T, and the other holding an attribute vector, x, to determine the class label of x, without revealing T and x to the other party in a semi-honest setting.
- The time and communication complexities of our protocol are linear to the number of nodes and the height of T and exclude any exponential factors.
- The DAG of the BDG in our protocol is represented by a look-up vector V, and the other party obliviously refers to V. We demonstrate how the length of V is reduced by using a succinct data structure called GLOUDS to achieve linear complexity.

- An oblivious evaluation of a split function in each internal node is conducted before graph traversal. We propose a novel protocol called eROT that enables the correct edges to obliviously be chosen during traversal.
- We implemented our protocol and tested it on BDGs of various sizes; we found that its actual runtime and communication size were concordant with the theoretical complexities. We also compared the runtime and communication size of our protocol to those in previous studies [4,6] to confirm that our protocol was an order of magnitude faster.

The rest of the paper is organized as follows. Section 2 describes important building blocks for the proposed method and the security model that was assumed for this study. We detail our method in Sect. 3 and evaluate it on various datasets in Sect. 4. Section 5 concludes the paper.

2   Preliminary

2.1   Notation

We denote vector v as $(v_0, \ldots, v_{n-1})$ and the i-th element of v as $v[i]$. The $\{a_0, \ldots, a_{n-1}\}$ represents a set of size n. The $\{a_i\}_{i=0}^{n-1}$ stands for $\{a_0, \ldots, a_{n-1}\}$. We define the "rotation" of a vector as: given n dimensional vector v, the r-rotation of v results in vector $\hat{v}$, each of whose elements is $\hat{v}[(i+r)_{\mathsf{mod}\ n}] = v[i]$. The $\langle P(x) \rangle$ returns 1 if predicate $P(x)$ is true given x, otherwise 0. The notation, $r \in_R A$, means r is a uniformly chosen random value from a set A. We define λ-bit unary representation of $x \in \{0, \ldots, \lambda - 1\}$ as a λ-bit vector that has 1 at x-th least significant bit and has 0 at the other bits, and denote it as $\mathsf{UNARY}_\lambda(x)$.

2.2   Additively Homomorphic Encryption

We used a semantically secure additively homomorphic public-key encryption scheme in our protocol and especially assumed a lifted-ElGamal cryptosystem [11] with plaintext space $\mathbb{Z}_p$ whose message in a ciphertext is located in the exponent. The public-key encryption scheme is equipped with three algorithms:

1. KeyGen: outputs a public/private key pair (pk, sk).
2. $\mathsf{Enc_{pk}}(m)$: outputs a ciphertext $[\![m]\!]$, by encrypting a plaintext m, with pk.
3. $\mathsf{Dec_{sk}}([\![m]\!])$: outputs a plaintext m, by decrypting a ciphertext $[\![m]\!]$, with sk.

$[\![m]\!]$ represents a ciphertext of a plaintext m. Likewise, $[\![v]\!]$ represents a ciphertext vector, each of whose elements is an encryption of each element of a vector v. A public key of AHE has $\oplus$, $\otimes$, $\ominus$ operations on ciphertexts described as follows. Given two plaintexts m_1, m_2, we can compute $[\![m_1 + m_2]\!] = [\![m_1]\!] \oplus [\![m_2]\!]$ by using $\oplus$ operation. We can also compute multiplication by a constant k ($[\![k \cdot m]\!] = k \otimes [\![m]\!]$). Negation on a ciphertext is represented by $\ominus [\![m]\!]$. In our setting, the user generates and holds a public/private key pair (pk, sk), and the server only receives a public key pk so that only the user can decrypt ciphertexts and the server can only conduct encryption and additively homomorphic computation.

2.3   Oblivious Transfer

Oblivious transfer (OT) is a secure two-party protocol between a sender and a chooser. A chooser in 1-out-of-N OT specifies an index $i \in \{0, \ldots, N-1\}$, and only obtains the i-th element of the sender's vector $\boldsymbol{v}$, without disclosing i to the sender. We denote the execution of OT with an index i, and a vector $\boldsymbol{v}$ by $OT_1^N(i, \boldsymbol{v})$. While there are several efficient implementations that achieve OT_1^N functionality, we use simple protocols based on additively homomorphic operation which require $O(N)$ computational cost and communication size. For the space limitation, we omit the implementation detail.

2.4   GLOUDS

The graph level order unary degree sequence (GLOUDS) [14] is the succinct data structure of a DAG, which is a query-time efficient data structure that uses the space close to information-theoretic lower bound. GLOUDS regards a DAG as an integration of a spanning tree and "non-tree" edges that are not included in the tree, and introduces the idea of "shadow nodes", which are duplicates of non-tree nodes (nodes with incoming edges > 1) to virtually treat a graph as a tree, while it avoids unnecessary copies of nodes. When we transform a DAG into an equivalent tree, it is necessary to repeat copying of a subtree rooted from a non-tree node for all the incoming edges to the node, which causes exponential growth in total tree size. Since GLOUDS generates as many shadow nodes as the number of non-tree edges, it is considered to be efficient when there are not too many non-tree edges.

More precisely, GLOUDS consists of a trit $(0, 1, 2)$ sequence B of length N and an auxiliary vector H, where N is the sum of the number of nodes and the number of edges $+ 1$. The nodes in the DAG are numbered in level order (from top to bottom and left to right) and the root is numbered 0. The nodes are visited in level order during construction of GLOUDS. When each node is visited, 0 is stored in B, and all the children of the node are stored in left-to-right order. If a child is already observed, a trit 2 is stored in B, and 1 otherwise. The root node is considered as a child of an unshown supernode, and hence 1 is stored in B as the first element (i.e., $B[0] = 1$). H memorizes numbers of shadow nodes in the order in which they appear in B as 2. For the case of the DAG in Fig. 1, after storing $B[0] = 1$, $B[1] = 0$ is stored when the node "0" is visited. Since the node "1" and "2" are the children of the node "0" and they are not observed, $B[2] = 1$ and $B[3] = 1$ are stored. Similarly, $B[4] = 0$, $B[5] = 1$ and $B[6] = 1$ are stored for the visit of the node "1", and $B[7] = 0$, $B[8] = 2$ and $B[9] = 1$ are stored for the visit of the node "2". Note that $B[8] = 2$ because the node "4", which is the left child of the node "2", is already observed. 4 is recorded in $H[0] = 4$. After visiting all the nodes, B and H are described as $B = 10110110210110210000$ and $H = [4, 7]$. Either 1 or 2 in vector B corresponds to any one of the nodes in the DAG, and 0 is regarded as a delimiter between groups of siblings. Note that 0 is also considered as a parent node of a right-neighbour group of siblings; therefore, the same node appears more than once in B.

Here, we define two operations on sequence B as:

Definition 1. *Operations on trit sequence B*
$\mathsf{Rank}_c(B, p)$: *returns the number of* $c \in \{0, 1, 2\}$ *in the prefix* $B[0, p)$ $(0 \le p < N)$
$\mathsf{Select}_c(B, i)$: *returns the position of the i-th* $c \in \{0, 1, 2\}$ *in B (i starts from 0.)*

One can move from a position p in B that stores a trit 1 or 2 (a parent node), to another position p' (x-th child of the node) by carrying out the equation below.

$$p' = \begin{cases} \mathsf{Select}_0(B, \mathsf{Rank}_1(B, p)) + x & (if\ B[p] = 1) \\ \mathsf{Select}_0(B, H[\mathsf{Rank}_2(B, p)]) + x & (if\ B[p] = 2) \end{cases} \tag{1}$$

For simplicity, we let $\mathsf{SelRan}(B, p)$ be the first term of Eq. 1. The $\mathsf{SelRan}(B, p)$ computes a position in B of the left delimiter of $B[p]$'s children. Since the siblings are stored sequentially, one can specify the x-th child by adding an offset x to $\mathsf{SelRan}(B, p)$. Figure 1 has an example of SelRan. For example, let us consider the case of $p = 2$. $B[2]$ corresponds to the node "1". The children of the node "1" are "3" and "4", and they correspond to $B[5]$ and $B[6]$. $\mathsf{SelRan}(B, 2)$ returns 4 and $B[4] = 0$ is the left delimiter of them. We define a map of a position in B and a node id, such that $\mathsf{ID}(p)$ returns id of the node that corresponds to $B[p]$. For example, $\mathsf{ID}(2) = 1$ and $\mathsf{ID}(4) = 1$. Note that $\mathsf{ID}(p) = \mathsf{ID}(\mathsf{SelRan}(B, p))$. GLOUDS can be regarded as a generalization of the level order unary degree sequence (LOUDS) [18], which is a succinct data structure for ordered trees; hence, our protocol can be immediately applied to DT.

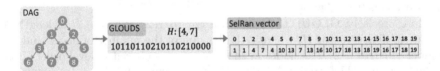

Fig. 1. Example of a BDG, corresponding GLOUDS and SelRan vector.

2.5   BDG and Efficient Design Principle of a Look-Up Vector by GLOUDS

BDG consists of a rooted DAG and a set of split functions $\{\mathsf{Split}_i\}_{i=0}^{m-1}$, where m is the number of internal nodes. Given an attribute $\mathcal{X} \in \mathbb{Z}$, a split function that is assigned to an internal node performs a greater than operation: $\mathsf{Split}_i(\mathcal{X}) = \langle t_i < \mathcal{X} \rangle$ to choose either a right or left child. BDG in our protocol is mainly represented by a look-up vector $\boldsymbol{v}$ of length N (also referred to as the SelRan vector), and a vector of ciphertexts $[\![\boldsymbol{o}]\!]$ that encrypts an offset vector $\boldsymbol{o}$ of length N. $\boldsymbol{v}$ represents the DAG, and $\boldsymbol{o}$ represents outputs of all the split functions taking a query (a set of attributes). Both $\boldsymbol{v}$ and $[\![\boldsymbol{o}]\!]$ are held by the party holding the BDG, and the other party traverses the BDG by obliviously referring to those vectors. $\boldsymbol{v}$ and $\boldsymbol{o}$ are described as:

$$v[i] = \begin{cases} \mathsf{SelRan}(B, i) & (if \; B[i] \neq 0) \\ i & (else) \end{cases} \text{ and } o[i] = \begin{cases} \langle \boldsymbol{\theta}[i] < \mathcal{X}_i \rangle + 1 & (if \; \boldsymbol{\tau}[i] = \mathsf{I}) \\ 0 & (else) \end{cases},$$

where $\mathcal{X}_i$ is a user's attribute for the split function that is associated with node $\mathsf{ID}(i)$, $\boldsymbol{\tau}$ is a **type vector** storing the types of each position and $\boldsymbol{\theta}$ is a **threshold vector**. $\boldsymbol{\tau}[i] = \mathsf{L}(\text{eaf})$ if node $\mathsf{ID}(i)$ is a leaf, $\boldsymbol{\tau}[i] = \mathsf{Z}(\text{ero})$ if node $\mathsf{ID}(i)$ is not a leaf and $B[i] = 0$. $\boldsymbol{\tau}[i] = \mathsf{I}(\text{nternal})$ otherwise. $\boldsymbol{\theta}[i]$ is a threshold of a split function that is associated with node $\mathsf{ID}(i)$ when $\boldsymbol{\tau}[i] = \mathsf{I}$. $\boldsymbol{\theta}[i]$ is set to empty otherwise. $v[p]$ returns the position of the left delimiter of node $\mathsf{ID}(p)$'s children and $o[p]$ returns the choice of a child. Therefore, one can compute the position of next node in B by:

$$p' = v[p] + o[p].$$

Note that the outputs of split functions include both parties' privacy; hence, the two parties need to jointly compute $[\![o]\!]$ without revealing their private parameters. We will describe how this is accomplished in Subsect. 3.4. v and o allow *self-loop* at positions $\{i \mid B[i] = 0, 0 \leq i < N\}$ by setting $v[i] = i$ and $o[i] = 0$. If one reaches such position i and node $\mathsf{ID}(i)$ is a leaf, one can stay on the same leaf to conceal the path length from the root toward each leaf. The self-loop at a non-leaf node can avoid incorrect traversal. The party holding BDG also prepares **label vector** z. $z[i]$ is set to a class label associated with node $\mathsf{ID}(i)$ if $\boldsymbol{\tau}[i] = \mathsf{L}$ and $B[i] = 0$. Otherwise, $z[i]$ is a random value within the possible range of class labels. Figure 2 has an example of these data structures that represent a BDG. The nodes and edges that are colored in orange show an example of a traversal from the root node to the node 7 when $\langle t_0 < x_0 \rangle = 1$, $\langle t_2 < x_2 \rangle = 0$ and $\langle t_4 < x_4 \rangle = 0$. The corresponding elements in the table in Fig. 2 are also colored in orange. The traversal starts by referring to $v[0] = 1$ to know the next position is $v[0] + d_0 = 3$. Similarly, one can know the next position by $v[3] + d_2 = 8$, and visits the node 7 by $v[8] + d_4 = 14$. Finally, one reaches the position $v[14] + 0 = 18$ where a self-loop is allowed, and stays at the position while computing $v[18] + 0 = 18$.

It is possible to design more space-efficient SelRan vector and auxiliary vectors, and we used it for the experiments in Sect. 4, however, we do not describe how we designed such vectors due to the space limitation. See the forthcoming full version of this paper for the details.

3   Method

3.1   Problem Setting

We assumed a party $\mathcal{A}$ had a private attribute vector x, and a party $\mathcal{B}$ had a private BDG T, in our protocol. $\mathcal{A}$ and $\mathcal{B}$ are referred to as a user and a server in previous sections. Both x and T must be concealed from the other party. $\mathcal{A}$ and $\mathcal{B}$ participate in the two-party secure BDG evaluation protocol. $\mathcal{A}$ only obtains an output of BDG $T(x)$, while he/she gains no information about $\mathcal{B}$'s private information except for $T(x)$. $\mathcal{B}$ obtains nothing. We assumed

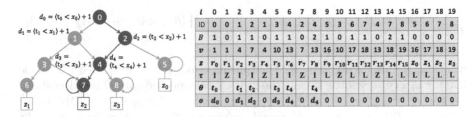

Fig. 2. Example of a BDG (left graph) and data structures for the BDG described in Subsect. 2.5 (right table). r_i is a random value. (Color figure online)

$\mathcal{A}$ and $\mathcal{B}$ shared three kinds of information: (1) length of the SelRan vector, (2) height of the BDG, and (3) length of the attribute vector. We considered a standard adversarial model in this work: a semi-honest model [8], in which even a corrupted party adheres to the specifications of a protocol.

3.2 Overview of Our Protocol

Our protocol is composed of following phases.

Comparison Phase (for Constructing Offset Vector): $\mathcal{B}$ eventually constructs and stores the encrypted offset vector without decrypting $\mathcal{A}$'s inputs in this phase. $\mathcal{A}$ and $\mathcal{B}$ securely calculate split functions associated with nodes to achieve this purpose. $\mathcal{B}$ stores all the decisions on which branch will be selected as ciphertexts. We used a secure comparison protocol to calculate split functions.

Evaluation Phase (for Computing Class Label on BDG): Two parties descend from the root to a leaf in the evaluation phase by recursively referring to the SelRan vector and $[\![o]\!]$ constructed in the comparison phase. After a leaf is reached, $\mathcal{A}$ retrieves a label associated with the leaf from z.

We will first describe several secure two party protocols that will be building blocks in Subsect. 3.3, and then explain how to construct the comparison phase and evaluation phase in detail in Subsect. 3.4. Our protocol can be seen as a sequential composition of the two protocols, Comparison Phase and Evaluation Phase. Therefore, security of our protocol is obvious if the underlying two protocols are secure.

3.3 Building Blocks

Comparison Protocol. A two-party secure comparison protocol that securely computes $\langle x < y \rangle$, is required to calculate split functions in Comparison phase. We used a variant [34] of the DGK comparison protocol [12] in our implementation, which is based on additively homomorphic encryption.

While problem settings of comparison protocols vary, we assumed the following setting: $\mathcal{A}$ and $\mathcal{B}$ had a plaintext input x and y. Only $\mathcal{B}$ acquired the encrypted comparison result $[\![\langle x < y \rangle]\!]$. Since we simply used the protocol and did not modify it, we will not go into details about the specification of the comparison protocol here. When x and y are ℓ bit integers, the time complexity and communication of both $\mathcal{A}$ and $\mathcal{B}$ are $O(\ell)$ in the DGK comparison protocol.

Recursive Oblivious Transfer. $\mathcal{A}$ recursively accesses $\mathcal{B}$'s SelRan vector $\boldsymbol{v}$, in the evaluation phase, i.e., $\mathcal{A}$ repeats querying an element of $\boldsymbol{v}$ and sets the next query depending on the query result. Not only queries but also intermediate results sent from $\mathcal{B}$ need to be hidden to protect private information for both parties. We used a known secure two-party protocol called recursive oblivious transfer (ROT) [28,31] for this problem.

Assuming $\mathcal{B}$ has a plaintext vector $\boldsymbol{v}$ of length N and $\mathcal{A}$ specifies a query p_0, ROT ensures that $\mathcal{A}$ obtains $\boldsymbol{v}[\boldsymbol{v}[\ldots \boldsymbol{v}[p_0]\ldots]]$ and $\mathcal{B}$ obtains nothing after an arbitrary number of iterations. ROT consists of σ steps, where σ is a common parameter between $\mathcal{A}$ and $\mathcal{B}$. Except for the initial and the last steps, rest of the steps repeat the same protocol. The initial step computes the next position starting from the initial position p_0 specified by $\mathcal{A}$. At the end of the initial step, $\mathcal{A}$ and $\mathcal{B}$ gain shares of the next position $p_1 = \boldsymbol{v}[p_0]$. The k-th step ($k = 1,\ldots,\sigma-2$) updates the position using the shares of the k-th position p_k. i.e., $\mathcal{A}$ and $\mathcal{B}$ gain shares of the next position $p_{k+1} = \boldsymbol{v}[p_k]$, p'_{k+1} and r_{k+1} where $r_{k+1} \in_R \mathbb{Z}$ is a random value. In the last step, the final value $\boldsymbol{v}[p_\sigma]$ is not divided into shares and only $\mathcal{A}$ knows the value.

For convenience, we denote $(p'_{k+1}, r_{k+1}) \leftarrow \mathsf{ROT}(p'_k, r_k, \boldsymbol{v})$ for the k-th step of ROT, which takes shares of a query p_k (i.e., p'_k, r_k), and a vector $\boldsymbol{v}$ as inputs, and outputs p'_{k+1} to $\mathcal{A}$ and r_{k+1} to $\mathcal{B}$. The initial step can also be represented by this notation by setting 0 to r_0. The time complexity and communication size of one round in ROT is $O(N)$ on both $\mathcal{A}$'s and $\mathcal{B}$'s sides due to the cost of OT.

eROT. The goal of eROT is recursive references to the offset vector $\boldsymbol{o}$ when it is encrypted. Specifying a query p_0, $\mathcal{A}$ obtains $\boldsymbol{o}[\boldsymbol{o}[\ldots \boldsymbol{o}[p_0]\ldots]]$ after an arbitrary number of iterations, and $\mathcal{B}$ obtains nothing. To achieve this goal, we assumed $\mathcal{B}$ had an $N \times \lambda$ ciphertext matrix $\boldsymbol{\Omega}$, instead of $[\![\boldsymbol{o}]\!]$. Each row $\boldsymbol{\Omega}[i]$ is meant to represent $\boldsymbol{o}[i]$. More concretely, $\boldsymbol{\Omega}[i]$ is a vector, each of whose elements is an encrypted bit of $\mathsf{UNARY}_\lambda(\boldsymbol{o}[i])$. For example, $\boldsymbol{o}[i] = 0$ is represented by $\boldsymbol{\Omega}[i][0] = [\![1]\!]$, $\boldsymbol{\Omega}[i][1] = [\![0]\!]$ and $\boldsymbol{\Omega}[i][2] = [\![0]\!]$ when $\lambda = 3$.

Initial (0-th) step:

1. $\mathcal{B}$ generates a random value $r_1 \in_R \mathbb{Z}$, and prepares a vector $\boldsymbol{u}'$ whose i-th element is masked by r_1: $\boldsymbol{u}'[i] = \bigoplus_{j=0}^{\lambda-1}(\boldsymbol{\Omega}[i][j] \otimes (j + r_1)_{\mathsf{mod}\ N})$ (namely, $\boldsymbol{u}[i] = [\![\boldsymbol{o}[i]]\!]$, $\boldsymbol{u}'[i] = [\![(\boldsymbol{o}[i] + r_1)_{\mathsf{mod}\ N}]\!]$.) $\mathcal{B}$ stores r_1.
2. $\mathcal{A}$ (chooser) and $\mathcal{B}$ (sender) engage in $OT_1^N(p_0, \boldsymbol{u}')$. $\mathcal{A}$ obtains $(p_1 + r_1)_{\mathsf{mod}\ N}$ decrypting $\boldsymbol{u}'[p_0] = [\![(p_1 + r_1)_{\mathsf{mod}\ N}]\!]$.

k-th ($k = 1, \ldots, \sigma - 2$) step:

$\mathcal{A}$ holds $p'_k = (p_k + r_k)_{\mathsf{mod}\ N}$ and $\mathcal{B}$ holds r_k.

1. $\mathcal{B}$ generates a random value $r_{k+1} \in_R \mathbb{Z}$. Then, $\mathcal{B}$ prepares a vector $\boldsymbol{u}'$ whose i-th element is masked by r_{k+1}: $\boldsymbol{u}'[i] = \bigoplus_{j=0}^{\lambda-1}(\boldsymbol{\Omega}[i][j] \otimes (j + r_{k+1})_{\mathsf{mod}\ N})$. $\mathcal{B}$ stores r_{k+1}.
2. $\mathcal{B}$ rotates $\boldsymbol{u}'$ by r_k elements to obtain $\hat{\boldsymbol{u}}'$.
3. $\mathcal{A}$ (chooser) and $\mathcal{B}$ (sender) engage in $OT_1^N(p'_k, \hat{\boldsymbol{u}}')$, and $\mathcal{A}$ obtains $(p_{k+1} + r_{k+1})_{\mathsf{mod}\ N}$ decrypting $\boldsymbol{u}'[p_k] = [\![(p_{k+1} + r_{k+1})_{\mathsf{mod}\ N}]\!]$.

Last step:

$\mathcal{B}$ does not mask $\boldsymbol{u}[i] (= \bigoplus_{j=0}^{\lambda-1}(\boldsymbol{\Omega}[i][j] \otimes j_{\mathsf{mod}\ N}))$ in the last step to send a true value to $\mathcal{A}$.

1. $\mathcal{B}$ rotates $\boldsymbol{u}$ by $r_{\sigma-1}$ elements to obtain $\hat{\boldsymbol{u}}$.
2. $\mathcal{A}$ (chooser) and $\mathcal{B}$ (sender) engage in $OT_1^N(p'_{\sigma-1}, \hat{\boldsymbol{u}})$, and $\mathcal{A}$ obtains $\boldsymbol{u}[p_{\sigma-1}] = [\![p_\sigma]\!]$. $\mathcal{A}$ obtains $\boldsymbol{o}[\boldsymbol{o}[\ldots \boldsymbol{o}[p_0]\ldots]\!] = p_\sigma$ by decrypting $[\![p_\sigma]\!]$.

For convenience, we denote $(p'_{k+1}, r_{k+1}) \leftarrow \mathsf{eROT}(p'_k, r_k, \boldsymbol{\Omega})$ for the k-th step of eROT, which takes shares of a query p_k (i.e., p'_k, r_k) and a matrix $\boldsymbol{\Omega}$ as inputs, and outputs p'_{k+1} to $\mathcal{A}$ and r_{k+1} to $\mathcal{B}$. The initial step can also be represented by this notation setting from 0 to r_0. Since the major part of the time complexity is the inner product and OT, the time complexity on $\mathcal{B}$'s side in one round is $O(N\lambda)$. The time complexity on $\mathcal{A}$'s side is $O(N)$ per iteration. The communication size per iteration is $O(N)$ due to the communication size for OT.

We state that the following security theorem is established for eROT.

Theorem 1. eROT *correctly outputs* $\boldsymbol{o}[\boldsymbol{o}[\ldots \boldsymbol{o}[p_0]\ldots]\!]$ *and is secure in the semi-honest setting.*

Proof. **Correctness:** Each row of $\boldsymbol{\Omega}$ is an unary representation of a value, and hence conducting $\boldsymbol{u}'[i] = \bigoplus_{j=0}^{\lambda-1}(\boldsymbol{\Omega}[i][j] \otimes (j+r)_{\mathsf{mod}\ N})$ correctly yields an encryption of $(p+r)_{\mathsf{mod}\ N}$, where p is the value stored at $\boldsymbol{\Omega}[i]$. Therefore, by performing the initial step of eROT, the two parties can obtain shares $(p_1 + r_1)_{\mathsf{mod}\ N}$ and r_1 of the true position p_1. In the k-th step, $\mathcal{A}$'s input $p'_k = (p_k + r_k)_{\mathsf{mod}\ N}$ to OT is a share of the true position p_k, and the two parties can obtain a correct element by rotating $\boldsymbol{u}'$ by r_k before conducting OT. After decrypting the encrypted value obtained by OT, $\mathcal{A}$ knows the share of the next position p_{k+1}. In the last step, $\mathcal{B}$ does not mask $\boldsymbol{u}'$ and, therefore by induction it holds that the protocol correctly outputs $\boldsymbol{o}[\boldsymbol{o}[\ldots \boldsymbol{o}[p_0]\ldots]\!]$ to $\mathcal{A}$.

Security: All the messages are exchanged by OT. Considering that secure OT is used, it is guaranteed that no information of $\mathcal{A}$ is leaked to $\mathcal{B}$. Security against a semi-honest user is established by secret sharing. Shares of intermediate results received by $\mathcal{A}$ are indistinguishable from uniformly distributed random values due to the property of modular addition. Thus, a semi-honest user cannot acquire any information from intermediate results. $\square$

Algorithm 1. Detailed description of comparison phase

- Public inputs: length of GLOUDS N; height d; length of attribute vector n
- Private input of $\mathcal{B}$: threshold vector $\boldsymbol{\theta}$; type vector $\boldsymbol{\tau}$
- Private input of $\mathcal{A}$: attribute vector $\boldsymbol{x}$

Step (1): $\mathcal{A}$ and $\mathcal{B}$ conduct comparison protocol coorperatively and $\mathcal{B}$ obtains $[\![\langle \boldsymbol{\theta}[j] < \mathcal{X}_j\rangle]\!]$. $\mathcal{X}_j$ is an element of attribute vector corresponding to the position j. $\mathcal{B}$ constructs a flag matrix $\boldsymbol{F}$.

    for $j \in \{0, \ldots, N-1\}$ **do**
      if $\boldsymbol{\tau}[j] = \mathsf{I}$ **then**
        $\boldsymbol{F}[j][0] \leftarrow [\![1]\!], \boldsymbol{F}[j][1] \leftarrow [\![\langle \boldsymbol{\theta}[j] < \mathcal{X}_j\rangle]\!], \boldsymbol{F}[j][2] \leftarrow [\![0]\!]$

Step (2): $\mathcal{B}$ constructs $\boldsymbol{W}$ from $\boldsymbol{F}$

    for $j \in \{0, \ldots, N-1\}, k \in \{1, 2\}$ **do**
      if $\boldsymbol{\tau}[j] = \mathsf{I}$ **then**
        $\boldsymbol{W}[j][k] \leftarrow \boldsymbol{F}[j][k-1] \oplus (\ominus\boldsymbol{F}[j][k])$ $\triangleright$ $[\![\mathsf{UNARY}_2(\langle \boldsymbol{\theta}[j] < \mathcal{X}_j\rangle)]\!]$

Step (3): $\mathcal{B}$ constructs an encrypted offset matrix $\boldsymbol{\Omega}$ based on $\boldsymbol{W}$.

    for $j \in \{0, \ldots, N-1\}$ **do**
      if $\boldsymbol{\tau}[j] = \mathsf{Z}$ or $\boldsymbol{\tau}[j] = \mathsf{L}$ **then**
        $\boldsymbol{\Omega}[j][0] \leftarrow [\![1]\!], \boldsymbol{\Omega}[j][1] \leftarrow [\![0]\!], \boldsymbol{\Omega}[j][2] \leftarrow [\![0]\!]$ $\triangleright$ $[\![\mathsf{UNARY}_3(0)]\!]$
      else if $\boldsymbol{\tau}[j] = \mathsf{I}$ **then**
        $\boldsymbol{\Omega}[j][0] \leftarrow [\![0]\!], \boldsymbol{\Omega}[j][1] \leftarrow \boldsymbol{W}[j][1], \boldsymbol{\Omega}[j][2] \leftarrow \boldsymbol{W}[j][2]$
                                    $\triangleright$ $[\![\mathsf{UNARY}_3(\langle \boldsymbol{\theta}[j] < \mathcal{X}_j\rangle + 1)]\!]$

3.4 Secure BDG Evaluation Using GLOUDS and AHE

Comparison Phase. $\mathcal{A}$ and $\mathcal{B}$ construct an encrypted matrix $\boldsymbol{\Omega}$, which corresponds to offset vector $\boldsymbol{o}$. The comparison phase ensures that no information from $\mathcal{B}$ or $\mathcal{A}$ will be disclosed, other than the number of comparisons. The following describes how we constructed $\boldsymbol{\Omega}$. The detailed algorithm is provided in Algorithm 1.

Construction of $\boldsymbol{F}$: Each split function is associated with one of positions $\{j \mid \boldsymbol{\tau}[j] = \mathsf{I} \wedge 0 \leq j < N\}$. $\mathcal{A}$ and $\mathcal{B}$ conduct a secure comparison protocol in Step (1) of Algorithm 1 to securely compute all the comparison results between attributes and thresholds. $\mathcal{B}$ finally constructs a flag matrix $\boldsymbol{F}$. We do not need to compute $\boldsymbol{F}[j]$ if $\boldsymbol{\tau}[j] \neq \mathsf{I}$.

Construction of $\boldsymbol{W}$: $\mathcal{B}$ constructs a matrix $\boldsymbol{W}$, each of whose rows is an encrypted 2-bit unary vector that represents an output of a split function.

Construction of $\boldsymbol{\Omega}$: $\mathcal{B}$ constructs $\boldsymbol{\Omega}$ in Step (3) based on $\boldsymbol{W}$ and $\boldsymbol{\tau}$. To make an encryption of $\mathsf{UNARY}_3(\langle \boldsymbol{\theta}[j] < \mathcal{X}_j\rangle + 1)$, we set $\boldsymbol{\Omega}[j][0] = [\![0]\!]$ if $\boldsymbol{\tau}[j] = \mathsf{I}$. If $\boldsymbol{\tau}[j] \neq \mathsf{I}$, $\boldsymbol{\Omega}[j]$ stores $\mathsf{UNARY}_3(0)$. Note that the lengths of rows of $\boldsymbol{\Omega}$ can be reduced to 1 when $\boldsymbol{\tau}[j] \neq \mathsf{I}$ (because the offset is 0). This is because $\mathcal{B}$ knows the offsets that do not rely on any user information and can minimize the bit length. To use the reduced form of the offset matrix, we modify the inner product in each step of eROT to $\bigoplus_{j=0}^{u_j}(\Omega[i][j] \cdot (j+r)_{\mathsf{mod}\ N})$, where $u_j \in \{1,3\}$ is the length of the row. As a result, we can also reduce the time complexity to $O(N)$, where N is the length of GLOUDS.

Also note that the calculation of an offset can be omitted when the node is a shadow node. A new type of position for shadow nodes should be defined to do that to distinguish them from other internal nodes.

We state that the following security theorem is established for Algorithm 1.

Theorem 2. *Algorithm 1 correctly outputs Ω and is secure in the semi-honest setting.*

Proof. **Correctness:** When an attribute $\mathcal{X}_j$ is less than an threshold $\theta[j]$, $(W[j][1], W[j][2])$ becomes $([\![1]\!], [\![0]\!])$, otherwise $([\![0]\!], [\![1]\!])$ assuming the correctness of the underlying secure comparison protocol. Therefore, all the rows of Ω satisfy the condition that they represent offsets in encrypted unary vectors.

Security: We have assumed that the underlying secure comparison protocol is secure in the semi-honest setting. Since the procedures after the secure comparison protocol only require server side operations on ciphertext, the security of the comparison phase is guaranteed by the security of the secure comparison protocol. $\qquad\square$

Evaluation Phase. This section describes the evaluation phase in which the participants securely descend a BDG using ROT and eROT.

We need to recursively refer to v and Ω by starting from an initial position $p_0 = 0$ to move from the root to the leaf. The next position p_{i+1}, given a starting position p_i on GLOUDS, which corresponds to the child, is calculated by adding the p_i-th elements of a SelRan vector v, and an encrypted offset matrix Ω. The next iteration will be executed after the next position p_{i+1} is set.

The private information of both parties must simultaneously be protected. There are two main security requirements: (1) $\mathcal{B}$ should not know the positions specified by $\mathcal{A}$ or the results of the protocol, and (2) v and Ω held by $\mathcal{B}$ should be concealed from $\mathcal{A}$. We used ROT and eROT to recursively refer to v and Ω concealing private information. Algorithm 2 describes the details of the evaluation phase satisfies the previously explained functionality and security requirements.

First, $\mathcal{A}$ and $\mathcal{B}$ start initialization in Step (1). $\mathcal{B}$ sets $r_1 = r_2 = r' = 0$. The r_1, r_2 store random values used in the previous iterations of ROT and eROT, and r' is sum of r_1 and r_2 modulo N. $\mathcal{A}$ sets the initial position $p'_0 = 0$. The two parties engage in ROT and eROT in Step (2) to update the position on GLOUDS by recursively referring to v and Ω. The (β'_{k+1}, r_1) and (ω'_{k+1}, r_2) correspond to random shares of $v[p_{k+1}]$ and an offset $o[p_{k+1}]$, which can be recovered by using $\mathcal{B}$'s random values r_1, r_2, and rotating v and Ω (without knowing these values due to the security of OT). Since the position of the x-th child is determined by $\mathsf{SelRan}(B, p) + x$, the next query is $p'_{k+1} = (\beta'_{k+1} + \omega'_{k+1})_{\mathsf{mod}\ N}$. This iteration is conducted d times regardless of the depth of a leaf, which should be reached. It should be noted that a position is fixed once it is reached at a position in B with

---

Algorithm 2. Detailed description of evaluation phase

- Public input: length of GLOUDS N; height d
- Private input of $\mathcal{B}$: SelRan vector v, encrypted offset matrix Ω, label vector z

Step (1): Initialization

$\qquad \mathcal{B}$ conducts: $r_1 \leftarrow 0,\ r_2 \leftarrow 0,\ r' \leftarrow r_1 + r_2$

$\qquad \mathcal{A}$ conducts: $p'_0 \leftarrow 0$

Step (2): Update the position in GLOUDS by iterating ROT, eROT.

$\qquad$ **for** $k = 0$ **to** $d - 1$ **do**

$\qquad\qquad \mathcal{A}$ and $\mathcal{B}$ engage in ROT and eROT.

$\qquad\qquad (\beta'_{k+1}, r_1) \leftarrow \mathsf{ROT}(p'_k, r', v),\ (\omega'_{k+1}, r_2) \leftarrow \mathsf{eROT}(p'_k, r', \Omega)$

$\qquad\qquad \mathcal{B}$ conducts: $r' \leftarrow (r_1 + r_2)_{\mathrm{mod}\ N}$

$\qquad\qquad \mathcal{A}$ conducts: $p'_{k+1} \leftarrow (\beta'_{k+1} + \omega'_{k+1})_{\mathrm{mod}\ N}$

Step (3): Get the output of BDG $T(x)$ from z using $OT_1^N(p'_d, z)$.

---

trit 0, which ensures that the last position is in the position that corresponds to the leaf due to the definition of v and Ω. Finally, $\mathcal{A}$ obtains the output of the BDG $T(x)$ from z using $OT_1^N(p'_d, z)$ in Step (3).

We state that the following security theorem is established for Algorithm 2.

Theorem 3. *Algorithm 2 correctly outputs $T(x)$ and is secure in the semi-honest setting.*

Proof. **Correctness:** Due to the way the look-up vector v and Ω are constructed, it is obvious that the evaluation phase can correctly compute $v[p_k] + o[p_k]$ in k-th step, if v and Ω are not randomized. v is randomized by r_1 and Ω (i.e., o) is randomized by r_2. Since following equation is established by considering the property of modular addition,

$$
\begin{aligned}
\{p'_{k+1} - r'\}_{\mathrm{mod}\ N} &= \{(\beta'_{k+1} - r_1) + (\omega'_{k+1} - r_2)\}_{\mathrm{mod}\ N} \\
&= v[p_k] + o[p_k] = p_{k+1},
\end{aligned}
$$

$p'_{k+1} = (\beta'_{k+1} + \omega'_{k+1})_{\mathrm{mod}\ N}$ and $r' = (r_1 + r_2)_{\mathrm{mod}\ N}$ are the shares of p_{k+1}, hence one can obtain shares of $v[p_{k+1}]$ and $o[p_{k+1}]$ by conducting ROT and eROT with the same arguments p'_{k+1} and r'. The label corresponding to the leaf is obtained by OT in Step (3). Therefore, by induction it holds that the evaluation phase correctly outputs $T(x)$.

Security: For the space limitation, we will only sketch out a proof. In Algorithm 2, all the messages are exchanged through OT in the subroutines ROT and eROT. Considering secure OT is used, it is guaranteed that no information of $\mathcal{A}$ is leaked to $\mathcal{B}$. All the messages sent from $\mathcal{B}$ are random share of $\mathcal{B}$'s information. Therefore, it is guaranteed that no information of $\mathcal{B}$ is leaked to $\mathcal{A}$ except for the final output $T(x)$. $\qquad\square$

3.5 Complexity

This subsection discusses the asymptotic time complexity and communication complexity of our protocol. The majority of computational and communication costs in the comparison phase are due to the comparison protocol and the construction of an encrypted offset matrix. The time complexity of the comparison protocol on the whole is $O(\ell m)$ on $\mathcal{B}$'s side. It is $O(\ell(n+m))$ on $\mathcal{A}$'s side by considering the encryption of an attribute vector and decryption of intermediate results. The construction of an encrypted offset matrix requires $O(N)$ computational cost as the computational cost is linear to the sum of the lengths of its rows. Therefore, the total time complexity of the comparison phase is $O(\ell m + N)$ on $\mathcal{B}$'s side and $O(\ell(n+m))$ on $\mathcal{A}$'s side. Since an encrypted offset matrix is constructed offline, all of the communication cost is required by the secure comparison protocol. The communication cost for $\mathcal{B}$ is $O(\ell m)$ and that for $\mathcal{A}$ is $O(\ell n + m)$ in the comparison phase. The time complexities of both $\mathcal{A}$ and $\mathcal{B}$ are $O(dN)$ for the evaluation phase. This is because ROT and eROT require $O(dN)$ computational cost. The communication cost is also $O(dN)$. We have summarized the time and communication complexity in Table 1.

Table 1. Time complexity and Communication of each phase of our method. d is the height of a DAG, ℓ is the bit length of $\mathcal{A}$'s and $\mathcal{B}$'s inputs, m is the number of split functions, n is the number of nodes in a DAG and N is the length of $\mathcal{B}$.

Phase	Time	Communication	# rounds
Comparison ($\mathcal{A}$)	$O(\ell(n+m))$	$O(\ell n)$	2
Comparison ($\mathcal{B}$)	$O(\ell m + N)$	$O(\ell m)$	
Evaluation ($\mathcal{A}$)	$O(dN)$	$O(dN)$	$d+1$
Evaluation ($\mathcal{B}$)	$O(dN)$	$O(dN)$	

4 Experiments

We evaluated the efficiency of our protocol with experiments under various settings. We implemented our protocol, which is secure against the semi-honest model using the C++ library of elliptic curve ElGamal encryption [24]. We used secp256k1 for the security parameters of lifted-ElGamal, which is secure at the 128-bit security level [9]. We used a standard desktop PC with a Xeon 3.40-GHz processor for a party $\mathcal{B}$ (a server) and a standard desktop PC with a Xeon 2.40-GHz processor for a party $\mathcal{A}$ (a user) (1 thread each). Both the server and the user were in the same local area network (LAN) in our experiments.

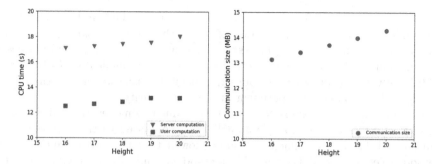

Fig. 3. CPU time (s) of server and user, and communication size (MB) on simulated datasets. We varied d from 16 to 20 while fixing other parameters ($m = 557, n = 95$, and $N = 1110$.)

4.1 Experiment on Simulated Dataset

First, we will present the results obtained from experiments on the simulated dataset. The dataset was composed of pairs of BDGs and attribute vectors. We varied the height d of the BDGs from 16 to 20 one by one, while fixing the number of nodes to 1110 (the number of comparisons m was 557.) and the lengths of attribute vectors n to 395. These parameters (except for d) were taken from Brickell et al. [6] to enable performance to be later compared in Sect. 4.2. Figure 3 plots the CPU time and communication size of the server and the user in our protocol. We observed that even when d was 20, our protocol finished within a practical timeframe and communication size (27 s and 14 MB). We also confirmed that the CPU time and communication size of both the server and the user were linear to d, which is consistent with theoretical complexity. We also confirmed that runtime overhead caused by network latency was not too large. When the round trip time (RTT) was 45 ms, which is regional RTT within North America [33], the increase in runtime was only 6 s.

4.2 Comparison to Conventional Methods

Brickell et al. [6] proposed a $O(n + \ell N + d)$ time BP (BDG) evaluation method based on AHE and Garbled Circuit. [6] reported the performance of their protocol on a BDG (1107 nodes and 395 attributes) as 302 s in CPU time and 25 MB in communication size. Since the exact topology of the BDG they used was unshown, we conducted an experiment on BDGs that have the same number of nodes and attributes with various heights for fair comparison. The results revealed that our method maintained 11 times better performance in runtime (26 s) than that of [6] and required 1.8 times less communication size (14 MB) even when the DAG is as high as $d = 20$. Barni et al. proposed a privacy-preserving evaluation method of LBP which is a generalization of BP. The time complexity of this method is $O(n + m\ell' + d)$ where $\ell'(> \ell)$ is bit length of threshold. Barni et al.'s [4] performance on the ECG dataset ($d = 4, m = 6$ and $n = 4$)

was 6.8 s in computation (without network communication) and 0.1122 MB in communication size. The performance of our method with the same parameters was about 8.85 times better (0.768 s) than [4] in terms of computational cost, although our method incurred slightly more cost in communication (0.156 MB). Additionally, the security level in our experimental setting was higher than that of [4,6]. (They conducted experiment at a 80 bit security).

We also conducted experiments on DTs trained using several real datasets used by conventional methods [5,30,35]. Even compared to the methods specialized in DT, the experimental results showed that the performance of our method exceeds that of Bost et al.'s method [5] and is almost equivalent to those of Wu et al.'s and Tai et al.'s methods [30,35]. Additionally, our protocol had an advantage on deep decision trees over [35] whose complexity is exponential to d. When $d = 17, m = 58$, and $n = 57$, our method achieved about a 4 fold faster runtime. The methods specialized in DTs can be used for BDGs by transforming a BDG into an equivalent DT. Their computational cost increases along with the increase of redundant nodes and edges incurred by the transformation. Therefore, while the state-of-the-art method by Tai et al. [30] performed slightly better than our method for DT evaluation, its runtime became worse than our method's runtime when it is tested on complex BDGs.

5   Conclusion

We proposed an efficient protocol for evaluating BDGs, which was designed by AHE and did not use heavy cryptographic primitives, such as fully homomorphic encryption. The protocol obliviously evaluated a look-up vector that was constructed based on GLOUDS to achieve linear time and communication complexities. We also proposed a design principle for the look-up vector to further reduce the vector size. The results obtained from the experiments indicated that the actual runtime and communication size were well concordant with theoretical complexities and that the runtime of our method was an order of magnitude faster than that in the previous approaches [4,6]. We also confirmed that our method was even faster for the DT evaluations compared to a previous approach that specialized in DT [35] when the tree was deep. Our method demonstrated a runtime in an experiment with BDGs that was faster than the state-of-the-art method of DT evaluation [30] that took advantage of the fact that a graph with information equivalent to that in a tree was much more compact than the tree. These results confirmed the efficiency of our method, and we also hope that it will contribute to secure utilization of valuable classifiers that aggregate knowledge extracted from abundant data resources. Another remarkable feature of our protocol is that it directly simulates step-by-step graph traversal, whereas other efficient methods [30,35] reformulate the graph traversal as an evaluation of polynomial equations. By using an additional look-up vector, our protocol enables traversals both to ascendant and to descendant. Such a feature could be useful for various applications that require more complex graph traversal (i.e. searching on DFA).

Acknowledgements. A part of this work is supported by Okawa Foundation Research Grant and JST CREST grant numbers: JPMJCR1503, JPMJCR1688, JPMJCR14D6.

References

1. Amazon machine learning - predictive analytics with AWS (2017). https://aws.amazon.com/aml/
2. Google cloud machine learning at scale — google cloud platform (2017). https://cloud.google.com/products/machine-learning/
3. Microsoft Azure: Machine learning (2017). https://azure.microsoft.com
4. Barni, M., Failla, P., Kolesnikov, V., Lazzeretti, R., Sadeghi, A.-R., Schneider, T.: Secure evaluation of private linear branching programs with medical applications. In: Backes, M., Ning, P. (eds.) ESORICS 2009. LNCS, vol. 5789, pp. 424–439. Springer, Heidelberg (2009). https://doi.org/10.1007/978-3-642-04444-1_26
5. Bost, R., Popa, R., Tu, S., Goldwasser, S.: Machine learning classification over encrypted data. In: NDSS, pp. 1–14 (2015)
6. Brickell, J., Porter, D.E., Shmatikov, V., Witchel, E.: Privacy-preserving remote diagnostics. In: CCS, pp. 498–507. ACM Press, New York (2007)
7. Bryant, R.E.: Graph-based algorithms for boolean function manipulation. IEEE TC **100**(8), 677–691 (1986)
8. Hazay, C., Lindell, Y.: Efficient Secure Two-Party Protocols: Information Security and Cryptography, 1st edn. Springer, Heidelberg (2010). https://doi.org/10.1007/978-3-642-14303-8
9. Certicom Research: Standards for Efficient Cryptography 2 (SEC 2): Recommended elliptic curve domain parameters (2010). http://www.secg.org/sec2-v2.pdf
10. Cock, M.D., et al.: Efficient and private scoring of decision trees, support vector machines and logistic regression models based on pre-computation. Cryptology ePrint Archive, Report 2016/736 (2016). https://eprint.iacr.org/2016/736
11. Cramer, R., Gennaro, R., Schoenmakers, B.: A secure and optimally efficient multi-authority election scheme. In: Fumy, W. (ed.) EUROCRYPT 1997. LNCS, vol. 1233, pp. 103–118. Springer, Heidelberg (1997). https://doi.org/10.1007/3-540-69053-0_9
12. Damgård, I., Geisler, M., Krøigaard, M.: Efficient and secure comparison for on-line auctions. In: Pieprzyk, J., Ghodosi, H., Dawson, E. (eds.) ACISP 2007. LNCS, vol. 4586, pp. 416–430. Springer, Heidelberg (2007). https://doi.org/10.1007/978-3-540-73458-1_30
13. Dowlin, N., Gilad-Bachrach, R., Laine, K., Lauter, K., Naehrig, M., Wernsing, J.: CryptoNets: applying neural networks to encrypted data with high throughput and accuracy. In: ICML, pp. 201–210. JMLR (2016)
14. Fischer, J., Peters, D.: GLOUDS: representing tree-like graphs. Discrete Algorithms **36**, 39–49 (2016)
15. Fredrikson, M., Jha, S., Ristenpart, T.: Model inversion attacks that exploit confidence information and basic countermeasures. In: CCS, pp. 1322–1333. ACM Press, New York (2015)
16. Fredrikson, M., Lantz, E., Jha, S., Lin, S.: Privacy in pharmacogenetics: an end-to-end case study of personalized warfarin dosing. In: USENIX, pp. 17–32 (2014)
17. Huang, C.L., Chen, M.C., Wang, C.J.: Credit scoring with a data mining approach based on support vector machines. Expert Syst. Appl. **33**(4), 847–856 (2007)

18. Jacobson, G.: Space-efficient static trees and graphs. In: FOCS, pp. 549–554. IEEE Press, New York (1989)
19. Kohavi, R., Li, C.H.: Oblivious decision trees graphs and top down pruning. In: IJCAI, pp. 1071–1077. Morgan Kaufmann, San Francisco (1995)
20. Lavecchia, A.: Machine-learning approaches in drug discovery: methods and applications. Drug Discov. Today **20**(3), 318–331 (2015)
21. Liu, J., Juuti, M., Lu, Y., Asokan, N.: Oblivious neural network predictions via MiniONN transformations. In: CCS, pp. 619–631. ACM Press, New York (2017)
22. Madabhushi, A., Lee, G.: Image analysis and machine learning in digital pathology: challenges and opportunities. Med. Image Anal. **33**, 170–175 (2016)
23. Meinel, C.: Modified branching programs and their computational power. LNCS, vol. 370. Springer, Heidelberg (1989). https://doi.org/10.1007/BFb0017563
24. Mitsunari, S.: C++ library implementing elliptic curve Elgamal crypto system. https://github.com/herumi/mcl
25. Mohassel, P., Niksefat, S.: Oblivious decision programs from oblivious transfer: efficient reductions. In: Keromytis, A.D. (ed.) FC 2012. LNCS, vol. 7397, pp. 269–284. Springer, Heidelberg (2012). https://doi.org/10.1007/978-3-642-32946-3_20
26. Oliveira, A.L., Sangiovanni-Vincentelli, A.: Using the minimum description length principle to infer reduced ordered decision graphs. Mach. Learn. **25**(1), 23–50 (1996)
27. Oliver, J.J.: Decision graphs: an extension of decision trees. Technical report, Department of Computer Science, Monash University (1992)
28. Shimizu, K., Nuida, K., Ratsch, G.: Efficient privacy-preserving string search and an application in genomics. Bioinformatics **32**(11), 1652–1661 (2016)
29. Shotton, J., Sharp, T., Kohli, P., Nowozin, S., Winn, J., Criminisi, A.: Decision jungles: compact and rich models for classification. In: Burges, C.J.C., Bottou, L., Welling, M., Ghahramani, Z., Weinberger, K.Q. (eds.) NIPS, pp. 234–242. Curran Associates Inc., New York (2013)
30. Tai, R.K.H., Ma, J.P.K., Zhao, Y., Chow, S.S.M.: Privacy-preserving decision trees evaluation via linear functions. In: Foley, S., Gollmann, D., Snekkenes, E. (eds.) ESORICS 2017. LNCS, vol. 10493, pp. 494–512. Springer, Cham (2017). https://doi.org/10.1007/978-3-319-66399-9_27
31. Troncoso-Pastoriza, J.R., Katzenbeisser, S., Celik, M.: Privacy preserving error resilient DNA searching through oblivious automata. In: CCS, pp. 519–528. ACM Press, New York (2007)
32. Vaidya, J., Yu, H., Jiang, X.: Privacy-preserving SVM classification. Knowl. Inf. Syst. **14**(2), 161–178 (2008)
33. Verizon Enterprise Solutions: Verizon: IP latency statistics (2017). http://www.verizonenterprise.com/about/network/latency/
34. Veugen, T.: Improving the DGK comparison protocol. In: WIFS, pp. 49–54. IEEE Press, New York (2012)
35. Wu, D.J., Feng, T., Naehrig, M., Lauter, K.: Privately evaluating decision trees and random forests. PoPETS **4**, 1–21 (2016)
36. Yao, A.C.C.: How to generate and exchange secrets. In: FOCS, pp. 162–167. IEEE Press, New York (1986)

Post-Quantum Cryptography

Key Reuse Attack on NewHope Key Exchange Protocol

Chao Liu[1]([✉]), Zhongxiang Zheng[2], and Guangnan Zou[3]

[1] Key Laboratory of Cryptologic Technology and Information Security,
Ministry of Education, Shandong University, Jinan, People's Republic of China
liu_chao@mail.sdu.edu.cn
[2] Department of Computer Science and Technology, Tsinghua University,
Beijing, People's Republic of China
zhengzx13@mails.tsinghua.edu.cn
[3] Space Star Technology Co., Ltd., Beijing, People's Republic of China
zouguangnan@spacestar.com.cn

Abstract. In recent years, Ring Learning with Errors (RLWE) key exchange has been recognized for its efficiency and is considered a potential alternative to Diffie-Hellman (DH) key exchange protocol. We focus on RLWE key exchange protocols in the context of key reuse. In 2016 (ePrint 085), Fluhrer firstly presented an attack aiming at the case where party B reuse his secret key. In key reuse attack, the adversary plays the role of A with the abilities to initiate any number of key exchange sessions with party B. The adversary initiates a sequence of key exchange sessions with a malformed key, then looks for the *signal* variations sent by party B. In this work, we describe a new key reuse attack against the NewHope key exchange protocol proposed by Alkim et al. in 2016. We give a detailed analysis of the *signal* function of the NewHope and describe a new key recovering technique based on the special property of NewHope's *signal*.

Keywords: RLWE · Key exchange · Post quantum · Key reuse · Active attack

1 Introduction

In 1994, Shor [16] devised that the discrete log problem can be solved in polynomial time by quantum computers. The fact that currently used Diffie-Hellman (DH) key exchange algorithms are mainly based on the hardness of the discrete log problem, leads to the search for cryptographic protocols with resistance to all known quantum algorithms. Lattice based constructions which appear to be resistant to attack by both classical and quantum computers, have been regarded as an important candidate for post-quantum cryptography. Recent progress in the development of Learning With Errors (LWE) and its variants puts lattice cryptography in an excellent position for use in practice. The Learning With

© Springer Nature Switzerland AG 2019
K. Lee (Ed.): ICISC 2018, LNCS 11396, pp. 163–176, 2019.
https://doi.org/10.1007/978-3-030-12146-4_11

Errors (LWE) problem was first introduced by Regev in [15] and then Lyuba-shevsky, Peikert and Regev [11] proposed an algebraic variant of LWE called Ring Learning With Errors (RLWE), which is of better efficiency.

Key exchange is a fundamental cryptographic primitive where cryptographic keys are allowed to exchanged between two or more participants. LWE and RLWE based key exchange protocols are considered to be a desirable replacement for DH protocols. Ding et al. firstly introduced such key exchange in [8] and several recent key exchange variants [2–5,14,18] that rely on the hardness of the LWE problem or RLWE problem have been proposed and implemented. In RLWE based key exchange, the two parties A and B firstly compute approximate values using the public key of the other's, then party B sends an additional information about the interval in which the approximately equal value lies, to party A. Finally, both two parties compute a shared secret using this additional information which we refer to as the *signal*. In this paper, we consider the case that party B reuses his secret key. Since key reuse is widely adopted for efficiency reasons (see [7], Sect. 4), attack analysis on such case will reveal some potential danger of the key exchange protocol.

In this work, we focus on NewHope key exchange proposed by Alkim et al. in [2]. NewHope implementation is very rapid, and is tested in Google Chrome Canary browser for its post quantum experiment [9]. And it was awarded the 2016 Internet Defense Prize [17].

Previous Work. The first attack on RLWE key exchange for reused public keys was described by Fluhrer in [10]. A detailed description of key reuse attack is presented by Ding et al. in [6] on key exchange proposed in [8]. The idea of their attacks is to deviate from the key exchange in generating the adversary's public key, then the adversary uses the variations of the *signal* to extract information about the party B's secret key.

Ding et al. [7] also described a new attack on Ding's one pass case key exchange [8]. This attack doesn't rely on the *signal* function output but use only the information of whether the final key of two parties agree.

Our Contribution. An error-reconciliation mechanism allows two parties with two "approximately agreeing" secret values to reach exact agreement. NewHope's error-reconciliation mechanism is more complex than Ding's key exchange's error-reconciliation mechanism presented in [8]. For details, its *signal* function is constructed based on a special lattice $\tilde{D}_4$ and the *signal* doesn't change regularly as Ding's key exchange does. Hence Fluhrer or Ding's attack can't be adopted directly. We analysis the *signal* function of NewHope and find that for every *signal*, there is one special vector corresponding to it. Let $\mathbf{a}_{k,l}[i]$ be one of dimension of this special vector, then we find that the sequence $(\mathbf{a}_{k,l}[i])_{k=0,1,...,q-1}$ have "periodic" property. We show that the number of period of $(\mathbf{a}_{k,l}[i])_{k=0,1,...,q-1}$ equals to the absolute value of one coefficient of the secret key. We introduce a technique to select sequence $(\mathbf{a}_{k,l}[i])_{k=0,1,...,q-1}$ and construct algorithm to compute the number of the period of this sequence, also we use a new way to eliminate the ambiguity of the $\pm$ sign of the coefficients. Experiments have been conducted to verify the correctness of our attack.

Organization. In Sect. 2 we discuss some notations used in this paper and background on RLWE. The NewHope key exchange and key reuse attack is reviewed in Sect. 3. And our attack is described in Sect. 4. The conclusion is presented in Sect. 5.

2 Preliminaries

2.1 Notation

Let n be an integer which is a power of 2. Define the ring of integer polynomials $R := \mathbb{Z}[x]/(x^n + 1)$. For any positive integer q, define $R_q := \mathbb{Z}_q[x]/(x^n + 1)$ as the ring of integer polynomials modulo $x^n + 1$ where every coefficient is reduced modulo q. We define for $x \in \mathbb{R}$ the rounding function $\lfloor x \rceil = \lfloor x + \frac{1}{2} \rfloor \in \mathbb{Z}$. For an even (resp. odd) positive integer α, we define $r' = r \bmod^{\pm}\alpha$ to be the unique element r' in the range $-\frac{\alpha}{2} < r' \leq \frac{\alpha}{2}$ (resp. $-\frac{\alpha-1}{2} \leq r' \leq \frac{\alpha-1}{2}$) such that $r' = r \bmod \alpha$. Sometimes for $v = v_0 + v_1 x + v_2 x^2 + \dots v_{n-1}x^{n-1} \in R$, we write $v[i]$ as v_i. Suppose χ is a probability distribution over R, $x \xleftarrow{\$} \chi$ means the sampling of $x \in R$ according to χ. For a probabilistic algorithm $\mathcal{A}$, we write $y \xleftarrow{\$} \mathcal{A}$ to represent that the output of $\mathcal{A}$ is assigned to y randomly.

We write (column) vectors in bold face as $\mathbf{v} = (v_0, v_1, \dots, v_{n-1})^T$, where $\mathbf{v}^T$ denotes the transpose of the vector, and matrices in bold face as $\mathbf{A}$. For a vector $\mathbf{v} = (v_0, \dots, v_{n-1})^T$ in $\mathbb{R}^n$, define the l_1 norm as $||\mathbf{v}||_1 = \sum_{i=0}^{n-1} |v_i|$ and the l_2 norm as $||\mathbf{v}||_2 = (\sum_{i=0}^{n-1} |v_i|^2)^{1/2}$. In this paper $|| \cdot ||$ denote the l_2 norm.

For any positive real $s \in \mathbb{R}$, we write $\rho_s(x) = exp(-\pi \frac{||x||^2}{s^2})$ as the Gaussian function which is scaled by a factor s. Let $\rho_s(\mathbb{Z}^n) = \sum_{\mathbf{x} \in \mathbb{Z}^n} \rho_s(\mathbf{x})$. Then for $\mathbf{x} \in \mathbb{Z}^n$, we let $D_{\mathbb{Z}^n, s}(\mathbf{x}) = \frac{\rho_s(\mathbf{x})}{\rho_s(\mathbb{Z}^n)}$ to indicate the n-dimensional discrete Gaussian distribution.

In applying the norms, we assume the coefficient embedding of elements from R to $\mathbb{R}^n$. For any element $y = \sum_{i=0}^{n-1} y_i x^i$ of R, we can embed this element into $\mathbb{R}^n$ as the vector $(y_0, \dots, y_{n-1})$. Also we define the ring of integer polynomials $\bar{R} := \mathbb{Z}[x]/(x^4 + 1)$. For any $y(x) = y_0 + y_1 x + \dots + y_{n-1}x^{n-1} \in R$, define a mapping:

$$\phi : R \quad \to \bar{R}^{\frac{n}{4}}$$
$$y(x) \quad \mapsto (\bar{y}_0(x), \bar{y}_1(x), \cdots, \bar{y}_{\frac{n}{4}-1}(x)) \tag{1}$$

where $\bar{y}_i(x) = y_i + y_{i+\frac{n}{4}}x + y_{i+2\cdot\frac{n}{4}}x^2 + y_{i+3\cdot\frac{n}{4}}x^3 \in \bar{R}$. Assume the coefficient embedding of $\bar{y}_i$ to $\mathbf{v}_i = (y_i, y_{i+\frac{n}{4}}, y_{i+2\cdot\frac{n}{4}}, y_{i+3\cdot\frac{n}{4}})^T$, then we call vector $\mathbf{v}_i$ a *split vector* of y.

2.2 Ring Learning with Errors

The learning with Errors (LWE) problem is first introduced by Regev [15] who shows that under a quantum reduction, solving LWE in the average cases is as hard as solving certain Lattice problems in the worst cases. But LWE based cryptosystems is not efficient for practical applications for its large key sizes of $O(n^2)$. In 2010, Lyubashevsky, Peikert, and Regev [11] introduced the Ring Learning with Errors (RLWE), which is the version of LWE in the ring setting and can drastically improve the efficiency. For a uniform random $a, s \overset{\$}{\leftarrow} R_q$ and error distribution χ, let $A_{s,\chi}$ denote the distribution of the RLWE pair $(a, as+e)$, where error $e \overset{\$}{\leftarrow} \chi$. Given $(a, as+e)$ for polynomial number of samples, the search version of RLWE is to find a secret s in R_q, and the decision version of the RLWE problem, denote R-DLWE$_{q,\chi}$ is to distinguish $A_{s,\chi}$ from the uniform distribution on $R_q \times R_q$. Like LWE, RLWE enjoys a worst case hardness guarantee, and we state in the following:

Theorem 1 ([11], **Theorem 3.6**). *Let $R = \mathbb{Z}[x]/(x^n + 1)$ where n is a power of 2. Let $\alpha = \alpha(n) < \sqrt{logn/n}$, and $q = 1 \mod 2n$ be a ploy(n)-bounded prime such that $\alpha q \geq \omega(\sqrt{logn})$. There exists a ploy(n)-time quantum reduction from $\tilde{O}(\sqrt{n}/\alpha)$-SIVP (Short Independent Vectors Problem) on ideal lattices in R to solving R-DLWE$_{q,\chi}$ with $l - 1$ samples, where $\chi = D_{\mathbb{Z}^n,s}$ is discrete Gaussian distribution with $s = \alpha q \cdot (nl/log(nl))^{1/4}$.*

3 The Protocol and Key Reuse Attack

The Newhope Key Exchange. The NewHope key exchange protocol proposed by Alkim in [2] is an instantiation of Peikert's RLWE based passively secure key-exchange protocol [14]. Firstly, we recall NewHope protocol. Let n be a power of 2, $R = \mathbb{Z}[X]/(X^n + 1)$ and $R_q = R/qR$. In NewHope key exchange parameter $(n, q) = (1024, 12289)$. NewHope protocol is listed in Table 1.

The RLWE secret and error is sampled from ψ_k which is a centered binomial. We note that one can sample from ψ_k by computing $\sum_{i=0}^{k} b_i - b_i'$ where the $b_i, b_i' \in \{0, 1\}$ are uniform independent bits, hence this way of sampling is very easy and efficient. NewHope uses the parameter $k = 16$.

HelpRec and Rec. In NewHope, party B firstly mapping $\phi(v_B) = (v_0, v_1, \ldots, v_{\frac{n}{4}-1})$, then coefficient embed v_i to $\mathbf{x}_i$. NewHope's error reconciliation is based on finding the closest vector in $\tilde{D}_4$ with basis

$$\mathbf{B} = \begin{pmatrix} 1 & 0 & 0 & 1/2 \\ 0 & 1 & 0 & 1/2 \\ 0 & 0 & 1 & 1/2 \\ 0 & 0 & 0 & 1/2 \end{pmatrix}$$

and we define $\mathbf{g}^t := (1/2, 1/2, 1/2, 1/2)$. HelpRec$(\mathbf{x}_i; b)$ function to compute 2-bit reconciliation information as:

$$\text{HelpRec}(\mathbf{x}_i; b) = \text{CVP}_{\tilde{D}_4} \left(\frac{4}{q} (\mathbf{x}_i + b\mathbf{g}) \right) \mod 4, \tag{2}$$

Table 1. NewHope scheme from [2].

Parameter: $q = 12289, n = 1024$
Error distribution: ψ_{16}

Party A	Party B
$seed \xleftarrow{\$} \{0,1\}^{\frac{n}{4}}$	
$a \leftarrow \text{Parse}(\text{SHAKE-128}(seed)) \in R_q$	
$s_A, e_A \xleftarrow{\$} \psi_{16}^n$	$s_B, e_1, e_2 \xleftarrow{\$} \psi_{16}^n$
$p_A \leftarrow as_A + e_A \in R_q \xrightarrow{(p_A, seed)}$	$a \leftarrow \text{Parse}(\text{SHAKE-128}(seed))$
	$p_B \leftarrow as_B + e_1 \in R_q$
	$v_B \leftarrow p_A s_B + e_2 \in R_q$
$v_A \leftarrow p_B s_A \in R_q \xleftarrow{(p_B, r)}$	$r \xleftarrow{\$} \text{FullHelpRec}(v_B) \in \mathbb{Z}_4^n$
$c \leftarrow \text{FullRec}(v_A, r)$	$c \leftarrow \text{FullRec}(v_B, r) \in \{0,1\}^{\frac{n}{4}}$
$\mu \leftarrow \text{SHA3-256}(c)$	$\mu \leftarrow \text{SHA3-256}(c)$

where $b \in \{0,1\}$ is a uniformly chosen random bit. Then compute $\mathbf{r} \xleftarrow{\$} \text{FullHelpRec}(v_B)$ where $\mathbf{r} = (r_0, r_1, \ldots, r_{n-1})^T$ by computing $\mathbf{r}_i \xleftarrow{\$} \text{HelpRec}(\mathbf{x}_i)$ where $\mathbf{r}_i = (r_i, r_{i+\frac{n}{4}}, r_{i+2 \cdot \frac{n}{4}}, r_{i+3 \cdot \frac{n}{4}})^T \in \{0,1,2,3\}^4$.

We call the output vector $\mathbf{r} \xleftarrow{\$} \text{FullHelpRec}(v_B)$ a *signal*. Vector $\mathbf{r} = (r_0, r_1, \ldots, r_{n-1})^T$ is split into vectors $\mathbf{r}_i = (r_i, r_{i+\frac{n}{4}}, r_{i+2 \cdot \frac{n}{4}}, r_{i+3 \cdot \frac{n}{4}})^T \in \{0,1,2,3\}^4$, and Rec function

$$\text{Rec}(\mathbf{x}_i, \mathbf{r}_i) = \text{Decode}\left(\frac{1}{q}\mathbf{x}_i - \frac{1}{4}\mathbf{B}\mathbf{r}_i\right) \tag{3}$$

computes one key bit from a vector $\mathbf{x}_i$ and a reconciliation vector $\mathbf{r}_i$. To compute $c \leftarrow \text{FullRec}(v_B, \mathbf{r})$ where $c = (c_0, \ldots, c_{\frac{n}{4}-1})^T$, one need compute $c_i \leftarrow \text{Rec}(\mathbf{x}_i, \mathbf{r}_i)$. $\text{CVP}_{\tilde{D}_4}$ and Decode are listed as Algorithms 1 and 2, respectively.

Algorithm 1. $\text{CVP}_{\tilde{D}_4}$

Require: $\mathbf{v} := (v_0, v_1, v_2, v_3) \in \mathbb{R}^4$
Ensure: An vector $\mathbf{z} \in \mathbb{Z}^4$ such that $\mathbf{Bz}$ is a closest vector to $\mathbf{v}$
1: **if** $(\|\mathbf{v} - \lfloor\mathbf{v}\rceil\|_1) < 1$ **then**
2:    **return** $(\lfloor v_0 \rceil, \lfloor v_1 \rceil, \lfloor v_2 \rceil, 0)^T + \lfloor v_3 \rceil \cdot (-1, -1, -1, 2)^T$
3: **else**
4:    **return** $(\lfloor v_0 \rceil, \lfloor v_1 \rceil, \lfloor v_2 \rceil, 1)^T + \lfloor v_3 \rceil \cdot (-1, -1, -1, 2)^T$
5: **end if**

Key Reuse Attack. For a key exchange protocol, suppose that an active adversary $\mathcal{A}$ has the ability to initiate any number of key exchange sessions with party

---

Algorithm 2. Decode

Require: $\mathbf{v} \in \mathbb{R}^4 / \mathbb{Z}^4$
Ensure: A bit c such that $c\mathbf{g}$ is a closest vector to $\mathbf{v} + \mathbb{Z}^4$
1: $\mathbf{x} = \mathbf{v} - \lfloor \mathbf{v} \rceil$
2: **return** 0 if $\|\mathbf{x}\|_1 \leq 1$ and 1 otherwise

B who reuses his secret key s_B. An adversary plays the role of party A in the protocol and aims to recover s_B. The adversary $\mathscr{A}$ can set p_A adaptively by deviating from the protocol. Then in a key reuse attack on NewHope, once $\mathscr{A}$ has collected enough sequences of $(p_B^{(i)}, \mathbf{r}^{(i)})$ from party B, he can recover s_B.

4   Key Reuse Attack on NewHope

In this Section, we describe our key reuse attack on NewHope. We firstly give a general overview of the attack in Sect. 4.1. In Sect. 4.2, we introduce a special sequence which will be used in our attack techniques. In Sect. 4.3, we describe the sequence selecting and period counting techniques which are aimed to recover the secret.

4.1   General Overview of Our Attack

We denote the deviated public key of the adversary as $p_{\mathscr{A}}$, and the secret and error terms of the adversary as $s_{\mathscr{A}}$ and $e_{\mathscr{A}}$ respectively. To explain our attack strategy, we firstly consider the simpler case when the error term e_2 is not added to the key computation of v_B of party B.

Choice of $s_{\mathscr{A}}$ and $e_{\mathscr{A}}$: The attacker chooses $s_{\mathscr{A}}$ to be 0 in R_q. The attacker chooses $e_{\mathscr{A}}$ to be the identity element 1 in R_q, and computes $p_{\mathscr{A}} = ke_{\mathscr{A}} = k$, where k takes values in $\mathbb{Z}_q$.

If we look at the key computation of B, we have $v_B = p_{\mathscr{A}} s_B = ks_B$. This results in the *signal* $\mathbf{r} \xleftarrow{\$} \text{FullHelpRec}(v_B)$ sent by the party B leaks the information of s_B, which is explained in the Sect. 4.2.

Oracle $\mathcal{S}$: There exists an oracle $\mathcal{S}$ which can be used to simulate party B's response. $\mathcal{S}$ performs the action of B and the adversary $\mathscr{A}$ has access to this oracle to make multiple queries. Assume that s_B is fixed for party B and $\mathcal{S}$ has access to the secret s_B. The input of $\mathcal{S}$ is $p_{\mathscr{A}}$ from $\mathscr{A}$. Oracle $\mathcal{S}$ computes $v_B = p_{\mathscr{A}} s_B$ and outputs $\mathbf{r} = \text{FullHelpRec}(v_B) \in \mathbb{Z}_4^n$.

We give the attack steps in the following.
Key reused attack by the adversary $\mathscr{A}$:

1. For every $k \in \{0, 1, \ldots, q-1\}$:
   a. Set $p_{\mathscr{A}} = k$.
   b. Invoke the oracle $\mathcal{S}$ with $p_{\mathscr{A}}$, and obtain $\mathcal{S}$'s output $\mathbf{r}_k = (r_{k,0}, \ldots, r_{k,n-1})$.

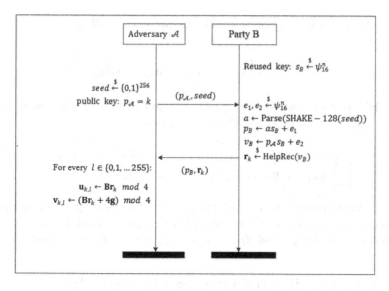

Fig. 1. One key exchange session of our key reused attack on NewHope key exchange.

2. For every $l \in \{0, 1, \ldots, \frac{n}{4} - 1\}$:
   a. For every $k \in \{0, 1, \ldots, q - 1\}$, set $\mathbf{r}_{k,l} := (r_{k,l}, r_{k,l+\frac{n}{4}}, r_{k,l+2 \cdot \frac{n}{4}}, r_{k,l+3 \cdot \frac{n}{4}})$
      and set vectors: $\mathbf{u}_{k,l} := \mathbf{Br}_{k,l} \bmod 4$; $\mathbf{v}_{k,l} := (\mathbf{Br}_{k,l} + 4\mathbf{g}) \bmod 4$.
   b. For every $i \in \{0, 1, 2, 3\}$:
      i. Set two sequences: $U_{l+\frac{n}{4} \cdot i} := \{\mathbf{u}_{0,l}[i], \mathbf{u}_{1,l}[i], \ldots, \mathbf{u}_{q-1,l}[i]\}$;
         $Z_{l+\frac{n}{4} \cdot i} := \{\mathbf{v}_{0,l}[i], \mathbf{v}_{1,l}[i], \ldots, \mathbf{v}_{q-1,l}[i]\}$.
      ii. Use $U_{l+\frac{n}{4} \cdot i}$ and $Z_{l+\frac{n}{4} \cdot i}$ to compute secret coefficient $s_B[l + \frac{n}{4} \cdot i]$.

Figure 1 is one instance of query of our attack. The most important step of our attack is in step (2.b.ii). In next several sections, we focus on how to compute secret's coefficient $s_B[l + \frac{n}{4} \cdot i]$ using sequences $U_{l+\frac{n}{4} \cdot i}$ and $Z_{l+\frac{n}{4} \cdot i}$.

4.2 Preparation

For fixed $l \in \{0, 1 \ldots, \frac{n}{4} - 1\}$ and $i \in \{0, 1, 2, 3\}$. In this section we describe that a special sequence can be selected from two sequences $U_{l+\frac{n}{4} \cdot i}$ and $Z_{l+\frac{n}{4} \cdot i}$. We show that this special sequence has "periodic" property, which is the key of our attack. And we define two periodic function to analysis the property of this special sequence.

The Special Sequence. Here, we describe a special sequence. If we look at oracle $\mathcal{S}$'s computation of the key v_B, we have $v_B = p_{\mathcal{A}} s_B = k \cdot s_B \bmod q$ where $k \in \mathbb{Z}$. This results in the *split vector* of v_B to be: $\mathbf{x}_{k,l} = (k s_B[l], k s_B[l + \frac{n}{4}], k s_B[l + 2 \cdot \frac{n}{4}], k s_B[l + 3 \cdot \frac{n}{4}])^T \bmod q$. For function HelpRec($\mathbf{x}_{k,l}; b$) in oracle $\mathcal{S}$, we firstly consider the case when $b = 0$ (The case when b is randomly selected from $\{0,1\}$ is described in Sect. 4.4). Then suppose $\mathbf{r}_{k,l} = \text{HelpRec}(\mathbf{x}_{k,l}; 0) \in \mathbb{Z}_4^4$, we can set

$$\hat{\mathbf{r}}_{k,l} := \text{CVP}_{\tilde{D}_4}(\frac{4}{q}\mathbf{x}_{k,l}) \in \mathbb{Z}^4 \text{ and } \mathbf{a}_{k,l} := \mathbf{B}\hat{\mathbf{r}}_{k,l} \in \mathbb{R}^4.$$

By the definition of Algorithm 1, it's easy to find that $\mathbf{a}_{k,l}$ is the closest vector of vector $\frac{4}{q}\mathbf{x}_{k,l}$ in lattice $\tilde{D}_4$ and there is equation:

$$\mathbf{a}_{k,l}[i] = \begin{cases} \lceil \frac{4}{q}(ks_B[l + \frac{n}{4} \cdot i] \bmod q) \rceil & \|\frac{4}{q}\mathbf{x}_{k,l} - \lfloor \frac{4}{q}\mathbf{x}_{k,l} \rceil\|_1 < 1 \\ \lfloor \frac{4}{q}(ks_B[l + \frac{n}{4} \cdot i] \bmod q) - \frac{1}{2} \rceil + \frac{1}{2} & \text{others.} \end{cases} \quad (4)$$

Obviously, the sequence $(\mathbf{a}_{k,l}[i] \bmod 4)_{k=0,1,\ldots,q-1}$ has some "periodic" property.

And here we describe the relationships between $(\mathbf{a}_{k,l}[i] \bmod 4)_{k=0,1,\ldots,q-1}$ and the two sequences $U_{l+\frac{n}{4}\cdot i}$ and $Z_{l+\frac{n}{4}\cdot i}$. Since $\mathbf{r}_{k,l} = \hat{\mathbf{r}}_{k,l} \bmod 4$, it's easy to verify that:

$$\mathbf{a}_{k,l} \bmod 4 = \mathbf{Br}_{k,l} \bmod 4 \text{ or } \mathbf{a}_{k,l} \bmod 4 = (\mathbf{Br}_{k,l} + 4\mathbf{g}) \bmod 4.$$

Note that in the step (2.a) of the attack, the adversary sets $\mathbf{u}_{k,l} := \mathbf{Br}_{k,l} \bmod 4$; $\mathbf{v}_{k,l} := (\mathbf{Br}_{k,l} + 4\mathbf{g}) \bmod 4$. Thus, one of vectors $\mathbf{u}_{k,l}$ and $\mathbf{v}_{k,l}$ is $\mathbf{a}_{k,l} \bmod 4$ and the other is $(\mathbf{a}_{k,l} + 4\mathbf{g}) \bmod 4$. Hence every element in $(\mathbf{a}_{k,l}[i] \bmod 4)_{k=0,1\ldots,q-1}$ is in one of sequences $U_{l+\frac{n}{4}\cdot i}$ and $Z_{l+\frac{n}{4}\cdot i}$.

Two Periodic Function. Here we say that $(\mathbf{a}_{k,l}[i] \bmod 4)_{k=0,1,\ldots,q-1}$ reveals some information of the secret s_B. To analysis the "periodic" property of $(\mathbf{a}_{k,l}[i] \bmod 4)_{k=0,1,\ldots,q-1}$, we can define the following two functions:

$$f_{0,x}(h) := \lceil \frac{4}{q} \cdot (h \cdot x \bmod q) \rceil \bmod 4;$$

$$f_{1,x}(h) := \left(\lfloor \frac{4}{q} \cdot (h \cdot x \bmod q) - \frac{1}{2} \rceil + \frac{1}{2} \right) \bmod 4.$$

where $x \in \mathbb{Z}$, $h \in \mathbb{R}$. When $x = 0$, $f_{0,0}(h) = 0$ and $f_{1,0}(h) = 0.5$. When $x \neq 0$, $f_{0,x}(h)$ and $f_{1,x}(h)$ are periodic functions with the fundamental period $N := \frac{q}{|x \bmod \pm q|}$. We denote the number of period for function $f(h)$ in domain $[0, q)$ as $P_{f,[0,q)}$. Then $P_{f_{0,x},[0,q)} = P_{f_{1,x},[0,q)} = |x \bmod^{\pm} q|$. One instance of function $f_{0,x}(k)$ and $f_{1,x}(k)$ is illustrated by Fig. 2. For sequence $(\mathbf{a}_{k,l}[i] \bmod 4)_{k=0,1,\ldots,q-1}$, equation

$$\mathbf{a}_{k,l}[i] \bmod 4 = f_{0,s_B[l+\frac{n}{4}\cdot i]}(k) \text{ or } \mathbf{a}_{k,l}[i] \bmod 4 = f_{1,s_B[l+\frac{n}{4}\cdot i]}(k) \quad (5)$$

holds for every $k \in \{0, 1, \ldots, q-1\}$. Then for such sequence $(\mathbf{a}_{k,l}[i] \bmod 4)_{k=0,1,\ldots,q-1}$, we define its number of period as $P_{\mathbf{a}_{k,l}[i] \bmod 4,[0,q)} := P_{f_{0,s_B[l+\frac{n}{4}\cdot i]},[0,q)} = |s_B[l + \frac{n}{4} \cdot i] \bmod^{\pm} q|$. Thus, the attacker can compute $|s_B[l + \frac{n}{4} \cdot i] \bmod^{\pm} q|$ by computing $P_{f_{0,s_B[l+\frac{n}{4}\cdot i]},[0,q)}$ or $P_{f_{1,s_B[l+\frac{n}{4}\cdot i]},[0,q)}$ once he has obtained $(\mathbf{a}_{k,l}[i] \bmod 4)_{k=0,1,\ldots,q-1}$.

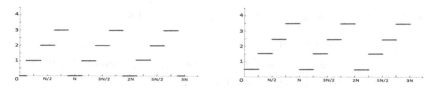

Fig. 2. Left figure represent $f_{0,3}(k)$ and the right figure represent $f_{1,3}(k)$, where $k \in [0, 12289]$, $N = \frac{12289}{3}$ which is defined in Sect. 4.2. The Vertical Axis for the value of function, and the Horizontal Axis for k.

4.3   Recover the Secret Key

In this section, we describe how to compute every coefficient of s_B using sequence $U_{l+\frac{n}{4}\cdot i}$ and $Z_{l+\frac{n}{4}\cdot i}$. This section is divided into two parts. The first part aims to select the sequence $(a_{k,l}[i] \bmod 4)_{k=0,1,...,q-1}$ described in above section, and the second part aims to compute every coefficient of s_B using the sequence we select.

Select Sequence. Here, we describe how to obtain $(a_{k,l}[i] \bmod 4)_{k=0,1,...,q-1}$ using $2q$ values $(u_{k,l}[i])_{k=0,1,...,q-1}$ and $(v_{k,l}[i])_{k=0,1,...,q-1}$.

In order to obtain $(a_{k,l}[i] \bmod 4)_{k=0,1,...,q-1}$, we have to find the relationships between the two adjacent elements $a_{k,l}[i] \bmod 4$ and $a_{k+1,l}[i] \bmod 4$. By Eq. (5), if $s_B[l + \frac{n}{4} \cdot i] \bmod^{\pm} q > 0$, there may exists some $k = \lfloor \frac{nN}{8} \rfloor$ where $n \in \mathbb{Z}$, such that:

$$(a_{k+1,l}[i] \bmod 4 - a_{k,l}[i] \bmod 4) \bmod 4 = 1; \tag{6}$$

and if $s_B[l + \frac{n}{4} \cdot i] \bmod^{\pm} q < 0$, there may exists some $k = \lfloor \frac{nN}{8} \rfloor$ where $n \in \mathbb{Z}$ such that:

$$(a_{k+1,l}[i] \bmod 4 - a_{k,l}[i] \bmod 4) \bmod 4 = -1; \tag{7}$$

and for others k in above two cases, there are:

$$-0.5 \le (a_{k+1,l}[i] \bmod 4 - a_{k,l}[i] \bmod 4) \bmod^{\pm} 4 \le 0.5. \tag{8}$$

We can select the sequence $(a_{k,l}[i] \bmod 4)_{k=0,1,...,q-1}$ using relations (6), (7) and (8). Firstly, to ensure the sequence we select is $(a_{k,l}[i] \bmod 4)_{k=0,1,...,q-1}$, we set the first number of the sequence:

$$w_0 = \begin{cases} u_{0,l}[i] & \text{if } |u_{0,l}[i] - 2| > |v_{0,l}[i] - 2|; \\ v_{0,l}[i] & \text{else.} \end{cases}$$

Here, we will select two sequences - the sequence W for the case when $s_B[l+\frac{n}{4}\cdot i] \bmod^{\pm} q > 0$ and the other sequence T for the case when $s_B[l+\frac{n}{4}\cdot i] \bmod^{\pm} q < 0$. Define a mapping:

$$\psi : \mathbb{R}^{2q} \qquad\qquad \to \mathbb{R}^{2q}, $$
$$(U_{l+\frac{n}{4}\cdot i}, Z_{l+\frac{n}{4}\cdot i}) \quad \mapsto (W, T) \tag{9}$$

where the elements in sequence $W = (w_k)_{k=0,\ldots,q-1}$ satisfy:

$$w_k = \begin{cases} w_0 & \text{when } k = 0; \\ \mathbf{u}_{k,l}[i] & \text{when } k > 0 \text{ and } (-0.5 \leq (\mathbf{u}_{k,l}[i] - w_{k-1}) \bmod^{\pm} 4 \leq 0.5 \text{ or} \\ & (\mathbf{u}_{k,l}[i] - w_{k-1}) \bmod 4 = 1); \\ \mathbf{v}_{k,l}[i] & \text{others.} \end{cases}$$

and the elements in sequence $T = (t_k)_{k=0,\ldots,q-1}$ satisfy:

$$t_k = \begin{cases} w_0 & \text{when } k = 0; \\ \mathbf{u}_{k,l}[i] & \text{when } k > 0 \text{ and } (-0.5 \leq (\mathbf{u}_{k,l}[i] - t_{k-1}) \bmod^{\pm} 4 \leq 0.5 \text{ or} \\ & (\mathbf{u}_{k,l}[i] - t_{k-1}) \bmod 4 = -1); \\ \mathbf{v}_{k,l}[i] & \text{others.} \end{cases}$$

Note that if Eq. (8) holds for every $k \in [0,q)$, $W = T = (\mathbf{a}_{k,l}[i] \bmod$

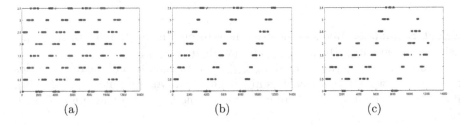

(a)   (b)   (c)

Fig. 3. An experimental data when one split vector of v_B is $\mathbf{x}_{k,l} = k \cdot (5, -7, 3, 4) \bmod q$. We put all values of $(\mathbf{u}_{k,l}[2])_{k=0,1,\ldots,q-1}$ and $(\mathbf{v}_{k,l}[2])_{k=0,1,\ldots,q-1}$ in the figure (a). Our selected sequence W is illustrated in the figure (b). And selected sequence T is illustrated in the figure (c). We can note that W has "period" property, while T is "out-of-order". The Vertical Axis for the value of the elements in sequence, and the Horizontal Axis for k.

$4)_{k=0,1,\ldots,q-1}$. If $W \neq T$, one of the two sequences W and T is $(\mathbf{a}_{k,l}[i] \bmod 4)_{k=0,1,\ldots,q-1}$. And if $s_B[l + \frac{n}{4} \cdot i] \bmod^{\pm} q > 0$, W is $(\mathbf{a}_{k,l}[i] \bmod 4)_{k=0,1,\ldots,q-1}$, while if $s_B[l + \frac{n}{4} \cdot i] \bmod^{\pm} q < 0$, T is $(\mathbf{a}_{k,l}[i] \bmod 4)_{k=0,1,\ldots,q-1}$. One instance of experimental data of our selecting technique is illustrated by Fig. 3, and we note that in this instance $s_B[l + \frac{n}{4} \cdot i] \bmod^{\pm} q = 3$, so $W = (\mathbf{a}_{k,l}[i] \bmod 4)_{k=0,1,\ldots,q-1}$.

Compute the Coefficients. Now, we show how to compute $s_B[l + \frac{n}{4} \cdot i] \bmod^{\pm} q$ using sequences W and T. In Sect. 4.2, we know that $P_{\mathbf{a}_{k,l}[i] \bmod 4, [0,q)} = |s_B[l + \frac{n}{4} \cdot i] \bmod^{\pm} q|$. Construct an algorithm Counting to compute the number of period of sequence W (or T). Algorithm Counting is listed as Algorithm 3.

We note that computing $P_{\mathbf{a}_{k,l}[i] \bmod 4, [0,q)}$ means that to compute $P_{f_{0,s_B[l+\frac{n}{4}\cdot i]}, [0,q)}$ or $P_{f_{1,s_B[l+\frac{n}{4}\cdot i]}, [0,q)}$. Since $|s_B[l + \frac{n}{4} \cdot i]| \leq 16$, for periodic functions $f_{0,s_B[l+\frac{n}{4}\cdot i]}$ and $f_{1,s_B[l+\frac{n}{4}\cdot i]}$, there exists h' in every period, such that

$f_{0,s_B[l+\frac{n}{4}\cdot i]}(h') = 0$ or $f_{1,s_B[l+\frac{n}{4}\cdot i]}(h') = 0.5$ or 3.5. In Algorithm 3, the count number c plus 1 at the point when the element in sequence W firstly equals to 0 or 0.5 or 3.5 in one period interval. Thus every period of function $f_{0,s_B[l+\frac{n}{4}\cdot i]}$ or $f_{1,s_B[l+\frac{n}{4}\cdot i]}$ in domain $[0,q)$ can be counted.

Algorithm 3. Counting

Require: $W = \{w_0, w_1, \ldots, w_{q-1}\} \in \mathbb{R}^q$
Ensure: The number of period of sequence W.
1: $k \leftarrow 0; c \leftarrow 0$
2: **while** ($w_k \neq 0$ or $w_k \neq 0.5$ or $w_k \neq 3.5$) **do**
3:     $k \leftarrow k+1$
4: **end while**
5: $c \leftarrow c+1$
6: **while** ($w_k \neq 2$ or $w_k \neq 2.5$) **do**
7:     $k \leftarrow k+1$
8: **end while**
9: **if** $k < q-1$ **then**
10:     **goto** step 2
11: **end if**
12: **return** c

Suppose $c_1 \leftarrow \text{Counting}(W)$ and $c_2 \leftarrow \text{Counting}(T)$. If $W = T$, $c_1 = c_2 = |s_B[l + \frac{n}{4} \cdot i] \bmod^{\pm} q|$. Then to determine the sign of $s_B[l + \frac{n}{4} \cdot i] \bmod^{\pm} q$, define:

$$sign = \sum_{\substack{k \in [0,q/c]; \\ |w_{k+1} - w_k| < 3}} (w_{k+1} - w_k).$$

where c is the output of Counting(W) and $w_k \in W$. Note that $[0, q/c]$ is one periodic interval. If $s_B[l + \frac{n}{4} \cdot i] \bmod^{\pm} q > 0$, it is easy to verified that $sign > 0$, and if $s_B[l + \frac{n}{4} \cdot i] \bmod^{\pm} q < 0$, there is $sign < 0$.

When $W \neq T$, we want to figure out that which one of c_1 and c_2 equals to $|s_B[l + \frac{n}{4} \cdot i] \bmod^{\pm} q|$. We need a parameter to measure the degree of approximation of the sequence we select and $(\mathbf{a}_{k,l}[i] \bmod 4)_{k=0,1,\ldots,q-1}$. Suppose the input sequence of Algorithm 3 is W. Let the value of k be k_j at the point when the Algorithm 3 loops to steps 5 for the j-th times ($0 < j \leq c_1$). Then the domina size of j-th periodic interval of W is $N_j = |k_{j+1} - k_j|$ ($j \in \{1, 2, \ldots, c_1 - 1\}$). Define parameter

$$var_W = \frac{\sum_{j=1}^{c_1-1} |N_j - \frac{\sum_{j=1}^{c_1-1} N_j}{c_1-1}|}{c_1 - 1}, \tag{10}$$

which is the variance of $N_1, \ldots, N_{c_1-1}$. Similarly, we can define var_T for sequence T. If $W = (\mathbf{a}_{k,l}[i] \bmod 4)_{k=0,1,\ldots,q-1}$, var_W should be very small because every two number in $\{N_1, \ldots, N_{c_1-1}\}$ are almost equal. Meanwhile sequence T hasn't "periodic" property (one instance is illustrated by Fig. 3c), which results in $var_T > var_W$. Similarly, the case when $T = (\mathbf{a}_{k,l}[i] \bmod 4)_{k=0,1,\ldots,q-1}$, $var_T <$

var_W. This property is verified in our experiments. Then if $var_W < var_T$, the sign of coefficient $s_B[l + \frac{n}{4} \cdot i] \bmod^{\pm} q$ is "+" and $|s_B[l + \frac{n}{4} \cdot i] \bmod^{\pm} q| = c_1$, otherwise the sign of $s_B[l + \frac{n}{4} \cdot i]$ is "−" and $|s_B[l + \frac{n}{4} \cdot i] \bmod^{\pm} q| = c_2$. Thus we can compute every coefficient of s_B.

4.4 Effect of e_2 and Parameter b

In our experiments, for the case where e_2 is added to v_B and b is chosen randomly from $\{0, 1\}$ in function $\text{HelpRec}(\mathbf{x}_{k,l}; b)$, we can still get the secret key correctly using algorithm described above. And the following analysis is only theoretical. Firstly, we analysis the case when e_2 is added to v_B. The key computation of B is $v_B = p_{\mathscr{A}} s_B + e_2$. Let $\tilde{\mathbf{x}}_{k,l} = (ks_B[l] + g_0, ks_B[l + \frac{n}{4}] + g_1, ks_B[l + 2 \cdot \frac{n}{4}] + g_2, ks_B[l + 3 \cdot \frac{n}{4}] + g_3)^T \bmod q$ where $g_i \overset{\$}{\leftarrow} \psi_{16}$ is one of *split vector* of v_B. Then if $s_B[l + \frac{n}{4} \cdot i] \bmod^{\pm} q > 0$, for Eq. (4), there may exists k such that

$$||\frac{4}{q}\tilde{\mathbf{x}}_{k,l} - \lfloor \frac{4}{q}\tilde{\mathbf{x}}_{k,l} \rceil ||_1 < 1; \quad ||\frac{4}{q}\tilde{\mathbf{x}}_{k+1,l} - \lfloor \frac{4}{q}\tilde{\mathbf{x}}_{k+1,l} \rceil ||_1 < 1$$

and

$$\lfloor \frac{4}{q}(ks_B[l + \frac{n}{4} \cdot i] + g_i \bmod q) \rceil - \lfloor \frac{4}{q}((k+1)s_B[l + \frac{n}{4} \cdot i] + g'_i \bmod q) \rceil = 1 \quad (11)$$

where $g_i, g'_i \overset{\$}{\leftarrow} \psi_{16}$. Equation (11) is equivalent to: $(\mathbf{a}_{k+1,l}[i] \bmod 4 - \mathbf{a}_{k,l}[i] \bmod 4) \bmod 4 = -1$. Thus the two sequences selected by mapping (9) will be both wrong. Similarly, if $s_B[l + \frac{n}{4} \cdot i] \bmod^{\pm} q < 0$, equation $(\mathbf{a}_{k+1,l}[i] \bmod 4 - \mathbf{a}_{k,l}[i] \bmod 4) \bmod 4 = 1$ may occurs. We hope to eliminate such wrong points in mapping (9). Our idea of improvement is to detect such wrong point and "delete" it. We add the following two steps to the mapping (9): (1) when selecting the sequence W, for k such that $|(\mathbf{u}_{k,l}[i] - w_{k-1}) \bmod^{\pm} 4| = 1$ and $|(\mathbf{u}_{k,l}[i] - \mathbf{u}_{k+1,l}[i]) \bmod^{\pm} 4| = 1$, let $w_k = w_{k-1}$; (2) when selecting the sequence T, for k such that $|(\mathbf{u}_{k,l}[i] - t_{k-1}) \bmod^{\pm} 4| = 1$ and $|(\mathbf{w}_{k,l}[i] - \mathbf{w}_{k+1,l}[i]) \bmod^{\pm} 4| = 1$, let $t_k = t_{k-1}$. Since such wrong points are sparse in $[0, q)$ (actually in our experiments we failed to found such wrong points), the "periodic" property of sequence we select by this way doesn't change, hence this improvement hasn't influence to the algorithm Counting. Other steps of the attack is all the same with the case when e_2 is not added to v_B.

Similarly, we consider the case when b is chosen randomly from $\{0, 1\}$ in function $\text{HelpRec}(\mathbf{x}_{k,l}; b)$. Firstly consider the case when $s_B[l + \frac{n}{4} \cdot i] \bmod^{\pm} q > 0$. Suppose two inputs for function $\text{HelpRec}(\mathbf{x}; b)$ is $(\mathbf{x}_{k,l}, 1)$ and $(\mathbf{x}_{k+1,l}, 0)$. Then for Eq. (4), there may exists k such that:

$$||\frac{4}{q}(\mathbf{x}_{k,l} + \mathbf{g}) - \lfloor \frac{4}{q}(\mathbf{x}_{k,l} + \mathbf{g}) \rceil ||_1 < 1; ||\frac{4}{q}\mathbf{x}_{k+1,l} - \lfloor \frac{4}{q}\mathbf{x}_{k+1,l} \rceil ||_1 < 1$$

and

$$\lfloor \frac{4}{q}((ks_B[l + \frac{n}{4} \cdot i] + 2) \bmod q)) \rceil - \lfloor \frac{4}{q}((k+1)s_B[l + \frac{n}{4} \cdot i] \bmod q) \rceil = -1$$

hold. This equals to $(\mathbf{a}_{k+1,l}[i] \bmod 4 - \mathbf{a}_{k,l}[i] \bmod 4) \bmod 4 = -1$. Similarly for the case when $s_B[l + \frac{n}{4} \cdot i] \bmod^{\pm} q < 0$, $(\mathbf{a}_{k+1,l}[i] \bmod 4 - \mathbf{a}_{k,l}[i] \bmod 4) \bmod 4 = 1$. Note that such wrong points is same with the case when e_2 is added to v_B described in above paragraph, and we can use the same way to eliminate such wrong points in mapping (9).

4.5   Adversary Time Complexity

From the above description of our attack, it is clear that the adversary needs q queries to recover every coefficient of s_B. The time complexity of selecting the sequence W and T for every coefficient of s_B is $2q$ and there needs about $2q$ times to compute the exact value of the coefficient using sequences W and T. Suppose times of once querying is t. Since q is $O(n)$, the query complexity of the complete attack is $q = O(n)$ and times complexity is $qt + n \cdot (2q + 2q) = O(n^2)$.

5   Conclusion

In this work, we have presented an detailed key reused attack on NewHope key exchange in recovering the secret of a reused key with q queries. We show that for NewHope key exchange, when the public key is fixed for a long term, an active adversary can collect a sequence of the *signal* and construct a sequence with "periodic" property, which reveals the information of the secret. The adversary can exploits the "periodic" property of the sequence and recovers the secret key of the honest party. We believe that such strategy of the key reuse attack can also be adapted to NISTPQC submission NewHope IND-CPA KEM [1]. But the NewHope IND-CCA KEM would stop the attack. This version of key exchange applies the Fujisaki-Okamoto transform and achieves CCA security.

Acknowledgments. This article is supported by The National Key Research and Development Program of China (Grant No. 2017YFA0303903). Authors thank Yang Yu for discussions and the anonymous ICISC'18 reviewers for helpful comments.

References

1. Alkim, E., Avanzi, R., Bos, J.W., Ducas, L.: NewHope, algorithm specifcations and supporting documentation. Version 1.0. Submission to NIST, 30 November 2017. https://newhopecrypto.org/data/NewHope_2017_12_21.pdf
2. Alkim, E., Ducas, L., Pöppelmann, T., Schwabe, P.: Post-quantum key exchange - a new hope. In: Proceedings Of the 25th USENIX Security Symposium, pp. 327–343. USENIX Association
3. Bos, J.W., et al.: Frodo: take off the ring! practical, quantum-secure key exchange from LWE. In: Proceedings Of the 2016 ACM SIGSAC Conference on Computer and Communications Security, pp. 1006–1018. ACM Press (2016)
4. Bos, J.W., Costello, C., Naehrig, M., Stebila, D.: Post-quantum key exchange for the TLS protocol from the ring learning with errors problem. In: 2015 IEEE Symposium on Security and Privacy, pp. 553–570. IEEE Computer Society Press, May 2015

5. Bos, J., et al.: CRYSTALS-Keyber: a CCA-secure module-lattice-base KEM. Cryptology ePrint Archive, Report 2017/634 (2017). http://eprint.iacr.org/2017/634

6. Ding, J.T., Alsayigh, S., Saraswathy, R.V., Fluhrer, S., Lin, X.D.: Leakage of Signal function with reused keys in RLWE key exchange. In: 2017 IEEE International Conference on Communications (ICC)

7. Ding, J., Fluhrer, S., Rv, S.: Complete attack on RLWE key exchange with reused keys, without signal leakage. In: Susilo, W., Yang, G. (eds.) ACISP 2018. LNCS, vol. 10946, pp. 467–486. Springer, Cham (2018). https://doi.org/10.1007/978-3-319-93638-3_27

8. Ding, J.T., Xie, X., Lin, X.: A simple provably secure key exchange scheme based on the learning with errors problem. Cryptology ePrint Archive, Report 2012/688 (2012). http://eprint.iacr.org//2012/688.pdf

9. Experimenting with post-quantum cryptography, July 2016. https://security.googleblog.com/2016/07/experimenting-with-post-quantum.html

10. Fluhrer, S.: Cryptanalysis of ring-LWE based key exchange with key share reuse. Cryptology ePrint Archive, Report 2016/085 (2016). http://eprint.iacr.org/2016/085

11. Lyubashevsky, V., Peikert, C., Regev, O.: On ideal lattices and learning with errors over rings. In: Gilbert, H. (ed.) EUROCRYPT 2010. LNCS, vol. 6110, pp. 1–23. Springer, Heidelberg (2010). https://doi.org/10.1007/978-3-642-13190-5_1

12. Micciancio, D., Regev, O.: Worst-case to average-case reductions based on Gaussian measures. SIAM J. Comput. 37, 267–302 (2007)

13. Peikert, C.: Public-key cryptosystems from the worst-case shortest vector problem. In: Proceedings of the 2009 ACM Symposium on Theory of Computing, Series. STOC 2009, pp. 333-342. ACM, New York (2009)

14. Peikert, C.: Lattice cryptography for the internet. In: Mosca, M. (ed.) PQCrypto 2014. LNCS, vol. 8772, pp. 197–219. Springer, Cham (2014). https://doi.org/10.1007/978-3-319-11659-4_12

15. Regev, O.: On lattices, learning with errors, random linear codes, and cryptography. In: Proceedings of the Thirty-Seventh Annual ACM Symposium on Theory of Computing. STOC 2005, pp. 84–93. ACM, New York (2005)

16. Shor, P.W.: Polynomial-time algorithms for prime factorization and discrete logarithms on a quantum computer. SIAM J. Comput. 1484–1509 (1997)

17. The Internet Defense Prize. https://internetdefenseprize.org/

18. Zhang, J., Zhang, Z., Ding, J., Snook, M., Dagdelen, Ö.: Authenticated key exchange from ideal lattices. In: Oswald, E., Fischlin, M. (eds.) EUROCRYPT 2015. LNCS, vol. 9057, pp. 719–751. Springer, Heidelberg (2015). https://doi.org/10.1007/978-3-662-46803-6_24

Supersingular Isogeny Diffie–Hellman Authenticated Key Exchange

Atsushi Fujioka[1(✉)], Katsuyuki Takashima[2], Shintaro Terada[3], and Kazuki Yoneyama[3]

[1] Kanagawa University, Kanagawa, Japan
fujioka@kanagawa-u.ac.jp
[2] Mitsubishi Electric, Kanagawa, Japan
Takashima.Katsuyuki@aj.MitsubishiElectric.co.jp
[3] Ibaraki University, Ibaraki, Japan
{17nm713n,kazuki.yoneyama.sec}@vc.ibaraki.ac.jp

Abstract. We propose two authenticated key exchange protocols from supersingular isogenies. Our protocols are the first post-quantum one-round Diffie–Hellman type authenticated key exchange ones in the following points: one is secure under the quantum random oracle model and the other resists against maximum exposure where a non-trivial combination of secret keys is revealed. The security of the former and the latter is proven under isogeny versions of the decisional and gap Diffie–Hellman assumptions, respectively. We also propose a new approach for invalidating the Galbraith–Vercauteren-type attack for the gap problem.

Keywords: One-round authenticated key exchange ·
Supersingular isogeny decisional Diffie–Hellman assumption ·
Degree-insensitive supersingular isogeny gap Diffie–Hellman
assumption · CK model · CK$^+$ model · Quantum adversary

1 Introduction

All conventional cryptosystems from discrete logarithm and/or factorization intractability assumptions would be totally broken by the emergence of quantum computers, i.e., by Shor's algorithm [27]. In the post-quantum era, it is important to confirm whether classical cryptographic techniques are still secure against quantum adversaries. Recently, strong security notions and constructions against quantum computers have been intensively studied (e.g., [1,3,10,32,33]). Moreover, National Institute of Standards and Technology has initiated a process to standardize quantum-resistant public-key cryptographic algorithms [24], so, to study quantum-resistant cryptosystems is a hot research area.

Key establishing over insecure channels is one of important cryptographic techniques. In a key establishing protocol, two parties exchange some messages, and then, they can share a key. Recent researches on this have lead to *authenticated key exchange* (AKE). In the post-quantum era, it is preferable to have an

© Springer Nature Switzerland AG 2019
K. Lee (Ed.): ICISC 2018, LNCS 11396, pp. 177–195, 2019.
https://doi.org/10.1007/978-3-030-12146-4_12

AKE protocol secure based on a problem which resists against quantum adversaries. We then propose two quantum-resistant AKE schemes from a (relatively) new mathematical foundation, i.e., supersingular isogenies.

Supersingular Isogeny Diffie–Hellman (SIDH). Computing a sequence of isogenies of elliptic curves is a new cryptographic basic operation in some applications. For example, a cryptographic hash function from expander graphs, proposed in [6], consists of computing an isogeny sequence, which is based on the hardness of constructing an isogeny between two (randomly chosen) isogenous curves. Diffie–Hellman (DH) type key exchange protocols based on isogenies are given by Rostovtsev and Stolbunov [26] and De Feo et al. [11], which were considered as candidates for post-quantum public-key primitives.

Childs et al. [7] considered the isogeny computation problem for *ordinary* elliptic curves, and obtained a subexponential-time quantum algorithm. In contrast, the algorithm cannot be applied to the supersingular case (because of noncommutativity of endomorphism rings). Therefore, both applications above, i.e., hash function and key exchange, need to employ *supersingular* elliptic curves (and the graph consisting of them). In particular, *supersingular isogeny Diffie–Hellman* (SIDH) protocol proposed by De Feo et al. [11] has short public keys compared to other post-quantum candidates, and has been intensively studied for serving as a drop-in replacement to existing Internet protocols [2,8,9].

Very recently, Petit [25] proposed a mathematical attack for the security of SIDH, but also showed that the security is not affected by the attack if we use appropriate public parameters as is given in Sect. 3.

Authenticated Key Exchange. In an AKE protocol, two parties have own static public keys, exchange ephemeral public keys, and compute a session key based on the public keys and the related secret keys. AKE protocols achieve that honest parties can establish a session key, and any malicious party cannot guess the session key. The latter condition is formulated in an indistinguishability game.

Regarding to this security game, several models have been invented, and the Canetti–Krawczyk (CK) model was proposed to capture leakage of the session state [5]. After the proposal, several security requirements have been indicated such as *key compromise impersonation* (KCI), *weak perfect forward secrecy* (wPFS), and *maximal exposure attacks* (MEX) (refer to [21] for KCI, wPFS, and MEX). The CK model has been integrated with KCI, wPFS, and MEX to the CK$^+$ model [13].

Recently, several SIDH AKE protocols have been proposed [14,22,23,31].

Galbraith proposed a one-round[1] protocol (SIDH TS2) in [14] based on the Unified Model DH protocol by Jeong, Katz, and Lee [18]. The protocol is CK-secure under a decisional problem in classical random oracle model (ROM).

---

[1] Galbraith claims that the protocol is one-round however the description shows that it is two-round as the responder generates the response after receiving the first message [14].

Longa shows a two-round SIDH AKE protocol (AKE-SIDH-SIKE) which is CK^+-secure from a KEM scheme [23]. However, it is based on a generic construction known already.

LeGrow, Jao, and Azarderakhsh defined a security model in which the adversary is allowed to make quantum queries, and proposed a *quantum* CK secure (qCK secure) protocol [22]. The protocol, we call it LJA, is secure in the quantum random oracle model (QROM) however it is two-round.

Xu et al. proposed a two-round protocol (AKE_{SIDH-2}) in [31], and the protocol is CK^+-secure under a decisional problem in classical random oracle model (ROM).

It is worth to note here that all the existing SIDH AKE protocols shown above *only* achieve two-pass protocols except the SIDH TS2 protocol. In a one-round protocol, two parties can simultaneously exchange their ephemeral keys, while in a two-pass one, a party has to wait for the ephemeral key from the other party. Moreover, a one-round AKE protocol has several advantages of efficiency, e.g., each party can pre-compute ephemeral keys in advance.

Supersingular Isogeny Gap DH Problem. Traditional DH AKE protocols have been constructed from several forms of DH assumptions, i.e., computational, decisional and gap DH assumptions, for attaining various trade-offs between security and efficiency. Recently, Galbraith and Vercauteren [16] and Thormarker [29] independently proposed attacks, called *GV-type attack* in this paper, on the supersingular isogeny computational DH (SI-CDH) problem with access to *decision degree* oracle, which determines whether two supersingular curves are isogenous of some *specific degree* or not. While the attack can be extended to *some* form of SI version of gap DH (SI-GDH) problem, still, there exist possible approaches to formulate a *secure* form of SI-GDH problem (and assumption) for which the above attack is ineffective. Therefore, it is important to find and establish such *secure* SI-GDH assumptions to rescue (a wide range of) SIDH-based AKE schemes on the gap assumptions. (For surveys on SIDH-related computational problems, refer to [16, 30].)

Contributions. We propose two one-round authenticated key exchange protocols from supersingular isogenies: one is a protocol secure in the CK model with a quantum adversary under a supersingular isogeny version of the DDH assumption, and the other is a protocol secure in the CK^+ model with a classical adversary under a supersingular isogeny version of the gap DH assumption.

We call the latter assumption *degree-insensitive (di-)SI-GDH* assumption in which an adversary has access to a degree-insensitive SI-DDH oracle, and then cannot employ the GV-type attack for which degree distinction is crucial. We expect that the new assumption is of independent interest. Then, both protocols have several advantages of efficiency and wide applicability in practical situations as they retain a simple one-round Diffie–Hellman structure, and are realized in exchanging a single elliptic curve with an auxiliary smooth-order torsion basis, which can be efficiently compressed [2,8]. We give a comparison table of the existing SIDH AKE protocols and our proposals in Table 1.

Table 1. Comparison of SIDH AKE protocols.

	Assumption	Model	Action	Proof
SIDH TS2 [14]	SI-CDH	CK	One-round (see footnote 1)	ROM
AKE-SIDH-SIKE [23]	SI-DDH	CK$^+$	Two-round	ROM
LJA [22]	SI-DDH	qCK	Two-round	QROM
AKE$_{SIDH\text{-}2}$ [31]	SI-DDH	CK$^+$	Two-round	ROM
SIDH UM	SI-DDH	CK	One-round	QROM
Biclique SIDH	di-SI-GDH	CK$^+$	One-round	ROM

Notations. When A is a set, $y \in_R A$ denotes that y is uniformly selected from A. When A is a random variable, $y \leftarrow_R A$ denotes that y is randomly selected from A according to its distribution. We denote the finite field of order q by $\mathbb{F}_q$.

2   Security Models: CK-Security and CK$^+$-Security

This section outlines the CK and CK$^+$ security definitions for two-pass PKI-based authenticated key exchange protocols. Note that, in our *post-quantum* CK and CK$^+$ models, all parties are modeled by probabilistic polynomial-time (ppt) Turing machines while the adversary is modeled by a polynomial time quantum machine. For further CK and CK$^+$ details and explanations, see [12,21]. It is worth to note here that the proposed protocols are one-round and thus, it is enough to describe the security model as for two-pass AKE because a two-pass model includes a one-round one.

We denote a party's identity $\hat{A}$, $\hat{B}$, $\hat{C}$, ..., where the ID space is **IDS**. A party honestly generates its own keys, static public and static secret ones, and the static public key is linked with the party's identity in some systems like PKI.[2] The maximum numbers of parties and sessions are polynomially bound in the security parameter.

We outline our models for a two-pass AKE protocol where parties, $\hat{A}$ and $\hat{B}$, exchange ephemeral public keys, X and Y, i.e., $\hat{A}$ sends X to $\hat{B}$ and $\hat{B}$ sends Y to $\hat{A}$, and thereafter derive a session key. The session key depends on the exchanged ephemeral keys, identifiers of the parties, the static keys, and the protocol instance that is used.

---

[2] Static public keys must be known to both parties in advance. They can be obtained by exchanging them before starting the protocol or by receiving them from a certificate authority. This situation is common for all PKI-based AKE protocols.

Keys. The public key owned by each party and its secret key are called *static public key* and *static secret key*, respectively. The one-time use session information exchanged in the protocol is called *ephemeral public key* as the information is generated from a temporary secret called *ephemeral secret key*.

Session. An invocation of a protocol is called a *session*. A session is activated via an incoming message of the forms $(\Pi, \mathcal{I}, \hat{A}, \hat{B})$ or $(\Pi, \mathcal{R}, \hat{A}, \hat{B}, Y)$, where $\Pi \in \mathbf{PRS}$ is a protocol identifier in the protocol ID space, $\mathbf{PRS}$. If $\hat{A}$ is activated with $(\Pi, \mathcal{I}, \hat{A}, \hat{B})$, then $\hat{A}$ is the session *initiator*, otherwise it is the session *responder*. We say that $\hat{A}$ is the *owner* (resp. *peer*) of session sid if the third (resp. fourth) coordinate of sid is $\hat{A}$. After activation, session initiator $\hat{A}$ creates ephemeral public key X and a new session identified with $(\Pi, \mathcal{I}, \hat{A}, \hat{B}, X, \perp)$, and sends $(\Pi, \mathcal{R}, \hat{B}, \hat{A}, X)$ to the session responder $\hat{B}$, who then prepares ephemeral public key Y and a new session identified with $(\Pi, \mathcal{R}, \hat{B}, \hat{A}, X, Y)$, computes the session key and sends $(\Pi, \mathcal{I}, \hat{A}, \hat{B}, X, Y)$ to $\hat{A}$. Upon receiving $(\Pi, \mathcal{I}, \hat{A}, \hat{B}, X, Y)$, $\hat{A}$ updates the session identifier $(\Pi, \mathcal{I}, \hat{A}, \hat{B}, X, \perp)$ with $(\Pi, \mathcal{I}, \hat{A}, \hat{B}, X, Y)$ and computes a session key for that session. We say that a session is *completed* if its owner computes a session key.

If $\hat{A}$ is the initiator of a session, the session is identified via sid $= (\Pi, \mathcal{I}, \hat{A}, \hat{B}, X, \perp)$ or sid $= (\Pi, \mathcal{I}, \hat{A}, \hat{B}, X, Y)$. If $\hat{B}$ is the responder of a session, the session is identified via sid $= (\Pi, \mathcal{R}, \hat{B}, \hat{A}, X, Y)$. The *matching session* of the session identified via $(\Pi, \mathcal{I}, \hat{A}, \hat{B}, X, Y)$ is a session with identifier $(\Pi, \mathcal{R}, \hat{B}, \hat{A}, X, Y)$ and vice versa.

Adversary. Adversary $\mathcal{M}$ is modeled as a probabilistic Turing machine that controls all communications including session activation. Activation is performed via a Send(MESSAGE) query. The MESSAGE has one of the following forms: $(\Pi, \mathcal{I}, \hat{A}, \hat{B})$, $(\Pi, \mathcal{R}, \hat{A}, \hat{B}, X)$, or $(\Pi, \mathcal{I}, \hat{A}, \hat{B}, X, Y)$. Each party submits its responses to adversary $\mathcal{M}$, who decides the global delivery order.

The secret information of a party is not accessible to adversary $\mathcal{M}$; however, leakage of secret information is obtained via the following adversary queries.

- SessionKeyReveal(sid): $\mathcal{M}$ obtains the session key for the session with session identifier sid, provided that the session is completed.
- SessionStateReveal(sid): $\mathcal{M}$ obtains the session state of the owner of session sid if the session is not completed (the session key is not established yet). The session state includes all ephemeral secret keys and intermediate computation results except for immediately erased information but does not include the static secret key.
- Corrupt($\hat{A}$): The query allows $\mathcal{M}$ to obtain all information of party $\hat{A}$. If a party, $\hat{A}$, is corrupted by a Corrupt($\hat{A}$) query issued by $\mathcal{M}$, then we call the party, $\hat{A}$, *dishonest*. If not, we call the party *honest*.

Definition 1 (Freshness). *Let* sid* *be the session identifier of a completed session, owned by an honest party $\hat{A}$ with an honest peer $\hat{B}$. If the matching* *session exists, then let* $\overline{\text{sid}}^*$ *be the session identifier of the matching session of* sid*. *Define* sid* *to be* fresh *if none of the following conditions hold:*

- $\mathcal{M}$ *issues* SessionKeyReveal(sid*), *or* SessionKeyReveal($\overline{\text{sid}}^*$) *if* $\overline{\text{sid}}^*$ *exists.*
- $\overline{\text{sid}}^*$ *exists and* $\mathcal{M}$ *makes either of the following queries*
  - SessionStateReveal(sid*) *or* SessionStateReveal($\overline{\text{sid}}^*$),
- $\overline{\text{sid}}^*$ *does not exist and* $\mathcal{M}$ *makes the following query*
  - SessionStateReveal(sid*).

Security Experiment. Initially, adversary $\mathcal{M}$ is given a set of honest parties, for whom $\mathcal{M}$ selects identifiers. Then the adversary makes any sequence of the queries described above. During the experiment, $\mathcal{M}$ makes a special query Test(sid*), where sid* is the session identifier of a fresh session, and is given with equal probability either the session key held by sid* or a random key; the query does not terminate the experiment. The experiment continues until $\mathcal{M}$ makes a guess whether the key is random or not. The adversary *wins* the game if the test session sid* is still fresh and if the guess by $\mathcal{M}$ was correct. The advantage of quantum adversary $\mathcal{M}$ in the AKE experiment with AKE protocol Π is defined as

$$\mathbf{Adv}_{\Pi}^{\text{AKE}}(\mathcal{M}) = \Pr[\mathcal{M} \text{ wins}] - \frac{1}{2}.$$

Definition 2 (Post-quantum CK security). *We say that an AKE protocol* Π *is post-quantum secure in the CK model if the following conditions hold:*

1. *If two honest parties complete matching sessions, then, except with negligible probability, they both compute the same session key.*
2. *For any polynomial-time quantum adversary* $\mathcal{M}$, $\mathbf{Adv}_{\Pi}^{\text{AKE}}(\mathcal{M})$ *is negligible in security parameter* λ *for the test session* sid*,
   (a) *if* $\overline{\text{sid}}^*$ *does not exist, or*
   (b) *if* $\overline{\text{sid}}^*$ *exists, and the static secret key of the owner of* sid* *and the static secret key of the owner of* $\overline{\text{sid}}^*$ *are given to* $\mathcal{M}$.

Definition 3 (Post-quantum CK$^+$ security). *We say that an AKE protocol* Π *is post-quantum secure in the CK$^+$ model if the following conditions hold:*

1. *If two honest parties complete matching sessions, then, except with negligible probability, they both compute the same session key.*
2. *For any polynomial-time quantum adversary* $\mathcal{M}$, $\mathbf{Adv}_{\Pi}^{\text{AKE}}(\mathcal{M})$ *is negligible in security parameter* λ *for the test session* sid*,
   (a) *if* $\overline{\text{sid}}^*$ *does not exist, and the static secret key of the owner of* sid* *is given to* $\mathcal{M}$,
   (b) *if* $\overline{\text{sid}}^*$ *does not exist, and the ephemeral secret key of the owner of* sid* *is given to* $\mathcal{M}$,
   (c) *if* $\overline{\text{sid}}^*$ *exists, and the static secret key of the owner of* sid* *and the static secret key of the owner of* $\overline{\text{sid}}^*$ *are given to* $\mathcal{M}$,
   (d) *if* $\overline{\text{sid}}^*$ *exists, and the ephemeral secret key of the owner of* sid* *and the ephemeral secret key of the owner of* $\overline{\text{sid}}^*$ *are given to* $\mathcal{M}$,

(e) if $\overline{\text{sid}}^*$ exists, and the static secret key of the owner of sid^* and the ephemeral secret key of the owner of $\overline{\text{sid}}^*$ are given to $\mathcal{M}$, or

(f) if $\overline{\text{sid}}^*$ exists, and the ephemeral secret key of the owner of sid^* and the static secret key of the owner of $\overline{\text{sid}}^*$ are given to $\mathcal{M}$.

The static and ephemeral public keys of our schemes include supersingular curves and points on them. We can test supersingularity of curves in polynomial time, e.g., [28]. We make an important remark: While Krawczyk mentions a strong adversary model where a corrupted party can choose to register any public key of its choice at any point during the protocol as a variant of the CK($^+$) model in [21], we do not allow the re-registration of static public key (similar to the CK($^+$) model), and the initial public key is honestly generated and has been used until the end of the protocol. It is because that an active attack which Galbraith et al. [15] proposed for revealing static keys might be considered as an effective attack when we adopt the above flexible key re-registration.

3   Supersingular Isogeny Diffie–Hellman (SIDH)

We describe the SIDH protocol, whose implementation is investigated in detail in [9] and subsequently in [2,4,8,19,20]. The security is studied in [15,25]. For making user secret keys short, we follow the description in the SIKE document [17], that is, the user key is given as just one scalar, e.g., $k_A \in \mathbb{Z}/\ell_A^{e_A}\mathbb{Z}$.

3.1   Original (Concrete) Description of SIDH

For two small primes ℓ_A, ℓ_B (e.g., $\ell_A = 2, \ell_B = 3$), we choose a large prime p such that $p \pm 1 = f \cdot \ell_A^{e_A}\ell_B^{e_B}$ for a small f and $\ell_A^{e_A} \approx \ell_B^{e_B} = 2^{\Theta(\lambda)}$, where λ is a security parameter. Then, we also choose a random supersingular elliptic curve E over $\mathbb{F}_{p^2}$ with $E(\mathbb{F}_{p^2}) \simeq (\mathbb{Z}/(p \pm 1)\mathbb{Z})^2 \supseteq (\mathbb{Z}/\ell_A^{e_A}\mathbb{Z})^2 \oplus (\mathbb{Z}/\ell_B^{e_B}\mathbb{Z})^2$. We use isogenies, ϕ_A and ϕ_B, with kernels of orders, $\ell_A^{e_A}$ and $\ell_B^{e_B}$, respectively, and the following commutative diagram for the SIDH key exchange between Alice and Bob.

$$
\begin{array}{ccc}
E & \xrightarrow{\ \phi_A\ } & E_A = E/\langle R_A \rangle \\
{\scriptstyle \phi_B}\downarrow & & \downarrow{\scriptstyle \phi_{AB}} \\
E_B = E/\langle R_B \rangle & \xrightarrow{\ \phi_{BA}\ } & E/\langle R_A, R_B \rangle
\end{array}
\qquad
\begin{aligned}
&\text{for } \ker \phi_A = \langle R_A \rangle \subset E[\ell_A^{e_A}], \\
&\ker \phi_B = \langle R_B \rangle \subset E[\ell_B^{e_B}], \\
&\ker \phi_{BA} = \langle \phi_B(R_A) \rangle \subset E_B[\ell_A^{e_A}], \\
&\ker \phi_{AB} = \langle \phi_A(R_B) \rangle \subset E_A[\ell_B^{e_B}].
\end{aligned}
$$

Below we first choose generators P_A, Q_A, P_B, Q_B such that $E[\ell_A^{e_A}] = \langle P_A, Q_A \rangle$, $E[\ell_B^{e_B}] = \langle P_B, Q_B \rangle$ and then set the random curve $E/\mathbb{F}_{p^2}$ and the above generators as public parameters, i.e., we define the generator as $\mathsf{pk}^{\text{sidh}} = (\mathfrak{g} = (E; P_A, Q_A, P_B, Q_B), \mathfrak{e} = (\ell_A, \ell_B, e_A, e_B)) \leftarrow_R \mathsf{Gen}^{\text{sidh}}(1^\lambda)$. Secret-key spaces for Alice and Bob are given as $SK_A = \mathbb{Z}/\ell_A^{e_A}\mathbb{Z}$ and $SK_B = \mathbb{Z}/\ell_B^{e_B}\mathbb{Z}$, respectively. DH-type key exchange is given as below (Fig. 1). Here, since $\langle \phi_B(P_A) + k_A \phi_B(Q_A) \rangle = \langle \phi_B(R_A) \rangle = \ker \phi_{BA}$ and $\langle \phi_A(P_B) + k_B \phi_A(Q_B) \rangle = \langle \phi_A(R_B) \rangle = \ker \phi_{AB}$ hold, we have the equality of the j-invariants $K_{\text{Alice}} = j(E_B/\ker \phi_{BA}) = j(E/\langle R_A, R_B \rangle) = j(E_A/\ker \phi_{AB}) = K_{\text{Bob}}$, and $K = K_{\text{Alice}} = K_{\text{Bob}}$ is a shared key. Alice's output includes $\phi_A(P_B)$ and $\phi_A(Q_B)$ as well as E_A, and the security is based on the hardness of isogeny problem with the auxiliary inputs.

Alice		**Bob**

$k_A \in_R SK_A :$

Alice's secret key,

$R_A = P_A + k_A Q_A,$

$\phi_A : E \to E_A = E/\langle R_A \rangle,$

$R_{BA} = \phi_B(P_A) + k_A \phi_B(Q_A),$

$K_{\text{Alice}} = j(E_B/\langle R_{BA} \rangle).$

$\xrightarrow{\quad E_A, \phi_A(P_B), \phi_A(Q_B) \quad}$

$\xleftarrow{\quad E_B, \phi_B(P_A), \phi_B(Q_A) \quad}$

$k_B \in_R SK_B :$

Bob's secret key,

$R_B = P_B + k_B Q_B,$

$\phi_B : E \to E_B = E/\langle R_B \rangle,$

$R_{AB} = \phi_A(P_B) + k_B \phi_A(Q_B),$

$K_{\text{Bob}} = j(E_A/\langle R_{AB} \rangle).$

Fig. 1. Outline of SIDH protocol (original description).

3.2 Crypto-Friendly Description of SIDH

We prepare an alternative crypto-friendly description of SIDH for a simple presentation of our proposed AKE.

We set

$$\mathfrak{g} = (E;\ P_A,\ Q_A,\ P_B,\ Q_B),\quad \mathfrak{a} = k_A,\ \text{and}\ \mathfrak{b} = k_B.$$

Let the sets of supersingular curves and those with an auxiliary torsion basis be

$$SSEC_p = \{\text{supersingular elliptic curve } E \text{ over } \mathbb{F}_{p^2}$$
$$\text{with } E(\mathbb{F}_{p^2}) \simeq (\mathbb{Z}/(p \pm 1)\mathbb{Z})^2 \supseteq (\mathbb{Z}/\ell_A^{e_A}\mathbb{Z})^2 \oplus (\mathbb{Z}/\ell_B^{e_B}\mathbb{Z})^2\},$$
$$SSEC_{p,A} = \{(E;\ P_B',\ Q_B') \,|\, E \in SSEC_p,\ (P_B',\ Q_B') : \text{basis of } E[\ell_B^{e_B}]\},$$
$$SSEC_{p,B} = \{(E;\ P_A',\ Q_A') \,|\, E \in SSEC_p,\ (P_A',\ Q_A') : \text{basis of } E[\ell_A^{e_A}]\}.$$

Thus, SIDH public keys of A and B are given elements of $SSEC_{p,A}$ and $SSEC_{p,B}$, respectively. Then, we define

$$\mathfrak{g}^{\mathfrak{a}} = (E_A;\ \phi_A(P_B),\ \phi_A(Q_B)) \in SSEC_{p,A},$$
$$\text{where } R_A = P_A + k_A Q_A,\ \phi_A : E \to E_A = E/\langle R_A \rangle,$$
$$\mathfrak{g}^{\mathfrak{b}} = (E_B;\ \phi_B(P_A),\ \phi_B(Q_A)) \in SSEC_{p,B},$$
$$\text{where } R_B = P_B + k_B Q_B,\ \phi_B : E \to E_B = E/\langle R_B \rangle,$$
$$\left(\mathfrak{g}^{\mathfrak{b}}\right)^{\mathfrak{a}} = j(E_{BA}),$$
$$\text{where } R_{BA} = \phi_B(P_A) + k_A\phi_B(Q_A),\ \phi_{BA} : E_B \to E_{BA} = E_B/\langle R_{BA} \rangle,$$
$$\left(\mathfrak{g}^{\mathfrak{a}}\right)^{\mathfrak{b}} = j(E_{AB}),$$
$$\text{where } R_{AB} = \phi_A(P_B) + k_B\phi_A(Q_B),\ \phi_{AB} : E_A \to E_{AB} = E_A/\langle R_{AB} \rangle.$$

We describe SIDH using this notation below (Fig. 2). Public parameters are $\mathfrak{g} = (E;\ P_A,\ Q_A,\ P_B,\ Q_B)$ and $\mathfrak{e} = (\ell_A, \ell_B, e_A, e_B)$. Here, shared secret is given as $K_{\text{Alice}} = \left(\mathfrak{g}^{\mathfrak{b}}\right)^{\mathfrak{a}} = \left(\mathfrak{g}^{\mathfrak{a}}\right)^{\mathfrak{b}} = K_{\text{Bob}}$, which shows correctness of the SIDH protocol.

4   Post-quantum Assumptions from SIDH

We define SI-CDH, SI-DDH, ds- and di-SI-GDH assumptions against quantum adversaries based on the notation in Sect. 3.2. The SI-DDH assumption is needed for indistinguishability security of SIDH shared keys. Moreover, all of the following assumptions excluding ds-SI-GDH (see Proposition 1) are considered reasonable at present.

Alice

$\mathfrak{a} \in_R SK_A$: Alice's secret key,

compute $\mathfrak{g}^a$,

$K_{\text{Alice}} = \left(\mathfrak{g}^b\right)^a$.

$\xrightarrow{\quad \mathfrak{g}^a \quad}$

$\xleftarrow{\quad \mathfrak{g}^b \quad}$

Bob

$\mathfrak{b} \in_R SK_B$: Bob's secret key,

compute $\mathfrak{g}^b$,

$K_{\text{Bob}} = \left(\mathfrak{g}^a\right)^b$.

Fig. 2. Outline of SIDH protocol (crypto-friendly description).

Definition 4 (SI-CDH Assumption). *Let S be a quantum machine adversary. For* $\mathsf{pk}^{\mathsf{sidh}} = (\mathfrak{g} = (E;\ P_A,\ Q_A,\ P_B,\ Q_B),\ \mathfrak{e} = (\ell_A, \ell_B, e_A, e_B)) \leftarrow_R \mathsf{Gen}^{\mathsf{sidh}}(1^\lambda)$ *and* $\mathfrak{a} \in_R SK_A$, $\mathfrak{b} \in_R SK_B$, S *receives (* $\mathsf{pk}^{\mathsf{sidh}}$, $\mathfrak{g}^a$, $\mathfrak{g}^b$*), and* S *outputs* $\mathfrak{h} \in \mathbb{F}_{p^2}$. *If* $\mathfrak{h} = \left(\mathfrak{g}^a\right)^b \left(= \left(\mathfrak{g}^b\right)^a\right)$, S *wins. We define the advantage of* S *for the SI-CDH problem as* $\mathbf{Adv}_{\mathfrak{g},\mathfrak{e}}^{\text{SI-CDH}}(S) = \Pr[S\ wins]$. *The SI-CDH assumption is: For any polynomial-time quantum machine adversary* S, *the advantage of* S *for the SI-CDH problem is negligible in security parameter* λ.

Definition 5 (SI-DDH Assumption). *Let S be a quantum machine adversary. For* $\mathsf{pk}^{\mathsf{sidh}} = (\mathfrak{g} = (E;\ P_A,\ Q_A,\ P_B,\ Q_B),\ \mathfrak{e} = (\ell_A, \ell_B, e_A, e_B)) \leftarrow_R \mathsf{Gen}^{\mathsf{sidh}}(1^\lambda)$ *and* $\mathfrak{a}, \mathfrak{r} \in_R SK_A$, $\mathfrak{b}, \mathfrak{s} \in_R SK_B$, S *receives* $\mathcal{X}_b$ *for* $b \in_R \{0,1\}$, *that is defined by*

$$\mathcal{X}_0 = (\ \mathsf{pk}^{\mathsf{sidh}},\ \mathfrak{g}^a,\ \mathfrak{g}^b,\ \left(\mathfrak{g}^a\right)^b\ ) \quad \text{and} \quad \mathcal{X}_1 = (\ \mathsf{pk}^{\mathsf{sidh}},\ \mathfrak{g}^a,\ \mathfrak{g}^b,\ \left(\mathfrak{g}^r\right)^s\ ),$$

S *outputs a guess bit* b'. *If* $b = b'$, S *wins. We define the advantage of* S *for the SI-DDH problem as* $\mathbf{Adv}_{\mathfrak{g},\mathfrak{e}}^{\text{SI-DDH}}(S) = \Pr[S\ wins] - 1/2$. *The SI-DDH assumption is: For any polynomial-time quantum machine adversary* S, *the advantage of* S *for the SI-DDH problem is negligible in security parameter* λ.

Definition 6 (ds- and di-SI-GDH Assumption). *Let S be a quantum machine adversary. For* $\mathsf{pk}^{\mathsf{sidh}} = (\mathfrak{g} = (E;\ P_A,\ Q_A,\ P_B,\ Q_B),\ \mathfrak{e} = (\ell_A, \ell_B, e_A, e_B)) \leftarrow_R \mathsf{Gen}^{\mathsf{sidh}}(1^\lambda)$ *and* $\mathfrak{a} \in_R SK_A$, $\mathfrak{b} \in_R SK_B$, S *receives* $(\mathsf{pk}^{\mathsf{sidh}},\ \mathfrak{g},\ \mathfrak{g}^a,\ \mathfrak{g}^b)$, *and* S *access SI-DDH oracle for any input* $\mathcal{X} = (\mathsf{pk}^{\mathsf{sidh}}, (E'_A; P'_{AB}, Q'_{AB}),\ (E'_B; P'_{BA},\ Q'_{BA}),\ \mathfrak{h}')$ *where* P'_{AB}, Q'_{AB} *(resp.* $P'_{BA}, Q'_{BA})$ *are points in* $E'_A(\mathbb{F}_{p^2})$ *(resp.* $E'_B(\mathbb{F}_{p^2}))$ *and* $\mathfrak{h}' \in \mathbb{F}_{p^2}$, *and then outputs* $\mathfrak{h} \in \mathbb{F}_{p^2}$. *If* $\mathfrak{h} = \left(\mathfrak{g}^a\right)^b \left(= \left(\mathfrak{g}^b\right)^a\right)$, S *wins. According to the behavior of SI-DDH oracle, we have two types of SI-GDH problem, i.e.,*

- **degree-sensitive SI-GDH (ds-SI-GDH) problem.** *The ds-SI-DDH oracle answers true if there exist a supersingular elliptic curve* E'_{AB} *and isogenies* $(\phi'_A, \phi'_B, \phi'_{AB}, \phi'_{BA})$ *among* E, E'_A, E'_B, E'_{AB} *which form a commutative diagram as in Fig. 3 such that*
  - *degree d'_A of ϕ'_A (and ϕ'_{BA}) is equal to $\ell_A^{e_A}$ and degree d'_B of ϕ'_B (and ϕ'_{AB}) is equal to $\ell_B^{e_B}$ and*

- $P'_{AB} = \phi'_A(P_B)$, $Q'_{AB} = \phi'_A(Q_B)$ and $P'_{BA} = \phi'_B(P_A)$, $Q'_{BA} = \phi'_B(Q_A)$ where points (P_A, Q_A, P_B, Q_B) are given in public key pk^{sidh}, and $\mathfrak{h}' = j(E'_{AB})$, and false otherwise. We call this case degree-sensitive SI-GDH (ds-SI-GDH) problem.

- **degree-insensitive SI-GDH (di-SI-GDH) problem.** The di-SI-DDH oracle answers true if there exist a supersingular elliptic curve E'_{AB} and isogenies $(\phi'_A, \phi'_B, \phi'_{AB}, \phi'_{BA})$ among E, E'_A, E'_B, E'_{AB} which form a commutative diagram as in Fig. 3 such that
  - degree d'_A of ϕ'_A (and ϕ'_{BA}) is a power of ℓ_A and degree d'_B of ϕ'_B (and ϕ'_{AB}) is a power of ℓ_B and
  - $P'_{AB} = \phi'_A(P_B)$, $Q'_{AB} = \phi'_A(Q_B)$ and $P'_{BA} = \phi'_B(P_A)$, $Q'_{BA} = \phi'_B(Q_A)$ where points (P_A, Q_A, P_B, Q_B) are given in public key pk^{sidh}, and $\mathfrak{h}' = j(E'_{AB})$, and false otherwise. We call this case degree-insensitive SI-GDH (di-SI-GDH) problem.

We define the advantage of adversary S for the ds–SI-GDH and di-SI-GDH problems as $\mathbf{Adv}_{g,e}^{ds\text{-}SI\text{-}GDH}(S) = \Pr[S\ wins]$ and $\mathbf{Adv}_{g,e}^{di\text{-}SI\text{-}GDH}(S) = \Pr[S\ wins]$, respectively. The ds-SI-GDH (resp. di-SI-GDH) assumption is: For any polynomial-time quantum machine adversary S, the advantage of S for the ds-SI-GDH (resp. di-SI-GDH) problem is negligible in security parameter λ.

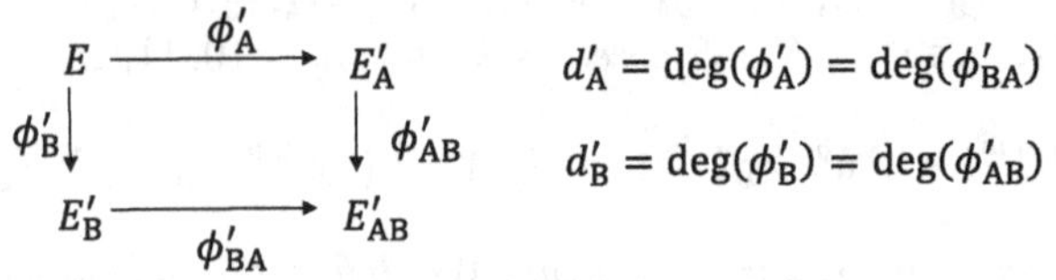

$d'_A = \deg(\phi'_A) = \deg(\phi'_{BA})$

$d'_B = \deg(\phi'_B) = \deg(\phi'_{AB})$

Fig. 3. Commutative diagram for true instances of SI-DDH oracles, in which it holds that $\ker(\phi'_{BA}) = \phi'_B(\ker(\phi'_A))$ and $\ker(\phi'_{AB}) = \phi'_A(\ker(\phi'_B))$.

Proposition 1 (adapted from [16]). The ds-SI-GDH assumption does not hold, i.e., there exists a ppt adversary against the ds-SI-GDH problem.

Proof Sketch. Very recently, Galbraith and Vercauteren proposed an attack on the SI-CDH problem with access to the decision degree (DD) oracle [16], which determines whether two supersingular curves are isogenous of some specific degree or not. As a basic building block, first, we describe an attack on the SI-CDH problem using the DD oracle. The input of the problem is $(\mathsf{pk}^{sidh} = (g = (E; P_A, Q_A, P_B, Q_B), \mathfrak{e} = (\ell_A, \ell_B, e_A, e_B)), E_A, P_{AB}, Q_{AB})$, where $\phi_A : E \to E_A$ is an $\ell_A^{e_A}$-isogeny, $P_{AB} = \phi_A(P_B)$, and $Q_{AB} = \phi_A(Q_B)$. The goal of the adversary S is to reveal ϕ_A. For that, S calculates integer u such that $u \cdot \ell_A \equiv 1$ (mod ℓ_B), and then one ℓ_A-isogeny $\psi : E_A \to E'$. S send

$$(\tilde{\mathsf{pk}}^{sidh} = (g, \tilde{\mathfrak{e}} = (\ell_A, \ell_B, e_A - 1, e_B), E', u \cdot \psi(P_{AB}), u \cdot \psi(Q_{AB}))$$

to the DD oracle. Here, we note that the exponent $e_A - 1$ is used instead of e_A for the implicitly defined ℓ_A-power isogeny. That is, the oracle distinguishes the degree (or length) of the isogeny, in other words, whether E' is $\ell_A^{e_A-1}$-isogenous to E or $\ell_A^{e_A+1}$-isogenous to E. See the left hand side of Fig. 4. Then, the adversary reveals all the isogeny by repeating this ℓ_A-backtracking decision.

Next, we extend the above strategy to solve the ds-SI-GDH problem. Namely, an ds-SI-GDH adversary obtains an input $(\mathsf{pk}^{\mathsf{sidh}} = (\mathfrak{g} = (E; P_A, Q_A, P_B, Q_B), \mathfrak{e} = (\ell_A, \ell_B, e_A, e_B)), E_A, P_{AB}, Q_{AB}, \ldots)$, where $\phi_A : E \to E_A$ is an $\ell_A^{e_A}$-isogeny, $P_{AB} = \phi_A(P_B)$, and $Q_{AB} = \phi_A(Q_B)$. The goal of the adversary $\mathcal{S}$ is to reveal ϕ_A. For that, $\mathcal{S}$ calculates one ℓ_A-isogeny $\psi : E_A \to E'$ as before. Moreover, $\mathcal{S}$ calculates degree $\ell_B^{e_B}$-isogenies $E \to E_B'$ and $E' \to E_{AB}'$ that makes commutative SIDH diagram (E, E', E_B', E_{AB}'). Then, $\mathcal{S}$ send

$$(\tilde{\mathsf{pk}}^{\mathsf{sidh}} = (\mathfrak{g}, \tilde{\mathfrak{e}} = (\ell_A, \ell_B, e_A - 1, e_B), E', E_B', \ldots, j(E_{AB}'))$$

to the ds-SI-DDH oracle and determine whether ψ is a backtracking step in ϕ_A or not. See the right hand side of Fig. 4. From here on, repeating this procedure, $\mathcal{S}$ can reveal ϕ_A. Also, $\mathcal{S}$ can compute E_{AB} by using E_B and ϕ_A, which solves the ds-SI-GDH problem. □

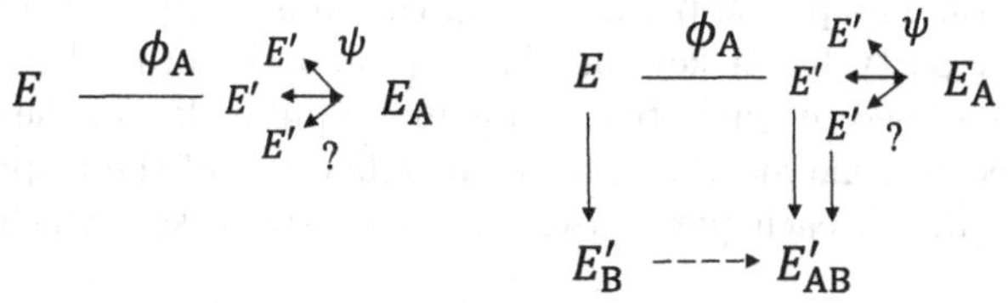

Fig. 4. Diagrams for the GV-type attack. The right (resp. left) hand side shows the strategy for the ds-SI-GDH problem (resp. the SI-CDH problem with access to the DD oracle). The attacker distinguishes which one of the $e_A + 1$ left arrows of ℓ_A-isogenies from E_A is backtracking by using the ds-SI-DDH (resp. the DD) oracle.

As described in the above proof, to distinguish the degree of isogeny (or distance between two elliptic curves in the ℓ_A-isogeny graph) is crucial for the GV-type attack. Since the ability for the distinction is given by the ds-SI-DDH oracle, the GV-type attack adversaries have *no advantages* in the di-SI-GDH problem. Therefore, in contrast to the ds-SI-GDH problem, we may assume that the di-SI-GDH problem cannot be solved by any efficient adversaries, and can be used for the basis of the security of our biclique scheme.

Note that auxiliary points $\phi_A'(P_B), \phi_A'(Q_B), \phi_B'(P_A), \phi_B'(Q_A)$ in true instance $\mathcal{X}$ for di-SI-DDH oracle impose some restrictions on implicitly defined isogenies ϕ_A', ϕ_B' (and ϕ_{AB}', ϕ_{BA}') used in Fig. 3. However, since degrees d_A' and d_B' of ϕ_A' and ϕ_B' can be chosen as *any* powers of ℓ_A and ℓ_B respectively, a wide range of tuples (E_A', E_B', E_{AB}') can be accepted for forming the commutative diagram in Fig. 3. Therefore, as an extreme possible case, *any* tuple of supersingular elliptic curves

(E'_A, E'_B, E'_{AB}) *might* form the commutative diagram in Fig. 3, that is, any tuple of such curves would be true instances in the hypothetical case. We cannot exclude such possibility from our present knowledge of the di-SI-GDH problem. A satisfiable analysis of the di-SI-GDH problem seems to need more understanding of the Ramanujan graph of ℓ-isogenies of supersingular curves.

Lemma 3.2 and Theorem 3.3 in [30] also show some interesting connection between computational and decisional SIDH problems. However, we notice that answers of all the oracles $(O_{E,1})_{\ell^e}, (O_{E,2})_{\ell^e}$ and $(O_{E,3})_{\ell^e}$ (for $\ell^e = \ell_1^{e_1}$ or $\ell_2^{e_2}$) are related to isogenies of degrees dividing ℓ^e, which is defined by public parameters. In particular, all the isogeny degrees have smaller or equal than ℓ^e. Our di-SI-GDH problem is related to unbounded degrees which are just a power of ℓ. Thus, Lemma 3.2 and Theorem 3.3 in [30] are now unrelated with our situation, but, we think seeking relationships between the di-SI-GDH problem and the results in [30] is an interesting research direction.

5   Proposed SIDH UM Protocol

In this section, we propose the SIDH UM protocol, where it can be proved in the quantum random oracle model under the SI-DDH assumption.

Before describing the protocol, we explain that each party needs to have two static public keys. The public parameter, $\mathfrak{g}$, contains two parameters, (P_1, Q_1) and (P_2, Q_2). A party has a key on (P_1, Q_1) and the other key on (P_2, Q_2). Then, (P_1, Q_1) is used to generate the ephemeral public key of the initiator and (P_2, Q_2) is used to generate the ephemeral public key of the responder. When the role is exchanged, each party uses the other static key which is not used before.

This double construction in public parameter and static public keys gives resistance to reflection attacks. To the best of our knowledge, the previous researches of key exchange on supersingular isogenies have lacked this consideration.

5.1   Useful Techniques for Quantum Random Oracle Model

A problem on security proofs in the quantum random oracle model is how to generate random values for exponentially many positions in order to simulate outputs of the hash function. For a hash function $H : Dom \rightarrow Rng$, in the quantum random oracle model, the adversary poses a superposition $|\phi\rangle = \Sigma\alpha_x|x\rangle$ and the oracle returns $\Sigma\alpha_x|H(x)\rangle$. If Rng is large for a quantum polynomial-time simulator, it is difficult to generate all random output values of H to compute $\Sigma\alpha_x|H(x)\rangle$. Zhandry [33] showed a solution with the notion of k-wise independent function.

A weight assignment on a set $\mathcal{X}$ is a function $D : \mathcal{X} \rightarrow \mathbb{R}$ such that $\Sigma_{x \in \mathcal{X}} D(x) = 1$. A distribution on $\mathcal{X}$ is a weight-assignment D such that $D(x) \geq 0$ for all $x \in \mathcal{X}$. Consider the set of functions $H : \mathcal{X} \rightarrow \mathcal{Y}$ for sets $\mathcal{X}$ and $\mathcal{Y}$, denoted by $H_{\mathcal{X},\mathcal{Y}}$. We define the marginal weight assignment $D_{\mathcal{W}}$ of

D on $H_{\mathcal{X},\mathcal{Y}}$ where the weight of a function $H_{\mathcal{W}} : \mathcal{W} \to \mathcal{Y}$ is equal to the sum of the weights of all $H \in H_{\mathcal{X},\mathcal{Y}}$ that agree with $H_{\mathcal{W}}$ on $\mathcal{W}$.

Definition 7 (*k*-wise equivalence). *We call two weight assignments D_1 and D_2 on $H_{\mathcal{X},\mathcal{Y}}$ k-wise equivalent if for all $\mathcal{W} \subseteq \mathcal{X}$ of size k, the marginal weight assignments $D_{1,\mathcal{W}}$ and $D_{2,\mathcal{W}}$ (of D_1 and D_2) over $H_{\mathcal{X},\mathcal{Y}}$ are identical.*

Definition 8 (*k*-wise independent function). *We call a function f k-wise independent function if f is k-wise equivalent to a random function.*

Lemma 1 (Theorem 3.1 in [33]). *Let A be a quantum algorithm making q quantum queries to an oracle $H : \mathcal{X} \to \mathcal{Y}$. If we draw H from some weight assignment D, then for every z, the quantity $\Pr_{H \leftarrow D}[A^H() = z]$ is a linear combination of the quantities $\Pr_{H \leftarrow D}[H(x_i) = r_i \forall i \in 1, \ldots, 2q]$ for all possible settings of the x_i and r_i.*

Lemma 2 (Theorem 6.1 in [33]). *If there exists $2q_i$-wise independent function, then any quantum algorithm A making q_i quantum queries to random oracles O_i can be efficiently simulated by a quantum algorithm B, which has the same output distribution, but makes no queries.*

Hence, a quantum algorithm B can simulate quantum random oracles in a polynomial-time. We use this simulation technique to simulate outputs of the hash function in the security proof of the SIDH UM protocol.

On the other hand, the other problem on security proofs in the quantum random oracle model is how to insert intended random values as the outputs of corresponding oracle inputs. Zhandry [33] showed a solution with the notion of semi-constant distributions $\mathbf{SC}_\omega$.

Definition 9 (Semi-constant distribution). *Define $\mathbf{SC}_\omega$, the semi-constant distribution, as the distribution over $H_{\mathcal{X},\mathcal{Y}}$ resulting from the following process:*

– *First, pick a random element y from $\mathcal{Y}$.*
– *For each $x \in \mathcal{X}$, do one of the following:*
  • *With probability ω, set $H(x) = y$. We call x a distinguished input to H.*
  • *Otherwise, set $H(x)$ to be a random element in $\mathcal{Y}$.*

Lemma 3 (Corollary 4.3 in [33]). *The distribution of outputs of a quantum algorithm making h queries to an oracle drawn from $\mathbf{SC}_\omega$ is at most a distance $\frac{3}{8}h^4\omega^2$ away from the case when the oracle is drawn from the uniform distribution.*

We suppose that the simulation succeeds with probability ϵ if the adversary uses an inserted random value as the outputs of corresponding oracle inputs. If the probability that the adversary uses one of the points is ω, then the simulation succeeds with probability $\epsilon\omega - \frac{3}{8}h^4\omega^2$. By choosing ω to maximize the success probability, the simulation succeeds with probability $O(\epsilon^2/h^4)$. We use this simulation technique to insert a SI-DDH instance into the hash function in the security proof of the SIDH UM protocol.

5.2   Description of SIDH UM Protocol

We give our SIDH UM protocol using the notation in Sect. 3.2. Public parameters are $\mathfrak{g} = (E;\, P_1,\, Q_1,\, P_2,\, Q_2)$ and $\mathfrak{e} = (\ell_1, \ell_2, e_1, e_2)$. We set $\Pi = \text{SIDHUM}$, that is, the protocol ID is "SIDHUM." Static and ephemeral keys are the same as our biclique SIDH protocol. Let two secret-key spaces for initiators and responders be given as $SK_1 = \mathbb{Z}/\ell_1^{e_1}\mathbb{Z}$ and $SK_2 = \mathbb{Z}/\ell_2^{e_2}\mathbb{Z}$, respectively.

User $\hat{A}$ has two static public keys, $A_1 = \mathfrak{g}^{\mathfrak{a}_1}$ and $A_2 = \mathfrak{g}^{\mathfrak{a}_2}$, where $\mathfrak{a}_1 = k_{A,1} \in_R SK_1$, $\mathfrak{a}_2 = k_{A,2} \in_R SK_2$, and $\mathfrak{a}_1$ and $\mathfrak{a}_2$ are $\hat{A}$'s static secret keys. User $\hat{B}$, also, has two static public keys, $B_1 = \mathfrak{g}^{\mathfrak{b}_1}$ and $B_2 = \mathfrak{g}^{\mathfrak{b}_2}$, where $\mathfrak{b}_1 = k_{B,1} \in_R SK_1$, $\mathfrak{b}_2 = k_{B,2} \in_R SK_2$, and $\mathfrak{b}_1$ and $\mathfrak{b}_2$ are $\hat{B}$'s static secret keys. Here, ephemeral secret keys for $\hat{A}$ and $\hat{B}$ are given as

$$\mathfrak{x} = k_X \in_R SK_1, \text{ and } \mathfrak{y} = k_Y \in_R SK_2,$$

respectively. $\hat{A}$ sends a ephemeral public key X as $X = \mathfrak{g}^{\mathfrak{x}}$ to $\hat{B}$, $\hat{B}$ sends back a ephemeral public key Y as $Y = \mathfrak{g}^{\mathfrak{y}}$ to $\hat{A}$.

$\hat{A}$ computes $Z_1 = B_2^{\mathfrak{a}_1}$, and $Z_2 = Y^{\mathfrak{x}}$, and then, obtains the session key K as $K = H(\Pi, Z_1, Z_2, \hat{A}, \hat{B}, X, Y)$, where H is a hash function.

$\hat{B}$ can computes the session key K as $K = H(\Pi, Z_1, Z_2, \hat{A}, \hat{B}, X, Y)$ from $Z_1 = A_1^{\mathfrak{b}_2}$, and $Z_2 = X^{\mathfrak{y}}$.

It is clear that the session keys of both parties are equal (Fig. 5).

$A_1 = \mathfrak{g}^{\mathfrak{a}_1}$	$B_1 = \mathfrak{g}^{\mathfrak{b}_1}$
$A_2 = \mathfrak{g}^{\mathfrak{a}_2}$	$B_2 = \mathfrak{g}^{\mathfrak{b}_2}$
$X = \mathfrak{g}^{\mathfrak{x}} \xrightarrow{X}$	$Y = \mathfrak{g}^{\mathfrak{y}}$
$\xleftarrow{Y}$	
$Z_1 = B_2^{\mathfrak{a}_1}$	$Z_1 = A_1^{\mathfrak{b}_2}$
$Z_2 = Y^{\mathfrak{x}}$	$Z_2 = X^{\mathfrak{y}}$
$K = H(\Pi, Z_1, Z_2, \hat{A}, \hat{B}, X, Y)$	

Fig. 5. Outline of SIDH UM protocol.

$A_1 = \mathfrak{g}^{\mathfrak{a}_1}$	$B_1 = \mathfrak{g}^{\mathfrak{b}_1}$
$A_2 = \mathfrak{g}^{\mathfrak{a}_2}$	$B_2 = \mathfrak{g}^{\mathfrak{b}_2}$
$X = \mathfrak{g}^{\mathfrak{x}} \xrightarrow{X}$	$Y = \mathfrak{g}^{\mathfrak{y}}$
$\xleftarrow{Y}$	
$Z_1 = Y^{\mathfrak{a}_1}$	$Z_1 = A_1^{\mathfrak{y}}$
$Z_2 = B_2^{\mathfrak{x}}$	$Z_2 = X^{\mathfrak{b}_2}$
$Z_3 = B_2^{\mathfrak{a}_1}$	$Z_3 = A_1^{\mathfrak{b}_2}$
$Z_4 = Y^{\mathfrak{x}}$	$Z_4 = X^{\mathfrak{y}}$
$K = H(\Pi, Z_1, Z_2, Z_3, Z_4, \hat{A}, \hat{B}, X, Y)$	

Fig. 6. Outline of Biclique SIDH protocol.

5.3   Security

Theorem 1. *Suppose that H is modeled as a quantum random oracle and that the SI-DDH assumption hold for $(\mathfrak{g}, \mathfrak{e})$. Then the SIDH UM protocol is a post-quantum CK-secure authenticated key exchange protocol in the quantum random oracle model.*

In particular, for any AKE quantum adversary $\mathcal{M}$ against the SIDH UM protocol that runs in time at most t, involves at most n honest parties and activates at most s sessions, and makes at most h queries to the quantum random oracle *and q* SessionKeyReveal *queries, there exists an SI-DDH quantum adversary $\mathcal{S}$ such that*

$$\mathbf{Adv}_{\mathfrak{g},\mathfrak{e}}^{\text{SI-DDH}}(\mathcal{S}) \geq \frac{2\mathbf{Adv}_{\text{SIDHUM}}^{\text{AKE}}(\mathcal{M})^2}{n^2 s^2 (8hq + 3(h + q + 1)^4)},$$

where $\mathcal{S}$ runs in time t plus time to perform $\mathcal{O}((n + s)\lambda)$ low-degree isogeny operations.

An intuition of the security proof is given in Sect. 5.1. The SI-DDH assumption used in Theorem 1 can be degree-sensitive. Hence, it implies security under the SI-CDH assumption by using the reduction in Proposition 1. However, an additional reduction cost is necessary. It is not trivial to directly prove security under the SI-CDH assumption because of the no-cloning theorem. Specifically, in the reduction to the CK security, the SI-CDH solver wants to extract the answer of the SI-CDH problem from a random oracle query by the AKE adversary. However, the query is a quantum state, and the solver cannot record a copy of the input. Thus, this proof strategy does not work. Recently, Zhandry [34] introduced a technique to record quantum queries. How to apply this technique to the proof is an open problem.

6   Proposed Biclique SIDH Protocol

In this section, we propose the biclique SIDH protocol, where it can be proved in the random oracle model under the di-SI-GDH assumption.

It is worth to note here that the SIDH UM protocol is secure in the quantum random oracle model under the SI-DDH assumption, and therefore, the SIDH UM protocol is superior than the biclique SIDH protocol in the following points: the computational model of adversaries and the assumption relaying to the security. However, the biclique SIDH protocol can be shown to be secure in the CK$^+$ model, that is, the protocol resists against maximum exposure where a non-trivial combination of secret keys is revealed. This shows that the biclique SIDH protocol is superior than the SIDH UM protocol in this sense.

As our SIDH UM protocol in Sect. 5, the public parameter, $\mathfrak{g}$, contains two parameters, (P_1, Q_1) and (P_2, Q_2) in our biclique SIDH protocol. A party has a key on (P_1, Q_1) and the other key on (P_2, Q_2).

6.1   Description of Biclique SIDH Protocol

We give our biclique SIDH protocol using the notation in Sect. 3.2. Public parameters are $\mathfrak{g} = (E; P_1, Q_1, P_2, Q_2)$ and $\mathfrak{e} = (\ell_1, \ell_2, e_1, e_2)$. We set $\Pi = \text{BCSIDH}$, that is, the protocol ID is "BCSIDH." Let two secret-key spaces for initiators and responders be given as $SK_1 = \mathbb{Z}/\ell_1^{e_1}\mathbb{Z}$ and $SK_2 = \mathbb{Z}/\ell_2^{e_2}\mathbb{Z}$, respectively.

User $\hat{A}$ has two static public keys, $A_1 = \mathfrak{g}^{\mathfrak{a}_1}$ and $A_2 = \mathfrak{g}^{\mathfrak{a}_2}$, where $\mathfrak{a}_1 = k_{A,1} \in_R SK_1$, $\mathfrak{a}_2 = k_{A,2} \in_R SK_2$, and $\mathfrak{a}_1$ and $\mathfrak{a}_2$ are $\hat{A}$'s static secret keys. User $\hat{B}$, also, has two static public keys, $B_1 = \mathfrak{g}^{\mathfrak{b}_1}$ and $B_2 = \mathfrak{g}^{\mathfrak{b}_2}$, where $\mathfrak{b}_1 = k_{B,1} \in_R SK_1$,

$\mathfrak{b}_2 = k_{\mathsf{B},2} \in_R SK_2$, and $\mathfrak{b}_1$ and $\mathfrak{b}_2$ are $\hat{B}$'s static secret keys. Here, ephemeral secret keys for $\hat{A}$ and $\hat{B}$ are given as

$$\mathfrak{x} = k_{\mathsf{X}} \in_R SK_1, \text{ and } \mathfrak{y} = k_{\mathsf{Y}} \in_R SK_2,$$

respectively. $\hat{A}$ sends an ephemeral public key X as $X = g^{\mathfrak{x}}$ to $\hat{B}$, $\hat{B}$ sends back an ephemeral public key Y as $Y = g^{\mathfrak{y}}$ to $\hat{A}$.

$\hat{A}$ computes the non-trivial combinations of the ephemeral and static public keys as $Z_1 = Y^{\mathfrak{a}_1}$, $Z_2 = B_2^{\mathfrak{x}}$, $Z_3 = B_2^{\mathfrak{a}_1}$, and $Z_4 = Y^{\mathfrak{x}}$, and then, obtains the session key K as $K = H(\Pi, Z_1, Z_2, Z_3, Z_4, \hat{A}, \hat{B}, X, Y)$, where H is a hash function.

$\hat{B}$ can computes the session key K as $K = H(\Pi, Z_1, Z_2, Z_3, Z_4, \hat{A}, \hat{B}, X, Y)$ from $Z_1 = A_1^{\mathfrak{y}}$, $Z_2 = X^{\mathfrak{b}_2}$, $Z_3 = A_1^{\mathfrak{b}_2}$, and $Z_4 = X^{\mathfrak{y}}$.

It is clear that the session keys of both parties are equal (Fig. 6).

Charles et al. [6] proposed a hash function secure against quantum adversaries from the isogeny computation intractability. Hence, we can use the isogeny-based hash function in the real implementation for H, however, H is modeled as a random oracle in the security proof below.

6.2   Security

Theorem 2. *Suppose that H is modeled as a random oracle and that the di-SI-GDH assumption hold for $(\mathfrak{g}, \mathfrak{e})$. Then the biclique SIDH protocol is a post-quantum CK^+-secure authenticated key exchange protocol in the random oracle model.*

In particular, for any AKE quantum adversary $\mathcal{M}$ against the biclique SIDH protocol that runs in time at most t, involves at most n honest parties and activate at most s sessions, and makes at most h queries to the random oracle, there exists a di-SI-GDH quantum adversary $\mathcal{S}$ such that

$$\mathbf{Adv}_{\mathfrak{g},\mathfrak{e}}^{\text{di-SI-GDH}}(\mathcal{S}) \geq \min\left\{\frac{1}{sn}, \frac{1}{n^2}, \frac{1}{s^2}\right\} \cdot \mathbf{Adv}_{\text{BCSIDH}}^{\text{AKE}}(\mathcal{M}),$$

where $\mathcal{S}$ runs in time t plus time to perform $\mathcal{O}((n+s)\lambda)$ low-degree isogeny operations and make $\mathcal{O}(h+s)$ queries to di-SI-DDH oracle.

As we consider a case where the security model is CK^+, an adversary may access to a non-trivial combination of secret keys. However, it means that the adversary cannot access to the other combination of the secret key. Thus, the di-SI-GDH solver can embedded an instance to the public keys where secret key are not revealed. As we assume the random oracle model, the adversary has to make a query which contains the di-SI-GDH answer, and then, the theorem can be proved. Note here that the di-SI-DDH oracle is necessary to keep consistency between the answers by the di-SI-GDH solver on adversary's questions.

We consider how to extend our security proof in the random oracle model to that in the *quantum* random oracle model as in the SIDH UM protocol. For completing the simulation, we need to extend the di-SI-GDH assumption

(Definition 6). Namely, in random oracle simulation, $\mathcal{S}$ first checks compatibility of input elements using di-SI-DDH oracle. Hence, in the quantum ROM situation, since inputs are given in quantum superposition form, we should extend the di-SI-DDH oracle to take as input the superpositions. If the di-SI-GDH quantum adversary allows the extended di-SI-DDH oracle access, then our security proof can be converted to quantum ROM secure one.

7   Conclusion

We proposed two authenticated key exchange protocols from supersingular isogenies: SIDH UM and biclique SIDH. We also discussed a new approach for invalidating the Galbraith–Vercauteren attack for the gap problem on the supersingular isogeny Diffie–Hellman, and defined the di-SI-GDH assumption.

The SIDH UM protocol is secure in the CK and quantum random oracle models under the SI-DDH assumption. The biclique SIDH protocol is secure in the CK^+ and random oracle models under the di-SI-GDH assumption.

Our protocols are the first post-quantum one-round Diffie–Hellman type authenticated key exchange ones in the following points: one is secure under the quantum random oracle model and the other resists against maximum exposure where a non-trivial combination of secret keys is revealed.

References

1. Ambainis, A., Rosmanis, A., Unruh, D.: Quantum attacks on classical proof systems: the hardness of quantum rewinding. In: FOCS 2014, pp. 474–483 (2014)
2. Azarderakhsh, R., Jao, D., Kalach, K., Koziel, B., Leonardi, C.: Key compression for isogeny-based cryptosystems. In: AsiaPKC 2016, pp. 1–10 (2016)
3. Boneh, D., Dagdelen, Ö., Fischlin, M., Lehmann, A., Schaffner, C., Zhandry, M.: Random oracles in a quantum world. In: Lee, D.H., Wang, X. (eds.) ASIACRYPT 2011. LNCS, vol. 7073, pp. 41–69. Springer, Heidelberg (2011). https://doi.org/10.1007/978-3-642-25385-0_3
4. Bos, J.W., Friedberger, S.: Fast arithmetic modulo $2^x\,p^y \pm 1$. In: ARITH 2017, pp. 148–155 (2017)
5. Canetti, R., Krawczyk, H.: Analysis of key-exchange protocols and their use for building secure channels. In: Pfitzmann, B. (ed.) EUROCRYPT 2001. LNCS, vol. 2045, pp. 453–474. Springer, Heidelberg (2001). https://doi.org/10.1007/3-540-44987-6_28
6. Charles, D., Lauter, K., Goren, E.: Cryptographic hash functions from expander graphs. J. Crypt. **22**(1), 93–113 (2009)
7. Childs, A., Jao, D., Soukharev, V.: Constructing elliptic curve isogenies in quantum subexponential time. J. Math. Crypt. **8**(1), 1–29 (2014)
8. Costello, C., Jao, D., Longa, P., Naehrig, M., Renes, J., Urbanik, D.: Efficient compression of SIDH public keys. In: Coron, J.-S., Nielsen, J.B. (eds.) EUROCRYPT 2017. LNCS, vol. 10210, pp. 679–706. Springer, Cham (2017). https://doi.org/10.1007/978-3-319-56620-7_24

9. Costello, C., Longa, P., Naehrig, M.: Efficient algorithms for supersingular isogeny Diffie-Hellman. In: Robshaw, M., Katz, J. (eds.) CRYPTO 2016. LNCS, vol. 9814, pp. 572–601. Springer, Heidelberg (2016). https://doi.org/10.1007/978-3-662-53018-4_21

10. Dagdelen, Ö., Fischlin, M., Gagliardoni, T.: The fiat–shamir transformation in a quantum world. In: Sako, K., Sarkar, P. (eds.) ASIACRYPT 2013. LNCS, vol. 8270, pp. 62–81. Springer, Heidelberg (2013). https://doi.org/10.1007/978-3-642-42045-0_4

11. De Feo, L., Jao, D., Plût, J.: Towards quantum-resistant cryptosystems from supersingular elliptic curve isogenies. J. Math. Crypt. **8**(3), 209–247 (2014)

12. Fujioka, A., Suzuki, K., Xagawa, K., Yoneyama, K.: Practical and post-quantum authenticated key exchange from one-way secure key encapsulation mechanism. In: ASIACCS 2013, pp. 83–94 (2013)

13. Fujioka, A., Suzuki, K., Xagawa, K., Yoneyama, K.: Strongly secure authenticated key exchange from factoring, codes, and lattices. Des. Codes Crypt. **76**(3), 469–504 (2015). A preliminary version appeared in PKC 2012 (2012)

14. Galbraith, S.D.: Authenticated key exchange for SIDH. IACR Cryptology ePrint Archive 2018, 266 (2018). http://eprint.iacr.org/2018/266

15. Galbraith, S.D., Petit, C., Shani, B., Ti, Y.B.: On the security of supersingular isogeny cryptosystems. In: Cheon, J.H., Takagi, T. (eds.) ASIACRYPT 2016. LNCS, vol. 10031, pp. 63–91. Springer, Heidelberg (2016). https://doi.org/10.1007/978-3-662-53887-6_3

16. Galbraith, S.D., Vercauteren, F.: Computational problems in supersingular elliptic curve isogenies. IACR Cryptology ePrint Archive 2017, 774 (2017). http://eprint.iacr.org/2017/774

17. Jao, D., et al.: Supersingular Isogeny Key Encapsulation (SIKE). Submission to NIST Post-Quantum Cryptography Standardization (2017)

18. Jeong, I.R., Katz, J., Lee, D.H.: One-round protocols for two-party authenticated key exchange. In: Jakobsson, M., Yung, M., Zhou, J. (eds.) ACNS 2004. LNCS, vol. 3089, pp. 220–232. Springer, Heidelberg (2004). https://doi.org/10.1007/978-3-540-24852-1_16

19. Koziel, B., Azarderakhsh, R., Kermani, M.M., Jao, D.: Post-quantum cryptography on FPGA based on isogenies on elliptic curves. IEEE Trans. Circuits Syst. **64**–**I**(1), 86–99 (2017)

20. Koziel, B., Jalali, A., Azarderakhsh, R., Jao, D., Mozaffari-Kermani, M.: NEON-SIDH: efficient implementation of supersingular isogeny Diffie-Hellman key exchange protocol on ARM. In: Foresti, S., Persiano, G. (eds.) CANS 2016. LNCS, vol. 10052, pp. 88–103. Springer, Cham (2016). https://doi.org/10.1007/978-3-319-48965-0_6

21. Krawczyk, H.: HMQV: a high-performance secure Diffie-Hellman protocol. In: Shoup, V. (ed.) CRYPTO 2005. LNCS, vol. 3621, pp. 546–566. Springer, Heidelberg (2005). https://doi.org/10.1007/11535218_33

22. LeGrow, J., Jao, D., Azarderakhsh, R.: Modeling quantum-safe authenticated key establishment, and an isogeny-based protocol. IACR Cryptology ePrint Archive 2018, 282 (2018). http://eprint.iacr.org/2018/282

23. Longa, P.: A note on post-quantum authenticated key exchange from supersingular isogenies. IACR Cryptology ePrint Archive 2018, 267 (2018). http://eprint.iacr.org/2018/267

24. National Institute of Standards and Technology: Post-Quantum crypto standardization: Call for Proposals Announcement, December 2016. http://csrc.nist.gov/groups/ST/post-quantum-crypto/cfp-announce-dec2016.html

25. Petit, C.: Faster algorithms for isogeny problems using torsion point images. In: Takagi, T., Peyrin, T. (eds.) ASIACRYPT 2017. LNCS, vol. 10625, pp. 330–353. Springer, Cham (2017). https://doi.org/10.1007/978-3-319-70697-9_12

26. Rostovtsev, A., Stolbunov, A.: Public-key cryptosystem based on isogenies. IACR Cryptology ePrint Archive 2006, 145 (2006). http://eprint.iacr.org/2006/145

27. Shor, P.W.: Polynomial-time algorithms for prime factorization and discrete logarithms on a quantum computer. SIAM J. Comput. 26(5), 1484–1509 (1997)

28. Sutherland, A.: Identifying supersingular elliptic curves. LMS J. Comput. Math. 15, 317–325 (2012)

29. Thormarker, E.: Post-quantum cryptography: supersingular isogeny Diffie-Hellman key exchange. Master's thesis, Stockholm University (2017)

30. Urbanik, D., Jao, D.: SoK: the problem landscape of SIDH. In: APKC 2018, pp. 53–60 (2018)

31. Xu, X., Xue, H., Wang, K., Tian, S., Liang, B., Yu, W.: Strongly secure authenticated key exchange from supersingular isogeny. IACR Cryptology ePrint Archive 2018, 760 (2018). http://eprint.iacr.org/2018/760

32. Zhandry, M.: How to construct quantum random functions. In: FOCS 2012, pp. 679–687 (2012)

33. Zhandry, M.: Secure identity-based encryption in the quantum random oracle model. In: Safavi-Naini, R., Canetti, R. (eds.) CRYPTO 2012. LNCS, vol. 7417, pp. 758–775. Springer, Heidelberg (2012). https://doi.org/10.1007/978-3-642-32009-5_44

34. Zhandry, M.: How to record quantum queries, and applications to quantum indifferentiability. IACR Cryptology ePrint Archive 2018, 276 (2018). http://eprint.iacr.org/2018/276

On the Complexity of the LWR-Solving BKW Algorithm

Hiroki Okada[1]([⊠]), Atsushi Takayasu[2], Kazuhide Fukushima[1],
Shinsaku Kiyomoto[1], and Tsuyoshi Takagi[2]

[1] KDDI Research, Inc., Saitama, Japan
ir-okada@kddi-research.jp
[2] The University of Tokyo, Tokyo, Japan

Abstract. Duc *et al.* applied the Blum-Kalai-Wasserman (BKW) algorithm to the learning with rounding (LWR) problem. The number of blocks is a parameter of the BKW algorithm. By optimizing the number of blocks, we can minimize the time complexity of the BKW algorithm. However, Duc *et al.* did not derive the optimal number of blocks theoretically, but they searched it for numerically. In this paper, we theoretically derive the asymptotically optimal number of blocks and show the minimum time complexity of the algorithm. Furthermore, we derive an equation that relates the Gaussian parameter σ of the LWE problem and the modulus p of the LWR problem. When σ and p satisfy the equation, the asymptotic time complexity of the BKW algorithm to solve the LWE and LWR problems are the same.

Keywords: Lattice · Learning with errors · Learning with rounding · Blum-Kalai-Wasserman algorithm

1 Introduction

Background. In December 2016, the National Institute of Standards and Technology (NIST) initiated post-quantum cryptography (PQC) standardization. In the list of the round 1 submissions [33], there are several lattice-based schemes whose security are based on learning with errors (LWE) problem (e.g., [5,6,13,17,18]) and learning with rounding (LWR) problem (e.g., [9,21,24,34]). Therefore, studies of the algorithm to solve the LWE and LWR problems are important for design and security analysis of post-quantum cryptosystems.

The LWE problem, which is an extension of the learning parity with noise (LPN) problem, is introduced by Regev [42]. An adversary of the LWE problem receives samples $(\boldsymbol{a}_j, \langle \boldsymbol{a}_j, \boldsymbol{s} \rangle + e_j) \in \mathbb{Z}_q^n \times \mathbb{Z}_q$ $(j = 1, 2, \dots)$ from the LWE oracle, where $\boldsymbol{a}_j$ is a uniformly random vector in $\mathbb{Z}_q^n$, $\boldsymbol{s}$ is a fixed secret vector in $\mathbb{Z}_q^n$, and $e_j \in \mathbb{Z}_q$ is a noise (usually, discrete Gaussian noise). The goal of the adversary is to recover secret vector $\boldsymbol{s}$. We note that the LPN problem has a fixed modulus $q = 2$ and the noise follows the Bernoulli distribution. In [42], Regev presents a reduction from worst-case lattice problems to the average-case LWE problem.

© Springer Nature Switzerland AG 2019
K. Lee (Ed.): ICISC 2018, LNCS 11396, pp. 196–214, 2019.
https://doi.org/10.1007/978-3-030-12146-4_13

We can classify algorithms for solving the LWE problem into two families. The first family uses lattice reduction techniques, which have been extensively studied (see, *e.g.*, [14,20,26,30,31,38–41]). The expected complexity of these algorithms is often considered when parameters for LWE-based schemes are discussed, such as in [3]. The second family is tailor-made for the LPN and LWE problems without lattice reduction, which includes the main subject of this paper: the Blum-Kalai-Wasserman (BKW) algorithm [15]. The BKW algorithm can be described as a "block-wise" and addition-only variant of the standard Gaussian elimination. First, we separate the vector $a_j \in \mathbb{Z}_q^n$ into a blocks: We can write $a_j = (a_{j,1}\|a_{j,2}\|\ldots\|a_{j,a})$, where $a_{j,1},\ldots,a_{j,a} \in \mathbb{Z}_q^{n/a}$, and then, by adding the samples together like the Gaussian elimination, we obtain "reduced" samples $a_j' = (a_{j,1}'\|\mathbf{0}\|\ldots\|\mathbf{0})$. As reported in [32], improved variants of the BKW algorithm such as [29,36] are some of the asymptotically fastest algorithms. Although some algorithm [37] based on lattice reduction outperforms these BKW algorithms for some parameter-sets (q,σ), it allows a heuristic [32].

The LWR problem is proposed by Banerjee *et al.* [12] with its reduction from the LWE problem. We can consider the LWR problem as a deterministic variant of the LWE problem in which the noise additions are replaced with deterministic rounding operations. An adversary of the LWR problem receives samples $(a_j, \lceil \frac{p}{q}\langle a_j, s\rangle\rfloor) \in \mathbb{Z}_q^n \times \mathbb{Z}_p$ $(j = 1, 2, \ldots)$ from the LWR oracle, where p is a rounding modulus such that $p < q$. Compared with LWE-based cryptographic schemes, LWR-based schemes can be simply implemented because they replace the rich Gaussian error sampling process of the LWE-based schemes with the rounding operations (which can be simply implemented by rounding off the lower-order bits). The LWR problem was initially applied to low-depth pseudorandom functions [10–12,16], and there have been a number of applications, cf. lossy trapdoor functions [7], public-key cryptosystems [9,21,43] and key exchange protocol [34].

However, few studies have been performed on the complexity of the LWR problem, while the complexity of the LWE problem has been extensively studied. The complexity of the LWR problem is often estimated by adopting the LWE-solving algorithms to the LWR problem. In [3], Albrecht *et al.* estimates the cost of running primal and dual lattice attacks, which is based on lattice reduction techniques, against lattice-based schemes including LWE-based schemes and LWR-based schemes in the list of the round 1 submissions for the NIST PQC standardization [33]. They consider that the cost of lattice attacks for the LWR and LWE problem are the same when an equation $\sigma = \frac{q}{2\sqrt{3}p}$ holds, as considered in [22,24]. This equation is simply derived by comparing the variance of the Gaussian noise of the LWE problem and the "rounding error" of the LWR problem. Note that the equation, which relates the hardness of the LWR problem and the LWE problem, is limited to attacks based on lattice reduction techniques, and it is not shown that the conversion equation can be applied for the BKW algorithm.

Previous Works. The BKW algorithm initially targeted the LPN problem, and its time complexity is sub-exponential in $2^{O(n/\log n)}$. Albrecht *et al.* [2] expanded it to solve the LWE problem whose time complexity is $q^{O(n/\log n)}$. Duc *et al.* improved Albrecht *et al.*'s BKW algorithm and also introduced its variant for the LWR problem, which was the first algorithmic analysis of the LWR problem. They showed that the time complexity of the LWR-solving algorithm is $q^{O(n/\log n)}$ when we choose the number of the block as $a = O(\log n)$. However, they did not show this choice of a is optimal; thus the minimum time complexity of the algorithm is not shown.

After the BKW algorithm proposed by Albrecht *et al.*, new variants of the BKW algorithm [1,4,28,29,36] for solving a *small-secret* LWE problem, whose secret vector s is extremely small (e.g. $s \in \{0,1\}^n$), are proposed. These algorithms can be applied to the general LWE problem, whose secret vector s is uniform in $\mathbb{Z}_q^n$, by transforming the general LWE problem to *small-secret* LWE problem with a technique called *secret-error switching* [8,19,36], and it is shown that some of these algorithms [29,36] solve the general LWE problem faster. However, it is not shown that these new type of the BKW algorithm can be applied to the LWR problem. In order to apply the *secret-error switching* technique to the LWR problem, we need to convert LWR samples into LWE samples with uniform error by substituting the LWR samples $(a_j, \lceil \frac{p}{q}\langle a_j, s\rangle \rceil)$ with $(a_j, \frac{q}{p}\lceil \frac{p}{q}\langle a_j, s\rangle \rceil)$, and, as mentioned in [36], solving this converted LWR problem with their algorithm is out of reach. On the other hand, Duc *et al.*'s LWR-solving BKW algorithm does not need to convert LWR samples into LWE samples; the algorithm is tailor-made for solving the LWR problem.

Our Contribution. We first review Duc *et al.*'s LWR-solving BKW algorithm, and then derive the time complexity in a simpler form. Subsequently, we theoretically derive the optimal choice of the number of blocks a that asymptotically minimize the time complexity, while Duc *et al.* searched numerically for the optimal a in [25]. Thus, an entirely theoretical analysis of the time complexity of the algorithm is shown in this paper: We show that the minimum time complexity of the BKW algorithm is $t = q^{O(n/\log n)}$ and the required number of samples is $m = q^{O(n/\log n)}$. We also confirm that the derived parameter is accurately optimal by showing the results of some concrete instances of the LWR problem, and that they fit the results given by Duc *et al.*

Furthermore, we derive a conversion equation between the Gaussian parameter σ in the LWE problem and the rounding modulus p in the LWR problem, by comparing the time complexity of the BKW algorithm for the LWE and LWR problems: We show that the time complexity of the BKW algorithm to solve the LWE problem and that to solve the LWR problem are the same when σ and p satisfy equation $\sigma = \frac{q}{2\sqrt{3}p}$. This equation coincides with the equation derived from the complexity analysis of the attacks based on lattice reduction techniques. Thus, our result means that the equation is applicable also for the complexity analysis based on the BKW algorithm.

2   Preliminaries

Notations. We denote the logarithm of base 2 and the natural logarithm as $\log(\cdot)$ and $\ln(\cdot)$, respectively. We denote the imaginary unit as i, and a real part of $x \in \mathbb{C}$ as $\mathrm{Re}(x)$. We let $\lceil \cdot \rfloor : \mathbb{R} \to \mathbb{Z}$ be the rounding function that rounds to the closest integer. (In the case of equality, we take the floor.) We define $\theta_q := e^{\frac{2\pi i}{q}}$ and also $\theta_p := e^{\frac{2\pi i}{p}}$. We write vectors in bold. By $\boldsymbol{a}_j$ we denote the j-th vector of the list of vectors. We denote a partial vector of a vector $\boldsymbol{a} = (a_1, a_2, \ldots, a_n)$ by $\boldsymbol{a}_{(k,l)} := (a_k, a_{k+1}, \ldots, a_l)$, where $1 \le k \le l \le n$. By $(\boldsymbol{a} \| \boldsymbol{b})$ we denote the concatenation of two vectors $\boldsymbol{a}$ and $\boldsymbol{b}$. We denote by $\langle \cdot, \cdot \rangle$ the usual dot product of two vectors, and we define $\langle \cdot, \cdot \rangle_q := \langle \cdot, \cdot \rangle \pmod{q}$. We write $\boldsymbol{s} \xleftarrow{U} \mathcal{S}$ to denote the process of sampling $\boldsymbol{s}$ uniformly at random over $\mathcal{S}$, and we write $e \leftarrow \chi$ to denote the process of sampling e according to a probability distribution χ.

2.1   LWE and LWR Problem

The LWE Problem. We define the LWE oracle and the LWE problem.

Definition 1 *(LWE oracle). Let n, q be positive integers. Learning with Error (LWE) oracle $\mathsf{LWE}_{s,\chi}$ for a fixed vector $\boldsymbol{s} \in \mathbb{Z}_q^n$ and probability distribution χ over $\mathbb{Z}_q$ is an oracle returning $\left\{ (\boldsymbol{a}, c) \,\middle|\, c = \langle \boldsymbol{a}, \boldsymbol{s} \rangle + e \bmod q, \ \boldsymbol{a} \xleftarrow{U} \mathbb{Z}_q^n, e \leftarrow \chi \right\}$.*

For the distribution of noise χ, variants of the Gaussian distribution that is discretized into $\mathbb{Z}_q$ are used. In this paper, we consider two types of Gaussian distributions that are considered in [25]; the *rounded Gaussian distribution* $\bar{\Psi}_{\sigma,q}$ and the *discrete Gaussian distribution* $D_{\sigma,q}$. The rounded Gaussian distribution $\bar{\Psi}_{\sigma,q}$ is proposed in the initial LWE problem by Regev [42], and is also considered in [2,27]. Its probability mass function for integer x in the interval $]-\frac{q}{2}, \frac{q}{2}]$, is given by $\Pr[x \leftarrow \bar{\Psi}_{\sigma,q}] = \int_{x-\frac{1}{2}}^{x+\frac{1}{2}} g(\theta; q, \sigma)d\theta$, where $g(\theta; q, \sigma)$ is the probability density function of the *wrapped Gaussian distribution* $\bar{\Psi}_{\sigma,q}$, which is defined by $g(\theta; q, \sigma) := \sum_{l=-\infty}^{\infty} \frac{1}{\sigma\sqrt{2\pi}} \exp\left(\frac{-(\theta+lq)^2}{2\sigma^2}\right)$, for $\theta \in]-\frac{q}{2}, \frac{q}{2}]$. The *discrete Gaussian distribution* $D_{\sigma,q}$ is used in most of the cryptographic applications of the LWE problem and in the classical LWE problem reduction [19]. This distribution is, for x an integer in $]-\frac{q}{2}, \frac{q}{2}]$, $\Pr[x \leftarrow D_{\sigma,q}] = \dfrac{\exp(-\frac{x^2}{2\sigma^2})}{\sum_{y \in]-\frac{q}{2}, \frac{q}{2}]} \exp(-\frac{y^2}{2\sigma^2})}$.

Definition 2 *(Search-LWE). The Search-LWE problem is the problem of recovering the hidden secret $\boldsymbol{s}$ given m samples $(\boldsymbol{a}_j, c_j) \in \mathbb{Z}_q^n \times \mathbb{Z}_q$ $(j = 1, 2, \ldots, m)$ received from $\mathsf{LWE}_{s,\chi}$.*

The LWR Problem. We define the LWR problem which is the main focus of this paper. For the purpose, we define an LWR Oracle in advance.

Definition 3 *(LWR oracle). Let n, q be natural numbers. Learning with Rounding (LWR) oracle $\mathsf{LWR}_{s,p}$ for a hidden vector $\boldsymbol{s} \in \mathbb{Z}_q^n$ and rounding modulus p is an oracle returning $\left\{ (\boldsymbol{a}, c) \,\middle|\, c = \left\lceil \frac{p}{q} \langle \boldsymbol{a}, \boldsymbol{s} \rangle_q \right\rfloor, \ \boldsymbol{a} \xleftarrow{U} \mathbb{Z}_q^n \right\}$.*

Definition 4 *(LWR problem). The LWR problem is the problem of recovering the hidden secret s given m samples $(a_j, c_j) \in \mathbb{Z}_q^k \times \mathbb{Z}_q$ $(j = 1, 2, \ldots, m)$ received from* $\mathsf{LWR}_{s,p}$.

The rounding calculation in the LWR sample generates a "rounding error," which is similar to the Gaussian noise added in the LWE sample. Duc *et al.* proved that "rounding error" follows a uniform distribution, in Lemma 19 in [25].

Lemma 1 *(Lemma 19 in [25]). Let n and $q > p \geq 2$ be positive integers, q prime. Let (a, c) be a random sample from an LWR oracle $\mathsf{LWR}_{s,p}$. Then, the "rounding error," given by*

$$\xi = \frac{p}{q}\langle a, s \rangle_q - c, \tag{1}$$

follows the uniform distribution in a discrete subset of $[-\frac{1}{2}, \frac{1}{2}]$ with mean zero. Furthermore, the characteristic function of ξ, for $t \in \mathbb{R}_{\neq 0}$, is

$$\phi_\xi(t) := E[e^{\pm it\xi}] = \frac{\sin(\frac{t}{2})}{q\sin(\frac{t}{2q})}. \tag{2}$$

Banerjee *et al.*, showed a reduction from the LWE problem to the LWR problem, in the paper [12] in which they first introduced the LWR problem. Note that the decision version of the LWE (or LWR) problem can be described as follows: given m samples of the form $(a, c) \in \mathbb{Z}_q^n \times \mathbb{Z}_q$ (or $\mathbb{Z}_p$), where $a \xleftarrow{U} \mathbb{Z}_q^n$, distinguish whether $c \xleftarrow{U} \mathbb{Z}_q$ (or $\mathbb{Z}_p$) or $c = \langle a, s \rangle + e$ (or $c = \left\lceil \frac{p}{q}\langle a, s \rangle_q \right\rfloor$), for a fixed secret $s \in \mathbb{Z}_q^n$.

Theorem 1 *(Theorem 3.2 in [12]). Let $\beta \in \mathbb{R}_+$, χ be any efficiently samplable distribution over $\mathbb{Z}$ such that $\Pr_{x \leftarrow \chi}[|x| > \beta]$ is negligible, and let $q \geq p \cdot \beta \cdot n^{\omega(1)}$. Then, solving decision-LWR with secrets of size n and parameters p and q is at least as hard as solving decision-LWE over $\mathbb{Z}_q^n$ with noise distribution χ.*

Alwen *et al.* [7] also showed the reduction without the super-polynomial parameters, but it limits the number of samples that the LWR oracle allows the adversary to receive.

Theorem 2 *(Theorem 4.1 from [7]). Let λ be the security parameter. Let n, l, m, p, γ be positive integers, p_{max} be the largest prime divisor of q, and $p_{max} \geq 2\beta\gamma nmp$. Let χ be the probability distribution over $\mathbb{Z}$ such that the average absolute value of $x \leftarrow \chi$ is less than β. Then, if $n \geq (l + \lambda + 1)\log(q)/\log(2\gamma) + 2\lambda$ and if $\gcd(q, q/p_{max}) = 1$, the decision-LWR with secrets of size n, parameters p and q and limited to m queries is at least as hard as solving decision-LWE over $\mathbb{Z}_q^l$ with noise distribution χ and limited to m queries.*

2.2 Duc *et al.*'s BKW Algorithm for the LWR Problem

We recall Duc *et al.*'s BKW algorithm to solve the LWR problem. The BKW algorithm consists of three stages: (1) Sample reduction, (2) Hypothesis testing,

---

Algorithm 1. The BKW algorithm to solve the LWR problem [25]

Input: natural numbers a, b $(ab = n), m$, and samples (a_j, c_j), $(j = 1, 2, \ldots, m)$. We represent the set of samples as $\mathcal{S} := \{(a_j, c_j)\}_{j=1}^{m}$.

Output: s.

(Stage 1: Sample reduction.)

  for $l = 0$ to $a - 2$ **do**

    $\mathcal{S}' \leftarrow \phi$: empty set

    $T_l \leftarrow \phi$

    repeat

      extract one sample (a, c) from $\mathcal{S}$.

      if $a_{(b(a-l-1)+1, b(a-l))} = \mathbf{0}$ **then**

        $\mathcal{S}' \leftarrow \mathcal{S}' \cup (a, c)$

      else if there is $(a', c') \in T_l$ such that $(a \pm a')_{(b(a-l-1)+1, b(a-l))} = \mathbf{0}$ **then**

        $\mathcal{S}' \leftarrow \mathcal{S}' \cup (a \pm a', c \pm c')$

      else

        $T_l \leftarrow T_l \cup (a, c)$

      end if

    until $\mathcal{S} = \phi$

    $\mathcal{S} \leftarrow \mathcal{S}'$

  end for

  $T_{a-1} \leftarrow \mathcal{S}$

  for $l = a - 1$ to 0 **do**

    (Stage 2: Hypothesis testing.)

    Let $m_l := \#T_l$, and we denote by $(\overline{a}_j^l, \overline{c}_j^l)$ the samples included in T_l.

    Calculate $f(y) := \sum_{j=1}^{m_l} \mathbf{1}_{\{\overline{a}_j^l = y\}} \theta_p^{\overline{c}_j^l}$ for all $y \in \mathbb{Z}_q^b$

    Calculate the DFT of f, which is $\hat{f}(z) = \sum_{j=1}^{m_l} \theta_p^{-(\frac{p}{q}\langle \overline{a}_j^l, z \rangle - \overline{c}_j^l)}$.

    Calculate $s_{(b(a-l-1)+1, b(a-l)))} \leftarrow \arg\max_z \operatorname{Re}(\hat{f}(z))$.

    (Stage 3: Back substitution.)

    Using the obtained $s_{(bl+1, b(l+1))))}$, update sets $T_{l'}$ for $0 \leq l' < l$.

  end for

  return $(s_{(1,b)} \| s_{(b+1, 2b)} \| \cdots \| s_{((a-1)b+1, ab)})$

---

and (3) Back substitution. For simplicity, in this paper, we consider only the case that the number of blocks a and the block length b satisfy $ab = n$. Algorithm 1 shows an overview of the algorithm.

Stage 1: Sample Reduction. We receive m samples $\{(a_j, c_j)\}_{j=1}^{m}$ from LWR oracle $\mathsf{LWR}_{s,p}$, and represent the set of samples as $\mathcal{S} := \{(a_j, c_j)\}_{j=1}^{m}$. We separate the vector $a_j \in \mathbb{Z}_q^n$ into a blocks whose length are b: We can write $a_j = (a_{j(1,b)} \| a_{j(b+1, 2b)} \| \cdots \| a_{j((a-1)b+1, ab)})$. In Stage 1, our goal is to produce samples whose elements are all zero except for the first block, with addition or subtraction of pairs of samples. For $l = 0$, we extract a sample (a, c) from $\mathcal{S}$, and search another sample (a', c') such that $(a \pm a')_{((a-1)b+1, ab)} = \mathbf{0}$, then we store the sample $(a \pm a', c \pm c')$ in the temporary set $\mathcal{S}'$. If a sample (a, c) already holds $a_{((a-1)b+1, ab)} = \mathbf{0}$, we directly store it in $\mathcal{S}'$. If we cannot find the sample

$(\boldsymbol{a}', c')$ such that $(\boldsymbol{a} \pm \boldsymbol{a}')_{((a-1)b+1,ab)} = \boldsymbol{0}$, we store the sample $(\boldsymbol{a}, c)$ in $\mathcal{T}_0$. After we finish extracting samples and empty the set $\mathcal{S}$, we renew $\mathcal{S} \leftarrow \mathcal{S}'$ and move on to the next step for $l = 1$. In this manner, we recursively generate the sets $\mathcal{T}_l$ for $0 \leq l \leq a - 2$, and then we set $\mathcal{T}_{a-1} \leftarrow \mathcal{S}$ in the end. Note that the samples $(\boldsymbol{a}, c)$ in $\mathcal{T}_l$ hold $\boldsymbol{a}_{((a-l)b+1,ab)} = \boldsymbol{0}$ (except for $l = 0$). In particular, the samples $(\boldsymbol{a}, c)$ in $\mathcal{T}_{a-1}$ hold $\boldsymbol{a}_{(b+1,ab)} = \boldsymbol{0}$. We may think of the reduced samples in $\mathcal{T}_{a-1}$ as the set of samples of the b-dimensional LWR problem, although the variance of their noise is larger than those of the original samples. Hereinafter, the samples in $\mathcal{T}_l$ are termed "reduced samples", and represent $\mathcal{T}_l = \{(\overline{\boldsymbol{a}}_j^l, \overline{c}_j^l)\}_{j=1}^{m_l}$, where $m_l := \#\mathcal{T}_l$. Note that the maximum number of samples whose $(a - l)$-th block cannot vanish is $\frac{q^b-1}{2}$, and the minimum (worst) number of reduced samples in $\mathcal{T}_{a-1}$ (i.e. minimum value of m_{a-1}) is

$$m' = m - (a-1)\frac{q^b - 1}{2}. \tag{3}$$

Stage 2: Hypothesis Testing. For simplicity, we explain Stage 2 and Stage 3 only for $l = a - 1$. (In Sect. 3.1, we only consider the time complexity to recover $\boldsymbol{s}_{(1,b)}$ because the whole time complexity of the algorithm is at most a positive constant multiple of it). For simplicity of notation, we define a b-dimensional vector $\overline{\boldsymbol{a}}_j := (\overline{\boldsymbol{a}}_j^{a-1})_{(1,b)}$, and denote $\overline{c}_j := \overline{c}_j^{a-1}$. The goal of this stage is to estimate the first b elements of $\boldsymbol{s}$, denoted as $\boldsymbol{s}_{(1,b)}$. We define the function $f(\boldsymbol{y}) := \sum_{j=1}^{m'} \mathbb{1}_{\{\overline{\boldsymbol{a}}_j = \boldsymbol{y}\}} \theta_p^{\overline{c}_j}$, where $\boldsymbol{y} \in \mathbb{Z}_q^b$, $\theta_p := e^{\frac{2\pi i}{p}}$, and $\mathbb{1}_{\{\overline{\boldsymbol{a}}_j = \boldsymbol{y}\}}$ is 1 when $\overline{\boldsymbol{a}}_j = \boldsymbol{y}$ is true and 0 otherwise. The discrete Fourier transform of f is

$$\hat{f}(\boldsymbol{z}) := \sum_{\boldsymbol{y} \in \mathbb{Z}_q^b} f(\boldsymbol{y})\theta_q^{-\langle \boldsymbol{y}, \boldsymbol{z}\rangle} = \sum_{j=1}^{m'} \theta_p^{-(\frac{p}{q}\langle \overline{\boldsymbol{a}}_j, \boldsymbol{z}\rangle - \overline{c}_j)}. \tag{4}$$

Then, we search the max $\mathrm{Re}(\hat{f}(\boldsymbol{z}))$, and output $\boldsymbol{s}_{(1,b)} = \mathrm{argmax}_{\boldsymbol{z}}\ \mathrm{Re}(\hat{f}(\boldsymbol{z}))$. We explain how the output $\mathrm{argmax}_{\boldsymbol{z}}\ \mathrm{Re}(\hat{f}(\boldsymbol{z}))$ estimates the secret vector. We define the "rounding error" of the reduced samples $\{(\overline{\boldsymbol{a}}_j, \overline{c}_j)\}_{j=1}^{m'}$ by $\overline{\xi}_j := \frac{p}{q}\langle \overline{\boldsymbol{a}}_j, \boldsymbol{s}_{(1,b)}\rangle - \overline{c}_j$, as like (1). Recall that the $\overline{\boldsymbol{a}}_j$ is produced by $a - 1$ times of the "tree-like" addition of the original samples in the process of Stage 1, i.e. $\overline{\boldsymbol{a}}_j$ is the sum of the 2^{a-1} original samples, thus we can write $\overline{\boldsymbol{a}}_j = (\boldsymbol{a}_{j,1} \pm \boldsymbol{a}_{j,2} \pm \cdots \pm \boldsymbol{a}_{j,2^{a-1}})_{(1,b)}$, where $\boldsymbol{a}_{j,1}, \ldots, \boldsymbol{a}_{j,2^{a-1}}$ are the original samples. Similarly, we can write $\overline{c}_j = c_{j,1} \pm c_{j,2} \pm \cdots \pm c_{j,2^{a-1}}$, and obtain

$$\overline{\xi}_j = \frac{p}{q}\langle \boldsymbol{a}_{j,1} \pm \boldsymbol{a}_{j,2} \pm \cdots \pm \boldsymbol{a}_{j,2^{a-1}}, \boldsymbol{s}\rangle - (c_{j,1} \pm c_{j,2} \pm \cdots \pm c_{j,2^{a-1}})$$

$$= \sum_{k=1}^{2^{a-1}} \frac{p}{q}\langle \boldsymbol{a}_{j,k}, \boldsymbol{s}\rangle - c_{j,k} = \sum_{k=1}^{2^{a-1}} \xi_{j,k},$$

where the $\xi_{j,k}$ are independent rounding errors from original samples. From the above equation and (4), when $\boldsymbol{z} = \boldsymbol{s}_{(1,b)}$, we obtain $\hat{f}(\boldsymbol{s}_{(1,b)}) = \sum_{j=1}^{m'} \theta_p^{-(\sum_{k=1}^{2^{a-1}} \xi_{j,k})}$. On the other hand, when $\boldsymbol{z} \neq \boldsymbol{s}_{(1,b)}$, $\frac{p}{q}\langle \overline{\boldsymbol{a}}_j, \boldsymbol{z}\rangle - \overline{c}_j$ distribute uniformly in $]0, p]$.

Thus, when we select an appropriate value of parameter a such that the sum of the rounding errors $\sum_{k=1}^{2^{a-1}} \xi_{j,k}$ does not grow too large, $\text{Re}(\hat{f}(s))$ is so much larger than $\text{Re}(\hat{f}(z))$ that the hypothesis test succeeds with high probability.

Stage 3: Back Substitution. Using the obtained $s_{(1,b)}$, update the sets T_l by zeroing-out b elements in each sample: Replace all $(a, c) \in T_{l'}$ for $0 \le l' < a - 1$ with (a', c'), where $a' = (0||a_{(b+1,n)}) \in \mathbb{Z}_q^n$, $c' = c - \frac{p}{q}\langle a_{(1,b)}, s_{(1,b)}\rangle_q \in \mathbb{Z}_p$. Then back to Stage 2 to obtain $s_{(b+1,2b)}$.

Repeating a rounds of Stages 2 to 3, we estimate $s_{(1,b)}, s_{(b+1,2b)}, \cdots,$ $s_{(a(b-1)+1,ab)}$, and obtain $s = (s_{(1,b)}||s_{(b+1,2b)}||\cdots||s_{((a-1)b+1,ab)})$.

3   Analysis of BKW Algorithm for the LWR Problem

We derive the minimum time complexity and the minimum number of required samples, by optimizing the number of blocks a which is a parameter of the BKW algorithm.

As with Duc *et al.* in this paper, we consider only the case that the block length b satisfy $n = ab$, for simplicity. Therefore, the block length b is determined by the number of blocks a, as $b = n/a$. Note that the complexity of the BKW algorithm for the general case, where $n = (a - 1) \cdot b + n'$ and $n' < b$, is asymptotically the same with that for the case where $ab = n$. We always consider q to be a prime, and $q > p > 4$ because we need the condition to prove Lemma 3.

In Sect. 3.1, we analyze the time complexity of the BKW algorithm for solving the LWR problem, using a as a parameter. In Sect. 3.2, we derive the optimal value of a that asymptotically minimizes the asymptotic time complexity. In Sect. 3.3, we calculate the concrete time complexity of the BKW algorithm for several LWR instances, and confirm that the optimal value of a minimizes the time complexity of the algorithm. Furthermore, in Sect. 4.1, we derive a equation that relates σ of the LWE problem and p of the LWR problem. When σ and p satisfy the equation, the asymptotic time complexity of the BKW algorithm to solve the LWE and LWR problems are the same.

3.1   Complexity Analysis

We analyze the time complexity and required number of samples to solve the LWR problem. We asymptotically analyze the time complexity and make it in a simple form so that we can theoretically derive the optimal number of blocks a in Sect. 3.2. We first refer to Lemma 2 (Theorem 23 in [25]), which is the analysis of the minimum number of samples needed to solve the LWR problem.

Lemma 2 *(Theorem 23 in [25]). We define the probability that the algorithm cannot recover the correct answer $\epsilon := \Pr\left[\text{argmax}_z \text{Re}\left(\hat{f}(z)\right) \ne s_{(1,b)}\right]$. Then, the number of samples required to solve the LWR problem with oracle $\mathsf{LWR}_{s,p}$ is*

$$m^{\mathsf{LWR}} = \frac{8n}{a} \ln\left(\frac{q}{\epsilon}\right) \left((R_{q,p})^{2^{a-1}} - \left(\frac{3}{p}\right)^{2^{a-1}}\right)^{-2} + (a-1)\frac{q^{n/a}-1}{2}, \qquad (5)$$

where $R_{q,p} := \frac{\sin\left(\frac{\pi}{p}\right)}{q\sin\left(\frac{\pi}{pq}\right)}$.

Note that $R_{q,p}$ is derived based on the characteristic function of the "rounding error" given in (2): $R_{q,p} = \phi_\xi(\frac{2\pi}{p})$ holds. As discussed later, this m^{LWR} in (5) is the main term of the time complexity of the algorithm.

In the following Lemma 3, we analyze the asymptotic behavior of the complicated part of the m^{LWR}. We describe it in a simpler form in order to enable the analysis of the minimum time complexity, which will be given later in Sect. 3.2. Note that we use the error rate of the LWR sample $\alpha_{\mathrm{lwr}} := \frac{1}{p}\sqrt{\frac{\pi}{6}}$ [21] to describe the time complexity for simplicity of notation.

Lemma 3. Let $\alpha_{\mathrm{lwr}} := \frac{1}{p}\sqrt{\frac{\pi}{6}}$. When $q > p > 4$, we have

$$\left((R_{q,p})^{2^{a-1}} - \left(\frac{3}{p}\right)^{2^{a-1}}\right)^{-2} = \exp(\pi\alpha_{\mathrm{lwr}}^2 2^a) + O\left(\frac{1}{p^2 q^2}\right). \qquad (6)$$

Proof. First, we prove that

$$\left((R_{q,p})^{2^{a-1}} - \left(\frac{3}{p}\right)^{2^{a-1}}\right)^{-2} = (R_{q,p})^{-2^a} + O\left(\frac{1}{p^{2^{a-1}}}\right) \qquad (7)$$

holds. Note that $q > p > 4$. We obtain $R_{q,p} = \frac{\frac{p}{\pi}\sin\left(\frac{\pi}{p}\right)}{\frac{pq}{\pi}\sin\left(\frac{\pi}{pq}\right)} \geq \frac{p}{\pi}\sin\left(\frac{\pi}{p}\right)$. Since $\frac{p}{\pi}\sin\left(\frac{\pi}{p}\right)$ is monotonically increasing when $p > 4$, we obtain $R_{q,p} > \frac{4}{\pi}\sin\left(\frac{\pi}{4}\right) = 0.9003\cdots$, and $R_{q,p} > \frac{3}{p}$. Let $x = (R_{q,p})^{2^{a-1}}$ and $y = \left(\frac{3}{p}\right)^{2^{a-1}}$, then we have $\frac{y}{x} < 1$. Using Taylor expansion, we obtain $(x-y)^{-2} = \frac{1}{x^2}\left(1 + O(\frac{y}{x})\right)$. Therefore, (7) holds.

Next, we derive (6) from (7). Using Taylor expansion, we obtain $R_{q,p} = 1 - \frac{\pi^2}{6p^2} + O\left(\frac{1}{p^2 q^2}\right) = 1 - \pi\alpha_{\mathrm{lwr}}^2 + O\left(\frac{1}{p^2 q^2}\right)$, and $R_{q,p} - \exp\left(-\pi\alpha_{\mathrm{lwr}}^2\right) = O\left(\frac{1}{p^2 q^2}\right)$. Consequently, from this equation, we obtain

$$(R_{q,p})^{-2^a} = \left(\exp\left(-\pi\alpha_{\mathrm{lwr}}^2\right) + O\left(\frac{1}{p^2 q^2}\right)\right)^{-2\cdot 2^{a-1}}$$

$$= \left(\exp\left(\pi\alpha_{\mathrm{lwr}}^2 \cdot 2\right)\left(1 + O\left(\frac{1}{p^2 q^2}\right)\right)\right)^{2^{a-1}}$$

$$= \exp\left(\pi\alpha_{\mathrm{lwr}}^2 2^a\right)\left(1 + O\left(\frac{1}{p^2 q^2}\right)\right)^{2^{a-1}}$$

$$= \exp\left(\pi\alpha_{\mathrm{lwr}}^2 2^a\right) + O\left(\frac{1}{p^2 q^2}\right). \qquad (8)$$

Thus, using (7) and (8), we have (6). □

We can now derive the number of required samples and the time complexity of the algorithm.

Theorem 3. *Let n and $q > p > 4$ be positive integers, q prime, and a be a natural number. Fix $\epsilon \in (0,1)$. When at least $m^{\mathsf{LWR}} = \mathrm{poly}(\exp(\pi\alpha_{\mathrm{lwr}}^2 2^a), q^{n/a})$ samples are given by LWR oracle $\mathsf{LWR}_{s,p}$, the time complexity of the BKW algorithm to recover secret s with a probability of at least $1 - \epsilon$ is $t^{\mathsf{LWR}} = \mathrm{poly}(\exp(\pi\alpha_{\mathrm{lwr}}^2 2^a), q^{n/a})$, where $\alpha_{\mathrm{lwr}} = \frac{1}{p}\sqrt{\frac{\pi}{6}}$.*

Proof. From Lemmas 2 and 3, the number of required samples to solve the LWR problem is $m^{\mathsf{LWR}} = \frac{8n}{a}\ln\left(\frac{q}{\epsilon}\right)\left(\exp(\pi\alpha_{\mathrm{lwr}}^2 2^a) + O\left(\frac{1}{p^2 q^2}\right)\right) + (a-1)\frac{q^{n/a}-1}{2}$. Recall that the number of the "reduced" samples we obtain after Stage 1 is $m' = m^{\mathsf{LWR}} - (a-1)\frac{q^{n/a}-1}{2} = \frac{8n}{a}\ln\left(\frac{q}{\epsilon}\right)\left(\exp(\pi\alpha_{\mathrm{lwr}}^2 2^a) + O\left(\frac{1}{p^2 q^2}\right)\right)$, which is defined in (3).

In Stage 1, since we apply the addition for $O(m^{\mathsf{LWR}})$ samples in $\mathbb{Z}_q^n$ for $a-1$ times, the time complexity is $t_1 = O(anm^{\mathsf{LWR}})$. In Stage 2, We first calculate $f(\boldsymbol{y}) := \sum_{j=1}^{m'} \mathbb{1}_{\{\bar{a}_j = y\}}\theta_p^{\bar{c}_j}$, for all $\boldsymbol{y} \in \mathbb{Z}_q^b$. Since we need only to calculate $f(\boldsymbol{y})$ for $\boldsymbol{y} \in \{\bar{a}_j\}_{j=1}^{m'}$, the time complexity for calculating $f(\boldsymbol{y})$ is $O(m') = O(\exp(\pi\alpha_{\mathrm{lwr}}^2 2^a)(n/a)\ln q)$. After that, we compute the DFT of f, the complexity of which is $O(q^{n/a}(n/a)\ln q)$. Finally, we search $\max \hat{f}(\boldsymbol{z})$ defined in (4) for all $\boldsymbol{z} \in \mathbb{Z}_q^{n/a}$, the time complexity of which is $O(q^{n/a}n/a)$. Therefore, the time complexity of Stage 2 is $t_2 = O(\exp(\pi\alpha_{\mathrm{lwr}}^2 2^a)(n/a)\ln q) + O(q^{n/a}(n/a)\ln q)$. In Stage 3, since we update all samples stored in $T_{l'}$ $(0 \le l' < a - 1)$ (the total number of these samples is $m^{\mathsf{LWR}} - m'$) with inner product calculation of the vectors in $\mathbb{Z}_q^{n/a}$, the time complexity of Stage 3 is $t_3 = O((m^{\mathsf{LWR}} - m')n/a) = O(q^{n/a}n/a)$. Therefore, the time complexity of the BKW algorithm is $t^{\mathsf{LWR}} = t_1 + t_2 + t_3 = O(\exp(\pi\alpha_{\mathrm{lwr}}^2 2^a)(n/a)\ln q) + O(q^{n/a}(n/a)\ln q) = \mathrm{poly}(\exp(\pi\alpha_{\mathrm{lwr}}^2 2^a), q^{n/a})$. $\square$

3.2   Parameter Optimization

We analyze the optimal choice for input parameter a to asymptotically minimize the asymptotic time complexity of the BKW algorithm to solve the LWR problem. Furthermore, we analyze the minimum time complexity.

Theorem 4 *(Optimal choice of a). The optimal parameter a that asymptotically minimizes the asymptotic time complexity of the algorithm to solve the LWR problem is*

$$a = \left\lfloor \frac{1}{\ln 2}W\left(\frac{n\ln q\ln 2}{\pi\alpha_{\mathrm{lwr}}^2}\right)\right\rfloor \tag{9}$$

where W is Lambert W function [23].

Proof. From Theorem 3, we obtain the time complexity $t = O(\exp(\pi\alpha_{lwr}^2 2^a)$ $(n/a)\ln q) + O(q^{n/a}(n/a)\ln q)$. Note that $\exp(\pi\alpha_{lwr}^2 2^a)$ monotonically increases and $q^{n/a}$ monotonically decreases, as a increases. Therefore, the time complexity is asymptotically minimized[1] when a satisfies

$$\exp(\pi\alpha_{lwr}^2 2^a) = q^{n/a}. \tag{10}$$

From (10), by simple arithmetic, we obtain $(\ln 2)ae^{(\ln 2)a} = \frac{n\ln q\ln 2}{\pi\alpha_{lwr}^2}$. To solve this equation for a, we use the Lambert W function, which satisfies $W(ze^z) = z$. We obtain $W((\ln 2)ae^{(\ln 2)a}) = (\ln 2)a$, and we obtain (9).   □

Since the Lambert function $W(x)$·has an asymptotic form as $W(x) = \ln(x) - \ln(\ln(x)) + o(1)$, we can evaluate $a = \frac{1}{\ln 2}\left(\ln\left(\frac{n\ln q\ln 2}{\pi\alpha_{lwr}^2}\right) - \ln\ln\left(\frac{n\ln q\ln 2}{\pi\alpha_{lwr}^2}\right)\right) + o(1)$. Furthermore, when we consider q to be at most exponential of n (this range of q includes most of q used in LWE cryptosystems), we obtain $\log q = O(n)$, and $a = O(\log n)$. Using this value, (10), and Theorem 3, we obtain the corollary below.

Corollary 1 *(Minimum time complexity). Let n and $q > p > 4$ be positive integers, q prime. Let $a = \left\lfloor\frac{1}{\ln 2}W\left(\frac{n\ln q\ln 2}{\pi\alpha_{lwr}^2}\right)\right\rfloor$, where $\alpha_{lwr} = \frac{1}{p}\sqrt{\frac{\pi}{6}}$. Fix $\epsilon \in (0,1)$. When at least $q^{O(n/\log n)}$ samples are given by LWR oracle $\mathsf{LWR}_{s,p}$, the time complexity of the BKW algorithm to recover secret s with a probability of at least $1 - \epsilon$ is $q^{O(n/\log n)}$.*

3.3   Concrete Analysis

Table 1 shows the concrete time complexity of the BKW algorithm. We denote the time complexity of the LWR-solving BKW algorithm by $\mathcal{C}^{\mathsf{LWR}}$. Then, similar to Theorem 17 in [25], we obtain

$$\mathcal{C}^{\mathsf{LWR}} = \frac{1}{4}(a-2)(a-1)(2n/a+1)(q^{n/a}-1) + nq^{n/a}\log(q)$$

$$+ \sum_{j=0}^{a-1} m_{j,\epsilon}^{\prime\mathsf{LWR}}\left(\frac{a-1-j}{2}(n+2)+2\right), \tag{11}$$

where $m_{j,\epsilon}^{\prime\mathsf{LWR}} := \frac{8n}{a}\ln\left(\frac{q}{\epsilon}\right)\left(R_{q,p}^{2^{a-1-j}} - \left(\frac{3}{p}\right)^{2^{a-1-j}}\right)^{-2}$. We use the same parameters n, q and p as in Table 2 in [25]: For type (a), $q = \mathrm{nextprime}(\lceil(2\sigma n)^3\rceil)$,

---

[1] Let $\tilde{a}$ satisfies $\exp(\pi\alpha_{lwr}^2 2^{\tilde{a}}) = q^{n/\tilde{a}}$, and Let $t_{\tilde{a}}$ be the time complexity with $a = \tilde{a}$, namely $t_{\tilde{a}} = O(\exp(\pi\alpha_{lwr}^2 2^{\tilde{a}})(n/a)\ln q)$. If we set $a > \tilde{a}$, then we obtain $t_a = O(\exp(\pi\alpha_{lwr}^2 2^a)(n/a)\ln q)$, and $t_a > t_{\tilde{a}}$ since $\exp(\pi\alpha_{lwr}^2 2^a) > \exp(\pi\alpha_{lwr}^2 2^{\tilde{a}})$. If we set $a < \tilde{a}$, then we obtain $t = O(q^{n/a}(n/a)\ln q)$, and $t_a > t_{\tilde{a}}$ since $q^{n/a} > q^{n/\tilde{a}}$. Therefore, $\tilde{a}$ is asymptotically optimal.

$p = \text{nextprime}(\lceil \sqrt[3]{q} \rceil)$ and for type (b), $p = 13$, $q = \text{nextprime}(\lceil 2\sigma np \rceil)$, where $\sigma = \frac{n^2}{\sqrt{2\pi n}(\log(n))^2}$. These parameters are selected based on Corollary 4.2 in [7], which follows Theorem 2. Type (a) parameters maximize the efficiency, and type (b) parameters minimize the modulus to error ratio (q/σ). Note that we also ignored the constraint on the number of samples m as Duc *et al.* did. We set $a = \left\lfloor \frac{1}{\ln 2} W \left(\frac{n \ln q \ln 2}{\pi \alpha_{\text{lwr}}^2} \right) \right\rfloor$ and calculate m^{LWR} and C^{LWR} in (5) and (11), respectively. We also set $\epsilon = 0.01$ in Table 1, following the setting given in Table 2 of [25].

Table 1. The worst case time complexity (C^{LWR}) and the number of required samples (m^{LWR}) for the LWR-solving BKW algorithm. We also provide the value of a theoretically derived in (9), which asymptotically approaches the optimal value that minimizes the complexity. In this table, "*" means the value is optimal, and "†" means the value is not optimal. The optimal values are shown in parenthesis.

(type)	n	q	p	a	$\log(C^{\text{LWR}})$	$\log(m^{\text{LWR}})$
(a)	32	6318667	191	19*	51.00	42.70
	40	23166277	293	20*	60.66	52.18
	64	383056211	733	24† (23)	92.70† (92.10)	83.08† (82.80)
	80	1492443083	1151	25*	110.82	101.11
	96	4587061889	1663	26*	132.17	122.15
	112	11942217841	2287	28*	148.00	137.68
	128	27498355153	3023	29*	167.44	156.88
(b)	32	2411	13	11*	44.53	37.00
	40	3709	13	11*	53.24	45.44
	64	9461	13	12*	81.48	72.92
	80	14867	13	12*	103.76	94.86
	96	21611	13	12*	126.83	117.66
	112	29717	13	13*	140.08	130.63
	128	39241	13	13*	162.50	152.84

In Table 1, we can observe that our choice of the number of blocks a asymptotically (but almost completely) minimizes the time complexity of the algorithm.

4   Comparison Between the LWE and LWR Problems

4.1   Relation Between σ and p

In this section, we compare the time complexity of the BKW algorithm to solve the LWE and LWR problem, and then derive a relation between p in the LWR problem and σ in the LWE problem.

In Theorem 3, we showed that the time complexity of the BKW algorithm to solve the LWR problem is $\text{poly}(\exp(\pi\alpha_{\text{lwr}}^2 2^a), q^{n/a})$. On the other hand, based on Theorem 16 in [25], Kaminakaya et al. [35] analyzed the time complexity of the BKW algorithm to solve the LWE problem, and showed that the complexity is $\text{poly}(\exp(\pi\alpha_{\text{lwe}}^2 2^a), q^{n/a})$, where $\alpha_{\text{lwe}} := \frac{\sqrt{2\pi}\sigma}{q}$. We will describe the result later in Lemma 5 and refer to the proof given in [35]. As a preparation, we refer to the Theorem 16 in [25], which shows the number of samples required to solve the LWE problem:

Lemma 4 *(Theorem 16 in [25])*. *Let $\epsilon := \Pr\left[\text{argmax}_z \text{Re}\left(\hat{f}(z)\right) \neq s_{(1,b)}\right]$ be the probability that the algorithm does not recover the correct answer. Then, the number of samples required to solve the LWE problem with oracle $\mathsf{LWE}_{s,\chi}$ is*

$$m^{\mathsf{LWE}} = \frac{8n}{a}\ln\left(\frac{q}{\epsilon}\right)(R_{q,\sigma,\chi})^{-2^a} + (a-1)\frac{q^{n/a}-1}{2}, \tag{12}$$

where

$$R_{q,\sigma,\chi} = \begin{cases} \frac{q}{\pi}\sin\left(\frac{\pi}{q}\right)e^{-2\pi^2\sigma^2/q^2} & \text{when } \chi = \bar{\Psi}_{q,\sigma}, \\ 1 - \frac{2\pi^2\sigma^2}{q^2} & \text{when } \chi = D_{q,\sigma}. \end{cases}$$

Based on this Lemma 4, we can show the time complexity of the BKW algorithm for the LWE problem:

Lemma 5 *([35])*. *Let a and b be natural numbers such that $ab = n$. There is an algorithm to solve the LWE problem whose oracle is $\mathsf{LWE}_{s,\chi}$, with the number of samples $m = \text{poly}(\exp(\pi\alpha_{\text{lwe}}^2 2^a), q^{n/a})$, and the time complexity $t = \text{poly}(\exp(\pi\alpha_{\text{lwe}}^2 2^a), q^{n/a})$, where $\alpha_{\text{lwe}} := \frac{\sqrt{2\pi}\sigma}{q}$, both when $\chi = D_{\sigma,q}$ and $\chi = \bar{\Psi}_{\sigma,q}$.*

Proof. Here, we refer the proof given in [35]. Similar to the proof of Theorem 3, using Lemma 4, we can prove that there is an algorithm to solve the LWE problem whose oracle is $\mathsf{LWE}_{s,\chi}$, with number of samples $m = \text{poly}((R_{q,\sigma,\chi})^{-2^a}, q^{n/a})$, and time complexity $t = \text{poly}((R_{q,\sigma,\chi})^{-2^a}, q^{n/a})$. Thus, we need only prove that

$$R_{q,\sigma,\chi} = O(\exp(-\pi\alpha_{\text{lwe}}^2)) \tag{13}$$

holds, both when $\chi = \bar{\Psi}_{\sigma,q}$ and when $\chi = D_{\sigma,q}$. When $\chi = \bar{\Psi}_{\sigma,q}$, since $\sin\left(\frac{\pi}{q}\right) < \frac{\pi}{q}$, we obtain $R_{q,\sigma,\chi} < e^{-2\pi^2\sigma^2/q^2} = \exp(-\pi\alpha_{\text{lwe}}^2)$, which means (13) holds. Next, we prove that (13) holds when $\chi = D_{\sigma,q}$. Using Taylor expansion, we obtain $R_{q,\sigma,\chi} - \exp(-\pi\alpha_{\text{lwe}}^2) = 1 - \pi\alpha_{\text{lwe}}^2 - \exp(-\pi\alpha_{\text{lwe}}^2) = -\frac{\alpha_{\text{lwe}}^4}{2} + O(\alpha_{\text{lwe}}^6)$, thus we obtain $R_{q,\sigma,\chi} = \exp(-\pi\alpha_{\text{lwe}}^2) + O(\alpha_{\text{lwe}}^4)$. □

We now can derive the relation between the parameters of the LWE problem and the LWR problem.

Corollary 2. *The time complexity of the BKW algorithm to solve the LWE problem over $\mathbb{Z}_q^n$ with Gaussian parameter σ and that to solve the LWR problem over $\mathbb{Z}_q^n$ with rounding modulus p are asymptotically the same, when q, p and σ satisfy*

$$\sigma = \frac{q}{2\sqrt{3}p}. \tag{14}$$

Proof. The time complexity of the BKW algorithm to solve the LWE and LWR problems are given in Theorem 3 and Lemma 5, respectively. Solving the equation $\pi\alpha_{\mathrm{lwe}} = \pi\alpha_{\mathrm{lwr}}$ for σ, we obtain (14).

4.2   Noise Distribution of Concrete Instances

We confirm that, when σ of the LWE problem and p of the LWR problem satisfy (14), the distribution of the Gaussian noise of LWE samples and the rounding error of LWR samples after sample reduction are similar by showing concrete examples. From the similarity of the LWR and LWE problems, the LWE-solving BKW algorithm in [25] is almost the same as the LWR-solving algorithm: Only the hypothesis testing stage is different. In the LWE-solving algorithm, (4) is replaced by $\hat{f}(z) = \sum_{j=1}^{m'} \theta_q^{-(\langle \overline{a}_j, z \rangle - \overline{c}_j)}$, where $\theta_q := e^{\frac{2\pi i}{q}}$. We define $\overline{e}_j := \langle \overline{a}_j, s \rangle - \overline{c}_j$, and denote $\overline{a}_j = a_{j,1} \pm a_{j,2} \pm \cdots \pm a_{j,2^{a-1}}$, $\overline{c}_j = c_{j,1} \pm c_{j,2} \pm \cdots \pm c_{j,2^{a-1}}$, then we obtain

$$\overline{e}_j = \langle a_{j,1} \pm a_{j,2} \pm \cdots \pm a_{j,2^{a-1}}, s \rangle - (c_{j,1} \pm c_{j,2} \pm \cdots \pm c_{j,2^{a-1}})$$

$$= \sum_{k=1}^{2^{a-1}} \langle a_{j,k}, s \rangle - c_{j,k} = \sum_{k=1}^{2^{a-1}} e_{j,k},$$

where $e_{j,k}$ is the independent Gaussian noise from the original LWE samples. When $z = s$, we obtain $\hat{f}(s) = \sum_{j=1}^{m'} \theta_q^{-(\langle \overline{a}_j, s \rangle - \overline{c}_j)} = \sum_{j=1}^{m'} \theta_q^{-(\sum_{k=1}^{2^{a-1}} e_{j,k})}$.

Figure 1 shows examples of the distribution of the Gaussian noise of LWE samples and the rounding error of LWR samples after sample reduction. The two figures on the left show histograms of $\theta_q^{-(\langle \overline{a}_j, s \rangle_q - \overline{c}_j)}$ on a complex plane, where $(\overline{a}_j, \overline{c}_j), j \in \{1, 2, \ldots, m'\}$ are LWE samples obtained after the sample reduction stage. The two figures on the right show histograms of $\theta_p^{-(\frac{p}{q}\langle \overline{a}_j, s \rangle - \overline{c}_j)}$, where $(\overline{a}_j, \overline{c}_j), j \in \{1, 2, \ldots, m'\}$ are LWR samples obtained after the sample reduction stage. In these figures, we used $n = 128, q = 16411, a = 8, m = 2^{20}, m' = 2^{12}$ and $l = 0$. In type (a) figure, we used $\sigma = q/(\sqrt{2\pi}n(\log(n))^2)$, which is for Regev cryptosystem [42]. For type (b), we used $\sigma = 4q/(\sqrt{2\pi}n(\log(n))^2)$. Parameter p is calculated from σ according to (14). From Fig. 1, we can observe that those distributions are similar when parameter σ and p satisfy (14).

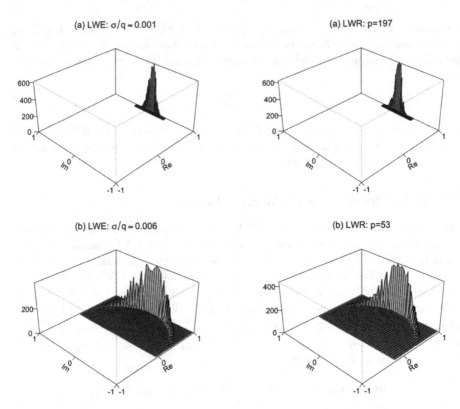

Fig. 1. Distribution of the noises of the LWE and LWR samples obtained after the sample reduction stage. Parameter p for the LWR sample is calculated from σ according to (14), which relates the complexity of the BKW algorithm for the LWE problem and the LWR problem.

4.3   Time Complexity of Concrete Instances

We denote the time complexity of the LWE-solving BKW algorithm by $\mathcal{C}^{\mathsf{LWE}}$. Then, similar to Theorem 17 in [25], we obtain

$$\mathcal{C}^{\mathsf{LWE}} = \frac{1}{4}(a-2)(a-1)(2n/a+1)(q^{n/a}-1) + nq^{n/a}\log(q)$$

$$+ \sum_{j=0}^{a-1} m'^{\mathsf{LWE}}_{j,\epsilon}\left(\frac{a-1-j}{2}(n+2)+2\right)$$

where $m'^{\mathsf{LWE}}_{j,\epsilon} := \frac{8n}{a}\ln\left(\frac{q}{\epsilon}\right) \cdot (R_{q,\sigma,\chi})^{-2^{a-j}}$. The number of samples of the LWE-solving BKW algorithm m^{LWE} is given in (12). Table 2 shows the time complexity of the LWE problem for various parameters of the Regev cryptosystem [42], and the time complexity of the LWR problem whose parameter p is calculated by (14). Concretely, in Table 2, $q = \text{nextprime}(n^2)$, $\sigma = \frac{n^2}{\sqrt{2\pi n}(\log(n))^2}$,

$p = \text{nextprime}\left(\frac{q}{2\sqrt{3}\sigma}\right)$, $\epsilon = 0.01$. From this table, we can confirm that the complexity of the LWE problem and the LWR problem whose parameters satisfy (14) are almost the same.

Table 2. The time complexity and the required number of samples of the LWE-solving BKW algorithm and the LWR-solving BKW algorithm, when $p = \frac{q}{2\sqrt{3}\sigma}$.

n	q	LWE (Regev [42])				LWR ($p = \frac{q}{2\sqrt{3}\sigma}$)			
		σ	a	$\log(\mathcal{C}^{\mathsf{LWE}})$	$\log(m^{\mathsf{LWE}})$	p	a	$\log(\mathcal{C}^{\mathsf{LWR}})$	$\log(m^{\mathsf{LWR}})$
64	4099	5.67	19	49.74	43.61	211	19	49.70	43.60
80	6421	7.14	20	60.22	53.85	263	20	60.20	53.84
96	9221	8.65	21	71.72	63.79	311	21	71.03	63.65
112	12547	10.20	21	82.73	75.94	359	21	82.73	75.94
128	16411	11.79	22	91.84	84.86	409	22	91.84	84.86

5 Conclusion

We analyzed the time complexity of the BKW algorithm for the LWR problem and theoretically derived the optimal number of blocks a that asymptotically (but almost completely) minimizes the time complexity of the algorithm, while Duc *et al.* numerically searched for the optimal value of a [25].

Furthermore, we derived the relation between the parameters of the LWE and LWR problems with the same time complexity of the BKW algorithm, which is $\sigma = \frac{q}{2\sqrt{3}p}$. This equation coincides with the equation derived by the complexity analysis of the lattice attacks: We showed that the conversion equation can also be applied for complexity analysis based on the BKW algorithm.

References

1. Albrecht, M.R.: On dual lattice attacks against small-secret LWE and parameter choices in HElib and SEAL. In: Coron, J.-S., Nielsen, J.B. (eds.) EUROCRYPT 2017. LNCS, vol. 10211, pp. 103–129. Springer, Cham (2017). https://doi.org/10.1007/978-3-319-56614-6_4
2. Albrecht, M.R., Cid, C., Faugère, J.C., Fitzpatrick, R., Perret, L.: On the complexity of the BKW algorithm on LWE. Des. Codes Cryptogr. **74**(2), 325–354 (2015)
3. Albrecht, M.R., et al.: Estimate all the {LWE, NTRU} schemes!. In: Catalano, D., De Prisco, R. (eds.) SCN 2018. LNCS, vol. 11035, pp. 351–367. Springer, Cham (2018). https://doi.org/10.1007/978-3-319-98113-0_19
4. Albrecht, M.R., Faugère, J.-C., Fitzpatrick, R., Perret, L.: Lazy modulus switching for the BKW algorithm on LWE. In: Krawczyk, H. (ed.) PKC 2014. LNCS, vol. 8383, pp. 429–445. Springer, Heidelberg (2014). https://doi.org/10.1007/978-3-642-54631-0_25

5. Albrecht, M.R., Orsini, E., Paterson, K.G., Peer, G., Smart, N.P.: Tightly secure ring-LWE based key encapsulation with short ciphertexts. In: Foley, S.N., Gollmann, D., Snekkenes, E. (eds.) ESORICS 2017. LNCS, vol. 10492, pp. 29–46. Springer, Cham (2017). https://doi.org/10.1007/978-3-319-66402-6_4

6. Alkim, E., Ducas, L., Pöppelmann, T., Schwabe, P.: Post-quantum key exchange - a new hope. In: USENIX Security Symposium, pp. 327–343 (2016)

7. Alwen, J., Krenn, S., Pietrzak, K., Wichs, D.: Learning with rounding, revisited. In: Canetti, R., Garay, J.A. (eds.) CRYPTO 2013. LNCS, vol. 8042, pp. 57–74. Springer, Heidelberg (2013). https://doi.org/10.1007/978-3-642-40041-4_4

8. Applebaum, B., Cash, D., Peikert, C., Sahai, A.: Fast cryptographic primitives and circular-secure encryption based on hard learning problems. In: Halevi, S. (ed.) CRYPTO 2009. LNCS, vol. 5677, pp. 595–618. Springer, Heidelberg (2009). https://doi.org/10.1007/978-3-642-03356-8_35

9. Baan, H., et al.: Round2: KEM and PKE based on GLWR. Cryptology ePrint Archive, Report 2017/1183 (2017). https://eprint.iacr.org/2017/1183

10. Banerjee, A., Fuchsbauer, G., Peikert, C., Pietrzak, K., Stevens, S.: Key-homomorphic constrained pseudorandom functions. In: Dodis, Y., Nielsen, J.B. (eds.) TCC 2015. LNCS, vol. 9015, pp. 31–60. Springer, Heidelberg (2015). https://doi.org/10.1007/978-3-662-46497-7_2

11. Banerjee, A., Peikert, C.: New and improved key-homomorphic pseudorandom functions. In: Garay, J.A., Gennaro, R. (eds.) CRYPTO 2014. LNCS, vol. 8616, pp. 353–370. Springer, Heidelberg (2014). https://doi.org/10.1007/978-3-662-44371-2_20

12. Banerjee, A., Peikert, C., Rosen, A.: Pseudorandom functions and lattices. In: Pointcheval, D., Johansson, T. (eds.) EUROCRYPT 2012. LNCS, vol. 7237, pp. 719–737. Springer, Heidelberg (2012). https://doi.org/10.1007/978-3-642-29011-4_42

13. Bansarkhani, R.E.: LARA - a design concept for lattice-based encryption. Cryptology ePrint Archive, Report 2017/049 (2017). https://eprint.iacr.org/2017/049

14. Becker, A., Gama, N., Joux, A.: A sieve algorithm based on overlattices. LMS J. Comput. Math. 17(A), 49–70 (2014)

15. Blum, A., Kalai, A., Wasserman, H.: Noise-tolerant learning, the parity problem, and the statistical query model. J. ACM 50(4), 506–519 (2003)

16. Boneh, D., Lewi, K., Montgomery, H., Raghunathan, A.: Key homomorphic PRFs and their applications. In: Canetti, R., Garay, J.A. (eds.) CRYPTO 2013. LNCS, vol. 8042, pp. 410–428. Springer, Heidelberg (2013). https://doi.org/10.1007/978-3-642-40041-4_23

17. Bos, J., et al.: CRYSTALS - Kyber: a CCA-secure module-lattice-based KEM. In: 2018 IEEE European Symposium on Security and Privacy (EuroS&P), pp. 353–367 April 2018

18. Bos, J., et al.: Frodo: take off the ring! practical, quantum-secure key exchange from LWE. In: Proceedings of the 2016 ACM SIGSAC Conference on Computer and Communications Security, CCS 2016, pp. 1006–1018. ACM (2016)

19. Brakerski, Z., Langlois, A., Peikert, C., Regev, O., Stehlé, D.: Classical hardness of learning with errors. In: Proceedings of the Forty-Fifth Annual ACM Symposium on Theory of Computing, STOC 2013, pp. 575–584. ACM (2013)

20. Chen, Y., Nguyen, P.Q.: BKZ 2.0: better lattice security estimates. In: Lee, D.H., Wang, X. (eds.) ASIACRYPT 2011. LNCS, vol. 7073, pp. 1–20. Springer, Heidelberg (2011). https://doi.org/10.1007/978-3-642-25385-0_1

21. Cheon, J.H., Kim, D., Lee, J., Song, Y.: Lizard: cut off the tail! a practical post-quantum public-key encryption from LWE and LWR. In: Catalano, D., De Prisco, R. (eds.) SCN 2018. LNCS, vol. 11035, pp. 160–177. Springer, Cham (2018). https://doi.org/10.1007/978-3-319-98113-0_9

22. Cheon, J.H., et al.: Lizard. Technical report, National Institute of Standards and Technology (2017). https://csrc.nist.gov/

23. Corless, R.M., Gonnet, G.H., Hare, D.E., Jeffrey, D.J., Knuth, D.E.: On the Lambert W function. Adv. Comput. Math. **5**, 329–359 (1996)

24. D'Anvers, J.-P., Karmakar, A., Sinha Roy, S., Vercauteren, F.: Saber: module-LWR based key exchange, CPA-secure encryption and CCA-secure KEM. In: Joux, A., Nitaj, A., Rachidi, T. (eds.) AFRICACRYPT 2018. LNCS, vol. 10831, pp. 282–305. Springer, Cham (2018). https://doi.org/10.1007/978-3-319-89339-6_16

25. Duc, A., Tramèr, F., Vaudenay, S.: Better algorithms for LWE and LWR. In: Oswald, E., Fischlin, M. (eds.) EUROCRYPT 2015. LNCS, vol. 9056, pp. 173–202. Springer, Heidelberg (2015). https://doi.org/10.1007/978-3-662-46800-5_8

26. Gama, N., Nguyen, P.Q., Regev, O.: Lattice enumeration using extreme pruning. In: Gilbert, H. (ed.) EUROCRYPT 2010. LNCS, vol. 6110, pp. 257–278. Springer, Heidelberg (2010). https://doi.org/10.1007/978-3-642-13190-5_13

27. Goldwasser, S., Kalai, Y.T., Peikert, C., Vaikuntanathan, V.: Robustness of the learning with errors assumption. In: Innovations in Computer Science (ICS 2010). Tsinghua University Press (2010)

28. Guo, Q., Johansson, T., Mårtensson, E., Stankovski, P.: Coded-BKW with sieving. In: Takagi, T., Peyrin, T. (eds.) ASIACRYPT 2017. LNCS, vol. 10624, pp. 323–346. Springer, Cham (2017). https://doi.org/10.1007/978-3-319-70694-8_12

29. Guo, Q., Johansson, T., Stankovski, P.: Coded-BKW: solving LWE using lattice codes. In: Gennaro, R., Robshaw, M. (eds.) CRYPTO 2015. LNCS, vol. 9215, pp. 23–42. Springer, Heidelberg (2015). https://doi.org/10.1007/978-3-662-47989-6_2

30. Hanrot, G., Pujol, X., Stehlé, D.: Algorithms for the shortest and closest lattice vector problems. In: Chee, Y.M., et al. (eds.) IWCC 2011. LNCS, vol. 6639, pp. 159–190. Springer, Heidelberg (2011). https://doi.org/10.1007/978-3-642-20901-7_10

31. Hanrot, G., Pujol, X., Stehlé, D.: Analyzing blockwise lattice algorithms using dynamical systems. In: Rogaway, P. (ed.) CRYPTO 2011. LNCS, vol. 6841, pp. 447–464. Springer, Heidelberg (2011). https://doi.org/10.1007/978-3-642-22792-9_25

32. Herold, G., Kirshanova, E., May, A.: On the asymptotic complexity of solving LWE. Des. Codes Cryptogr. **86**(1), 55–83 (2018)

33. Information Technology Laboratory, National Institute of Standards and Technology: Post-Quantum Cryptography. https://csrc.nist.gov/Projects/Post-Quantum-Cryptography. Accessed 31 Jan 2018

34. Jin, Z., Zhao, Y.: Optimal key consensus in presence of noise. CoRR abs/1611.06150 (2016)

35. Kaminakaya, K., Kunihiro, N., Takayasu, A.: BKW algorithm for solving LWE Problem. In: Symposium on Cryptography and Information Security, SCIS 2016. IEICE (2016 in Japanese)

36. Kirchner, P., Fouque, P.-A.: An improved BKW algorithm for LWE with applications to cryptography and lattices. In: Gennaro, R., Robshaw, M. (eds.) CRYPTO 2015. LNCS, vol. 9215, pp. 43–62. Springer, Heidelberg (2015). https://doi.org/10.1007/978-3-662-47989-6_3

37. Laarhoven, T.: Sieving for shortest vectors in lattices using angular locality-sensitive hashing. In: Gennaro, R., Robshaw, M. (eds.) CRYPTO 2015. LNCS, vol. 9215, pp. 3–22. Springer, Heidelberg (2015). https://doi.org/10.1007/978-3-662-47989-6_1

38. Lindner, R., Peikert, C.: Better key sizes (and attacks) for LWE-based encryption. In: Kiayias, A. (ed.) CT-RSA 2011. LNCS, vol. 6558, pp. 319–339. Springer, Heidelberg (2011). https://doi.org/10.1007/978-3-642-19074-2_21

39. Liu, M., Nguyen, P.Q.: Solving BDD by enumeration: an update. In: Dawson, E. (ed.) CT-RSA 2013. LNCS, vol. 7779, pp. 293–309. Springer, Heidelberg (2013). https://doi.org/10.1007/978-3-642-36095-4_19

40. Nguyen, P.Q.: Lattice reduction algorithms: theory and practice. In: Paterson, K.G. (ed.) EUROCRYPT 2011. LNCS, vol. 6632, pp. 2–6. Springer, Heidelberg (2011). https://doi.org/10.1007/978-3-642-20465-4_2

41. Nguyen, P.Q., Stehlé, D.: Low-dimensional lattice basis reduction revisited. In: Buell, D. (ed.) ANTS 2004. LNCS, vol. 3076, pp. 338–357. Springer, Heidelberg (2004). https://doi.org/10.1007/978-3-540-24847-7_26

42. Regev, O.: On lattices, learning with errors, random linear codes, and cryptography. J. ACM **56**(6), 34:1–34:40 (2009)

43. Xie, X., Xue, R., Zhang, R.: Deterministic public key encryption and identity-based encryption from lattices in the auxiliary-input setting. In: Visconti, I., De Prisco, R. (eds.) SCN 2012. LNCS, vol. 7485, pp. 1–18. Springer, Heidelberg (2012). https://doi.org/10.1007/978-3-642-32928-9_1

Secret Sharing and Searchable Encryption

A Hierarchical Secret Sharing Scheme Based on Information Dispersal Techniques

Koji Shima$^{(\boxtimes)}$ and Hiroshi Doi

Institute of Information Security, Yokohama, Japan
{dgs164101,doi}@iisec.ac.jp

Abstract. Hierarchical secret sharing schemes are known for how they share a secret among a group of participants partitioned into levels. In this study, we consider using a systematic information dispersal algorithm (IDA). We then apply the general concept of hierarchy to the generator matrix used in a systematic IDA and propose an *ideal* hierarchical secret sharing scheme applicable at any level. For perfect privacy, secret sharing schemes depend on the fact that an adversary can only pool at most $k - 1$ shares. However, in our hierarchical scheme, we need to consider an adversary can also pool k or more shares of lower-level participants. Moreover, considering practical use, we present our evaluation of our software implementation.

Keywords: Secret sharing scheme · Hierarchical access structure · IDA · Ideal scheme · Software implementation

1 Introduction

In today's modern information society, there is a strong need to securely store large amounts of secret information and both prevent information theft or leakage and avoid information loss. Secret sharing schemes are known to simultaneously satisfy the need to distribute and manage secret information so as to prevent such information theft and loss. [1] and [2] independently introduced the basic idea of a (k, n) threshold secret sharing scheme almost four decades ago in 1979. In Shamir's (k, n) threshold scheme, n shares are generated from the secret, and each of these shares is distributed to a participant. Next, the secret can be recovered using any subset k of the n shares, but it cannot be recovered with fewer than k shares. Furthermore, every subset comprising less than k participants cannot obtain any information regarding the secret. Therefore, the original secret is secure even if some of the shares are leaked or exposed. Conversely, the secret can be recovered even if a few of the shares are missing.

Several hierarchical secret sharing schemes are known for how they share the given secret among a group of participants who are partitioned into levels. In such schemes, often, a minimal number of higher-level participants are required

© Springer Nature Switzerland AG 2019
K. Lee (Ed.): ICISC 2018, LNCS 11396, pp. 217–232, 2019.
https://doi.org/10.1007/978-3-030-12146-4_14

to recover the secret. For example, opening a bank vault may require, say, three employees, at least one of whom must be a department manager. In this scenario, we have what is called a $(\{1,3\}, n)$ hierarchical secret sharing scheme. In [3,4], Tassa introduced polynomial derivatives to generate shares and focused on questions related to Birkhoff interpolation problems.

1.1 Secret Sharing Schemes and Hierarchical Schemes

Kurihara et al. [5–7] proposed $(3, n)$ and (k, n) threshold schemes that use only XOR operations to distribute and recover the secret. In [8], they then presented a faster technique for realizing field operations not over $GF(q^L)$ but over $GF(q)$ by using the construction mechanisms of Feng et al. [9] and Blömer et al. [10] for the matrix representation of finite fields. Chen et al. [11] proposed a (k, n) threshold scheme that constructs shares based on a systematic IDA. All above-mentioned schemes are *ideal*.

Next, in [12] and [13], Yamamoto and Blakley et al. each introduced a *ramp* secret sharing scheme, which exhibited a trade-off between security and space efficiency. In [14], Krawczyk proposed a secret sharing scheme called Secret Sharing Made Short, which provides *computational security*, meaning that it encrypts data with a key-based encryption algorithm, then distributes the encrypted data using an IDA and the key via a secret sharing scheme. In [7], Kurihara et al. briefly introduced a *ramp* scheme based on their XOR-based (k, n) threshold scheme. In [15], they then proposed a fast (k, L, n) *ramp* scheme. In [16], Resch et al. proposed a dispersal scheme that provides *computational security*; this scheme enriches Rabin's IDA [17], then combines the All-or-Nothing Transform [18] with the systematic Reed-Solomon code. In [19], Béguin et al. showed how to realize *computational* secret sharing schemes for general access structure. Their approach reduced the problem to an optimization problem.

Tassa [3,4] proposed a $(\mathbf{k}, n)$ hierarchical secret sharing scheme in which a minimal number of higher-level participants are required for recovering the secret. Tassa's scheme is *ideal*. Tassa used the derivative of a polynomial to achieve hierarchy and recover the secret via Birkhoff interpolation. In [20], Selçuk et al. proposed a function called the truncated version to achieve the described hierarchy. This truncated version truncates the polynomial from to the lowest-order term depending on the level. In [29], Shima et al. proposed a hierarchical secret sharing scheme over finite fields of characteristic two.

In addition, Tassa [3,4] showed other hierarchical settings studied by other authors. Shamir [2] suggested accomplishing a hierarchical scheme by assigning capable participants a large number of shares. However, when any subset of lower-level participants is sufficiently large, only the lower-level participants are needed to recover the secret. Simmons [21] and Brickell [22] considered other hierarchical settings. However, the necessary number of participants is the highest of the thresholds associated with the various levels. Therefore, their hierarchical settings are unsuitable for the scenario in which a minimal number of higher-level participants must be involved in recovery of the secret.

Tassa then defined a $(\mathbf{k}, n)$ hierarchical secret sharing scheme as follows.

Definition 1. *Let* $\mathbf{k} = \{k_i\}_{i=0}^m$, $0 < k_0 < \cdots < k_m$, *and let* $k = k_m$ *be the maximal threshold. A* $(\mathbf{k}, n)$ *hierarchical secret sharing scheme where a minimal number of higher-level participants are required for any recovery of the secret is defined as the following access structure* $\varGamma$:

$$\varGamma = \left\{ \mathcal{V} \subset \mathcal{U} : \left| \mathcal{V} \cap \left(\bigcup_{j=0}^{i} \mathcal{U}_j \right) \right| \geq k_i, \forall i \in \{0, 1, \cdots, m\} \right\}.$$

Here, let $\mathcal{U}$ *be a set of* n *participants and assume that* $\mathcal{U}$ *is composed of levels, that is,* $\mathcal{U} = \bigcup_{i=0}^m \mathcal{U}_i$, *where* $\mathcal{U}_i \cap \mathcal{U}_j = \emptyset$ *for all* $0 \leq i < j \leq m$. *The scheme then generates each share of the participants* $u \in \mathcal{U}$ *to satisfy the access structure.*

Given $\mathbf{k} = \{1, 3\}$ as an example, we have a $(\{1, 3\}, n)$ hierarchical scheme that consists of two levels and requires at least one indispensable participant from $\mathcal{U}_0$ and three or more participants from $\mathcal{U}_0 \bigcup \mathcal{U}_1$ to recover the secret.

1.2   Example Scenarios of Hierarchical Schemes

Tassa presented the opening of a bank vault as an example scenario. In this scenario, a fast $(\{1, 3\}, n)$ hierarchical secret sharing scheme is required. Castiglione et al. [23, 24] presented other scenarios; the project manager and team members can access a project workspace according to their levels of authority; nurses may access a subset of patients' clinical data, while a doctor can access all data.

Here, we present a file management system as an example scenario. We store the indispensable participant's share in local storage such as smartphones, and we store the remaining two shares in external storage such as USB mass storage and cloud storage. Only the owner of the smartphone can recover this data by using either of the two external storage devices, and data cannot be recovered using only the two external storage devices. Considering practical use in this scenario, there is a need for a fast $(\{1, 2\}, 3)$ hierarchical secret sharing scheme. In general, a fast $(\{1, k\}, k + 1)$ hierarchical secret sharing scheme would be useful.

1.3   Our Contributions

In this paper, we introduce our hierarchical IDA, which is a hierarchical secret sharing scheme applicable to any level. Our scheme is both *perfect* and *ideal*. We use operations with $\mathrm{GF}(2^L)$. For perfect privacy, Chen et al.'s (k, n) threshold scheme [11] depends on the fact that an adversary can only pool at most $k - 1$ shares. However, in our hierarchical scheme, we need to consider an adversary can also pool k or more shares of lower-level participants. Therefore, Chen et al.'s scheme cannot be directly applied to hierarchical schemes, which we present in more detail in Sect. 4.2. Our overall contributions are summarized as follows:

- We apply hierarchy to the generator matrix used in an IDA and realize a hierarchical secret sharing scheme applicable to any level. We solve the aforementioned issues and provide mathematical proof.

- In a single hierarchy, or a non-hierarchical secret sharing scheme, our scheme is more efficient in implementation than Chen et al.'s scheme [11] because in our scheme, all matrices $\mathbf{G}'$ used by $Recover^{\mathrm{IDA}}$ of the corresponding rows are the same.
- We achieve a $(\{1, k\}, k + 1)$ hierarchical scheme using only XOR operations. As a result, we can use simple 64-bit XOR operations instead of $\mathrm{GF}(2^L)$. Then, this scheme is much faster than an approach by Tassa [3,4].

2   Preliminaries

2.1   Notations and Definitions

We use the following notations and definitions throughout this paper.

- $\oplus$ denotes a bitwise XOR operation.
- $\oplus_{j=a}^{b} c_j$ denotes $c_a \oplus \cdots \oplus c_b$.
- $\|$ denotes a concatenation of bit sequences.
- $\|_{j=a}^{b} c_j$ denotes $c_a \| \cdots \| c_b$.
- $H(X)$ denotes the Shannon entropy of a random variable X.
- $\mathbf{v}[j]$ denotes the j-th element in vector $\mathbf{v}$.
- $\mathbf{v}[0][1] \cdots [n-1]$ denotes vector $\mathbf{v}$ with exactly n elements.
- Elements in $\mathrm{GF}(2^L)$ can be identified with polynomials $f_L(X) = \sum_{i=0}^{L-1} f_i X^i$. $f_i \in \mathrm{GF}(2)$. They can also be represented by decimal numbers or hexadecimal numbers of $f_{L-1} \cdots f_1 f_0$ binary. For example, $f_8(X) = X^5 + 1 \in \mathrm{GF}(2)[X]$ can be represented by 33 or 21 h of 00100001 binary.

2.2   Perfect and Ideal Secret Sharing Schemes

In this subsection, we refer to Beimel [25] for a *perfect* secret sharing scheme and refer to Blundo et al. [26,27] and Kurihara et al. [5–7] for an *ideal* secret sharing scheme. Let S be a random variable in a given probability distribution on the secret, S_B be a random variable in a given probability distribution on the shares in each authorized set B, and S_T be a random variable in a given probability distribution on the shares of each unauthorized set T. A *perfect* secret sharing scheme would satisfy the following conditions:

Correctness, Accessibility $H(S|S_B) = 0$.
Perfect privacy, Perfect security $H(S|S_T) = H(S)$.

A secret sharing scheme is called *ideal* if it is *perfect* and the information rate equals one. In other words, if the size of every bit of the shares equals the bit size of the secret, the scheme is *ideal*. As Tassa [4] mentioned in Definition 1.1, we may apply the information rate to a hierarchical secret sharing scheme.

2.3 Systematic IDA

A (k, n) systematic IDA constitutes two more specific algorithms, i.e., $Share^{\text{IDA}}$ and $Recover^{\text{IDA}}$.

$Share^{\text{IDA}}$ takes as input data message D and outputs a codeword to distribute D among n participants. D is parsed into column vector $\mathbf{D}$ that has k elements, with each element in $\text{GF}(2^L)$. Generator matrix or dispersal matrix $\mathbf{G} = [g_{(i,j)}]_{i=1,j=1}^{n,\ k}$ is a publicly known $n \times k$ matrix with the following conditions:

- The first k rows form the $k \times k$ identity matrix.
- Any subset k of the n rows of $\mathbf{G}$ is linearly independent.

Column vector $\mathbf{C}$ with n elements is then output as codeword $\mathbf{C} = \mathbf{G} \cdot \mathbf{D}$. Since the first k rows of $\mathbf{G}$ form the identity matrix, we obtain

$$
\mathbf{C} = \mathbf{G} \cdot \begin{pmatrix} \mathbf{D}[0] \\ \vdots \\ \mathbf{D}[k-1] \end{pmatrix} = \begin{pmatrix} \mathbf{D}[0] \\ \vdots \\ \mathbf{D}[k-1] \\ \mathbf{C}[k] \\ \vdots \\ \mathbf{C}[n-1] \end{pmatrix},
$$

where each element $\mathbf{D}[i], \mathbf{C}[i] \in \text{GF}(2^L)$. Then, $\mathbf{G}$ is derived from a Vandermonde matrix using a sequence of elementary matrix transformations. In [28], Plank et al. describe how to prepare an $n \times k$ Vandermonde matrix in which $g_{(i,j)} = (i-1)^{j-1}$ and we turn the first k rows into the identity matrix using a sequence of elementary matrix transformations. This satisfies the conditions of $\mathbf{G}$ since elementary matrix operations do not change the rank of a matrix. Furthermore, a square Vandermonde matrix with $g_{(i,j)} = x^{j-1}$ is invertible if all x are distinct.

$Recover^{\text{IDA}}$ takes as input the remaining k elements $\mathbf{C}'$ for codeword $\mathbf{C}$ and outputs data message D. Here, $\mathbf{C}'$ is a column vector that has k elements. Through this process, new $k \times k$ matrix $\mathbf{G}'$ is formed from $\mathbf{G}$ and corresponds to the remaining k elements. After $\mathbf{G}'$ is inverted, we obtain D via $\mathbf{D} = (\mathbf{G}')^{-1} \cdot \mathbf{C}'$.

Employing a (k, n) systematic IDA instead of a (k, n) non-systematic IDA improves performance because it does not need to encode the first k codeword elements and similarly does not need to decode codeword elements that are equal to message data elements.

3 Related Work

From [11], Chen et al. presented a scheme that constructs an *ideal* threshold scheme with a systematic IDA. Since an IDA is essentially a *ramp* scheme, their scheme generates dummy keys at random, passing both these dummy keys and the secret values XORed with these dummy keys to the systematic IDA. Their scheme then applies some cyclic shifts to each of the outputs to generate shares.

Let P_x for $x = 0, \cdots, n-1$ be n participants for the (k, n) threshold scheme over $F = GF(2^L)$. Then, generator matrix $\mathbf{G}$ is publicly known and has elements in $GF(2^L)$, as remarked in Sect. 3.3. Furthermore, let the secret be given by $s \in \{0, 1\}^\lambda, \lambda = L \cdot k$. Secret s is parsed into $\mathbf{s} \in F^k$ with k elements of length L bits. Here, s must be padded to λ bits if s is less than λ bits.

3.1   Chen et al.'s Distribution Algorithm

Table 1 shows their distribution algorithm. Step 1 generates random values called dummy keys $r_1, \cdots, r_{k-1} \in \{0, 1\}^\lambda$. These values are parsed into $\mathbf{r}_1, \cdots, \mathbf{r}_{k-1} \in F^k$. In Step 2, s and the dummy keys are XORed to produce $s' \in \{0, 1\}^\lambda$, then s' is parsed into $\mathbf{r}_0 \in F^k$. In Step 3, each $\mathbf{r}_i$ is passed into $Share^{IDA}$. As a result, we obtain each column vector $\mathbf{R}_0, \cdots \mathbf{R}_{k-1} \in F^n$, each of which has n elements. In Steps 4 and 5, each participant P_x securely receives share $w_x \in \{0, 1\}^\lambda$. To shed further light on this algorithm, we detail Steps 3 and 4. In Step 3, we illustrate $k \times n$ matrix

$$\mathbf{M} = \begin{pmatrix} \mathbf{R}_0^T \\ \mathbf{R}_1^T \\ \vdots \\ \mathbf{R}_{k-1}^T \end{pmatrix} = \begin{pmatrix} \mathbf{R}_0[0][1]\cdots[n-1] \\ \mathbf{R}_1[0][1]\cdots[n-1] \\ \vdots \\ \mathbf{R}_{k-1}[0][1]\cdots[n-1] \end{pmatrix}.$$

Next, in Step 4, we illustrate matrix

$$\mathbf{M}' = \begin{pmatrix} \mathbf{R}_0[0]\,[1] \cdots [n-2]\,[n-1] \\ \mathbf{R}_1[1]\,[2] \cdots [n-1]\quad[0] \\ \vdots \;\; \vdots \qquad \vdots \qquad \vdots \\ \mathbf{R}_{k-1}[k-1]\,[k] \cdots [k-3]\,[k-2] \end{pmatrix},$$

applying $j-1$ left cyclic shifts to elements of the j-th row of $\mathbf{M}$ for $j = 1, \cdots, k$. Further, w_x concatenates the elements in the $x + 1$-th column of $\mathbf{M}'$.

3.2   Chen et al.'s Recovery Algorithm

Table 2 shows the corresponding recovery algorithm. The algorithm takes as input shares of participants P_i for $i = t_0, \cdots, t_{k-1}$ that cooperate to recover the secret. Step 1 parses each participant's share w_i into its k elements. More specifically, k of the n elements for each row are given in the distribution algorithm. Since elements in each row are cyclically shifted when the shares are generated, indexes of the k elements are different in each row, implying that all matrices $\mathbf{G}_i'$ passed into $Recover^{IDA}$ for the corresponding row differ. In Step 2, each column vector $\mathbf{R}_i'$ for $i = 0, \cdots, k-1$ has the k elements of the $i + 1$-th row in Step 1, and these column vectors are passed into $Recover^{IDA}$. In Steps 3 and 4, s' and $r_1, \cdots, r_{k-1}$ are then recovered from the available shares. Finally, in Steps 5 and 6, these recovered values are XORed to retrieve secret s.

Table 1. Distribution algorithm	**Table 2.** Recovery algorithm
Input: $s \in \{0,1\}^{\lambda}$ Output: $(w_0, \cdots w_{n-1})$	Input: $(w_{t_0}, \cdots w_{t_{k-1}})$ Output: s

1: for $i \leftarrow 1$ to $k-1$:
      $\mathbf{r}_i \leftarrow r_i \overset{\$}{\leftarrow} \{0,1\}^{\lambda}$
 2: $\mathbf{r}_0 \leftarrow s' \leftarrow s \oplus \{\oplus_{j=1}^{k-1} r_j\}$
 3: for $i \leftarrow 0$ to $k-1$:
      $\mathbf{R}_i \leftarrow Share^{\mathrm{IDA}}(\mathbf{r}_i, \mathbf{G})$
 4: for $i \leftarrow 0$ to $n-1$:
      $w_i \leftarrow \|_{j=0}^{k-1} \mathbf{R}_j[i+j \ (\mathrm{mod}\ n)]$
 5: return $(w_0, \cdots w_{n-1})$
	1: for $i \leftarrow 0$ to $k-1$:
	      $\|_{j=0}^{k-1} \mathbf{R}_j[t_i + j \ (\mathrm{mod}\ n)] \leftarrow w_{t_i}$
	 2: for $i \leftarrow 0$ to $k-1$:
	      $r_i \leftarrow Recover^{\mathrm{IDA}}(\mathbf{R}'_i, \mathbf{G}'_i)$
	 3: for $i \leftarrow 0$ to $k-1$:
	      $r_i \leftarrow \mathbf{r}_i$
	 4: $s' \leftarrow r_0$
	 5: $s \leftarrow s' \oplus \{\oplus_{j=1}^{k-1} r_j\}$
	 6: return s

3.3 Remark

Chen et al. showed generator matrix $\mathbf{G}$ as an $n \times k$ binary matrix. Any subset k of the n rows of $\mathbf{G}$ should be linearly independent, but not all of the parameters with k and n can satisfy the condition. Given $k = 3$ and $n = 5$ as an example, we cannot find any combinations of $g_0, g_1, g_2 \in \mathrm{GF}(2)$ in

$$\mathbf{G} = \begin{pmatrix} 1 & 0 & 0 \\ 0 & 1 & 0 \\ 0 & 0 & 1 \\ 1 & 1 & 1 \\ g_0 & g_1 & g_2 \end{pmatrix}.$$

In general, $\mathbf{G}$ has elements in $\mathrm{GF}(2^L)$. With $n = k, k+1$, $\mathbf{G}$ has elements in $\mathrm{GF}(2)$.

4 Our Proposed Scheme

In this section, we describe our proposed $(\mathbf{k}, n)$ hierarchical secret sharing scheme that satisfies Definition 1. We use operations with $F = \mathrm{GF}(2^L)$. Let the secret be given by $s \in \{0,1\}^{\lambda}, \lambda = L \cdot k$. Secret s is parsed into $\mathbf{s} \in F^k$ with k elements of length L bits. Note that s must be padded to λ bits if s is less than λ bits. We use $n \times k$ generator matrix $\mathbf{G}$ with the following properties:

- $\mathbf{G}$ has a defined hierarchy such that only an authorized subset can recover the secret.
- $\mathbf{G}$ does not have any row of $(y \cdots y)$, where $y \in F$.

$\mathbf{G}$ is publicly known. Since it is a uniquely determined table in a fixed system, we are able to recover the secret only from shares. We may describe $\mathbf{G}$ before using elementary matrix transformations required for the systematic IDA. This $\mathbf{G}$ is hereinafter referred to as a hierarchical generator matrix, and the IDA using this $\mathbf{G}$ is also hereinafter referred to as a hierarchical IDA.

4.1   Participant Identities and Hierarchical Generator Matrix

Let $P_x \in \mathcal{U}$ for $x = 0, \cdots, n-1$ define n participants and let $0 \leq l_0 \leq \cdots \leq l_m = n$. Each participant P_x has identity $x \in F$. Without loss of generality, we may assume that each P_x belongs to the following levels:

$$P_0, \cdots, P_{l_0-1} \in \mathcal{U}_0, \quad P_{l_0}, \cdots, P_{l_1-1} \in \mathcal{U}_1, \quad \cdots, \quad P_{l_{m-1}}, \cdots, P_{l_m-1} \in \mathcal{U}_m.$$

For example, $l_0 = 0$ means no participants belong to $\mathcal{U}_0$. Furthermore, let $0 \in \mathcal{U}_0$ be the phantom participant and let u_x (described later) always be assigned to zero.

P_x corresponds to the $x + 1$-th row of $\mathbf{G}$. In other words, we can view the $n \times k$ matrix as $\mathbf{G} = [g_{(x,j)}]_{x=0,\ j=1}^{n-1,\ k}$. Then, we introduce a hierarchy to $\mathbf{G}$, constructing $\mathbf{G}$ such that $k \times k$ matrix $\mathbf{G}'$ satisfies $\det(\mathbf{G}') = 0$ for any unauthorized k participants, where $\mathbf{G}'$ is generated from the rows of $\mathbf{G}$ corresponding to the k participants. Here, $P_x \in \mathcal{U}_i$ for $i = 0, \cdots, m$ generates $\mathbf{G}$ with

$$g_{(x,j)} = \begin{cases} u_x^{j-1-k_{i-1}} & (j > k_{i-1}) \\ 0 & (j \leq k_{i-1}) \end{cases},$$

where $g_{(x,j)} \in F$ and $k_{-1} = 0$. Given a $(\{2,3,5\}, n)$ hierarchical secret sharing scheme as an example, we obtain

$$\mathbf{G} = \begin{pmatrix} 1 & u_0 & u_0^2 & u_0^3 & u_0^4 \\ \vdots & \vdots & \vdots & \vdots & \vdots \\ 1 & u_{l_0-1} & u_{l_0-1}^2 & u_{l_0-1}^3 & u_{l_0-1}^4 \\ 0 & 0 & 1 & u_{l_0} & u_{l_0}^2 \\ \vdots & \vdots & \vdots & \vdots & \vdots \\ 0 & 0 & 1 & u_{l_1-1} & u_{l_1-1}^2 \\ 0 & 0 & 0 & 1 & u_{l_1} \\ \vdots & \vdots & \vdots & \vdots & \vdots \\ 0 & 0 & 0 & 1 & u_{n-1} \end{pmatrix}.$$

Intuitively, shares of lower-level participants are not generated from the secret and some random values. Here, u_x for $\mathbf{G}$ can be assigned from $2^L - 1$ values except zero over F. However, depending on the assignment of u_x, there is a specific case in which the secret cannot be recovered in spite of the presence of an authorized subset. Given $\mathbf{G}$ for a $(\{1,3\}, 5)$ hierarchical scheme over $\mathrm{GF}(2^8)$ as an example to understand the meaning of $\det(\mathbf{G}') = 0$, one of the matrices $\mathbf{G}_1'$ derives $\det(\mathbf{G}_1') = 0$ when $\mathbf{G}_1$ is given by $u_x = 1, 2, 3, 4, 5$. $\mathbf{G}_2$ given by $u_x = 1, 2, 4, 5, 6$ is required. Note that elements are represented by decimal numbers following Sect. 2. It is sufficient to find one $\mathbf{G}$, such as $\mathbf{G}_2$ for this scheme, i.e.,

$$\mathbf{G}_1 = \begin{pmatrix} 1 & 1 & 1 \\ 1 & 2 & 4 \\ 0 & 1 & 3 \\ 0 & 1 & 4 \\ 0 & 1 & 5 \end{pmatrix}, \mathbf{G}_1' = \begin{pmatrix} 1 & 1 & 1 \\ 1 & 2 & 4 \\ 0 & 1 & 3 \end{pmatrix}, \mathbf{G}_2 = \begin{pmatrix} 1 & 1 & 1 \\ 1 & 2 & 4 \\ 0 & 1 & 4 \\ 0 & 1 & 5 \\ 0 & 1 & 6 \end{pmatrix}.$$

Therefore, there is a case in which $\det(\mathbf{G}') = 0$, where the corresponding $\mathbf{G}'$ to the authorized subset. As Tassa described in Sect. 3.2 of [4] regarding probability, we can make the same argument for this issue. In other words, Tassa stated that the p used to allocate participant identities over $GF(p)$ should be usually very large. Similarly, the L used to allocate u_x of $\mathbf{G}$ over $GF(2^L)$ can be large in our scheme. To keep the paper compact, we do not discuss it further.

4.2   An Issue with Applying Hierarchy to IDA

A (k, n) IDA recovers not only data message D, but also all n elements if any k elements are given because the remaining $n - k$ elements can be calculated from publicly known generator matrix $\mathbf{G}$. A $(\mathbf{k}, n)$ hierarchical IDA also has a similar property. We consider a $(\{1, 3\}, 8)$ hierarchical IDA as an example. Let d_1, d_2, d_3 be passed into the systematic hierarchical IDA and let the eight codeword elements be $d_1, d_2, d_3, c_1, \cdots, c_5$. Furthermore, let d_1, d_2 be used for $\mathcal{U}_0$ and let $d_3, c_1, \cdots, c_5$ be used for $\mathcal{U}_1$. If three elements in $\mathcal{U}_1$ are given, such as c_1, c_2, c_3, we need to consider all elements not only in $\mathcal{U}_1$ but also in $\mathcal{U}_0$ can be calculated. As we described in Sect. 1.3, secret sharing schemes depend on the fact that an adversary can only pool at most $k-1$ shares. In Chen et al.'s scheme, the cyclic shifts after $Share^{IDA}$ work well to satisfy perfect privacy. However, in our hierarchical scheme, the cyclic shifts after $Share^{IDA}$ have no effect to satisfy perfect privacy because we need to consider no elements in $\mathcal{U}_0$ can be calculated when three elements in $\mathcal{U}_1$ are given in the example.

4.3   Distribution Algorithm

The dealer securely distributes each share $w_x \in \{0, 1\}^\lambda$ to participant P_x. Table 3 shows this specific algorithm. The underlined portions show differences between the algorithm of Table 1 and our algorithm. More specifically, with each column vector $\mathbf{r}_0, \cdots, \mathbf{r}_{k-1}$ transposed, Step 3 constructs matrix

$$\mathbf{M}_r = \begin{pmatrix} \mathbf{r}_0[0] & \cdots & \mathbf{r}_0[k-1] \\ \vdots & \ddots & \vdots \\ \mathbf{r}_{k-1}[0] & \cdots & \mathbf{r}_{k-1}[k-1] \end{pmatrix} = \begin{pmatrix} \mathbf{r}'_0[0] & \cdots & \mathbf{r}'_{k-1}[0] \\ \vdots & \ddots & \vdots \\ \mathbf{r}'_0[k-1] & \cdots & \mathbf{r}'_{k-1}[k-1] \end{pmatrix}$$

and defines $\mathbf{r}'_0, \cdots, \mathbf{r}'_{k-1}$ as column vectors. In Step 5, no cyclic shifts are used after $Share^{IDA}$. Through Steps 1 through 5 of our algorithm, we can illustrate $k \times n$ matrix

$$\mathbf{M} = \begin{pmatrix} \mathbf{R}_0^T \\ \vdots \\ \mathbf{R}_{k-1}^T \end{pmatrix} = \begin{pmatrix} \overbrace{\mathbf{R}_0[0] \cdots [l_0 - 1]}^{\mathcal{U}_0} \cdots \overbrace{[l_{m-1}] \cdots [n-1]}^{\mathcal{U}_m} \\ \vdots \quad \vdots \quad \vdots \quad \vdots \\ \mathbf{R}_{k-1}[0] \cdots [l_0 - 1] \cdots [l_{m-1}] \cdots [n-1] \end{pmatrix}.$$

Each participant P_x securely receives share w_x concatenating the elements in the $x + 1$-th column of $\mathbf{M}$.

Table 3. Our distribution algorithm

Table 4. Our recovery algorithm

Input: $s \in \{0,1\}^{\lambda}$	Input: $(w_{t_0}, \cdots w_{t_{k-1}})$
Output: $(w_0, \cdots w_{n-1})$	Output: s

Table 3:

1: for $i \leftarrow 1$ to $k-1$:

   $r_i \leftarrow r_i \xleftarrow{\$} \{0,1\}^{\lambda}$

2: $\mathbf{r}_0 \leftarrow s' \leftarrow s \oplus \{\oplus_{j=1}^{k-1} r_j\}$

3: $\mathbf{M}_r = (\mathbf{r}'_0 \cdots \mathbf{r}'_{k-1}) \leftarrow (\mathbf{r}_0 \cdots \mathbf{r}_{k-1})^{\mathrm{T}}$

4: for $i \leftarrow 0$ to $k-1$:

   $\mathbf{R}_i \leftarrow Share^{\mathrm{IDA}}(\mathbf{r}'_i, \mathbf{G})$

5: for $i \leftarrow 0$ to $n-1$:

   $w_i \leftarrow ||_{j=0}^{k-1} \mathbf{R}_j[i]$

6: return $(w_0, \cdots w_{n-1})$

Table 4:

1: for $i \leftarrow 0$ to $k-1$:

   $||_{j=0}^{k-1} \mathbf{R}_j[t_i] \leftarrow w_{t_i}$

2: for $i \leftarrow 0$ to $k-1$:

   $\mathbf{r}'_i \leftarrow Recover^{\mathrm{IDA}}(\mathbf{R}'_i, \underline{\mathbf{G}'})$

3: $(\mathbf{r}_0 \cdots \mathbf{r}_{k-1})^{\mathrm{T}} \leftarrow \mathbf{M}_r = (\mathbf{r}'_0 \cdots \mathbf{r}'_{k-1})$

4: for $i \leftarrow 0$ to $k-1$:

   $r_i \leftarrow \mathbf{r}_i$

5: $s' \leftarrow r_0$

6: $s \leftarrow s' \oplus \{\oplus_{j=1}^{k-1} r_j\}$

7: return s

4.4 Recovery Algorithm

Table 4 shows our recovery algorithm. This algorithm takes as input shares of participants P_i for $i = t_0, \cdots, t_{k-1}$ that cooperate to recover the secret. The underlined portions show differences between the algorithm of Table 2 and our algorithm. Since there are no cyclic shifts after $Share^{\mathrm{IDA}}$, all matrices $\mathbf{G}'$ used by $Recover^{\mathrm{IDA}}$ of the corresponding rows are the same. Step 1 parses each participant's share w_i into its k elements. In Step 2, each column vector $\mathbf{R}'_i$ for $i = 0, \cdots, k-1$ has the k elements of the $i+1$-th row from Step 1, and these column vectors are passed into $Recover^{\mathrm{IDA}}$. In Steps 3 through 5, s' and $r_1, \cdots, r_{k-1}$ are recovered from the available shares. Finally, in Steps 6 and 7, these recovered values are XORed to retrieve secret s.

4.5 Security Analysis

Theorem 1 proves the correctness and perfect privacy of our proposed scheme. We obtain secret s if all elements of $\mathbf{M}_r$ are recovered with $\mathbf{G}'$. Without loss of generality, we may focus on k elements of each j-th row of $\mathbf{M}_r^{\mathrm{T}}$, recovered from the j-th row of $\mathbf{M}$ with $Recover^{\mathrm{IDA}}$, because each j-th row can be processed independently from the others. We then apply this argument to every other row.

Next, the j-th row of $\mathbf{M}_r^{\mathrm{T}}$ has $k-1$ random values $\mathbf{r}'_{j-1}[1], \cdots, \mathbf{r}'_{j-1}[k-1] \in F$ and the value $\mathbf{r}'_{j-1}[0]$ XORed with those random values and secret $\mathbf{s}[j-1] \in F$. Therefore, any $k-1$ of the k values cannot reveal anything about the secret. We then pass the k values into $Share^{\mathrm{IDA}}$ and obtain $\mathbf{R}_{j-1}$. Lemma 1 proves that $\mathbf{R}_{j-1}$ cannot reveal anything about the secret.

Lemma 1. *Assume that any set of L' participants $T = \{P_{t_0}, \cdots, P_{t_{L'-1}}\} \notin \Gamma$ agrees to recover the secret. Let $y \in F \backslash \{0\}$. Then, if $n \times k$ generator matrix $\mathbf{G}$ whose j-th row is $(y \cdots y)$ is not used, we receive no information regarding the secret. In information theoretic terms, $H(S|S_T) = H(S)$, shown in Sect. 2.2.*

Proof. Suppose that s and $r_1, \cdots, r_{k-1}$ are mutually independent and that $r_1, \cdots, r_{k-1}$ are selected from the finite set $\{0,1\}^\lambda$ with uniform probability $1/2^\lambda$. We define $k \times k$ matrices $\mathbf{X}$, $\mathbf{A}$, and $\mathbf{A}'$ as

$$
\mathbf{X} = \begin{pmatrix} 1\,1\,\cdots\,1 \\ 0\,1\,\cdots\,0 \\ \vdots\,\vdots\,\ddots\,\vdots \\ 0\,0\,\cdots\,1 \end{pmatrix}, \quad \mathbf{A} = \begin{pmatrix} \mathbf{s}^{\mathrm{T}} \\ \mathbf{r}_1^{\mathrm{T}} \\ \vdots \\ \mathbf{r}_{k-1}^{\mathrm{T}} \end{pmatrix}, \quad \mathbf{A}' = \begin{pmatrix} 0\,\cdots\,0 \\ \mathbf{r}_1^{\mathrm{T}} \\ \vdots \\ \mathbf{r}_{k-1}^{\mathrm{T}} \end{pmatrix},
$$

respectively. Here, $\mathbf{X}^{-1} = \mathbf{X}$. Steps 1 through 3 of Table 3 can be represented as $\mathbf{M}_r = \mathbf{X} \cdot \mathbf{A}$. With $Share^{\mathrm{IDA}}$, we obtain $n \times k$ matrix $\mathbf{M}^{\mathrm{T}} = \mathbf{G} \cdot \mathbf{M}_r$. We then define $n \times k$ matrix $\mathbf{Y} = \mathbf{G} \cdot \mathbf{X}$, we obtain

$$
\mathbf{M}^{\mathrm{T}} = \mathbf{G} \cdot \mathbf{M}_r = \mathbf{G} \cdot \mathbf{X} \cdot \mathbf{A} = \mathbf{Y} \cdot \mathbf{A}.
$$

Here, let $\mathbf{G}_{(i)}$ be a matrix constructed by any subset i of the n rows of $\mathbf{G}$ for $i = 1, \cdots, L'$ and let $\mathbf{Y}_{(i)} = \mathbf{G}_{(i)} \cdot \mathbf{X}$. Furthermore, let $\mathbf{M}_{(i)}^{\mathrm{T}} = \mathbf{Y}_{(i)} \cdot \mathbf{A}$. Without loss of generality, we may assume that $\mathrm{rank}(\mathbf{G}_{(i)}) = i$ because we can consider $\mathbf{G}_{(\alpha)}$ such that $\mathrm{rank}(\mathbf{G}_{(i)}) = \alpha < i$. Then, $\mathrm{rank}(\mathbf{Y}_{(i)}) = \mathrm{rank}(\mathbf{G}_{(i)})$ because $\mathbf{X}$ is regular. Consider $i = 1$. It is apparent that the corresponding participant receives share $w \in \{0,1\}^\lambda$ regarding secret s if $\mathbf{Y}_{(1)} = (y\ 0\ \cdots\ 0)$, i.e.,

$$
w \leftarrow \mathbf{M}_{(1)}^{\mathrm{T}} = \mathbf{Y}_{(1)} \cdot \mathbf{A} = \left(y \cdot \mathbf{s}^{\mathrm{T}} \right), \quad \mathbf{G}_{(1)} = \mathbf{Y}_{(1)} \cdot \mathbf{X}^{-1} = \left(y \cdots y \right).
$$

If $\mathbf{Y}_{(1)} \neq (y\ 0\ \cdots\ 0)$, the vector $\mathbf{Y}_{(1)} \cdot \mathbf{A}'$ is uniformly distributed over $\{0,1\}^\lambda$ because all elements of the vector are L-bit random numbers that are mutually independent and distributed uniformly over $\{0,1\}^L$. Then, we suppose w' denotes a fixed value of w. $w = w'$ can be obtained with uniform probability $1/2^\lambda$ from any selected s. Because s is independent of w, we have $H(S|S_T) = H(S)$.

Next, we prove that vector $(y, 0, \cdots, 0)$ is not expressed by linear combination of the row vectors of $\mathbf{Y}_{(i)}$ for $i \geq 2$. Here, we may assume that none of the rows of $\mathbf{G}_{(i)}$ are equal to $(y\ \cdots\ y)$ because we already prove $\mathbf{G}_{(1)} = (y\ \cdots\ y)$. Matrix

$$
\mathbf{Y}_{(i)} = \begin{pmatrix} y_1\ a_{(1,1)}\ \cdots\ a_{(1,k-1)} \\ \vdots\ \vdots\ \ddots\ \vdots \\ y_i\ a_{(i,1)}\ \cdots\ a_{(i,k-1)} \end{pmatrix} = \mathbf{G}_{(i)} \cdot \mathbf{X}
$$

can be represented. If $y_j = 0$ for $j = 1, \cdots, i$, we receive no information regarding the secret because the j-th row of $\mathbf{Y}_{(i)}$ is formed by $(0\ *)$. Consider $y_j \neq 0$ for all $j = 1, \cdots, i$. There exist matrices $\mathbf{A}_{(i)}$ and $\mathbf{B}_{(i)}$ such that

$$
\mathbf{G}_{(i)} = \mathbf{Y}_{(i)} \cdot \mathbf{X}^{-1} = \begin{pmatrix} y_1\ \cdots\ y_1 \\ \vdots\ \ddots\ \vdots \\ y_i\ \cdots\ y_i \end{pmatrix} + \begin{pmatrix} 0\ a_{(1,1)}\ \cdots\ a_{(1,k-1)} \\ \vdots\ \vdots\ \ddots\ \vdots \\ 0\ a_{(i,1)}\ \cdots\ a_{(i,k-1)} \end{pmatrix} = \mathbf{A}_{(i)} + \mathbf{B}_{(i)}.
$$

Since rank is subadditive and $\mathrm{rank}(\mathbf{A}_{(i)}) = 1$,

$$
\mathrm{rank}(\mathbf{A}_{(i)} + \mathbf{B}_{(i)}) \leq \mathrm{rank}(\mathbf{A}_{(i)}) + \mathrm{rank}(\mathbf{B}_{(i)}) = 1 + \mathrm{rank}(\mathbf{B}_{(i)})
$$

and the rank of $\mathbf{B}_{(i)}$ is either $i-1$ or i. If $\operatorname{rank}(\mathbf{B}_{(i)}) = i$, $\sum_{j=1}^{i} a_j \cdot a_{(j,t)} \neq 0$ for at least one of $t = 1, \cdots k-1$, where $a_1, \cdots a_i$ are scalars. Therefore, the vector expressed by linear combination of the row vectors of $\mathbf{Y}_{(i)}$ is not $(y, 0, \cdots, 0)$. Next, if $\operatorname{rank}(\mathbf{B}_{(i)}) = i-1$, there exists matrix $\mathbf{G}_{(i)}$ of rank $i-1$. Furthermore, the remaining row of $\mathbf{G}_{(i)}$ should be assigned for the highest level $\mathcal{U}_0$ because $y_j \neq 0$. However, in our scheme, rows of $\mathbf{G}$ in the highest level are derived from a Vandermonde matrix. In such a condition, $\operatorname{rank}(\mathbf{G}_{(i)})$ should be i, i.e., $\operatorname{rank}(\mathbf{B}_{(i)}) \neq i-1$ by proof by contradiction. The proof is thus complete.   □

Theorem 1. *Assume that corresponding square matrix $\mathbf{M}_{\mathcal{V}}$, or namely, the $\mathbf{G}'$ required to recover the secret, is regular for any minimal authorized subset $\mathcal{V} \in \Gamma$, i.e., $|\mathcal{V}| = k$. Then both correctness and perfect privacy hold.*

Proof. Theorem 1 is based on an approach by Tassa [3,4]. We first consider correctness. Each participant P_x receives a part of the share $\sigma(x)_j$ generated by elements of the j-th row of $\mathbf{M}_r^{\mathrm{T}}$ that are passed into the IDA. Here, let $\mathbf{G}(x)$ denote the $x+1$-th row of $\mathbf{G}$. We then obtain $\sigma(x)_j = \mathbf{G}(x) \cdot \mathbf{r}'_{j-1}$. When all participants $\mathcal{V} = \{P_{t_0}, \cdots, P_{t_{k-1}}\}$ pool their shares $\boldsymbol{\sigma}_j$ together, they need to solve $\boldsymbol{\sigma}_j = \mathbf{M}_{\mathcal{V}} \cdot \mathbf{r}'_{j-1}$, i.e.,

$$
\boldsymbol{\sigma}_j = \begin{pmatrix} \sigma(t_0)_j \\ \vdots \\ \sigma(t_{k-1})_j \end{pmatrix} = \begin{pmatrix} \mathbf{G}(t_0) \\ \vdots \\ \mathbf{G}(t_{k-1}) \end{pmatrix} \begin{pmatrix} \mathbf{r}'_{j-1}[0] \\ \vdots \\ \mathbf{r}'_{j-1}[k-1] \end{pmatrix} = \mathbf{G}' \cdot \mathbf{r}'_{j-1}
$$

in unknown vector $\mathbf{r}'_{j-1}$. Since $\mathbf{M}_{\mathcal{V}}$ is regular by the assumption noted above, we are able to uniquely solve unknown vector $\mathbf{r}'_{j-1}$ for every j-th row of $\mathbf{M}_r^{\mathrm{T}}$.

Next, we consider perfect privacy. For the j-th row of $\mathbf{M}_r^{\mathrm{T}}$, we may view $\mathbf{r}'_{j-1}[0]$ in the unknown vector $\mathbf{r}'_{j-1}$ as a secret. Furthermore, a square matrix is regular if and only if its determinant is nonzero. Equivalently, the rows of $\mathbf{M}_{\mathcal{V}}$ are linearly independent. Let $\mathcal{V}_u \notin \Gamma$ be an unauthorized subset and $\mathbf{M}_{\mathcal{V}_u}$ be the corresponding matrix. We aim to show that even if all participants in $\mathcal{V}_u$ pool their shares together, they cannot reveal anything regarding the secret. This also implies that any value of the secret is accepted from their shares. The proof here is that the secret is not included in the row space of $\mathbf{M}_{\mathcal{V}_u}$, in the set of all possible linear combinations of the rows of $\mathbf{M}_{\mathcal{V}_u}$.

Without loss of generality, we may assume that $\mathcal{V}_u$ is missing only one participant to become authorized, and we may simplify the process by adding to $\mathcal{V}_u$ phantom participant $0 \in \mathcal{U}_0$ such that we can obtain an authorized subset. Then the square matrix corresponding to the authorized subset is regular by the assumption, and the rows of the square matrix are linearly independent. As a result, the share of participant $0 \in \mathcal{U}_0$ cannot be generated from $\mathcal{V}_u$, and the share is equivalent to $\mathbf{r}'_{j-1}[0]$. In addition, even when $\mathcal{V}_u$ is missing only one participant at the j-th level, the access structure holds by adding one higher-level participant, i.e., the highest-level participant $0 \in \mathcal{U}_0$.

The proof is thus complete and we conclude that $\det(\mathbf{M}_{\mathcal{V}}) \neq 0$ is required for both correctness and perfect privacy.   □

5 Software Implementation

We evaluated our scheme using one general purpose machine, as described in Table 5. We then used a file size of 888,710 bytes as an example. For operations with $GF(2^L)$, the additive operation is replaced by the XOR operation, the multiplication operation uses the Russian peasant multiplication method, the division operation uses $x^{-1} = x^{2^L-2}$, and the shift operation uses only the shift operation by one bit. In our experiments, we only used $GF(2^8)$ and a lookup table that was precomputed for the multiplication and division operations over $GF(2^8)$. More specifically, all results of the multiplication operations were pre-stored in an array of 2^{16} bytes, while those for the division operations were stored in another array of 2^{16} bytes. When each of the multiplication and division operations actually took place, the operation consisted of a lookup in each array. No cryptographic libraries were used. Then, the primitive polynomial used for $GF(2^8)$ is $x^8 + x^4 + x^3 + x^2 + 1$. Table 6 shows our experimental results.

Table 5. Test environment

CPU/RAM	Intel ® Celeron ® Processor G1820 2.70 GHz × 2, 2 MB cache/3.6 GB
OS	CentOS 7 Linux 3.10.0-229.20.1.el7.x86_64
Programing language/Compiler	C/gcc 4.8.3 (-O3 -flto -DNDEBUG)

Table 6. Results of our experiments for recovery

Level k	Number of participants $(|\mathcal{U}_0|, \cdots, |\mathcal{U}_m|)$	Throughput (Mbps)
$\{1, 3\}$	$(2,3)$	857
$\{2, 4\}$	$(3,4)$	562
$\{2, 3, 5\}$	$(3,3,3)$	373
$\{2, 4, 6, 10\}$	$(2,3,6,4)$	108
$\{3, 7, 11, 14, 17\}$	$(3,4,5,4,4)$	34.9

In terms of optimization, we can construct a scheme using only XOR operations with $(k+1) \times k$ binary generator matrix

$$
\mathbf{G} = \begin{pmatrix} 1 & 0 & \cdots & 0 \\ 0 & 1 & \cdots & 0 \\ \vdots & \vdots & \ddots & \vdots \\ 0 & 0 & \cdots & 1 \\ 0 & 1 & \cdots & 1 \end{pmatrix}.
$$

This matrix requires no multiplication operations. We then used simple 64-bit XOR operations. With $k = 3$, we achieved approximately 6.3 Gbps for recovery.

5.1 Computational Costs

Tassa's Approach. Table 7 shows the computational costs of the recovery algorithm. In general, the size of the secret, for example, a file size of 1 MB, exceeded L bits. In our analysis, we refer to such an initial computation processed once for that recovery as a precomputation. Here, a $t \times t$ determinant required $\mathcal{O}(t^3)$ when we used LU decomposition. Converting a matrix to a triangular matrix required some division operations. Furthermore, a $t \times t$ matrix inverse also required $\mathcal{O}(t^3)$ and some division operations when we used Gaussian elimination. Finally, $t \times t$ matrix multiplication required $\mathcal{O}(t^3)$ if carried out naively. Note that Tassa's scheme can be applied only to GF(p), where p is a prime number. Arithmetic operations, using multiple-precision arithmetic, required higher computational costs than operations over GF(2^L). Therefore, our scheme was much faster than Tassa's approach.

Table 7. Computational costs for recovery

Tassa's approach	
Precomputation	One $k \times k$ determinant
Recovery per time	One $k \times k$ determinant, one division operation
Our scheme	
Precomputation	One $k \times k$ matrix inverse
Recovery per time	One $k \times k$ matrix product, $k - 1$ XOR operations

Blömer et al.'s Technique. We consider applying Blömer et al.'s technique to our scheme because Kurihara et al. [8] reported that any secret sharing scheme over GF(2^L) could be converted to a scheme over GF(2). As Blömer et al. [10] mentioned in Sect. 4, we stored all coefficient vectors of field elements in a table and used table look-ups to compute τ. This operation required $\mathcal{O}(1)$. Furthermore, matrix-vector multiplication required at most L XOR operations. In our implementation using a lookup table for the multiplication operation over GF(2^8), a multiplication operation only required $\mathcal{O}(1)$.

6   Conclusions

In this paper, we focused on a fast $(\mathbf{k}, n)$ hierarchical secret sharing scheme applicable to any level. To achieve this, we applied a hierarchy to the generator matrix used in an IDA. Our scheme is both *perfect* and *ideal*. Given the implementation used in our experiments applicable to any level, we found our implementation on a general purpose PC was able to recover the given secret with $\mathbf{k} = \{1, 3\}$ at a processing speed of approximately 850 Mbps. With our $(\{1, 3\}, 4)$ optimized scheme, we achieved approximately 6.3 Gbps for recovery.

Acknowledgments. The authors thank the anonymous reviewers for their helpful comments. This work was supported by JSPS KAKENHI Grant Number JP18K11306.

References

1. Blakley, G.R.: Safeguarding cryptographic keys. In: AFIPS, vol. 48, pp. 313–317 (1979)
2. Shamir, A.: How to share a secret. Commun. ACM **22**(11), 612–613 (1979)
3. Tassa, T.: Hierarchical threshold secret sharing. In: Naor, M. (ed.) TCC 2004. LNCS, vol. 2951, pp. 473–490. Springer, Heidelberg (2004). https://doi.org/10.1007/978-3-540-24638-1_26
4. Tassa, T.: Hierarchical threshold secret sharing. J. Cryptol. **20**(2), 237–264 (2007)
5. Kurihara, J., Kiyomoto, S., Fukushima, K., Tanaka, T.: A fast $(3, n)$-threshold secret sharing scheme using Exclusive-OR operations. IEICE Trans. Fundam. **E91-A**(1), 127–138 (2008)
6. Kurihara, J., Kiyomoto, S., Fukushima, K., Tanaka, T.: On a fast (k, n)-threshold secret sharing scheme. IEICE Trans. Fundam. **E91-A**(9), 2365–2378 (2008)
7. Kurihara, J., Kiyomoto, S., Fukushima, K., Tanaka, T.: A new (k, n)-threshold secret sharing scheme and its extension. In: Wu, T.-C., Lei, C.-L., Rijmen, V., Lee, D.-T. (eds.) ISC 2008. LNCS, vol. 5222, pp. 455–470. Springer, Heidelberg (2008). https://doi.org/10.1007/978-3-540-85886-7_31
8. Kurihara, J., Uyematsu, T.: A novel realization of threshold schemes over binary field extensions. IEICE Trans. Fundam. **E94-A**(6), 1375–1380 (2011)
9. Feng, G.-L., Deng, R.-H., Bao, F.: Packet-loss resilient coding scheme with only XOR operations. IEE Proc. Commun. **151**(4), 322–328 (2004)
10. Blömer, J., Kalfane, M., Karp, R., Karpinski, M., Luby, M., Zuckerman, D.: An XOR-based erasure-resilient coding scheme. ICSI Technical report TR-95-048 (1995)
11. Chen, L., Laing, T.M., Martin, K.M.: Efficient, XOR-based, ideal (t, n)−threshold schemes. In: Foresti, S., Persiano, G. (eds.) CANS 2016. LNCS, vol. 10052, pp. 467–483. Springer, Cham (2016). https://doi.org/10.1007/978-3-319-48965-0_28
12. Yamamoto, H.: On secret sharing system using (k, L, n) threshold scheme. IEICE Trans. Fundam. (Jpn. Ed.) **J68-A**(9), 945–952 (1985). Secret sharing system using (k, L, n) threshold scheme. Electron. Commun. Jpn. (Engl. Ed.) Part I **69**(9), 46–54 (1986)
13. Blakley, G.R., Meadows, C.: Security of ramp schemes. In: Blakley, G.R., Chaum, D. (eds.) CRYPTO 1984. LNCS, vol. 196, pp. 242–268. Springer, Heidelberg (1985). https://doi.org/10.1007/3-540-39568-7_20
14. Krawczyk, H.: Secret sharing made short. In: Stinson, D.R. (ed.) CRYPTO 1993. LNCS, vol. 773, pp. 136–146. Springer, Heidelberg (1994). https://doi.org/10.1007/3-540-48329-2_12
15. Kurihara, J., Kiyomoto, S., Fukushima, K., Tanaka, T.: A fast (k, L, n)-threshold ramp secret sharing scheme. IEICE Trans. Fundam. **E92-A**(8), 1808–1821 (2009)
16. Resch, J.K., Plank, J.S.: AONT-RS: blending security and performance in dispersed storage systems. In: 9th USENIX Conference on File and Storage Technologies, pp. 191–202 (2011)
17. Rabin, M.O.: Efficient dispersal of information for security, load balancing, and fault tolerance. J. ACM **36**(2), 335–348 (1989)
18. Rivest, R.L.: All-or-nothing encryption and the package transform. In: Biham, E. (ed.) FSE 1997. LNCS, vol. 1267, pp. 210–218. Springer, Heidelberg (1997). https://doi.org/10.1007/BFb0052348
19. Béguin, P., Cresti, A.: General short computational secret sharing schemes. In: Guillou, L.C., Quisquater, J.-J. (eds.) EUROCRYPT 1995. LNCS, vol. 921, pp. 194–208. Springer, Heidelberg (1995). https://doi.org/10.1007/3-540-49264-X_16

20. Selçuk, A.A., Kaşkaloğlu, K., Özbudak, F.: On hierarchical threshold secret sharing. IACR Cryptology ePrint Archive 2009, 450 (2009)
21. Simmons, G.J.: How to (really) share a secret. In: Goldwasser, S. (ed.) CRYPTO 1988. LNCS, vol. 403, pp. 390–448. Springer, New York (1990). https://doi.org/10.1007/0-387-34799-2_30
22. Brickell, E.F.: Some ideal secret sharing schemes. In: Quisquater, J.-J., Vandewalle, J. (eds.) EUROCRYPT 1989. LNCS, vol. 434, pp. 468–475. Springer, Heidelberg (1990). https://doi.org/10.1007/3-540-46885-4_45
23. Castiglione, A., De Santis, A., Masucci, B.: Hierarchical and shared key assignment. In: NBiS-2014, pp. 263–270 (2014)
24. Castiglione, A., et al.: Hierarchical and shared access control. IEEE Trans. Inf. Forensics Secur. 11(4), 850–865 (2016)
25. Beimel, A.: Secret-sharing schemes: a survey. In: Chee, Y.M., Guo, Z., Ling, S., Shao, F., Tang, Y., Wang, H., Xing, C. (eds.) IWCC 2011. LNCS, vol. 6639, pp. 11–46. Springer, Heidelberg (2011). https://doi.org/10.1007/978-3-642-20901-7_2
26. Blundo, C., De Santis, A., Gargano, L., Vaccaro, U.: On the information rate of secret sharing schemes. TCS 154, 283–306 (1996)
27. Blundo, C., De Santis, A., Gargano, L., Vaccaro, U.: On the information rate of secret sharing schemes. In: Brickell, E.F. (ed.) CRYPTO 1992. LNCS, vol. 740, pp. 148–167. Springer, Heidelberg (1993). https://doi.org/10.1007/3-540-48071-4_11
28. Plank, J.S., Ding, Y.: Note: correction to the 1997 tutorial on reed-solomon coding. Softw.-Pract. Exp. 35(2), 189–194 (2005)
29. Shima, K., Doi, H.: A hierarchical secret sharing scheme over finite fields of characteristic 2. J. Inf. Process. 25, 875–883 (2017)

Cheating-Immune Secret Sharing Schemes from Maiorana-McFarland Boolean Functions

Romar B. dela Cruz[1]([⊠])[iD] and Say Ol[2]

[1] Institute of Mathematics, University of the Philippines Diliman,
Quezon City, Philippines
rbdelacruz@math.upd.edu.ph
[2] Teacher Education College, Phnom Penh, Cambodia
say_ol@yahoo.com

Abstract. We consider cheating-immune secret sharing schemes proposed by Pieprzyk and Zhang. This type of secret sharing scheme keeps dishonest participants from having a better chance (over the honest ones) of knowing the secret using their incorrect shares. We show that the class of Maiorana-McFarland Boolean functions can be used to construct such schemes. Consequently, new cheating-immune secret sharing schemes are presented.

Keywords: Secret sharing scheme · Boolean function ·
Error-correcting code

1 Introduction

A *secret sharing scheme* (SSS) is a technique of allocating access to a secret among a set of participants in such a way that only certain subsets are allowed to determine the secret. It was introduced independently by Shamir [30] and Blakley [1] for the protection of cryptographic keys. It is now a fundamental primitive as it is used to construct cryptographic protocols such as for secure multiparty computation [12] and oblivious transfer [32].

In general, an SSS starts with the *share distribution* phase followed by the *secret reconstruction* phase. In the share distribution phase, there is a dealer who produces the shares to be given to the participants. In the secret reconstruction phase, a subset of participants attempt to determine the secret using their shares. We consider the setting wherein the participants submit their shares to a trusted combiner who reconstructs the secret. We assume that the dealer and combiner are honest but the participants can cheat during secret reconstruction.

Tompa and Woll [33] showed that if a secret sharing scheme is linear, then it can be subjected to an attack from dishonest participants. During the secret reconstruction phase, the cheaters can submit invalid shares to the combiner. As a result, the combiner returns an invalid secret and the cheaters are able to

© Springer Nature Switzerland AG 2019
K. Lee (Ed.): ICISC 2018, LNCS 11396, pp. 233–247, 2019.
https://doi.org/10.1007/978-3-030-12146-4_15

compute the valid secret using their valid shares and the invalid secret. This attack also prevents the honest participants from knowing the valid secret. Several approaches to counter this attack can be found in the literature (for instance see [2,6,11,19,21,25,29]). The survey article by Martin [24] is a comprehensive analysis of the different types of SSS that handle dishonest participants, dealers and combiners.

This work deals with *cheating-immune secret sharing schemes* proposed by Pieprzyk and Zhang [28]. This type of SSS is capable of preventing the dishonest participants of gaining an advantage over the honest ones in the attack described above. In a cheating-immune scheme, the cheaters will not be able to determine the secret even if they submit invalid shares during reconstruction. If we compare with the other SSS that deal with cheating, there is no detection or identification of cheaters in a cheating-immune scheme. There is also no correction of the submitted invalid shares which means that the honest participants also do not recover the secret. A nice property of cheating-immune schemes is that the share size is the same as the secret size (in other schemes, either we have large shares or the recovery of the secret requires more than the minimum number of shares). The main problem in the theory of cheating-immune schemes is the construction of such schemes for any access structure. Properties and constructions of such schemes are studied in [4,13,14,22,23,26,27].

In this paper, we show that the class of Maiorana-McFarland Boolean functions can be used to construct cheating-immune schemes and we present new schemes. We used the techniques in the work of Carlet [9]. Our method of construction can be seen as a generalization of the method in [14]. This paper is organized as follows. Section 2 contains the definition and model of binary cheating-immune schemes. In Sect. 3, we present the relation between cheating-immune schemes and Boolean functions. Sections 4, 5 and 6 contain the main results of the paper. We summarize the work in the last section.

2   Cheating-Immune Secret Sharing Schemes

Let $\mathcal{P} = \{P_1, P_2, \ldots, P_n\}$ be the set of n participants. The set of all authorized or qualified subsets $\Gamma \subseteq 2^{\mathcal{P}}$ is called the *access structure*.

Definition 1. *A secret sharing scheme realizing an access structure Γ is a method to distribute shares of a secret K such that*

i. *if a subset of participants $A \in \Gamma$ then A can reconstruct the secret K*
ii. *if a subset of participants $B \notin \Gamma$ then B cannot reconstruct the secret K.*

We say that a secret sharing scheme is *perfect* if unauthorized subsets obtain no information about the secret. Otherwise, the scheme is *non-perfect*, that is, it is possible for an unauthorized subset to obtain partial information about the secret. A measure of efficiency of a secret sharing scheme is the so-called *information rate* which is the ratio of the size of the secret and the size of the shares. We assume that the dealer selects the secrets uniformly at random.

We use the model of a cheating-immune (n, n)-SSS over $\mathbb{F}_2$ introduced in [26]. The scheme is represented by a *defining function* $f : \mathbb{F}_2^n \to \mathbb{F}_2$ that maps each possible vector of shares $\alpha = (\alpha_1, \alpha_2, \ldots, \alpha_n)$ to a secret K. All participants must submit their shares to the combiner in order to reconstruct the secret. Let $\alpha, \beta \in \mathbb{F}_2^n$. We say that β *covers* α, denoted by $\alpha \preceq \beta$, if whenever $\alpha_i \neq 0$ then $\beta_i \neq 0$, $1 \leq i \leq n$. The *Hamming weight* of a vector α will be denoted by $wt(\alpha)$.

We represent the cheaters by a *cheating vector* δ with $\delta_i = 1$ if P_i is a cheater and 0 otherwise. Hence, $wt(\delta)$ gives the number of cheaters. Given two vectors x and δ, we distinguish the shares of the cheaters from the honest participants using the following vectors:

i. $x_\delta^+ = (x_1^+, \ldots, x_n^+)$ with $x_i^+ = x_i$ if $\delta_i = 1$ and $x_i^+ = 0$ if $\delta_i = 0$
ii. $x_\delta^- = (x_1^-, \ldots, x_n^-)$ with $x_i^- = x_i$ if $\delta_i = 0$ and $x_i^- = 0$ if $\delta_i = 1$

The vector x_δ^+ represents the cheaters' valid shares while x_δ^- represents the honest participants' shares.

Recall that when cheaters submit invalid shares during reconstruction, they will use the secret (sent by the combiner) to determine the true secret. Consider now the following sets of shares:

$$R(\delta, \alpha_\delta^+, K) = \{x_\delta^- \mid f(x_\delta^- \oplus \alpha_\delta^+) = K\}$$
$$R(\delta, \alpha_\delta^+ \oplus \delta, K^*) = \{x_\delta^- \mid f(x_\delta^- \oplus \alpha_\delta^+ \oplus \delta) = K^*\}$$

The set $R(\delta, \alpha_\delta^+, K)$ consists of all possible shares of honest participants such that combined with the cheaters' valid shares, will produce the original secret K. On the other hand, the set $R(\delta, \alpha_\delta^+ \oplus \delta, K^*)$ contains all the possible shares of honest participants such that combined with the cheaters' incorrect shares, will produce the secret K^*. The *probability of successful cheating with respect to* δ, α is given by

$$\rho_{\delta,\alpha} = |R(\delta, \alpha_\delta^+ \oplus \delta, K^*) \cap R(\delta, \alpha_\delta^+, K)| / |R(\delta, \alpha_\delta^+ \oplus \delta, K^*)|.$$

We now define a k-cheating-immune (n, n)-SSS or k-CI (n, n)-SSS. Note that we assume that all cheaters submit invalid shares during reconstruction.

Definition 2. *An (n, n)-SSS over $\mathbb{F}_2$ is k-cheating-immune if for every $\alpha, \delta \in \mathbb{F}_2^n$ with $1 \leq wt(\delta) \leq k$, we have $\rho_{\delta,\alpha} = 1/2$.*

The general case where not all cheaters submit invalid shares is handled by the so-called *strictly cheating-immune secret sharing schemes*. In this type of scheme, we use two vectors $\delta, \tau \in \mathbb{F}_2^n$ such that δ represents the cheaters while τ represents the cheaters who submitted fake shares. Note that $\tau \preceq \delta$. The value

$$\rho_{\delta,\tau,\alpha} = |R(\delta, \alpha_\delta^+ \oplus \tau, K^*) \cap R(\delta, \alpha_\delta^+, K)| / |R(\delta, \alpha_\delta^+ \oplus \tau, K^*)|$$

is the *probability of successful cheating with respect to* δ, τ, α.

Definition 3. *An (n, n)-SSS over $\mathbb{F}_2$ is strictly k-cheating-immune if, for every $\alpha, \delta, \tau \in \mathbb{F}_2^n$ such that $1 \leq wt(\delta) \leq k$ and $\tau \preceq \delta$, we have that $\rho_{\delta,\tau,\alpha} = 1/2$.*

3   Cheating-Immune Schemes and Boolean Functions

We now describe the connection between Boolean functions and cheating-immune secret sharing schemes. The defining function of an (n,n)-SSS over $\mathbb{F}_2$ is a Boolean function on $\mathbb{F}_2^n$. We recall some basic concepts in the theory of Boolean functions (for reference, see [8,10]).

A Boolean function f is *affine* if $f(x_1, x_2, \ldots, x_n) = a_1 x_1 \oplus a_1 x_2 \oplus \ldots \oplus a_n x_n \oplus c$. The affine function f is *linear* if $c = 0$. We say that f is *balanced* on $\mathbb{F}_2^n$ if $|f^{-1}(0)| = |f^{-1}(1)| = 2^{n-1}$. A nonconstant affine function is balanced.

A Boolean function f is said to be *k-resilient* if for every subset $\{j_1, j_2, \ldots, j_k\}$ of $\{1, 2, \ldots, n\}$ and every $(a_1, a_2, \ldots, a_k) \in \mathbb{F}_2^k$, the restricted function

$$f(x_1, x_2, \ldots, x_n)|_{x_{j_1} = a_1, x_{j_2} = a_2, \ldots, x_{j_k} = a_k}$$

is balanced on $\mathbb{F}_2^{n-k}$. We note that if f is k-resilient then it is also l-resilient for $0 \leq l \leq k$.

We say that a Boolean function f satisfies the *strengthened propagation of degree k* or $SP(k)$ if for any $\delta \in \mathbb{F}_2^n$ such that $1 \leq wt(\delta) \leq k$, and for any $\tau \preceq \delta$, the function $f(x_{\bar{\delta}} \oplus \tau) \oplus f(x_{\bar{\delta}} \oplus \tau \oplus \delta)$ is balanced. A function f satisfying $SP(k)$ also satisfies $SP(l)$ for $1 \leq l \leq k$.

The following theorems characterize cheating-immune (n,n)-SSS:

Theorem 1 ([27]). *An (n,n)-SSS over $\mathbb{F}_2$ with defining function f is k-CI if and only if f is k-resilient and satisfies $SP(k)$.*

Theorem 2 ([27]). *An (n,n)-SSS over $\mathbb{F}_2$ with defining function f is strictly k-CI if and only if the following conditions are satisfied:*

1. *f is k-resilient.*
2. *For any integer l with $0 \leq l \leq k - 1$, every function obtained from f by fixing any l input variables satisfies $SP(k - l)$.*

A bound on the number of cheaters is given by the following result:

Theorem 3 ([4]). *An (n,n)-SSS over $\mathbb{F}_2$ with defining function f can be k-cheating-immune only if $2k \leq n - 2$.*

4   Cheating-Immune SSS from Maiorana-McFarland Boolean Functions

Theorem 1 states that constructing a k-CI (n,n)-SSS over $\mathbb{F}_2$ is equivalent to constructing a Boolean function satisfying resiliency and strengthened propagation. In this section, we will show that a class of *Maiorana-McFarland* Boolean functions can be used to construct cheating-immune SSS. The Maiorana-McFarland Boolean functions are well-studied and these functions are used to build Boolean functions with cryptographic properties.

Let s, t be positive integers and ϕ be a vectorial Boolean function from $\mathbb{F}_2^t$ to $\mathbb{F}_2^s$, or a (t, s)-vectorial function given by

$$\phi(x_1, x_2, \ldots, x_t) = (\phi_1(x_1, x_2, \ldots, x_t), \phi_2(x_1, x_2, \ldots, x_t), \ldots, \phi_s(x_1, x_2, \ldots, x_t))$$

where its coordinate functions $\phi_1, \phi_2, \ldots, \phi_s$ are t-variable Boolean functions. Let g be a t-variable Boolean function. An $(s + t)$-variable Boolean function f defined by

$$f(x, y) = x \cdot \phi(y) \oplus g(y),$$

where $x \in \mathbb{F}_2^s, y \in \mathbb{F}_2^t$ is said to be of *Maiorana-McFarland* form. We call f an *MM function* for short.

The next theorem gives a condition under which a Maiorana-McFarland function satisfies resiliency.

Theorem 4 ([7]). *An MM function $f(x, y) = x \cdot \phi(y) \oplus g(y)$ is k-resilient if for every $y \in \mathbb{F}_2^t$, we have $wt(\phi(y)) \geq k + 1$.*

We now show that this class of functions also satisfies strengthened propagation criterion. We will use the following lemma.

Lemma 1. *A Boolean function f is balanced on $\mathbb{F}_2^n$ if there exists a subset $\{i_1, i_2, \ldots, i_k\} \subseteq \{1, 2, \ldots, n\}$ such that for every $a = (a_1, a_2, \ldots, a_k) \in \mathbb{F}_2^k$, the restricted function f_a obtained from f by substituting $x_{i_1} = a_1, x_{i_2} = a_2, \ldots, x_{i_k} = a_k$ is balanced.*

Proof. For every $a \in \mathbb{F}_2^k$, by assumption, f_a is balanced on $\mathbb{F}_2^{n-k}$. Hence, we have $|f_a^{-1}(0)| = |f_a^{-1}(1)| = 2^{n-k-1}$. Since there are 2^k possibilities for a, then $|f^{-1}(0)| = |f^{-1}(1)| = 2^k \times 2^{n-k-1} = 2^{n-1}$. Thus, f is balanced on $\mathbb{F}_2^n$.

A modification of the construction of Boolean functions satisfying propagation criterion using *Maiorana-McFarland* Boolean functions considered by [9] gives us a construction of Boolean functions satisfying strengthened propagation.

Theorem 5. *An MM function $f(x, y) = x \cdot \phi(y) \oplus g(y)$ satisfies $SP(k)$ if the following conditions are satisfied:*

1. *For any $a \in \mathbb{F}_2^s$ such that $1 \leq wt(a) \leq k$, the function $a \cdot \phi(y)$ is balanced on $\mathbb{F}_2^t$.*
2. *For any $y, z \in \mathbb{F}_2^t$ such that $wt(y \oplus z) \geq 1$, we have $wt(\phi(y) \oplus \phi(z)) \geq k$.*

Proof. Let $z = (x, y) = (x_1, x_2, \ldots, x_s, y_1, y_2, \ldots, y_t)$. For any $\delta, \tau \in \mathbb{F}_2^{s+t}$ such that $1 \leq wt(\delta) \leq k$ and $\tau \preceq \delta$, we denote by $\delta = (\delta^x, \delta^y)$ and $\tau = (\tau^x, \tau^y)$ where $\delta^x = (\delta_1, \delta_2, \ldots, \delta_s), \tau^x = (\tau_1, \tau_2, \ldots, \tau_s), \delta^y = (\delta_{s+1}, \delta_{s+2}, \ldots, \delta_{s+t})$ and $\tau^y = (\tau_{s+1}, \tau_{s+2}, \ldots, \tau_{s+t})$.

Define $h(z_\delta^-) = f(z_\delta^- \oplus \tau) \oplus f(z_\delta^- \oplus \tau \oplus \delta)$. Then,

$$h(z_\delta^-) = f(x_{\delta^x}^- \oplus \tau^x, y_{\delta^y}^- \oplus \tau^y) \oplus f(x_{\delta^x}^- \oplus \tau^x \oplus \delta^x, y_{\delta^y}^- \oplus \tau^y \oplus \delta^y)$$

$$= a(y_{\delta^y}^-) \cdot x_{\delta^x}^- \oplus b(y_{\delta^y}^-)$$

where

$$a(y_{\delta v}^-) = \phi(y_{\delta v}^- \oplus \tau^y) \oplus \phi(y_{\delta v}^- \oplus \tau^y \oplus \delta^y)$$

$$b(y_{\delta v}^-) = \tau^x \cdot \phi(y_{\delta v}^- \oplus \tau^y) \oplus (\tau^x \oplus \delta^x) \cdot \phi(y_{\delta v}^- \oplus \tau^y \oplus \delta^y) \oplus$$
$$g(y_{\delta v}^- \oplus \tau^y) \oplus g(y_{\delta v}^- \oplus \tau^y \oplus \delta^y)$$

Case 1. If $\delta^y = 0$ then $\tau^y = 0, y_{\delta v}^- = y$ and $wt(\delta^x) = wt(\delta)$. Hence, $h(z_\delta^-) = \delta^x \cdot \phi(y)$ is balanced by the first condition.

Case 2. If $\delta^y \neq 0$ then $0 \leq wt(\delta^x) \leq k - 1$. In other words, the number of constant coordinates of $x_{\delta x}^-$ is less than or equal to $k-1$. For every substitution of $t - wt(\delta^y)$ variables in $y_{\delta v}^-$, by the second condition, $wt(a(y_{\delta v}^-)) \geq k$. Hence, the function obtained from $h(z_\delta^-)$ by the substitution is a non-constant affine function which is balanced. Therefore, $h(z_\delta^-)$ is balanced by Lemma 1.

In conclusion, the function f satisfies $SP(k)$.

5   Construction of CI-SSS Using Binary Systematic Codes

Similar to what was done on [9], we use binary systematic codes to come up with concrete examples of functions satisfying the conditions in Theorems 4 and 5. This method of construction is a generalization of [14] which uses linear codes. The technique used here allows us to use nonlinear codes. We start with a discussion of some basic concepts on binary codes (the reader is referred to [18, 20] for a complete treatment of codes).

A nonempty subset $C \subseteq \mathbb{F}_2^n$ is called a *binary code* of length n. The *Hamming distance* between two vectors $x, y \in \mathbb{F}^n$, denoted by $d(x, y)$, is the number of positions where x and y differ. The *minimum distance* of C is defined as

$$d(C) = \min\{d(x, y) \mid x, y \in C, x \neq y\}.$$

A binary code of length n having M codewords and minimum distance d is called an (n, M, d)-*code*. The distance from a vector $\alpha \in \mathbb{F}_2^n$ to a code C is given by $d(\alpha, C) = \min\{d(\alpha, c) \mid c \in C\}$. The *covering radius* of C is defined to be $\rho = \max\{d(x, C) \mid x \in \mathbb{F}_2^n\}$.

Definition 4. *A binary code C is said to be k-systematic if there exists k positions $i_1, \ldots, i_k$ such that every element of $\mathbb{F}_2^k$ appears in exactly one codeword of C in the specified positions. The set $\{i_1, i_2, \ldots, i_k\}$ is called an information set of C.*

Let C be a binary k-systematic code. It follows from the definition that C has 2^k codewords. Let $c = (c_1, c_2, \ldots, c_n) \in C$. The coordinates $c_{i_1}, c_{i_2}, \ldots, c_{i_k}$ are called *information bits* and the remaining coordinates are called *parity-check bits*. Hence, if all of the parity-check bits of a binary k-systematic code are deleted, we obtain the code $\mathbb{F}_2^k$.

A *binary linear code* C is a k-dimensional subspace of $\mathbb{F}_2^n$. A binary linear code of length n, dimension k and minimum distance d is called an $[n, k, d]$-code.

A $k \times n$ matrix whose rows form a basis of C is called a *generator matrix*. The *dual code* of C is its $(n-k)$-dimensional dual space $C^{\perp} = \{x \in \mathbb{F}_2^n \mid c \cdot x = 0, \forall c \in C\}$. Note that a binary linear code is k-systematic.

Let C be an (n, M, d) binary code and let

$$B_i = \frac{1}{|C|} \sum_{c \in C} |\{x \in C \mid d(c, x) = i\}|.$$

The list $B_1, B_2, \ldots, B_n$ is called the *distance distribution* of C. The homogeneous polynomial $D_C(x, y) = \sum_{i=0}^{n} B_i x^{n-i} y^i$ is called the *distance enumerator* of C. The *dual distance* of an (n, M, d) binary code C is the smallest positive integer d' such that the coefficient of $x^{n-d'} y^{d'}$ of $D_C(x+y, x-y)$ is nonzero. In the case that C is linear, the dual distance is the same as the minimum distance of $C^{\perp}$.

Due to the notion of equivalence of codes, we can assume that the information set of a given systematic code is the set $\{1, \ldots, k\}$. We also assume that the generator matrix of a given linear code is in standard form, i.e. $[I_k \mid A]$ where I_k is the identity matrix of order k and A is a $k \times (n-k)$ binary matrix.

We now proceed with the construction of cheating-immune schemes using binary systematic codes.

Lemma 2 ([9]). *Let C be a binary code of length s with dual distance d'. Then, for every $a \in \mathbb{F}_2^s$ such that $1 \leq wt(a) \leq d'-1$, the s-variable Boolean function $\psi(x) = a \cdot x$ is still balanced when its domain is restricted to C.*

Theorem 6. *An MM function $f(x, y) = x \cdot \phi(y) \oplus g(y)$ satisfies $SP(k)$ if the (t, s)-vectorial function ϕ is injective and the code $\phi(\mathbb{F}_2^t)$ has minimum distance $d \geq k$ and dual distance $d' \geq k+1$.*

Proof. We will show that the conditions of Theorem 5 are satisfied. For any $a \in \mathbb{F}_2^s$ such that $1 \leq wt(a) \leq k$, the s-variable Boolean function $\psi(x) = a \cdot x$ is balanced on $\phi(\mathbb{F}_2^t)$ (because $k \leq d'-1$) thanks to Lemma 2. Since ϕ is injective, for any $z \in \phi(\mathbb{F}_2^t)$, there is a unique $y \in \mathbb{F}_2^t$ such that $z = \phi(y)$. Thus, the composition $(\psi \circ \phi)(y) = a \cdot \phi(y)$ is balanced on $\mathbb{F}_2^t$. For any $y, z \in \mathbb{F}_2^t$ with $wt(y \oplus z) \geq 1$, $wt(\phi(y) \oplus \phi(z)) \geq k$ because ϕ is injective and $\phi(\mathbb{F}_2^t)$ has minimum distance $d \geq k$.

Theorem 7. *Let C be an $(s, 2^t, d)$ binary t-systematic code with dual distance d' and covering radius ρ. Let $k = \min\{d, d'-1, \rho-1\}$ and $\alpha \in \mathbb{F}_2^s$ such that $d(\alpha, C) \geq k+1$. Define the (t, s)-vectorial function $\phi(x) = \alpha \oplus (x, u(x))$ where $u(x)$ is vector of $(s-t)$ parity-check bits of a codeword of C whose t information bits are represented by the vector x. Let g be an arbitrary t-variable Boolean function. Then the MM function $f(x, y) = x \cdot \phi(y) \oplus g(y)$ defines a k-CI $(s+t, s+t)$-SSS.*

Proof. For every $y \in \mathbb{F}_2^t$, $wt(\phi(y)) \geq d(\alpha, C) = k+1$. By Theorem 4, f is k-resilient. The code $\phi(\mathbb{F}_2^t) = \alpha \oplus C$ has the same minimum distance $d \geq k$ and the same dual distance $d' \geq k+1$ as C. By Theorem 5, f satisfies $SP(k)$. Due to Theorem 1, the $(s+t, s+t)$-SSS defined by f is k-CI.

We present examples of schemes obtained using the preceding theorem. The computations were performed using Magma [5]. In case that C is linear with generator matrix G, the function ϕ can be written as $\phi(y) = \alpha \oplus yG$. Then the defining function f will be $f(x, y) = x \cdot (\alpha \oplus yG) \oplus g(y)$.

Example 1 (**a new scheme**). Let C be the $[12, 5, 4]$ binary linear code with dual distance $d' = 4$, covering radius $\rho = 4$ and generator matrix

$$G = \begin{bmatrix} 1 & 0 & 0 & 0 & 0 & 0 & 1 & 1 & 0 & 1 & 1 & 1 \\ 0 & 1 & 0 & 0 & 0 & 1 & 0 & 1 & 1 & 0 & 1 & 1 \\ 0 & 0 & 1 & 0 & 0 & 1 & 1 & 0 & 1 & 1 & 0 & 1 \\ 0 & 0 & 0 & 1 & 0 & 0 & 0 & 0 & 1 & 1 & 1 & 0 \\ 0 & 0 & 0 & 0 & 1 & 1 & 1 & 1 & 0 & 0 & 0 & 0 \end{bmatrix}.$$

Using $\alpha = (0, 0, 0, 0, 0, 0, 1, 1, 1, 0, 1, 0)$ with $d(\alpha, C) = 4$, we obtain a 3-CI $(17, 17)$-SSS.

Example 2 (**using some classes of linear codes**)

a. 1-CI $(m + 1, m + 1)$-SSS: For $m \geq 4$, let $C = \mathcal{R}_m$, the $[m, 1, m]$ binary repetition code with dual distance $d^{\perp} = 2$ and covering radius $\rho = \lfloor \frac{n}{2} \rfloor$. Choose α such that $d(\alpha, C) = 2$.

b. 2-CI $(2^m + m - 1, 2^m + m - 1)$-SSS: For $m \geq 3$, let $C = \mathcal{S}_m$, the $[2^m - 1, m, 2^{m-1}]$ binary *Simplex* code with dual distance $d^{\perp} = 3$ and covering radius $\rho = 2^{m-1} - 1$. Choose α such that $d(\alpha, C) = 3$.

c. 3-CI $(2^m + m + 1, 2^m + m + 1)$-SSS: For $m \geq 4$, let $C = \mathcal{R}(1, m)$, the $[2^m, m + 1, 2^{m-1}]$ first-order *Reed-Muller* code with dual distance $d^{\perp} = 4$ and covering radius $2^{m-1} - 2^{\lceil m/2 \rceil - 1} \leq \rho \leq 2^{m-1} - 2^{m/2-1}$ [17]. Choose α such that $d(\alpha, C) \geq 4$.

Example 3 (**using nonlinear codes**). For even integer $m \geq 4$, there exists two well-known classes of binary nonlinear systematic codes [20]:

i. $(2^m, 2^{2m}, 2^{m-1} - 2^{\frac{m}{2}-1})$ *Kerdock* code $\mathcal{K}(m)$ with dual distance 6 and covering radius $2^{m-1} - 2^{\frac{m}{2}-1}$

ii. $(2^m, 2^{2^m - 2m}, 6)$ *Preparata* code $\mathcal{P}(m)$ with dual distance $2^{m-1} - 2^{\frac{m}{2}-1}$ and covering radius 3.

We use these codes to obtain the following schemes:

a. 2-CI $(2^{m+1} - 2m, 2^{m+1} - 2m)$-SSS: For even integer $m \geq 4$, let $C = \mathcal{P}(m)$ and choose $\alpha \in \mathbb{F}_2^m$ such that $d(\alpha, \mathcal{P}(m)) = 3$.

b. 5-CI $(2^m + 2m, 2^m + 2m)$-SSS: For even integer $m \geq 6$, let $C = \mathcal{K}(m)$ and choose $\alpha \in \mathbb{F}_2^m$ such that $d(\alpha, \mathcal{K}(m)) = 6$.

6   Strictly Cheating-Immune SSS

Here we consider the construction of strictly cheating-immune SSS from the class of Maiorana-McFarland Boolean functions. The goal is to construct functions satisfying the conditions given by Theorem 2. The next theorem talks about the strengthened propagation property.

Theorem 8. *Let $f(x, y) = x \cdot \phi(y) \oplus g(y)$ be an MM function satisfying the following conditions:*

1. *for any $a \in \mathbb{F}_2^s$ with $1 \le wt(a) \le k$, the function $a \cdot \phi(y)$ on $\mathbb{F}_2^t$ is $(k-1)$-resilient;*
2. *for any $y, z \in \mathbb{F}_2^t$ if $1 \le wt(y \oplus z) \le k$, we have $wt(\phi(y) \oplus \phi(z)) \ge k$.*

Then, for any integer l with $0 \le l \le k - 1$, every function obtained from f by keeping any l input variables constant satisfies $SP(k - l)$.

Proof. Let $z = (x, y)$. For any integer l with $0 \le l \le k - 1$, we denote by $\underline{x}$ and $\underline{y}$ the vectors obtained from x and y by fixing u and v coordinates constant such that $u + v = l$. If we let $\underline{z} = (\underline{x}, \underline{y})$ then $f(\underline{z})$ is the $(s + t - l)$-variable Boolean function obtained from f by fixing l input variables.

Now we show that $f(\underline{z})$ satisfies $SP(k - l)$. Let $\delta, \tau \in \mathbb{F}_2^n$, $n = s + t$, such that $\tau \preceq \delta, 1 \le wt(\delta) \le k - l$ and the set of nonzero coordinates of δ is a subset of the nonconstant coordinates of $\underline{z}$. We write $\delta = (\delta^{\underline{x}}, \delta^{\underline{y}})$ and $\tau = (\tau^{\underline{x}}, \tau^{\underline{y}})$ where $\delta^{\underline{x}}$ and $\tau^{\underline{x}}$ are the first s coordinates of δ and τ, and $\delta^{\underline{y}}$ and $\tau^{\underline{y}}$ are the remaining t coordinates of δ and τ, respectively.

Define $h(\underline{z}_\delta^-) = f(\underline{z}_\delta^- \oplus \tau) \oplus f(\underline{z}_\delta^- \oplus \tau \oplus \delta)$. Then $h(\underline{z}_\delta^-) = \underline{x}_{\delta^{\underline{x}}}^- \cdot a(\underline{y}_{\delta^{\underline{y}}}^-) \oplus b(\underline{y}_{\delta^{\underline{y}}}^-)$ where

$$a(\underline{y}_{\delta^{\underline{y}}}^-) = \phi(\underline{y}_{\delta^{\underline{y}}}^- \oplus \tau^{\underline{y}}) \oplus \phi(\underline{y}_{\delta^{\underline{y}}}^- \oplus \tau^{\underline{y}} \oplus \delta^{\underline{y}})$$

$$b(\underline{y}_{\delta^{\underline{y}}}^-) = \tau^{\underline{x}} \cdot \phi(\underline{y}_{\delta^{\underline{y}}}^- \oplus \tau^{\underline{y}}) \oplus (\tau^{\underline{x}} \oplus \delta^{\underline{x}}) \cdot \phi(\underline{y}_{\delta^{\underline{y}}}^- \oplus \tau^{\underline{y}} \oplus \delta^{\underline{y}}) \oplus$$

$$g(\underline{y}_{\delta^{\underline{y}}}^- \oplus \tau^{\underline{y}}) \oplus g(\underline{y}_{\delta^{\underline{y}}}^- \oplus \tau^{\underline{y}} \oplus \delta^{\underline{y}})$$

Case 1. If $\delta^{\underline{y}} = 0$ then $\underline{y}_{\delta^{\underline{y}}}^- = \underline{y}$, $\tau^{\underline{y}} = 0$ and $1 \le wt(\delta^{\underline{x}}) = wt(\delta) \le k - l \le k$. By the first condition, $\delta^{\underline{x}} \cdot \phi(y)$ is $(k-1)$-resilient. In addition, $v = l - u \le l \le k - 1$. Hence, $h(\underline{z}_\delta^-) = \delta^{\underline{x}} \cdot \phi(\underline{y})$ is balanced because it is obtained from the function $\delta^{\underline{x}} \cdot \phi(y)$ by fixing v input variables constant.

Case 2. If $\delta^{\underline{y}} \ne 0$ then $0 \le wt(\delta^{\underline{x}}) < wt(\delta) \le k - l$. Hence, the number of constant coordinates of $\underline{x}_{\delta^{\underline{x}}}^-$ is $u + wt(\delta^{\underline{x}}) \le l + (k - l - 1) = k - 1$. For every substitution of the $t - v - wt(\delta^{\underline{y}})$ variables in $\underline{y}_{\delta^{\underline{y}}}^-$, by the second condition, $wt(a(\underline{y}_{\delta^{\underline{y}}}^-)) \ge k$. Hence, the function obtained from $h(\underline{z}_\delta^-)$ by the substitution is a non-constant affine function which is balanced. Therefore, $h(\underline{z}_\delta^-)$ is balanced by Lemma 1.

In conclusion, the function $f(\underline{z})$ satisfies $SP(k - l)$.

An (s, t)-vectorial function ϕ is balanced if for every $y \in \mathbb{F}_2^t$, $|\phi^{-1}(y)| = 2^{s-t}$. The function ϕ is said to be k-resilient if it is balanced and every function obtained from ϕ by keeping k input variables constant is balanced.

Lemma 3 ([3]). *Let ϕ be a (t,r)-vectorial k-resilient function and ψ be an (r,s)-vectorial balanced function. Then the (t,s)-vectorial function $\psi \circ \phi$ is k-resilient.*

We now look at the construction of a function ϕ satisfying the conditions of Theorem 8. Similar to [9], we split ϕ into a composition of two simpler vectorial functions.

Theorem 9. *Suppose that $\phi = \phi_2 \circ \phi_1$ where ϕ_1 is a (t,r)-vectorial function and ϕ_2 is an (r,s)-vectorial function with the following properties:*

1. *(a) ϕ_1 is $(k-1)$-resilient;*
   (b) for any $y, z \in \mathbb{F}_2^t$ with $1 \leq wt(y \oplus z) \leq k$, we have $wt(\phi_1(y) \oplus \phi_1(z)) \geq 1$;
2. *(a) for any $a \in \mathbb{F}_2^s$ with $1 \leq wt(a) \leq k$, the function $a \cdot \phi_2(y)$ is balanced;*
   (a) for any $y, z \in \mathbb{F}_2^r$ with $wt(y \oplus z) \geq 1$, we have $wt(\phi_2(y) \oplus \phi_2(z)) \geq k$.

Then ϕ satisfies the condition of Theorem 8.

Proof. From $1(a)$ and $2(a)$, for any $a \in \mathbb{F}_2^s$ with $1 \leq wt(a) \leq k$, $a \cdot \phi(y) = a \cdot (\phi_2 \circ \phi_1)(y) = (a \cdot \phi_2) \circ \phi_1(y)$ is $(k-1)$-resilient thanks to Lemma 3. Hence, the first condition of Theorem 8 is satisfied. The $1(b)$ and $2(b)$ trivially imply the second condition of Theorem 8.

Next, we use binary systematic codes to construct ϕ_1 and ϕ_2. First we recall a connection between codes and orthogonal arrays. A *binary (n, k, λ)-orthogonal array* is a $\lambda 2^k \times n$ array such that for any k columns, every element of $\mathbb{F}_2^k$ appears in exactly λ rows. A binary orthogonal array is said to be simple if no two rows are identical. A *large set of binary (n, k, λ)-orthogonal arrays* is a set of $2^{n-k}/\lambda$ simple (n, k, λ)-orthogonal arrays such that every element of $\mathbb{F}_2^n$ appears in exactly one of the (n, k, λ)-orthogonal arrays in the set.

Lemma 4 [15]. *An $(n, 2^k, d)$-binary k-systematic code C with dual distance d' is also a binary $(n, d'-1, 2^{k-d'+1})$-orthogonal array.*

A relation between resilient functions and orthogonal arrays is given by the following lemma:

Lemma 5 [31]. *A k-resilient (t,r)-vectorial function is equivalent to a large set of binary $(t, k, 2^{t-r-k})$-orthogonal arrays.*

The next two results concern the functions ϕ_1 and ϕ_2.

Theorem 10. *Let C_1 be a $(t, 2^{t-r}, d_1)$-binary $(t-r)$-systematic code with $d_1 \geq k+1$ and dual distance $d_1' \geq k$. Let $\phi_1(x,y) = u(x) \oplus y$ be a (t,r)-vectorial function where $x \in \mathbb{F}_2^{t-r}, y \in \mathbb{F}_2^r$ and $u(x)$ is vector of parity-check bits of a codeword of C_1 whose information bits are represented by the vector x. Then ϕ_1 has the following properties:*

1. *ϕ_1 is $(k-1)$-resilient; and*
2. *for any $y, z \in \mathbb{F}_2^t$ with $1 \leq wt(y \oplus z) \leq k$, we have $wt(\phi_1(y) \oplus \phi_1(z)) \geq 1$.*

Proof. For any $z \in \mathbb{F}_2^r$, consider $\phi_1^{-1}(z) = \{(x,y) \mid \phi_1(x,y) = z, x \in \mathbb{F}^{t-r}$ and $y \in \mathbb{F}^r\}$. Since $\phi_1(x,y) = z \Leftrightarrow y = u(x) \oplus z$, we get $\phi_1^{-1}(z) = \{(x, u(x) \oplus z) \mid x \in \mathbb{F}_2^{t-r}\}$. Let $\mathbf{0} \in \mathbb{F}_2^{t-r}$ be the zero vector of length $t-r$. Then $\phi_1^{-1}(z) = (\mathbf{0}, z) \oplus C_1$ is a $(t, 2^{t-r}, d_1)$-binary $(t-r)$-systematic code with dual distance d_1'. By Lemma 4, $\phi_1^{-1}(z)$ is a binary $(t, d_1' - 1, 2^{t-r-d_1'+1})$-orthogonal array. It is also a binary $(t, k-1, 2^{t-r-k+1})$-orthogonal array since $k \leq d_1'$. By Lemma 5, ϕ_1 is $(k-1)$-resilient.

For any $y, z \in \mathbb{F}_2^t$ with $1 \leq wt(y \oplus z) \leq k$, suppose that $wt(\phi_1(y) \oplus \phi_1(z)) = 0$. It follows that $y, z \in \phi_1^{-1}(w)$ for some $w \in \mathbb{F}_2^r$. Since $\phi_1^{-1}(w) = (\mathbf{0}, w) \oplus C_1$ has minimum distance $d_1 \geq k + 1$, we obtain $wt(y \oplus z) \geq k + 1$, a contradiction. Consequently, $wt(\phi_1(y) \oplus \phi_1(z)) \geq 1$. $\square$

Theorem 11. *Let C_2 be an $(s, 2^r, d_2)$-binary r-systematic code with $d_2 \geq k$ and dual distance $d_2' \geq k + 1$. Let $\phi_2(y) = \alpha \oplus (y, v(y))$ where $y \in \mathbb{F}_2^r, \alpha \in \mathbb{F}_2^s$ and $v(y)$ is a vector of parity-check bits of a codeword of C_2 whose information bits are represented by the vector y. Then ϕ_2 has the following properties:*

1. *for any $a \in \mathbb{F}_2^s$ with $1 \leq wt(a) \leq k$, the function $a \cdot \phi_2(y)$ is balanced; and*
2. *for any $y, z \in \mathbb{F}_2^r$ with $wt(y \oplus z) \geq 1$, we have $wt(\phi_2(y) \oplus \phi_2(z)) \geq k$.*

Proof. For an arbitrary $\alpha \in \mathbb{F}_2^s$, ϕ_2 is injective (see the proof of Theorem 6). $\square$

We now present a construction of strictly cheating-immune schemes from Maiorana-McFarland functions.

Theorem 12. *Let $C_1 = \{(x, u(x)) \mid x \in \mathbb{F}_2^{t-r}\}$ be a $(t, 2^{t-r}, d_1)$-binary $(t-r)$-systematic code with dual distance d_1' and let $\phi_1(x,y) = u(x) \oplus y$ be a (t,r)-vectorial function where $x \in \mathbb{F}_2^{t-r}, y \in \mathbb{F}_2^r$. Suppose that $C_2 = \{(x, v(x)) \mid x \in \mathbb{F}_2^r\}$ is an $(s, 2^r, d_2)$-binary r-systematic code with dual distance d_2' and covering radius ρ. Let $k = \min\{d_1 - 1, d_1' d_2, d_2' - 1, \rho - 1\}$ and let $\phi_2(y) = \alpha \oplus (y, v(y))$ be a (r, s)-vectorial function where $y \in \mathbb{F}_2^r, \alpha \in \mathbb{F}_2^s$ such that $d(\alpha, C_2) \geq k + 1$. Define $\phi = \phi_2 \circ \phi_1$ and $f(x,y,z) = x \cdot \phi(y,z) \oplus g(y,z)$ where $x \in \mathbb{F}_2^s$ and g is an arbitrary t-variable Boolean function. Then the MM function f defines a k-CI $(s+t, s+t)$-SSS.*

Proof. Since $\phi(\mathbb{F}_2^t) = \alpha \oplus C_2$ then for any $(y, z) \in \mathbb{F}_2^t$ we must have $wt(\phi(y,z)) \geq k + 1$. By Theorem 4, f is k-resilient. The functions ϕ_1 and ϕ_2 satisfy the conditions of Theorems 10 and 11 respectively. Hence, they also satisfy the conditions of Theorem 9. Thus, ϕ satisfies the conditions of Theorem 8. Due to Theorem 2, the $(s+t, s+t)$-SSS defined by f is k-CI. $\square$

If C_1 and C_2 are linear codes with generator matrices $G_1 = [I_{t-r} \mid A]$ and G, respectively, then $\phi_1(y, z) = yA \oplus z$ and $\phi_2(y) = yG$. Thus, the defining function f can be written as $f(x, y, z) = x \cdot (\alpha \oplus (yA \oplus z)G) \oplus g(y, z)$.

Example 4 (**new schemes**)

a. Strictly 2-CI (13, 13)-SSS: Let C_1 be a $[6, 3, 3]$ binary self-dual code and C_2 be the $[7, 3, 4]$ binary Simplex code with $d_2^\perp = 3$ and covering radius $\rho = 3$. Consider a generator matrix $G_1 = [I_3 \mid A]$ of C_1 and a generator matrix G of C_2 where

$$A = \begin{bmatrix} 0 & 1 & 1 \\ 1 & 0 & 1 \\ 1 & 1 & 1 \end{bmatrix} \text{ and } G = \begin{bmatrix} 1 & 0 & 0 & 0 & 1 & 1 & 1 \\ 0 & 1 & 0 & 1 & 0 & 1 & 1 \\ 0 & 0 & 1 & 1 & 1 & 0 & 1 \end{bmatrix}.$$

Choose $\alpha = (0, 0, 1, 0, 1, 1, 0)$, then $d(\alpha, C_2) = 3$.

b. Strictly 3-CI (21, 21)-SSS: Let C_1 be a $[9, 4, 4]$ binary linear code with $d_1^\perp = 3$ and C_2 be a $[12, 5, 4]$ binary linear code with $d_2^\perp = 4$ and covering radius $\rho = 4$. We use generator matrices $G_1 = [I_4 \mid A]$ and G where

$$A = \begin{bmatrix} 0 & 1 & 1 & 1 & 1 \\ 1 & 1 & 0 & 0 & 1 \\ 1 & 1 & 0 & 1 & 0 \\ 1 & 1 & 1 & 0 & 0 \end{bmatrix} \text{ and } G = \begin{bmatrix} 1 & 0 & 0 & 0 & 0 & 0 & 1 & 1 & 0 & 1 & 1 & 1 \\ 0 & 1 & 0 & 0 & 0 & 1 & 0 & 1 & 1 & 0 & 1 & 1 \\ 0 & 0 & 1 & 0 & 0 & 1 & 1 & 0 & 1 & 1 & 0 & 1 \\ 0 & 0 & 0 & 1 & 0 & 0 & 0 & 0 & 1 & 1 & 1 & 0 \\ 0 & 0 & 0 & 0 & 1 & 1 & 1 & 1 & 0 & 0 & 0 & 0 \end{bmatrix}.$$

Choose $\alpha = (0, 0, 0, 0, 0, 0, 1, 1, 1, 0, 1, 0)$, then $d(\alpha, C_2) = 4$.

c. Strictly 3-CI (22, 22)-SSS: Let C_1 be a $[10, 5, 4]$ binary self-dual code and C_2 be a $[12, 5, 4]$ binary linear code with $d_2^\perp = 4$ and covering radius $\rho = 4$. We use generator matrices $G_1 = [I_5 \mid A]$ and G where

$$A = \begin{bmatrix} 0 & 1 & 1 & 1 & 1 \\ 1 & 0 & 1 & 1 & 1 \\ 1 & 1 & 0 & 1 & 0 \\ 1 & 1 & 1 & 0 & 0 \\ 1 & 1 & 0 & 0 & 1 \end{bmatrix} \text{ and } G = \begin{bmatrix} 1 & 0 & 0 & 0 & 0 & 0 & 1 & 1 & 0 & 1 & 1 & 1 \\ 0 & 1 & 0 & 0 & 0 & 1 & 0 & 1 & 1 & 0 & 1 & 1 \\ 0 & 0 & 1 & 0 & 0 & 1 & 1 & 0 & 1 & 1 & 0 & 1 \\ 0 & 0 & 0 & 1 & 0 & 0 & 0 & 0 & 1 & 1 & 1 & 0 \\ 0 & 0 & 0 & 0 & 1 & 1 & 1 & 1 & 0 & 0 & 0 & 0 \end{bmatrix}.$$

Choose $\alpha = (0, 0, 0, 0, 0, 0, 1, 1, 1, 0, 1, 0)$, then $d(\alpha, C_2) = 4$.

d. Strictly 3-CI (23, 23)-SSS: Let C_1 be an $[11, 6, 4]$ binary linear code with $d_1^\perp = 3$ and generator matrix $G_1 = [I_6 \mid A]$ and C_2 be a $[12, 5, 4]$ binary linear code with $d_2^\perp = 4$, covering radius $\rho = 4$ and generator matrix G where

$$A = \begin{bmatrix} 0 & 0 & 1 & 1 & 1 \\ 0 & 1 & 0 & 1 & 1 \\ 0 & 1 & 1 & 0 & 1 \\ 1 & 0 & 1 & 1 & 0 \\ 1 & 1 & 0 & 1 & 0 \\ 1 & 1 & 1 & 0 & 0 \end{bmatrix} \text{ and } G = \begin{bmatrix} 1 & 0 & 0 & 0 & 0 & 0 & 1 & 1 & 0 & 1 & 1 & 1 \\ 0 & 1 & 0 & 0 & 0 & 1 & 0 & 1 & 1 & 0 & 1 & 1 \\ 0 & 0 & 1 & 0 & 0 & 1 & 1 & 0 & 1 & 1 & 0 & 1 \\ 0 & 0 & 0 & 1 & 0 & 0 & 0 & 0 & 1 & 1 & 1 & 0 \\ 0 & 0 & 0 & 0 & 1 & 1 & 1 & 1 & 0 & 0 & 0 & 0 \end{bmatrix}.$$

Choose $\alpha = (0, 0, 0, 0, 0, 0, 1, 1, 1, 0, 1, 0)$, then $d(\alpha, C_2) = 4$.

Example 5 (**using nonlinear codes**)

a. Strictly 2-CI $(2^{m+1}, 2^{m+1})$-SSS: For even integer $m \geq 4$, let $C_1 = \mathcal{K}(m)$ and $\overline{C_2} = \mathcal{P}(m)$. Choose $\alpha \in \mathbb{F}_2^m$ such that $d(\alpha, C_2) = 3$.
b. Strictly 5-CI $(2^{m+1}, 2^{m+1})$-SSS: For even integer $m \geq 6$, let $C_1 = \mathcal{P}(m)$ and $\overline{C_2} = \mathcal{K}(m)$. Choose $\alpha \in \mathbb{F}_2^m$ such that $d(\alpha, C_2) \geq 6$.

7   Concluding Remarks

We showed that cheating-immune secret sharing schemes can be obtained from the class of Maiorana-MacFarland Boolean functions. We presented one new cheating-immune scheme, $k = 3$ for $n = 17$ and four new strictly cheating-immune schemes, $k = 2$ for $n = 13$ and $k = 3$ for $n = 21, 22, 23$. We also gave constructions of (strictly) cheating-immune secret sharing schemes from some well-known classes of binary nonlinear codes. There are still open cases in the construction of (n, n) cheating-immune secret sharing schemes. Another open problem is the construction of cheating-immune schemes for other access structures.

Acknowledgments. The authors would like to thank the reviewers for their comments and suggestions. The first author would like to thank the University of the Philippines Diliman for the financial support. The second author's work is supported by CIMPA and IMU.

References

1. Blakley, G.: Safeguarding cryptographic keys. In: Proceedings of AFIPS 1979 National Computer Conference, New York, vol. 48, pp. 313–317 (1979)
2. Bellare, M., Rogaway, P.: Robust computational secret sharing and a unified account of classical secret-sharing goals. In: ACM Conference on Computer and Communications Security, pp. 172–184. ACM (2007)
3. Bierbrauer, J., Gopalakrishnan, K., Stinson, D.R.: Bounds for resilient functions and orthogonal arrays. In: Desmedt, Y.G. (ed.) CRYPTO 1994. LNCS, vol. 839, pp. 247–256. Springer, Heidelberg (1994). https://doi.org/10.1007/3-540-48658-5_24
4. Braeken, A., Nikov, V., Nikova, S.: On cheating immune secret sharing. In: Proceedings of 25th Symposium on Information Theory in the Benelux, pp. 113–120 (2004)
5. Bosma, W., Cannon, J., Playoust, C.: The Magma algebra system. I. The user language. J. Symb. Comput. **24**, 235–265 (1997)
6. Cabello, S., Padró, C., Sáez, G.: Secret sharing schemes with detection of cheaters for general access structures. Des. Codes Cryptogr. **25**, 175–188 (2002)
7. Camion, P., Carlet, C., Charpin, P., Sendrier, N.: On Correlation-immune functions. In: Feigenbaum, J. (ed.) CRYPTO 1991. LNCS, vol. 576, pp. 86–100. Springer, Heidelberg (1992). https://doi.org/10.1007/3-540-46766-1_6

8. Carlet, C.: Boolean functions for cryptography and error-correcting codes. In: Boolean Models and Methods in Mathematics, Computer Science, and Engineering (Encyclopedia of Mathematics and its Applications), pp. 257–397. Cambridge University Press (2010)

9. Carlet, C.: On the propagation criterion of degree l and order k. In: Nyberg, K. (ed.) EUROCRYPT 1998. LNCS, vol. 1403, pp. 462–474. Springer, Heidelberg (1998). https://doi.org/10.1007/BFb0054146

10. Carlet, C.: Vectorial Boolean functions for cryptography. In: Boolean Models and Methods in Mathematics, Computer Science, and Engineering (Encyclopedia of Mathematics and its Applications), pp. 398–470. Cambridge University Press (2010)

11. Chor, B., Goldwasser, S., Micali, S., Awerbuch, B.: Verifiable secret sharing and achieving simultaneity in the presence of faults. In: FOCS 1985, pp. 383–395 (1985)

12. Cramer, R., Damgård, I., Maurer, U.: General secure multi-party computation from any linear secret-sharing scheme. In: Preneel, B. (ed.) EUROCRYPT 2000. LNCS, vol. 1807, pp. 316–334. Springer, Heidelberg (2000). https://doi.org/10.1007/3-540-45539-6_22

13. D'Arco, P., Kishimoto, W., Stinson, D.: Properties and constraints of cheating-immune secret sharing schemes. Discret. Appl. Math. **154**, 219–233 (2006)

14. dela Cruz, R., Wang, H.: Cheating-immune secret sharing schemes from codes and cumulative arrays. Cryptogr. Commun. **5**, 67–83 (2013)

15. Delsarte, P.: Four fundamental parameters of a code and their combinatorial significance. Inf. Control **23**, 407–438 (1973)

16. Guo-Zhen, X., Massey, J.: A spectral characterization of correlation-immune combining functions. IEEE Trans. Inf. Theory **34**(3), 569–571 (1988)

17. Helleseth, T., Klove, T., Mykkeltveit, J.: On the covering radius of binary codes. IEEE Trans. Inf. Theory **24**(5), 627–628 (1978)

18. Huffman, W.C., Pless, V.: Fundamentals of Error-Correcting Codes. Cambridge University Press, Cambridge (2003)

19. Kurosawa, K., Obana, S., Ogata, W.: t-Cheater identifiable (k, n) threshold secret sharing schemes. In: Coppersmith, D. (ed.) CRYPTO 1995. LNCS, vol. 963, pp. 410–423. Springer, Heidelberg (1995). https://doi.org/10.1007/3-540-44750-4_33

20. MacWilliams, F., Sloane, N.: The Theory of Error-Correcting Codes. North-Holland Publishing Company, Amsterdam (1977)

21. McEliece, R., Sarwate, D.: On sharing secrets and Reed-Solomon codes. Commun. ACM **24**, 583–584 (1981)

22. Ma, W.P., Lee, M.H.: New methods to construct cheating immune functions. In: Lim, J.-I., Lee, D.-H. (eds.) ICISC 2003. LNCS, vol. 2971, pp. 79–86. Springer, Heidelberg (2004). https://doi.org/10.1007/978-3-540-24691-6_7

23. Ma, W.P., Zhang, F.T.: New methods to construct cheating-immune multisecret sharing scheme. In: Feng, D., Lin, D., Yung, M. (eds.) CISC 2005. LNCS, vol. 3822, pp. 384–394. Springer, Heidelberg (2005). https://doi.org/10.1007/11599548_33

24. Martin, K.: Challenging the adversary model in secret sharing schemes. In: Coding and Cryptography II, Proceeidngs of the Royal Flemish Academy of Belgium for Science and the Arts, pp. 45–63 (2008)

25. Ogata, W., Kurosawa, K., Stinson, D.: Optimum secret sharing scheme secure against cheating. SIAM J. Discret. Math. **20**, 79–95 (2006)

26. Pieprzyk, J., Zhang, X.-M.: Cheating Prevention in Secret Sharing over $GF(p^t)$. In: Rangan, C.P., Ding, C. (eds.) INDOCRYPT 2001. LNCS, vol. 2247, pp. 79–90. Springer, Heidelberg (2001). https://doi.org/10.1007/3-540-45311-3_8

27. Pieprzyk, J., Zhang, X.-M.: Constructions of cheating-immune secret sharing. In: Kim, K. (ed.) ICISC 2001. LNCS, vol. 2288, pp. 226–243. Springer, Heidelberg (2002). https://doi.org/10.1007/3-540-45861-1_18

28. Pieprzyk, J., Zhang, X.M.: On cheating immune secret sharing. Discret. Math. Theor. Comput. Sci. **6**, 253–264 (2004)

29. Rabin, T., Ben-Or, M.: Verifiable secret sharing and multiparty protocols with honest majority. In: Proceedings of 21st ACM Symposium on Theory of Computing, pp. 73–85 (1989)

30. Shamir, A.: How to share a secret. Commun. ACM **22**, 612–613 (1979)

31. Stinson, D., Massey, J.: An infinite class of counterexamples to a conjecture concerning nonlinear resilient functions. J. Cryptol. **8**(3), 167–173 (1995)

32. Tassa, T.: Generalized oblivious transfer by secret sharing. Des. Codes Cryptogr. **58**(1), 11–21 (2011)

33. Tompa, M., Woll, H.: How to share a secret with cheaters. J. Cryptol. **1**, 133–138 (1988)

34. Wei, Y., Hu, Y.: New Construction of resilient functions with satisfying multiple cryptographic criteria. In: Proceedings of the 3rd International Conference on Information Security InfoSecu 2004, pp. 175–180. ACM (2004)

A New Privacy-Preserving Searching Model on Blockchain

Meiqi He, Gongxian Zeng, Jun Zhang, Linru Zhang, Yuechen Chen,
and SiuMing Yiu[✉]

The University of Hong Kong, Pok Fu Lam, Hong Kong
{mqhe,gxzeng,jzhang3,lrzhang,ycchen,smyiu}@cs.hku.hk

Abstract. It will be convenient for users if there is a market place that sells similar products provided by different suppliers. In physical world, this may not be easy, in particular, if the suppliers are from different regions or countries. On the other hand, this is more feasible in the virtual world. The Global Big Data Exchange in Guiyang, China, which provides a market place for traders to buy and sell data, is a typical example. However, these virtual market places are owned by third parties. The security/privacy is a concern in addition to the expensive service charges. In this work, we propose a new privacy-preserving searching model on blockchain which enables a decentralized and secure virtual search-and-match market place. The core technical contribution is a new searchable encryption scheme for blockchain. We adopt the similarity preserving hash and leverage smart contracts to protect the system from the forgery attack and double-rewarding attack. We formally prove the security and privacy of our protocol, and evaluate our scheme on the private net of Ethereum platform. Our experimental results show that our protocol can work efficiently.

Keywords: Security and privacy · Privacy-preserving searching · Blockchain

1 Introduction

As a customer, we all have the experience of looking for a product or a service. It would be convenient if there is a market place where we can find multiple suppliers of the same or similar product/service. In physical world, there are examples of this market place in different parts of the world (e.g. there is a building in Hong Kong selling wedding accessories, a tea street in Beijing, China selling different kinds of tea, and an Italian region in New York with many Italian restaurants etc.). However, it is not easy to have one that allows suppliers from different regions or countries to participate, except those organized by a third party (e.g. governments, trade organizations) which only open for a short period of time as the traders need to physically attend the event. On the other hand, it is more feasible to have such a market place in the virtual world. However, these virtual market places are usually owned by third parties. The security and

K. Lee (Ed.): ICISC 2018, LNCS 11396, pp. 248–266, 2019.
https://doi.org/10.1007/978-3-030-12146-4_16

privacy may be a major concern to the suppliers and customers in addition to expensive membership fees and service charges.

To further motivate our study, let us consider the following remarkable application: love matching service. There are multiple service providers. In order to use the actual service to find potential candidates for dating, in most cases, a customer needs to pay membership fees and other service charges. If we want to increase the chance of having a match, one may need to join multiple providers. The privacy of the customers (e.g. customers may not want others to know their criteria for choosing a partner) will totally rely on service providers. A more convenient and secure scenario is as follows. We have a market place allowing traders (service providers or product suppliers, we refer them as data owners in the rest of the paper) and customers to join. The market place is not owned by any third party. In our new model, customers can issue (or broadcast) a search query in this platform. Data owners can make a profit by finding a match for the customers. Instead of having the customer to join multiple providers or check the products from every supplier, now the customer only needs to issue one query, all data owners in the platform can help to locate the appropriate product/service for him. Ideally, the criteria in the search query is not revealed to the data owner to protect the privacy of the customers. The customers are only charged if matches are found. To realize this virtual market place, in this paper, we plan to explore a new "search-and-match" model on blockchains. The use of blockchains has the fundamental benefit that it does not require a trusted third party nor a centralized authority while it can provide a transparent and trusted platform for trading with low transaction fees (and service charges). Note that this platform can be extended for many applications such as data trading, job matching, finding rental apartment, and searching matched marrow.

The Abstract Problem: We model the problem as a privacy-preserving keyword matching problem on blockchains. The data of a data owner is represented as documents (e.g. product descriptions). Each document is characterized by a set of keywords. The user (or customer) query is in the form of a keyword. A data owner can earn a reward from a user for each document that matches the keyword in the query. Since all transactions occur in blockchain, which is transparent, the platform needs to satisfy the following basic requirements to guarantee the privacy of transactions.

1. The query is hidden from all data owners and other users. But then, the same query can be used by all data owners to search their documents.
2. The documents returned by a data owner are revealed only to the user who issued the query.
3. Miners of the blockchain (see Sect. 2 for a description of miners) are able to verify if the returned document indeed contains the keyword of the query, but they are not able to reveal the content of the document.

Besides the basic requirements, we also consider the following risks/attacks.

1. Double-rewarding attack: Greedy data owner may try to generate dupli-cate/similar documents to get double rewards once he knows which document can match the query.
2. Forgery attack: Only the data owner of the document that matches a query can reply to the query and get the reward.
3. Fair exchange: A customer cannot skip paying the reward as promised after receiving the document.

Highlights of Our Solutions and Contributions: At a first glance, the problem is similar to the traditional searchable encryption (SE) problem [6, 7,10,13,20], which allows a data owner to outsource a dataset onto a server allowing other customers to search it with a token while preserving the privacy of both the data and the query. However, there are major differences between our problem and the traditional SE problem. In traditional SE, the token for the customer is generated by the data owner. In our case, the customer will generate the token without any help from the data owner as the token needs to be used by multiple data owners. For traditional SE, the search is done by a third party, while in our case the search is done by the owner while the results need to be checked by a miner. Documents provided by different data owners are encrypted using different keys, i.e., the same encrypted query is required to search multiple documents encrypted by different keys.

To solve this problem, we propose a new multi-key searchable encryption scheme in the public key setting that enables a user to provide one search token, but allows multiple data owners to search documents encrypted with different keys. To achieve this, we are motivated by the multi-key searchable encryption scheme proposed in [18] to let each data owner transform the token to match its own documents. Each document is parsed into a set of distinct keywords and a *proof* for each keyword will be produced that can be verified by a miner. Note that [18] cannot be used directly in our case as [11,21] have shown that [18] has a leakage problem in keyword access pattern (for details, see Sect. 4). We thus propose new encryption and matching methods in our scheme. In partic-ular, to eliminate this leakage problem, we enhance our encryption method so that the data owner can update the ciphertext of the keyword every time the corresponding document has been replied to a query.

To make sure that others cannot reply to a query if the document does not belong to him, we embed the owner's public key into the encryption of the document. For the problems of forgery and double-rewarding attacks, we propose the followings. We adopt smart contracts to regulate the behaviors of owner and user. We use hash value to record the existence of documents while creating. Miners of the blockchain will only consider the document created before query. In brief, contracts, hash values and search tokens are all treated as transactions to be recorded into blockchain as undeniable proof. To avoid dishonest data owners conducting double-rewarding attack, we adopt TLSH [15], which is a similarity preserving hash used in digital forensics. In this way, data owners are required to provide the similarity hash values of their documents when replying a query and miners can detect duplication before mining it into blockchain[1]

In summary, the contributions of our paper are listed as follows.

- We propose a new privacy-preserving searching model for searching over encrypted data. The model can be used on blockchains. To realize this model, we design a novel multi-key searching encryption scheme.
- We identify several possible attacks in the new model and provide solutions to prevent these attacks, e.g. we carefully design smart contracts to handle the forgery attack and adopt similarity preserving hash to deal with double-rewarding problem.
- We formally prove the security of our scheme and evaluate our protocol in the private net of the existing Ethereum platform. Experimental results show that our protocol can run efficiently.

2 Background

2.1 Blockchain and Smart Contract

Please refer to [14] for an overview of blockchain. Here, we want to highlight an important role in blockchain, which is *miner*. They are responsible for verifying and adding new transactions, creating a new block by solving a puzzle (referred as the Proof of Work (PoW)) and receive some benefit in return. Once a block is added to the chain, it is extremely difficult for anyone to modify it, the correctness of the network can be guaranteed. In our case, all queries and replies are treated as transactions. Miners will be responsible to check if a reply from a data owner actually contains the keyword of the query before the reply can be added into the chain.

Smart contract is supported in some blockchain based platforms (*e.g.* Ethereum [4]). It is widely used in many complicated functions [8,9,12,17]. Smart contracts are computer programs that act as agreements where the terms of the agreement can be preprogrammed with the ability to be executed and enforced. With the decentralized setting, smart contracts are public to all users in the network. Once deployed, it is difficult to modify the contract, even by the creator, due to the Proof of Work. We leverage smart contracts in our construction to make sure that all parties would follow the defined actions.

2.2 Basics of Multi-key Searchable Encryption (MKSE)

The multi-key keyword matching scheme used in our system consists of the following steps.

---

[1] Note that in case two honest data owners have the same document or two very similar documents, it depends on the probability, the one whose document is confirmed by the miner first will get the reward.

- *Setup*: Generate the system parameters.
- *Encryption*: For each file, the data owner creates an encrypted keyword index so that it can be used in matching a query.
- *Retoken*: For each file, data owner computes a value for the user based on his public key if he is authorized to search the data. This protocol works as an access control so that token from the authorized user can be transformed to match the encrypted index.
- *Documents encryption*: Data owner encrypts the documents.
- *Search token*: User generates the query token for a keyword.
- *Match*: Data owner searches the query over the encrypted index.
- *Documents decryption*: User can decrypt the result documents if he pays for it.

2.3 Bilinear Map

Let G_1, G_2, G_T be groups of prime order p and $e : G_1 \times G_2 \rightarrow G_T$ be a bilinear map. g_1, g_2 are the generators of G_1, G_2. The map satisfies the following properties:

1. Computable: given $x \in G_1, y \in G_2$ there is a polynomial time algorithms to compute $e(x, y) \in G_T$.
2. Bilinear: for any integers $a, b \in [1, p]$, we have $e(g_1^a, g_2^b) = e(g_1, g_2)^{ab}$.
3. Non-degenerate: if g_1, g_2 are the generators of G_1, G_2 then $e(g_1, g_2)$ is a generator of G_T.

2.4 Similarity Preserving Hash

Hash functions are well-known and commonly used for proving integrity and file identification. Traditional hashes (MD5, SHA-1, SHA-2) are used to check if a file has been modified or tampered (even one bit change). However, for some other applications such as identifying new versions of documents and software, locating variants of malware families, finding similar infringing copies, deduplication on storage system, we cannot use traditional hashes. Similarity-preserving hash [1–3,15] was developed to handle these applications. In our case, we will adopt this technique to help miners to check duplicate/similar documents and only accept the first one to avoid the user double-paying for the same similar document.

Similarity-preserving hash aims at detecting similarity between objects by creating a digest in a way that similar objects will produce similar digests. When comparing two digests, a score related to the amount of content shared between them is given. There are several criteria to reflect the effectiveness and efficiency of different hash functions: (1) Efficiency: efficiency includes the comparison efficiency and space efficiency. (2) Sensitivity and robustness: sensitivity refers to the granularity at which an algorithm can detect similarity; robustness is a metric of how an algorithm can be in the midst of noise and plain transformations such as insertion/deletion.

The final decision on selecting an algorithm depends on the applications. In our case, we use the TLSH [15] algorithm for approximate matching as we pay more attention to the sensitiveness and robustness. Sdhash [3] and mvhash-B [1] suffer from active manipulation or anti-blacklisting. According to the authors [15], TLSH is more robust. It is reported that file can be deliberately modified by an adversary using randomization so that Ssdeep and Sdhash may fail, but TLSH still has a high chance to identify similar files. Additional experiments [16] showed that TLSH can detect strings which have been manipulated with adversarial intentions.

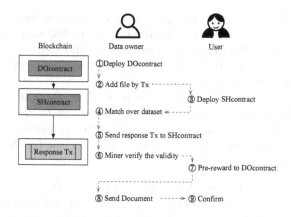

Fig. 1. System model

3 System Overview

In Fig. 1, we outline the architecture of our model. We summarize the challenges within these steps and present our ideas of how to resolve them in the following sections.

3.1 Method to Ensure Fair Exchange

To ensure fair data exchange, we regulate the allocation of deposits and rewards by issuing DOcontract. It works as a fair intermediary to ensure that the processing of exchanging money with data between an owner and a user is carried out properly like a safe remote purchase between a seller and a buyer. To publish the search token and collect the search results, we design the SHcontract. The followings are the two smart contracts we proposed:

(1) DOcontract: Smart contract to exchange data with money
To guarantee that no one can cheat for money during the trading, we have the following consideration. First, when data owners initialize the contract, they have to transfer some deposit to the contract. Before each response, the data owner

are required to pledge an amount of money from the funding of the contract. Therefore, we define the contract state to control the procedure of contract initialization. Specifically, when the DOcontract is sent to the blockchain, the state is set to *Created* and will change after each operation. As long as the data owner transfers the deposit, the contract state will be set to *Active*. The other functions can be executed if and only if the contract is active. Also, similar idea can be used to make sure they have pledge when response and the user has frozen the rewards before sending file. We restrict the access control of each function to guarantee that no one can cheat for money. The contract are designed to have the following functions.

- **Funding:** The owner must pledge a certain amount of money as an initial fund so that it can be used as deposit when answering search queries.
- **Deposit:** After sending response transactions, data owner pledges a certain amount of fund so that if they fail to provide valid document after response, the user who searches can get the deposit.
- **Reward:** This function allows the user to send the rewards to the contract. The money is *locked* until user receives the document from data owner.
- **Terminate:** As long as the owner would like to terminate the contract and there is no search work in process, he can activate the function to get back the funding in the contract.

(2) SHcontract: Smart contract to maintain search results
This smart contract is deployed by the user to collect and maintain the search results. SHcontract publishes the information about the query including search token, rewards for each response and maximum response number.

Data owners that would like to answer the query can interact with the contract by sending a response transaction. The function **Response** embedded in this contract requires the validation proof as input involving:

- Proof information indicating the correctness of matching;
- Similarity hash of the document.
- Block number and transaction index that contain the hash record.

Sending transactions to call a function in smart contract will embed the input arguments in the transaction. With this property, miners can easily extract and verify the proof provided by the data owner and determine whether to include it into block. And the SHcontract can receive and accept the response only when the miners accept and include the transaction.

3.2   Method to Resist Double-Rewarding

As introduced before, in our model, malicious data owner can produce a set of identical or similar documents to gain the rewards multiple times. We refer this as double-rewarding attack. To deal with the problem above, we propose to use duplication detection techniques in digital forensics called similarity-preserving hash to identify identical and similar documents to decrease the probability of having double-rewarding attack. In our design, data owners are required to include the similarity hash values in the response transaction. The miners who want to include this transaction have the responsibility to do the verification and then the SHcontract can receive and record the response after mining is completed. During the verification, miners compare the distance between this response with all the previous accepted responses. To generate and compare the hash values, we use the TLSH algorithm [15]. TLSH uses the following 4 steps to construct the digest.

- Process the input using a sliding window to populate an array of bucket counts;
- Calculate the quartile points;
- Construct the digest header values based on the quartile points, the length of the file and a checksum;
- Construct the digest body by generating a sequence of bit pairs, which depend on each bucket's value in relation to the quartile points.

4   Construction with SE

In this section, we show how to search on blockchain with a specific keyword search scheme. We first highlight our design technique for this part.

Hiding Keyword Access Pattern: Recently, the problem of leakage-abuse attack over searchable encryption has been widely investigated [5,11,21]. In [5], Cash *et al.* presented a characterization of the leakage profiles of in-the-wild searchable encryption products and SE schemes in the literature. In [11], Grubbs *et al.* presented many ways of attacking a software framework named Mylar [19] whose building blocks include a SE scheme protocol [18]. However, most of the attacks only leverage implementation issues of Mylar and are not related to the cryptographic protocols. In [21], the authors also mentioned that the leakage abuse attack in [11] could not be extended to cover any application using the underlying SE scheme and gave a deeper analysis to the MUSE (multi-user SE, also known as MKSE) scheme, in particular for [18]. Their work showed that in the case of MUSE, there is a new leakage problem named keyword access pattern and can result in serious attack if some users can collude with the server. In their following work [22], they proposed a secure and scalable MUSE scheme without this leakage. However, their scheme is based on two non-colluding parties which cannot be used in our setting. In our model, given that blockchain is a public network which is transparent to any users, without a careful design, publishing index ciphertext is likely to reveal such leakage so that attacks using keyword access pattern leakage are easier to be conducted. It is essential to get rid of the leakage of keyword access pattern. We thus require the data owner to update the ciphertext of the keyword after response, with low computational overhead, so that there will not be keyword access pattern leakage.

Unforgeable Identity: Another innovative elements in our keyword matching scheme is that, each data owner is linked with a secrete parameter x and publish $y = g_2^x$ as unforgeable identity. And x is involved in the index encryption protocol, y is included in the matching scheme. In this way, even if other user obtains your proof information, they cannot make use of it to gain profit. Let us consider the following case. Data owner A has a pre-knowledge about B's database that their data set have similar background. Then A can stay passive and observe B's responses. Upon B's releasing a proof for response, A will copy it and answer the query as well. Without our unforgeable identity x, y, the *proof* definitely can match the query, and A has high possibility to gain rewards without search.

4.1 System Setup

(1) At the beginning of setup, $\mathsf{Param}(\lambda)$ is called to input the security parameter and output system parameters.

$\mathsf{Param}(\lambda)$: Input the security parameter λ. Let G_1, G_2, G_T be groups of prime order p and $e : G_1 \times G_2 \to G_T$ be a bilinear map. g_1, g_2, g_T are the generators of G_1, G_2 and G_T respectively. Let $H_1 : \{0,1\}^* \to G_1$ be a collision resistant hash function. Let $sp = (p, G_1, G_2, G_T, g_1, g_2, H_1)$ be the system parameters.

---

(2) User who wants to publish search queries calls $\mathsf{SearchKey}(sp)$ to generate a pair of public key and secrete key pk, sk. Keep sk as a secret key and publish pk.

$\mathsf{SearchKey}(sp)$: User chooses a random number $\alpha \in \mathbb{Z}_p^*$ and computes $pk = g_2^\alpha$. Keep $sk = \alpha^{-1} \bmod p$ as a secret key and publish pk.

---

(3) Data owner deploys a DOcontract and generates a secret parameter $x \in \mathbb{Z}_p^*$ and publish $y = g_2^x$ as unforgeable identity which we mentioned before.

4.2 Add File

(1) Data owner parses document d into distinct keywords: $\{w_1, \cdots, w_m\}$. For each keyword, Data owner runs $\mathsf{Enc}(d, w_i)$ and outputs (k_{d,w_i}, c_{d,w_i}).

$\mathsf{Enc}(d, w_i)$:
(1) Data owner chooses a new key for his new files: $k_d \in \mathbb{Z}_p^*$.
(2) Data owner chooses a random number $t \in \mathbb{Z}_p^*$, and computes $r = g_2^t$.
(3) Data owner chooses a new key for w_i: k_{d,w_i}, and computes $s = (k_d k_{d,w_i} - xh)t^{-1} \bmod p$, where $h \in \mathbb{Z}_p^*$ is a random number.
(4) $c_{d,w_i} = (H_1(w_i)^h, H_1(w_i)^s, r)$.
(5) Output (k_{d,w_i}, c_{d,w_i}).

---

(2) Data owner chooses new keys for his new files and if the file is authorized to user u with public key pk, Data owner will compute $\mathsf{Delta}(d, pk, k_d){=}\Delta_{d,u}$.

$\mathsf{Delta}(d, pk, k_d)$:

1) Data owner computes $\Delta_{d,u} = (pk)^{k_d}$.
2) Output $\Delta_{d,u}$.

---

(3) Finally, DO encrypts the document d using $\mathsf{SKE}(d)$ and outputs the ciphertext.

$\mathsf{SKE}(d, k_d)$:

(1) Encrypt d by computing $C_d = d \oplus f(ID(d), k_d)$, where f is a pseudorandom function, $ID(d)$ is the identity of the document.
(2) Output C_d.

---

(4) Data owner publishes a transaction including the new hash value $hash\_new =$ SHA-1$(hash\_old||C_d)$. $hash\_old$ denotes the hash value of the data set before uploading this document.

4.3   Keyword Search

(1) As long as the user u wants to search for a keyword w, he computes the search token $tk_w = H_1(w)^{sk}$. and setup a SHcontract to collect the response.
(2) Data owner follows $\mathsf{Match}(tk_w, k_{d,w_i}, c_{d,w_i})$ to checks the equality and gives the output.

$\mathsf{Match}(tk_w, k_{d,w_i}, c_{d,w_i})$:

(1) Data owner parses c_{d,w_i} into three parts as (u_1, u_2, u_3).
(2) Check if $e(u_1, y)e(u_2, u_3)$ equals to $e(tk_w, \Delta_{d,u}^{k_{d,w_i}})$.
(3) If so, output 'match', if not, output 'no'.

---

(4) Data owner sends a response transaction to the SHcontract with similarity hash of this file and $proof = (\Delta_{d,u}^{k_{d,w_i}}, c_{d,w_i}, BlockNum, Index)$, where $BlockNum$ and $Index$ are the block number and transaction index that contains the hash value when uploading. After posting the $proof$, Data owner updates c_{d,w_i} by running $\mathsf{Enc}(d, w_i)$ with new parameters.
(5) Finally, Data owner is required to give deposit in the DOcontract.

4.4   Response Verification

After the response transaction is sent into the mining pool, the miners can get the $proof$ and do verifications to determine whether to include it into his blocks. (1) The response verification mainly consists of 3 steps:

- Check if $e(tk_w, \Delta_{d,u}^{k_{d,w_i}}) = e(H_1(w_i)^h, y)e(H_1(w_i)^s, r)$.
- Analyze the hash value in transaction (using $BlockNum$ and $Index$ to locate) to test if the new hash value is equal to $\mathsf{SHA}\text{-}1(hash\_old\|C_d)$.
- Check the similarity with previous responses.

(2) If the transaction passes the verification, miners will include it into his block so that the response is published on the blockchain and SHcontract will receive the transaction.

4.5   File Retrieval and Decryption

(1) After receiving the response, the user will call **Reward()** function in DOcontract to pledge the rewards, then the user computes $C_d' = C_d \oplus f(ID(d), sk)$ and sends it to the data owner to decrypt.
(2) Data owner computes $C_d'' = C_d' \oplus f(ID(d), k_d)$ and sends it back to the user. User can obtain the plaintext of d by $C_d'' \oplus f(ID(d), sk) = d$.
(3) As long as the deal is completed successfully, Data owner will get back the deposit and rewards, otherwise, the user will resume the rewards and get the deposit.

5   Security Analysis

We define the privacy of our keyword matching scheme using the simulation paradigm in searchable encryption, following [6,7], that is based on the notion of *leakage*. We first clarify the difference between keyword matching in our setting and previous SE setting. In previous SE scheme, data owners outsource the index and encrypted database onto a cloud server where the server is believed to have the most knowledge over the protocol and encrypted dataset. However, in our setting, there is no centralized server while all queries and responses are published on the blockchain which is a globally visible ledger. The users in this blockchain are supposed to be the ones who can obtain all the leakage and be curious to the dataset of the data owners and queries of the other users. Also, there is no "beginning leakage", including the length of the index and the size of the database compared to the previous SE scheme. All the leakage are query-revealed. In order to characterize the leakage in our scheme, we give the following definitions:

Definition 1. *(Leakage   function $\mathcal{L}$)   Given   a   search   input w,*
$\mathcal{L}=\{\mathsf{ap}(w, D), \mathsf{sp}(w), \, rl(w), \, tk_u(w), \, proof = (\Delta_{d,u}^{k_{d,w_i}}, c_{d,w_i}, \, BlockNum, \, Index),$

$C_d\}$, where $\mathsf{ap}(w, D)$ denotes the access pattern, $\mathsf{sp}(w)$ denotes the search pattern and $\mathsf{rl}(w)$ represents the number of documents matching each query. $tk_u(w)$ is the search token post by user u for keyword w, and proof is leaked in each response, where d is the matched document.

Theorem 1. *Our keyword match scheme has leakage profile $\mathcal{L}$ against the adversary if there exits a polynomial-time simulator $\mathcal{S}$, that for all polynomial-time adversary $\mathcal{A}$, the output of real execution and simulated execution are computationally indistinguishable.*

Due to the page limit, the formal proof is included in the Appendix A.

6 Experiments

While the construction of the previous sections gives an overview of our model and approach, we have yet to describe how our techniques integrate with existing blockchain platform. In this section, we show the evaluation results of our scheme on the private net of Ethereum platform, which is known as an open-source blockchain-based distributed computing platform and supports smart contract. We set up the go-ethereum from https://github.com/ethereum/go-ethereum, which is the official golang implementation of the Ethereum protocol. We built private Ethereum chain on a single server node with Intel(R) Core(TM) i5-3570 CPU, using single core processor.

6.1 Ethereum Platform

The basic functions are provided by the Ethereum, such as creating an account and sending transactions. Besides, it provides Ethereum Virtual Machine (EVM), which is part of the block verification protocol and can run the functions defined in the contracts. From the view point of developers, a contract has a specific address on the Ethereum blockchain, where we can store the contract code and data. Thus, we only need to send the Ethereum-specific binary format code onto the chain. When doing practically Turing complete computation, we can pass messages (contained in a transaction) to the contracts, which is exactly what we need. To reach consensus, all nodes in Ethereum would go through the transactions listed in the blocks and runs codes in the EVM. The Ethereum protocol charges a fee per computational step that is executed in a contract or transaction to prevent deliberate attacks and abuse on the Ethereum network. Every transaction is required to include a gas limit and a fee that it is willing to pay per gas. If the total amount of gas used by the computational steps spawned by the transaction, including the original message and any sub-messages that may be triggered, is less than or equal to the gas limit, then the transaction is processed.

6.2   Simulation Design

To measure the performance of our scheme, we deployed the DOcontract and SHcontract and implemented the verification protocol including the process of similarity comparison, hash checking and checking if the response matches the query. Also, since the new response transactions require verification before being mined into a block, we studied the impact of such kind of transactions on the normal Ethereum network.

For the two smart contracts, the functions overview has been introduced in Sect. 3.1. In this section, we address the specific challenges that we come up when we deploy them in Ethereum. Due to page limitation, we will not show the detailed pseudocode of DOcontract and SHcontract here.

As a matter of fact, in live Ethereum, the number of transactions a miner decide to include into his block depends on many factors. For instance, each miner will set a minimum gas unit before mining. Only transactions with gas above this level will be accepted. To simplify the experiment and simulate our protocol in the same standards, we assume that the miners all adopt greedy algorithm and the transaction with higher transaction fee has priority. In other words, they would mine a block that contains as many transactions as possible and the transaction that offers higher gas fee would be firstly to be included in a block.

In our experiment, the contracts are developed in Solidity language and the verification is implemented with Go language. We use the "crypto" package of golang and package "bn256" to implement the bilinear group.

6.3   Metrics

We test our protocols mainly according to the following criteria:

- *Response transaction verification time* is the average running time, over 1000 tests, required to verify a response transaction. Since the verification step is set before the miners mine the transaction into blocks, if the verification is not efficient, it will slow the miner down in comparison with mining normal transactions. We will also include the time for every sub-step. (You can refer the verification algorithm to Sect. 4.4 and one more step (*i.e.* check deposit) introduced in the above simulation design.)
- *Transaction waiting time* in blockchain denotes the interval between a transaction's creation and its inclusion in a block. It is an important measurement when designing a blockchain application. We want to know that how much influence our response transactions have on the average waiting time of all types of transactions and on the average waiting time of current existing normal transactions. We simulate the relationship between the waiting time and the instantaneous transactions number.

For different blockchain platforms, they have different block interval (i.e. the time to generate a new block), e.g. 10 min for Bitcoin and 12 s for Ethereum. Thus, we count the verification time by block numbers so that it is easier for readers to assess our scheme across different platforms.

6.4 Results

From Table 1, we can see that it takes about 50 ms to verify a response transaction and most of time is spent on performing bilinear map computation. From the websites that have the statistical performance of current live Ethereum (e.g. ETH Gas Station[2], Etherscan[3], etherchain[4]), we can know that each block usually contains about 70 transactions. Thus, it would be about 0.35 s to verify these transactions even all of them are our response transactions. We have known that, the block interval for Ethereum is around 12 s and 10 min for Bitcoin, thus, the verification time is negligible compared to the mining time.

Table 1. Verification time

Verification steps	Sub-steps	Time
Read Tx	Get Tx from pending pool	361.206 μs
	Analyze data from Tx	11.987 μs
Verification details	Check Match	52.234123 ms
	Check similarity (100 times)	6.4 μs
	Get Tx with new hash record	500.109 μs
	Extract new hash from Tx	58.483 μs
	Get Tx with old hash record	376.23 μs
	Extract old hash from Tx	300.333 μs
	Check file hash	350.162 μs
Check deposit	Check state in DOcontract	547.93 μs

Total: 54.746963 ms

Table 2. Transaction input size

Transaction	Size (Bytes)	Estimated gas usage
Tx to deploy DOcontract	4825	735044
Tx to deploy SHcontract	1543	268391
Response Tx	970	120190

For the reason that the contracts implement lots of functions in our scheme and require many parameters to ensure the protocol works fairly and correctly from setup to completion, it can be inferred that the response transactions in our

---

[2] https://ethgasstation.info/index.php.
[3] https://etherscan.io/.
[4] https://etherchain.org/.

Table 3. Running time of keyword search scheme

Algorithm	Token	Delta	Enc	ReToken	Match
Time (ms)	2.2528	5.0599	9.8592	23.0500	31.0791

scheme are larger than the normal transfer transactions. To make it clearer, we further investigate the contracts size, transaction size and gas usage in Ethereum. The results are shown in Table 2.

The transactions to deploy smart contracts are one-time consuming so we pay more attention to the response transactions, which need 4825 Bytes for each one in Table 2. At the time of writing this paper, the average block size in Ethereum is 16192 Bytes so the response transactions would occupy much space in a block. Also, a normal transaction uses about 43000 units gas on average but our response transaction costs about 120190 units gas. If comparing to an ether transfer transaction, which costs only 21000 units gas, the gas usage of our response transaction is about 6 times. However, from etherchain (See footnote 4), we know that, the current gas limit for each block is about 6800000 units, thus it has enough resource for our application while one block would contain less transactions and it takes longer waiting time when dealing with our response transactions. We do the following experiments to evaluate how much influence our response transactions have on the average waiting time of all types of transactions and on the average waiting time of current existing normal transactions.

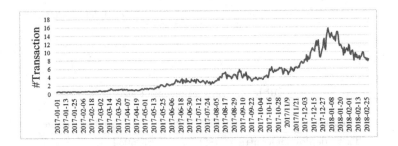

Fig. 2. Number of transactions per second

The evaluation on waiting time is displayed in Fig. 3(a) and (b). From Fig. 3(a), we can know that the average waiting time of all transactions increases with the increase of the percentage of response transactions among all transactions. Even there are 500 response transactions, the waiting time is no more than 6 blocks from Fig. 3(a). And from Fig. 3(b), we can see that the lines go up first and drop later, of which the reason is that we adopt greedy algorithm for miners and there are less normal transactions when the percentage is higher. In fact, in live Ethereum, we can estimate that the peak value of transactions generated

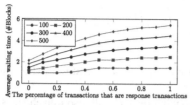

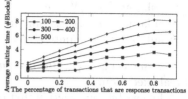

(a) Average waiting time of all types of transactions

(b) Average waiting time of normal transactions

Fig. 3. Waiting time

within a block interval is about 100 to 200. From etherchain (See footnote 4), the number of transactions generated per second since 01/01/2017 is plotted in Fig. 2. It is clearly that, even the quantity increases rapidly, the average level recently is about 7.4 transaction per second (88.8 per block interval). Under this background, the two figures present quite good results, that is, the average waiting time is about 2 to 3 blocks. In other words, our response transactions can integrate well with the existing blockchain-base network.

Last but not least, we evaluate the running time of algorithms in the keyword search scheme using go language. The results are summarized in Table 3. The time is averaged over 1000 tests with randomly generated keywords. We can conclude that the scheme has a modest overhead.

7   Conclusions

In this work, we propose and formulate a new model for privacy-preserving searching on blockchains. We present a keyword search scheme for searching over text data. We focus on single keyword search. It is desirable to extend our scheme to handle multiple keyword search with boolean operators, approximate matching for the keywords, and non-text files.

Acknowledgement. This project is partially supported by a RGC Project (CityU C1008-16G) funded by the HK Government.

A   Proof of Theorem 1

Proof. The simulator $\mathcal{S}$ is given leakage $\mathcal{L}$ to simulate the view of the adversary via imitating the real protocol. We generate the proof string $proof' = ((\Delta_{d,u}^{k_{d,w_i}})', (c_{d,w_i})', BlockNum, Index)$ and $(C_d)'$ as follows:

1. Simulating $(\Delta_{d,u}^{k_{d,w_i}})'$: Given $\mathcal{L}$, $\mathcal{A}$ choose a composite random key k_s that $k_s = k_{s_1} \cdot k_{s_2}$, and compute $(\Delta_{d,u}^{k_{d,w_i}})' = (\Delta_{d,u}^{k_{d,w_i}})^{k_s}$.

2. Simulating $(c_{d,w_i})'$: For each keyword query w_i, $(c_{d,w_i})'$ is consist of three parts.

- Simulating $(H(w_i)^h)'$: compute $(H(w_i)^h)' = (H(w_i)^h)^{k_s}$.
- Simulating $(H(w_i)^s)'$: compute $(H(w_i)^s)' = (H(w_i)^s)^{k_{s_1}}$.
- Simulating r': compute $r' = r^{k_{s_2}}$.

3. Simulating $(C_d)', BlockNum', Index'$: Use the same value in the response of real execution.

It follows by construction that response with $proof'$ will also match the search token tk_w if $proof$ does because:

$$e(tk_w, (\Delta_{d,u}^{k_{d,w_i}})') \stackrel{?}{=} e((H(w_i)^h)', y)e((H(w_i)^s)', r')$$

$$\text{Left} = e(H(w_i), (\Delta_{d,u}^{k_{d,w_i}})^{k_s})$$

$$= e(H(w_i), g_2)^{k_d k_{d,w_i} k_s}$$

$$\text{Right} = e((H(w_i)^h)', y)e((H(w_i)^s)', r')$$

$$= e(H(w_i)^{hk_s}, g_2^x)e(H(w_i)^{sk_{s_1}}, g_2^{tk_{s_2}})$$

$$= e(H(w_i), g_2)^{xhk_s + stk_{s_1} k_{s_2}}$$

$$= e(H(w_i), g_2)^{(st + xh)k_s}$$

Therefore, Left = Right if $proof$ matches tk_w.

We now claim that no polynomial-size distinguisher can distinguish between the distributions $proof'$ and $proof$. Note that in the simulation above, $\Delta_{d,u}^{k_{d,w_i}}$ and $(\Delta_{d,u}^{k_{d,w_i}})'$ as well as all the components in c_{d,w_i} and $(c_{d,w_i})'$ can be regarded as the problem to distinguish between g^{ab} and g^{abc}. For example, c_1 in c_{d,w_i} equals to $H(w_i)^h = g_1^{ah}$ and c_1' in $(c_{d,w_i})'$ equals to $H(w_i)^{hk_s} = g_1^{ahk_s}$. Then, we make the following Lemma 1.

Lemma 1. *If g^{ab} and g^{abc} are indistinguishable from random numbers in the same groups respectively, then g^{ab} and g^{abc} are indistinguished from each other.*

Since g^{ab} and g^{abc} are of the same structure, we only need to prove the distinguishability of anyone of them. For contradiction, we assume that there is a PPT adversary $\mathcal{D}$ that distinguishes g^{ab} and $R \stackrel{\$}{\leftarrow} G$, then we show how to construct a PPT reduction $\mathcal{B}$ that can use Exp to break the DDH assumption. breaks DDH.
Experiment

- Given a (multiplicative) cyclic group G of order p, and with generator g.
- $\mathcal{B}$ receives (g^a, g^b, g^{ab}) and (g^a, g^b, R), $R \stackrel{\$}{\leftarrow} G$. $\mathcal{B}$ passes g^{ab} and R to $\mathcal{D}$.
- $\mathcal{B}$ guesses the same as $\mathcal{D}$.

Finally, $(C_d), BlockNum, Index$ and $(C_d)', BlockNum', Index'$ are identical, therefore, no polynomial-size distinguisher can distinguish between the outputs of real execution and simulated execution.

References

1. Breitinger, F., Astebøl, K.P., Baier, H., Busch, C.: mvHash-B-A new approach for similarity preserving hashing. In: 2013 Seventh International Conference on IT Security Incident Management and it Forensics (IMF), pp. 33–44. IEEE (2013)
2. Breitinger, F., Baier, H.: Similarity preserving hashing: eligible properties and a new algorithm mrsh-v2. In: Rogers, M., Seigfried-Spellar, K.C. (eds.) ICDF2C 2012. LNICST, vol. 114, pp. 167–182. Springer, Heidelberg (2013). https://doi.org/10.1007/978-3-642-39891-9_11
3. Breitinger, F., Baier, H., Beckingham, J.: Security and implementation analysis of the similarity digest sdhash. In: First International Baltic Conference on Network Security & Forensics (nesefo) (2012)
4. Buterin, V.: Ethereum: a next-generation smart contract and decentralized application platform (2014). https://github.com/ethereum/wiki/wiki/%5BEnglish%5D-White-Paper
5. Cash, D., Grubbs, P., Perry, J., Ristenpart, T.: Leakage-abuse attacks against searchable encryption. In: Proceedings of the 22nd ACM SIGSAC Conference on Computer and Communications Security, pp. 668–679. ACM (2015)
6. Cash, D., et al.: Dynamic searchable encryption in very-large databases: data structures and implementation. In: NDSS, vol. 14, pp. 23–26 (2014)
7. Curtmola, R., Garay, J., Kamara, S., Ostrovsky, R.: Searchable symmetric encryption: improved definitions and efficient constructions. J. Comput. Secur. **19**(5), 895–934 (2011)
8. Decker, C., Wattenhofer, R.: A fast and scalable payment network with bitcoin duplex micropayment channels. In: Pelc, A., Schwarzmann, A.A. (eds.) SSS 2015. LNCS, vol. 9212, pp. 3–18. Springer, Cham (2015). https://doi.org/10.1007/978-3-319-21741-3_1
9. Delmolino, K., Arnett, M., Kosba, A.E., Miller, A., Shi, E.: Step by step towards creating a safe smart contract: lessons and insights from a cryptocurrency lab. IACR Cryptology ePrint Archive, p. 460 (2015)
10. Goh, E.J., et al.: Secure indexes. IACR Cryptology ePrint Archive, p. 216 (2003)
11. Grubbs, P., McPherson, R., Naveed, M., Ristenpart, T., Shmatikov, V.: Breaking web applications built on top of encrypted data. In: Proceedings of the 2016 ACM SIGSAC Conference on Computer and Communications Security, pp. 1353–1364. ACM (2016)
12. Heilman, E., Baldimtsi, F., Goldberg, S.: Blindly signed contracts: anonymous on-blockchain and off-blockchain bitcoin transactions. In: Clark, J., Meiklejohn, S., Ryan, P.Y.A., Wallach, D., Brenner, M., Rohloff, K. (eds.) FC 2016. LNCS, vol. 9604, pp. 43–60. Springer, Heidelberg (2016). https://doi.org/10.1007/978-3-662-53357-4_4
13. Kamara, S., Papamanthou, C., Roeder, T.: Dynamic searchable symmetric encryption. In: Proceedings of the 2012 ACM Conference on Computer and Communications Security, pp. 965–976. ACM (2012)
14. Nakamoto, S.: Bitcoin: a peer-to-peer electronic cash system (2008)
15. Oliver, J., Cheng, C., Chen, Y.: TLSH-a locality sensitive hash. In: 2013 Fourth Cybercrime and Trustworthy Computing Workshop (CTC), pp. 7–13. IEEE (2013)
16. Oliver, J., Forman, S., Cheng, C.: Using randomization to attack similarity digests. In: Batten, L., Li, G., Niu, W., Warren, M. (eds.) ATIS 2014. CCIS, vol. 490, pp. 199–210. Springer, Heidelberg (2014). https://doi.org/10.1007/978-3-662-45670-5_19

17. Poon, J., Dryja, T.: The bitcoin lightning network (2015)
18. Popa, R.A., Zeldovich, N.: Multi-key searchable encryption. IACR Cryptology ePrint Archive, p. 508 (2013)
19. Popa, R.A., et al.: Building web applications on top of encrypted data using Mylar. In: NSDI, pp. 157–172 (2014)
20. Song, D.X., Wagner, D., Perrig, A.: Practical techniques for searches on encrypted data. In: 2000 IEEE Symposium on Security and Privacy, S&P 2000, Proceedings, pp. 44–55. IEEE (2000)
21. Van Rompay, C., Molva, R., Önen, M.: A leakage-abuse attack against multi-user searchable encryption. Proc. Priv. Enhancing Technol. **2017**(3), 168–178 (2017)
22. Van Rompay, C., Molva, R., Önen, M.: Secure and scalable multi-user searchable encryption (2018)

Storage Security and Information Retrieval

ELSA: Efficient Long-Term Secure Storage of Large Datasets

Matthias Geihs[(✉)] and Johannes Buchmann

TU Darmstadt, Darmstadt, Germany
mgeihs@cdc.tu-darmstadt.de

Abstract. An increasing amount of information today is generated, exchanged, and stored digitally. This also includes long-lived and highly sensitive information (e.g., electronic health records, governmental documents) whose integrity and confidentiality must be protected over decades or even centuries. While there is a vast amount of cryptography-based data protection schemes, only few are designed for long-term protection. Recently, Braun et al. (AsiaCCS'17) proposed the first long-term protection scheme that provides renewable integrity protection and information-theoretic confidentiality protection. However, computation and storage costs of their scheme increase significantly with the number of stored data items. As a result, their scheme appears suitable only for protecting databases with a small number of relatively large data items, but unsuitable for databases that hold a large number of relatively small data items (e.g., medical record databases).

In this work, we present a solution for *efficient* long-term integrity and confidentiality protection of large datasets consisting of relatively small data items. First, we construct a renewable vector commitment scheme that is information-theoretically hiding under selective decommitment. We then combine this scheme with renewable timestamps and information-theoretically secure secret sharing. The resulting solution requires only a single timestamp for protecting a dataset while the state of the art requires a number of timestamps linear in the number of data items. We implemented our solution and measured its performance in a scenario where 12 000 data items are aggregated, stored, protected, and verified over a time span of 100 years. Our measurements show that our new solution completes this evaluation scenario an order of magnitude faster than the state of the art.

1 Introduction

1.1 Motivation and Problem Statement

Today, huge amounts of information are generated, exchanged, and stored digitally and these amounts will further grow in the future. Much of this data con-

This work has been co-funded by the DFG as part of project S6 within CRC 1119 CROSSING. This is the proceedings version as published at ICISC'18. An extended version can be found at arXiv.org [8].

K. Lee (Ed.): ICISC 2018, LNCS 11396, pp. 269–286, 2019.
https://doi.org/10.1007/978-3-030-12146-4_17

tains sensitive information (e.g., electronic health records, governmental documents, enterprise documents) and requires protection of *integrity* and *confidentiality*. Integrity protection means that illegitimate and accidental changes of data can be discovered. Confidentiality protection means that only authorized parties can access the data. Depending on the use case, protection may be required for several decades or even centuries. Databases that require protection are often complex and consist of a large number of relatively small data items that require continuous confidentiality protection and whose integrity must be verifiable independent from the other data items.

Today, integrity of digitally stored information is most commonly ensured using digital signatures (e.g., RSA [21]) and confidentiality is ensured using encryption (e.g., AES [19]). The commonly used schemes are secure under certain computational assumptions. For example, they require that computing the prime factors of a large integer is infeasible. However, as computing technology and cryptanalysis advances over time, computational assumptions made today are likely to break at some point in the future (e.g., RSA will become insecure once quantum computers are available [24]). Consequently, *computationally secure* cryptographic schemes have a limited lifetime and are insufficient to provide *long-term security*.

Several approaches have been developed to mitigate long-term security risks. Bayer et al. [1] proposed a technique for prolonging the validity of a digital signature by using digital timestamps. Based on their idea, a variety of long-term integrity protection schemes have been developed. An overview of existing long-term integrity schemes is given by Vigil et al. in [25]. In contrast to integrity protection, confidentiality protection cannot be prolonged. There is no protection against an adversary that stores ciphertexts today, waits until the encryption is weakened, and then breaks the encryption and obtains the plaintexts. Thus, if long-term confidentiality protection is required, then strong confidentiality protection must be applied from the start. A very strong form of protection can be achieved by using *information theoretically secure* schemes, which are invulnerable to computational attacks. For example, key exchange can be realized using quantum key distribution [11], encryption can be realized using one-time pad encryption [23], and data storage can be realized using proactive secret sharing [14]. An overview of information theoretically secure solutions for long-term confidentiality protection is given by Braun et al. [4].

Recently, Braun et al. proposed LINCOS [3], which is the first long-term secure storage architecture that combines long-term integrity with long-term confidentiality protection. While their system achieves high protection guarantees, it is only designed for storing and protecting a single large data object, but not databases that consist of a large number of small data items. One approach to store and protect large databases with LINCOS is to run an instance of LINCOS for each data item in parallel. However, with this construction the amount of work scales linearly with the number of stored data items. Especially, if the database consists of a large number of relatively small data items, this introduces a large communication and computation overhead.

1.2   Contribution

In this paper we propose an efficient solution to storing and protecting large and complex datasets over long periods of time. Concretely, we present the long-term secure storage architecture ELSA that uses renewable vector commitments in combination with renewable timestamps and proactive secret sharing to achieve this.

Our first contribution (Sect. 3) is to construct an extractable-binding and statistically hiding vector commitment scheme. Such a scheme allows for committing to a large number of data items by a single short commitment. The extractable binding property of the scheme enables renewable integrity protection [6] while the statistical hiding property ensures information theoretic confidentiality. Our construction is based on statistically hiding commitments and hash trees [18]. We prove that our construction is extractable binding given that the employed commitment scheme and hash function are extractable binding. Furthermore, we prove that our construction is statistically hiding under selective opening, which guarantees that by opening the commitments to some of the data items no information about unopened data items is leaked. The construction of extractable-binding and statistically hiding vector commitments may be of independent interest, for example, in the context of zero knowledge protocols [10].

Our second contribution (Sect. 4) is the construction of the long-term secure storage architecture ELSA, which uses our new vector commitment scheme construction to achieve efficient protection of large datasets. While protecting a dataset with LINCOS requires the generation of a commitment and a timestamp for each data item separately, ELSA requires only a single vector commitment and a single timestamp to protect the same dataset. Hence, the number of timestamps is decreased from linear in the number of data items to constant and this drastically reduces the communication and computation complexity of the solution. Moreover, as the vector commitment scheme is hiding under selective decommitment, the integrity of stored data items can still be verified individually without revealing information about unopened data items. ELSA uses a separate service for storing commitments and timestamps, which allows for renewing the timestamp protection without access to the stored confidential data. The decommitments are stored together with the data items at a set of shareholders using proactive secret sharing. We show that the long-term integrity security of ELSA can be reduced to the unforgeability security of the employed timestamp schemes and the binding security of the employed commitment schemes within their usage period. Long-term confidentiality security is based on the statistical hiding security of the employed commitment and secret sharing schemes.

Finally, we experimentally demonstrate (Sect. 5) the performance improvements achieved by ELSA in a scenario where 12 000 data items of size 10 kB are aggregated, stored, protected, retrieved, and verified during a timespan of 100 years. For this, we implemented ELSA and the state of the art long-term secure storage architecture LINCOS. Our measurements show that ELSA completes the evaluation scenario $17x$ faster than LINCOS and integrity protection consumes $101x$ less memory. In particular, protection renewal is significantly faster with ELSA. Renewing the timestamps for approximately 12 000 data items takes 21.89 min with LINCOS and only 0.34 s with ELSA. Furthermore, storage of the timestamps and commitments consumes 1.75 GB of storage space with LINCOS and only 17.27 MB with ELSA at the end of the experiment. These improvements are achieved at slightly higher storage costs for the shareholders. Each shareholder consumes 559 MB with LINCOS and 748 MB with ELSA. The storage costs for integrity protection are independent of the size of the data items. Storage, retrieval, and verification of a data item takes less than a second. Overall, our evaluation shows that ELSA provides practical performance and is suitable for storing and protecting large and complex databases that consist of relatively small data items over long periods of time (e.g., health record or governmental document databases).

1.3   Related Work

Our notion of vector commitments is reminiscent of the one proposed by Catalano and Fiore [7]. However, they do not consider the hiding property and therefore do not analyze hiding under selective opening security. Also, they do not consider extractable binding security. Hofheinz [15] studied the notion of selective decommitment and showed that schemes can be constructed that are statistically hiding under selective decommitment. However, they do not consider constructions of vector commitments where a short commitment is given for a set of messages. In [2], Bitansky et al. propose the construction of a SNARK from extractable collision-resistant hash functions in combination with Merkle trees. While their construction is similar to the extractable-binding vector commitment scheme proposed in Sect. 3.2, our construction relies on a weaker property (i.e., extractable-binding hash functions) and our security analysis provides concrete security estimates.

Weinert et al. [26] recently proposed a long-term integrity protection scheme that also uses hash trees to reduce the number of timestamps. However, their scheme does not support confidentiality protection, lacks a formal security analysis, and is less efficient than our construction. Only few work has been done with respect to combining long-term integrity with long-term confidentiality protection. The first storage architecture providing these two properties and most efficient to date is LINCOS [3]. Recently, another long-term secure storage architecture has been proposed by Geihs et al. [9] that provides access pattern hiding security in addition to integrity and confidentiality. On a high level, this is achieved by combining LINCOS with an information theoretically secure ORAM. While access pattern hiding security is an interesting property in certain scenarios where meta information about the stored data is known, it is achieved at the cost of additional computation and communication and it is out of the scope of this work.

2   Preliminaries

2.1   Notation

For a probabilistic algorithm $\mathcal{A}$ and input x, we write $\mathcal{A}(x) \to_r y$ to denote that $\mathcal{A}$ on input x produces y using random coins r. For a vector $V = (v_1, \ldots, v_n)$, $n \in \mathbb{N}$, and set $I \subseteq [n]$, define $V_I := (v_i)_{i \in I}$, and for $i \in [n]$, define $V_i := v_i$. For a pair of random variables (A, B), we define the statistical distance of A and B as $\Delta(A, B) := \sum_x |\Pr_A(x) - \Pr_B(x)|$.

2.2   Cryptographic Primitives

We briefly introduce the cryptographic primitives that are used in this paper. A more extensive description can be found in the full version [8].

Digital Signature Schemes. A digital signature scheme SIG is defined by a message space $\mathcal{M}$ and algorithms Setup, Sign, and Verify. Algorithm Setup $\to$ (sk, pk) for generating a secret signing key sk and a public verification key pk. Algorithm Sign$(sk, m) \to s$ gets as input a secret key sk and a message $m \in \mathcal{M}$ and outputs a signature s. Algorithm Verify$(pk, m, s) \to b$ gets as input a public key pk, a message m, and a signature s, and outputs $b = 1$, if the signature is valid, and 0, if it is invalid. A signature scheme is ϵ-unforgeable-secure if the probability of a t-bounded adversary $\mathcal{A}$ forging a signature is bounded by $\epsilon(t)$.

Timestamp Schemes. A timestamp scheme [12] is a protocol between a client and a timestamp service. The timestamp service initializes itself using algorithm Setup. The client uses protocol Stamp to request a timestamp from the timestamp service. Furthermore, there exists an algorithm Verify that allows anybody to verify the validity of a message-timestamp-tuple. Here, we consider signature-based timestamping, where the timestamp is a signature on the timestamped document and the current time. In this case, the timestamp service generating the signature must be trusted to use the correct time value.

Commitment Schemes. A (non-interactive) commitment scheme COM is defined by a message space $\mathcal{M}$ and algorithms Setup, Commit, and Verify. Algorithm Setup $\to pk$ generates a public commitment key pk. Algorithm Commit$(pk, m) \to (c, d)$ gets as input a public key pk and a message $m \in \mathcal{M}$ and outputs a commitment c and a decommitment d. Algorith Verify$(pk, m, c, d) \to b$ gets as input a public key pk, a message m, a commitment c, and a decommitment d, and outputs $b = 1$, if the decommitment is valid, and 0, if it is invalid. A commitment scheme is considered secure if it is hiding (i.e., a commitment does not leak information) and binding (i.e., the committer cannot change his mind about the committed message). There exist different flavors of defining binding security. Here, we are interested in extractable binding commitments as this enables renewable and long-term secure commitments [6].

Keyed Hash Functions. A keyed hash function is a tuple of algorithms (K, H) where K is a probabilistic algorithm that generates a key k and H is a deterministic algorithm that on input a key k and a message $x \in \{0, 1\}^*$ outputs a short fixed length hash $y \in \{0, 1\}^l$, for some $l \in \mathbb{N}$. We say a keyed hash function (K, F) is ϵ-extractable-binding if for any t_1-bounded algorithm $\mathcal{A}_1$, there exists a $t_\mathcal{E}$-bounded algorithm $\mathcal{E}$, such that for any t_2-bounded algorithm $\mathcal{A}_2$,

$$\Pr_{K \to k} \left[\begin{array}{c} H(k, x) = H(k, x^*) \wedge x \neq x^* : \\ \mathcal{A}_1(k) \to_r y, \mathcal{E}(k, r) \to x^*, \mathcal{A}_2(k, r) \to x \end{array} \right] \leq \epsilon(t_1, t_\mathcal{E}, t_2).$$

Secret Sharing Schemes. A proactive secret sharing scheme [14] is protocol between a data owner and a set of shareholders. It has a protocol Setup for generating system parameters, a protocol Share for sharing a data object, a protocol Reshare for refreshing the shares, and a protocol Reconstruct for reconstructing a data object from a given set of shares. In this work, we consider threshold secret sharing schemes, for which there exists a threshold parameter t (chosen by the data owner) such that any set of t shareholders can reconstruct the secret, but any set of less than t shareholders has no information about the secret.

3   Statistically Hiding and Extractable Binding Vector Commitments

In this section, we define statistically hiding and extractable binding vector commitments, describe a construction, and prove the construction secure. This construction is the basis for our performance improvements that we achieve with our new storage architecture presented in Sect. 4. The proofs of the presented theorems can be found in the full version [8].

3.1   Definition

A vector commitment scheme allows to commit to a vector of messages $[m_1, \dots, m_n]$. It is extractable binding, if the message vector can be extracted from the commitment and the state of the committer and it is hiding under partial opening if an adversary cannot infer any valuable information about unopened messages, even if some of the committed messages have been opened. Our vector commitments are reminiscent of the vector commitments introduced by Catalano and Fiore [7]. However, neither do they require their commitments to be extractable binding nor do they consider their hiding property.

Definition 1 (Vector commitment scheme). *A vector commitment scheme is a tuple* $(L, \mathcal{M}, \mathsf{Setup}, \mathsf{Commit}, \mathsf{Open}, \mathsf{Verify})$, *where* $L \in \mathbb{N}$ *is the maximum vector length,* $\mathcal{M}$ *is the message space, and* Setup, Commit, Open, *and* Verify *are algorithms with the following properties.*

Setup() $\to k$: *This algorithm generates a public key* k.

Commit$(k, [m_1, \ldots, m_n]) \to (c, D)$: *On input key k and message vector $[m_1, \ldots, m_n] \in \mathcal{M}^n$, where $n \in [L]$, this algorithm generates a commitment c and a vector decommitment D.*

Open$(k, D, i) \to d$: *On input key k, vector decommitment D, and index i, this algorithm outputs a decommitment d for the i-th message corresponding to D.*

Verify$(k, m, c, d, i) \to b$: *On input key k, message m, commitment c, decommitment d, and an index i, this algorithm outputs $b = 1$, if d is a valid decommitment from position i of c to m, and otherwise outputs $b = 0$.*

A vector commitment scheme is correct, if a decommitment produced by Commit and Open will always verify for the corresponding commitment and message.

Definition 2 (Correctness). *A vector commitment scheme $(L, \mathcal{M}, \mathsf{Setup}, \mathsf{Commit}, \mathsf{Open}, \mathsf{Verify})$ is correct if for all $n \in [L]$, $M \in \mathcal{M}^n$, $k \in \mathsf{Setup}$, $i \in [n]$,*

$$\Pr\left[\begin{array}{c} \mathsf{Verify}(k, M_i, c, d) = 1 : \\ \mathsf{Commit}(k, M) \to (c, D), \mathsf{Open}(k, D, i) \to d \end{array}\right] = 1.$$

A vector commitment scheme is statistically hiding under selective opening, if the distribution of commitments and openings does not depend on the unopened messages. For any public key k and message m, define $C_k(m)$ as the random variable that takes the value of c when sampling $\mathsf{Commit}(k, m) \to (c, d)$. A commitment scheme is ϵ-statistically-hiding if for any $k \in \mathsf{Setup}$, any pair of messages (m_1, m_2), $\Delta(C_k(m_1), C_k(m_2)) \le \epsilon$.

Definition 3 (Statistically hiding (under selective opening)). *Let $S = (L, \mathcal{M}, \mathsf{Setup}, \mathsf{Commit}, \mathsf{Open}, \mathsf{Verify})$ be a vector commitment scheme. For $n \in [L], I \subseteq [n], M \in \mathcal{M}^n, k \in \mathsf{Setup}$, we denote by $\mathsf{CD}_k(M, I)$ the random variable $(c, \bar{D}_I)$, where $(c, D) \leftarrow \mathsf{Commit}(k, M)$ and $\bar{D} \leftarrow (\mathsf{Open}(D, i))_{i \in [n]}$. Let $\epsilon \in [0, 1]$. We say S is ϵ-statistically-hiding, if for all $n \in \mathbb{N}, I \subseteq [n], M_1, M_2 \in \mathcal{M}^n$ with $(M_1)_I = (M_2)_I$, $k \in \mathsf{Setup}$,*

$$\Delta(\mathsf{CD}_k(M_1, I), \mathsf{CD}_k(M_2, I)) \le \epsilon.$$

A vector commitment scheme is extractable binding, if for every efficient committer, there exists an efficient extractor, such that for any efficient decommitter, if the committer gives a commitment that can be opened by a decommitter, then the extractor can already extract the corresponding messages from the committer at the time of the commitment.

Definition 4 (Extractable binding). *Let $\epsilon : \mathbb{N}^3 \to [0, 1]$. We say a vector commitment scheme $(L, \mathcal{M}, \mathsf{Setup}, \mathsf{Commit}, \mathsf{Open}, \mathsf{Verify})$ is ϵ-extractable-binding, if for all t_1-bounded algorithms $\mathcal{A}_1$, $t_{\mathcal{E}}$-bounded algorithms $\mathcal{E}$, and t_2-bounded algorithms $\mathcal{A}_2$,*

$$\Pr\left[\begin{array}{c} \mathsf{Verify}(p, m, c, d, i) = 1 \land m_i \ne m : \\ \mathsf{Setup}() \to k, \mathcal{A}_1(k) \to_r c, \\ \mathcal{E}(k, r) \to [m_1, m_2, \ldots], \mathcal{A}_2(k, r) \to (m, c, d, i) \end{array}\right] \le \epsilon(t_1, t_{\mathcal{E}}, t_2).$$

3.2 Construction: Extractable Binding

In the following, we show that the Merkle hash tree construction [18] can be casted into a vector commitment scheme and that this construction is extractable binding if the used hash function is extractable binding.

Construction 1. *Let (K, H) denote a keyed hash function and let $L \in \mathbb{N}$. The following is a description of the hash tree scheme by Merkle cast into the definition of vector commitments.*

Setup() $\rightarrow k$: *Run $K \rightarrow k$ and output k.*
Commit($k, [m_1, \ldots, m_n]$) $\rightarrow (c, D)$: *Set $l \leftarrow \min\{i \in \mathbb{N} : n \leq 2^i\}$. For $i \in \{0, \ldots, n-1\}$, compute $H(k, m_i) \rightarrow h_{i,l}$, and for $i \in \{n, \ldots, 2^l - 1\}$, set $h_{i,l} \leftarrow \perp$. For $i \in \{l-1, \ldots, 0\}$, $j \in \{0, \ldots, 2^i - 1\}$, compute $H(k, [h_{i-1,2j}, h_{i-1,2j+1}])$. Compute $H(k, [l, h_{0,0}]) \rightarrow c$. Set $D \leftarrow [h_{i,j}]_{i \in \{0, \ldots, l\}, j \in \{0, \ldots, 2^i - 1\}}$, and output (c, D).*
Open(k, D, i^*) $\rightarrow d$: *Let $D \rightarrow [h_{i,j}]_{i \in \{0, \ldots, l\}, j \in \{0, \ldots, 2^i - 1\}}$. Set $a_l \leftarrow i^*$. For $j \in \{l, \ldots, 1\}$, set $b_j \leftarrow a_j + 2(a_j \bmod 2) - 1$, $g_j \leftarrow h_{j, b_j}$, and $a_{j-1} \leftarrow \lfloor a_j / 2 \rfloor$. Set $d = [g_1, \ldots, g_l]$ and output d.*
Verify(k, m, c, d, i^*) $\rightarrow b^*$: *Let $d = [g_1, \ldots, g_l]$. Set $a_l \leftarrow i^*$ and compute $H(k, m) \rightarrow h_l$. For $i \in \{l, \ldots, 1\}$, if $a_i \bmod 2 = 0$, set $b_i \leftarrow [h_i, g_i]$, and if $a_i \bmod 2 = 1$, set $b_i \leftarrow [g_i, h_i]$, and then compute $H(k, b_i) \rightarrow h_{i-1}$ and set $a_{i-1} \leftarrow \lfloor a_i / 2 \rfloor$. Compute $H(k, [l, h_0]) \rightarrow c'$. Set $b^* \leftarrow (c = c')$. Output b^*.*

Theorem 1. *The vector commitment scheme described in Construction 1 is correct.*

Theorem 2. *Let (K, H) be an ϵ-extractable-binding hash function. The vector commitment scheme described in Construction 1 instantiated with (K, H) is ϵ'-extractable-binding with $\epsilon'(t_1, t_\mathcal{E}, t_2) = 2L * \epsilon(t_1 + t_\mathcal{E}/L, t_\mathcal{E}/L, t_2)$.*

3.3 Construction: Extractable Binding and Statistically Hiding

We now combine a statistically hiding and extractable binding commitment scheme with the vector commitment scheme from Construction 1 to obtain a statistically hiding (under selective opening) and extractable binding vector commitment scheme. The idea is to first commit with the statistically hiding scheme to each message separately and then produce a vector commitment to these individually generated commitments.

Construction 2. *Let COM be a commitment scheme and VC be a vector commitment scheme.*

Setup() $\rightarrow k$: *Run COM.Setup() $\rightarrow k_1$, VC.Setup() $\rightarrow k_2$, set $k \leftarrow (k_1, k_2)$, and output k.*
Commit($k, [m_1, \ldots, m_n]$) $\rightarrow (c, D)$: *Let $k \rightarrow (k_1, k_2)$. For $i \in \{1, \ldots, n\}$, compute COM.Commit(k_1, m_i) $\rightarrow (c_i, d_i)$. Then compute VC.Commit($k_2, [c_1, \ldots, c_n]$) $\rightarrow (c, D')$, set $D \leftarrow ([(c_1, d_1), \ldots, (c_n, d_n)], D')$, and output (c, D).*

Open$(k, D, i) \rightarrow d$: Let $k \rightarrow (k_1, k_2)$ and $D \rightarrow ([(c_1, d_1), \ldots, (c_n, d_n)], D')$. Compute VC.Open$(k_2, D', i) \rightarrow d'$, set $d \leftarrow (c_i, d_i, d')$, and output d.

Verify$(k, m, c, d, i) \rightarrow b$: Let $k \rightarrow (k_1, k_2)$ and $d \rightarrow (c', d', d'')$. Compute COM.Verify$(k_1, m, c', d') \rightarrow b_1$ and then compute VC.Verify$(k_2, c', c, d'', i) \rightarrow b_2$, set $b \leftarrow (b_1 \wedge b_2)$, and output b.

Theorem 3. *The vector commitment scheme described in Construction 2 is correct if* COM *and* VC *are correct.*

Theorem 4. *The vector commitment scheme described in Construction 2 is $L\epsilon$-statistically-hiding (under selective opening) if the commitment scheme* COM *is ϵ-statistically-hiding.*

Theorem 5. *If* COM *and* VC *of Construction 2 are ϵ-extractable-binding, Construction 2 is an ϵ'-extractable-binding vector commitment scheme with* $\epsilon'(t_1, t_\mathcal{E}, t_2) = L * \epsilon(t_1 + t_\mathcal{E}/L, t_\mathcal{E}/L, t_2)$.

4 ELSA: Efficient Long-Term Secure Storage Architecture

Now we present ELSA, a long-term secure storage architecture that efficiently protects large datasets. It provides long-term integrity and long-term confidentiality protection of the stored data. ELSA uses statistically-hiding and extractable-binding vector commitments (as described in Sect. 3) in combination with timestamps to achieve renewable and privacy preserving integrity protection. The confidential data is stored using proactive secret sharing to guarantee confidentiality protection secure against computational attacks. The data owner communicates with two subsystems (Fig. 1), where one is responsible for data storage with confidentiality protection and the other one is responsible for integrity protection. The evidence service is responsible for integrity protection updates and the secret share holders are responsible for storing the data and maintaining confidentiality protection. The evidence service also communicates with a timestamp service that is used in the process of evidence generation.

Fig. 1. Overview of the components of ELSA.

4.1 Construction

We now describe the storage architecture ELSA in terms of the algorithms Init, Store, RenewTs, RenewCom, RenewShares, and Verify. Algorithm Init initializes the architecture, Store allows to store new files, RenewTs renews the protection if the timestamp scheme security is weakened, RenewCom renews the protection if the commitment scheme security is weakened, RenewShares renews the shares to protect against a mobile adversary who collects multiple shares over time, and Verify verifies the integrity of a retrieved file.

We use the following notation. When we write SH.Store(name, dat) we mean that the data owner shares the data dat among the shareholders using protocol SHARE.Share associated with identifier name. If the shared data dat is larger then the size of the message space of the secret sharing scheme, dat is first split into chunks that fit into the message space and then the chunks are shared individually. Each shareholder maintains a database that describes which shares belong to which data item name. When we write SH.Retrieve(name), we mean that the data owner retrieves the shares associated identifier name from the shareholders and reconstructs the data using protocol SHARE.Reconstruct.

Initialization. The data owner uses algorithm ELSA.Init to initialize the storage system. The algorithm gets as input a proactive secret sharing scheme SHARE, a set of shareholder addresses $(\text{shURL}_i)_{i \in [N]}$, a sharing threshold T, and an evidence service address esURL. It then initializes the storage module SH by running protocol SHARE.Setup and the evidence service module ES by setting ES.evidence as an empty table and ES.renewLists as an empty list.

Data Storage. The client uses algorithm ELSA.Store (Algorithm 1) to store a set of data files $[\text{file}_i]_{i \in [n]}$, which works as follows. First a signature scheme SIG, a vector commitment scheme VC, and a timestamp scheme TS are chosen. Here, we assume that SIG is supplied with the secret key necessary for signature generation and VC is supplied with the public parameters necessary for commitment generation. The algorithm first signs each of the data objects individually. It then stores the file data, the public key certificate of the signature scheme instance, and the generated signature at the secret sharing storage system. Afterwards, the algorithm generates a vector commitment (c, D) to the file data vector and the signatures. For each file, the corresponding decommitment is extracted and stored at the shareholders. The file names filenames, the commitment scheme instance VC, the commitment c, and the chosen timestamp scheme instance TS are sent to the evidence service.

When the evidence service receives (filenames, VC, c, TS), it does the following in algorithm AddCom (Algorithm 2). It first timestamps the commitment (VC, c) and thereby obtains a timestamp ts. Then, it starts a new evidence list $l = [(\text{VC}, c, \text{TS}, \text{ts})]$ and assigns this list with all the file names in filenames. Also, it adds l to the list renewLists, which contains the lists that are updated on a timestamp renewal.

Algorithm 1. ELSA.Store($[\text{file}_i]_{i \in [n]}$, SIG, VC, TS)

---

filenames ← {};
for $i \in [n]$ do
  | SIG.Sign(file_i.dat) → s_i;
  | SH.Store(['data', file_i.name], [file_i.dat, SIG.$Cert$, s_i]);
  | filenames += file_i.name;
VC.Commit($[\text{file}_i.\text{dat}, \text{SIG}.Cert, s_i]_{i \in [n]}$) → (c, D);
for $i \in [n]$ do
  | VC.Open(D, i) → d;
  | SH.Store(['decom', file_i.name, i], d);
ES.AddCom(filenames, VC, c, TS);

Algorithm 2. ES.AddCom(filenames, VC, c, TS)

TS.Stamp((VC, c)) → ts;
$l \leftarrow [(\text{VC}, c, \text{TS}, \text{ts})]$;
for name ∈ filenames do
  | evidence[name] ← l;
  | renewLists += l;

Timestamp Renewal. Algorithm ES.RenewTs (Algorithm 3) is performed by the evidence service regularly in order to protect against the weakening of the currently used timestamp scheme. The algorithm gets as input a vector commitment scheme instance VC$'$ and a timestamp scheme instance TS. It first creates a vector commitment (c', D') for the list of renewal items renewLists. Here, we only require the extractable-binding property of VC$'$, while the hiding property is not required as all of the data stored at the evidence service is independent of the secret data due to the use of unconditionally hiding commitments by the data owner. For each updated list item i, the freshly generated timestamp, commitment, and extracted decommitment are added to the corresponding evidence list renewLists[i].

Algorithm 3. ES.RenewTs(VC$'$, TS)

VC$'$.Commit(renewLists) → (c', D');
TS.Stamp((VC', c')) → ts;
for $i \in [|\text{renewLists}|]$ do
  | VC$'$.Open(D', i) → d';
  | renewLists[i] += $(\text{VC}', c', d', \text{TS}, \text{ts})$;

Commitment Renewal. The data owner runs algorithm ELSA.RenewCom (Algorithm 4) to protect against a weakening of the currently used commitment scheme. It chooses a new commitment scheme instance VC and a new timestamp scheme instance TS and proceeds as follows. First the table of evidence lists ES.evidence are retrieved from the evidence service and complemented with the decommitment values stored at the shareholders. Next, a list with the data items, the signatures, and the current evidence for each data item is constructed. This list is then committed using the vector commitment scheme VC. The decommitments are extracted and stored at the shareholders, and the commitment is added to the evidence at the evidence service using algorithm ES.AddComRenew.

Algorithm 4. ELSA.RenewCom(VC, TS)

$\text{comIndices} \leftarrow \{\}; \text{comCount} \leftarrow \{\}; L \leftarrow [];$
for $\text{name} \in \text{ES.evidence}$ **do**
    $\text{SH.Retrieve}(['\text{data}', \text{name}]) \rightarrow (\text{dat}, \text{SIG}, s);$
    $\text{ES.evidence}[\text{name}] \rightarrow e;$
    for $i \in |e|$ **do**
        if $e_i.\text{VC} \neq \perp$ **then**
            $\text{SH.Retrieve}(['\text{decom}', \text{name}, i]) \rightarrow e_i.d;$

    $L \mathrel{+}= (\text{dat}, \text{SIG}, s, e);$
    $\text{comIndices}[\text{name}] \leftarrow |L|;$
    $\text{comCount}[\text{name}] \leftarrow |e|;$
$\text{VC.Commit}(L) \rightarrow (c, D);$
for $\text{name} \in \text{ES.evidence}$ **do**
    $\text{VC.Open}(D, \text{comIndices}[\text{name}]) \rightarrow d;$
    $\text{SH.Store}(['\text{decom}', \text{name}, \text{comCount}[\text{name}]], d);$
$\text{ES.AddComRenew}(\text{VC}, c, \text{TS});$

Algorithm 5. ES.AddComRenew(VC, c, TS)

$\text{TS.Stamp}((\text{VC}, c)) \rightarrow \text{ts};$
$l \leftarrow [(\text{VC}, c, \text{TS}, \text{ts})];$
$\text{renewLists} \leftarrow [l];$
for $\text{name} \in \text{evidence}$ **do**
    $\text{evidence}[\text{name}] \mathrel{+}= l;$

Secret Share Renewal. There are two types of share renewal supported by ELSA. The first type (ELSA.RenewShares()) triggers the share renewal protocol of the secret sharing system (i.e., the protocol SHARE.Reshare). This interactive protocol refreshes the shares at the shareholders so that old shares, which may have leaked already, cannot be combined with the new shares, which are obtained after the protocol has finished, to reconstruct the stored data. The second type (Algorithm 6) replaces the proactive sharing scheme entirely. This may be necessary if the scheme has additional security properties like verifiability (see proactive verifiable secret sharing [14]), whose security may be weakened. In this case, the data is retrieved, shared to the new shareholders, and finally the old shareholders are shutdown.

Algorithm 6. ELSA.RenewSharing(SHARE, $(\text{shURL}_i)_{i \in [N]}, T$)

SH'.Init(SHARE, $(\text{shURL}_i)_{i \in [N]}, T$);
$I \leftarrow$ ES.itemInfos;
for name $\in I$ **do**
   | SH.Retrieve('data/' + name) $\rightarrow$ dat;
   | SH'.Store('data/' + name, dat);
SH.Shutdown();
SH $\leftarrow$ SH';

Data Retrieval. The algorithm ELSA.Retrieve (Algorithm 7) describes the data retrieval procedure of ELSA. It gets as input the name of the data file that is to be retrieved. It then collects the evidence from the evidence service and the data from the shareholders. Next, the evidence is complemented with the decommitments and then the algorithm outputs the data with the corresponding evidence.

Algorithm 7. ELSA.Retrieve(name)

$e \leftarrow$ ES.evidence[name];
for $i \in [|e|]$ **do**
   if $e_i.\text{VC} \neq \perp$ **then**
      | SH.Retrieve(['decom', name, i]) $\rightarrow e_i.d$;
SH.Retrieve(['data', name]) $\rightarrow$ (dat, SIG, s);
$E \leftarrow$ (SIG, s, e);
return (dat, E);

Verification. Algorithm ELSA.Verify (Algorithm 8) describes how a verifier can check the integrity of a data item using the evidence produced by ELSA. Here we denote by $\mathrm{NTT}(i, e, t_{\mathsf{verify}})$ the time of the next timestamp after entry i of e and by $\mathrm{NCT}(i, e, t_{\mathsf{verify}})$ the time of the timestamp corresponding to the next commitment after entry i, and we set $\mathrm{NTT}(i, e, t_{\mathsf{verify}}) = t_{\mathsf{verify}}$ if i is the last timestamp and $\mathrm{NCT}(i, e, t_{\mathsf{verify}}) = t_{\mathsf{verify}}$ if i is the last commitment in e. The algorithm gets as input a reference to the considered PKI (e.g., a trust anchor), the current verification time t_{verify}, the data to be checked dat, the storage time t_{store}, and the corresponding evidence $E = (\mathsf{SIG}, s, e)$. The algorithm returns true, if dat is authentic and has been stored at time t_{store}.

In more detail, the verification algorithm works as follows. It first checks whether the signature s is valid for the data object dat under signature scheme instance SIG at the time of the first timestamp of the evidence list e. It also checks whether the corresponding commitment is valid for $(\mathsf{dat}, \mathsf{SIG}, s)$ at the time of the next commitment and the timestamp is valid at the next timestamp. Then, for each of the remaining $|e| - 1$ entries of e, the algorithm checks whether the corresponding timestamp is valid at the time of the next timestamp and whether the corresponding commitments are valid at the time of the next commitments. The algorithm outputs 1 if all checks return valid, and it outputs 0 in any other case.

Algorithm 8. ELSA.Verify$(\mathsf{PKI}, t_{\mathsf{verify}} : \mathsf{dat}, t_{\mathsf{store}}, E) \to b$

$(\mathsf{SIG}, s, e) \leftarrow E$;
$((\mathsf{VC}, c, d), (\mathsf{VC}', c', d'), (\mathsf{TS}, \mathsf{ts})) \leftarrow e_1$;
$t_{\mathrm{nt}} \leftarrow \mathrm{NTT}(1, e, t_{\mathsf{verify}})$; $t_{\mathrm{nc}} \leftarrow \mathrm{NCT}(1, e, t_{\mathsf{verify}})$;
$b \leftarrow \mathsf{SIG}.\mathsf{Verify}(\mathsf{PKI}, \mathsf{ts}.t : \mathsf{dat}, s)$;
$b \wedge= \mathsf{VC}.\mathsf{Verify}(\mathsf{PKI}, t_{\mathrm{nc}} : (\mathsf{dat}, \mathsf{SIG}, s), c, d)$;
$b \wedge= \mathsf{TS}.\mathsf{Verify}(\mathsf{PKI}, t_{\mathrm{nt}} : c, \mathsf{ts}, t_{\mathsf{store}})$;
$L \leftarrow (\mathsf{VC}, c, \mathsf{TS}, \mathsf{ts})$;

for $i \in [2, \dots, |e|]$ **do**
    $((\mathsf{VC}, c, d), (\mathsf{VC}', c', d'), (\mathsf{TS}, \mathsf{ts})) \leftarrow e_i$;
    $t_{\mathrm{nt}} \leftarrow \mathrm{NTT}(i, e, t_{\mathsf{verify}})$; $t_{\mathrm{nc}} \leftarrow \mathrm{NCT}(i, e, t_{\mathsf{verify}})$;
    if $\mathsf{VC} = \bot$ **then**
        $b \wedge= \mathsf{VC}'.\mathsf{Verify}(\mathsf{PKI}, t_{\mathrm{nt}} : L, c', d')$;
        $b \wedge= \mathsf{TS}.\mathsf{Verify}(\mathsf{PKI}, t_{\mathrm{nt}} : c', \mathsf{ts}, \mathsf{ts}.t)$;
        $L += (\mathsf{VC}', c', d', \mathsf{TS}, \mathsf{ts})$;
    else
        $\mathsf{dat}' \leftarrow (\mathsf{dat}, Cert, s, e[1, i-1])$;
        $b \wedge= \mathsf{VC}.\mathsf{Verify}(\mathsf{PKI}, t_{\mathrm{nc}} : \mathsf{dat}', c, d)$;
        $b \wedge= \mathsf{TS}.\mathsf{Verify}(\mathsf{PKI}, t_{\mathrm{nt}} : c, \mathsf{ts}, \mathsf{ts}.t)$;
        $L \leftarrow (\mathsf{VC}, c, \mathsf{TS}, \mathsf{ts})$;

return b;

4.2 Security Analysis

The security analysis can be found in the full version [8].

5 Performance Evaluation

We compare the performance of our new architecture ELSA with the performance of the storage architecture LINCOS [3], which is the fastest existing storage architecture that provides long-term integrity and long-term confidentiality.

5.1 Evaluation Scenario

For our evaluation we consider a scenario inspired by the task of securely storing electronic health records in a medium sized doctor's office. The storage time frame is 100 years. Every month, 10 new data items of size 10 kB (e.g., prescription data of patients) are added. Every year, one document from each of the previous years is retrieved and verified (e.g., historic prescription data is read from the archives).

We assume the following renewal schedule for protecting the evidence against the weakening of cryptographic primitives. The signatures are renewed every 2 years, as this is a typical lifetime of a public key certificate, which is needed to verify the signatures. While signature scheme instances can only be secure as long as the corresponding private signing key is not leaked to an adversary, commitment scheme instances do not involve the usage of any secret parameters. Therefore, their security is not threatened by key leakage and we assume that they only need to be renewed every 10 years in order to adjust the cryptographic parameter sizes or to choose a new and more secure scheme. Secret shares are renewed every 5 years, which we believe could be a typical shareholder life cycle.

In our architecture we instantiate the cryptographic schemes as follows. As signature scheme, we first use the RSA Signature Scheme [21] and then switch to the post-quantum secure XMSS signature scheme [5] by 2030, as we anticipate the development of large-scale quantum computers. As the vector commitment scheme we use Construction 2 with the statistically hiding commitment scheme by Halevi and Micali [13] whose security is based on the security of the used hash function which we instantiate with members of the SHA-2 hash function family [20]. If we model the hash functions as random oracles, they provide the necessary extractable-binding property. We adjust the signature and commitment scheme parameters over time as proposed by Lenstra and Verheul [16,17]. The resulting parameter sets are shown in Table 1. For the storage system, we use the secret sharing scheme by Shamir [22]. We run this scheme with 4 shareholders and a threshold of 3 shareholders are required for reconstruction. Resharing is carried out centrally by the data owner.

Table 1. Overview of the used commitment and signature scheme instances and their usage period.

Years	Signatures	Commitments
2018–2030	RSA-2048	HM-256
2031–2090	XMSS-256	HM-256
2091–2118	XMSS-512	HM-512

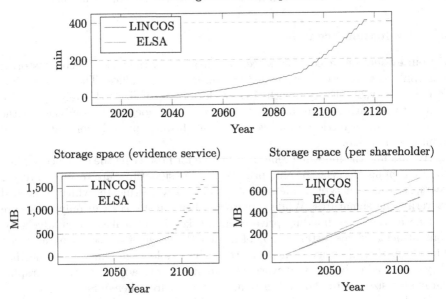

Fig. 2. Running time of the experiment and storage space consumption of the evidence service and per shareholder.

5.2   Results

We now present the results of our performance analysis. Figure 2 compares the computation time and storage costs of the two systems, ELSA and LINCOS. Our implementation was done using the programming language Java. The experiments were performed on a computer with a quad-core AMD Opteron CPU running at 2.3 GHz and the Java virtual machine was assigned 32 GB of RAM.

We observe that ELSA is much more computationally efficient compared to LINCOS. Completing the experiment using LINCOS took approximately 6.81 h, while it took only 24 min using ELSA. The biggest difference in the timings is observed when renewing timestamps. Timestamp renewal with LINCOS for year 2116 takes 21.89 min, while it takes only 0.34 s with ELSA. Data storage performance is also considerably faster with ELSA than with LINCOS. The same holds for the commitment renewal procedure. Data retrieval and verification performance is similar for the two systems.

Next, we observe that ELSA is also more efficient compared to LINCOS when it comes to the consumed storage space at the evidence service. This is, again, because ELSA requires fewer timestamps to be generated and stored than LINCOS. After running for 100 years, the evidence service of ELSA consumes only 17.27 MB while the evidence service of LINCOS consumes 1.75 GB of storage space. We observe by Fig. 2 that ELSA consumes slightly more storage space at the shareholders than LINCOS. This is because additional decommitment information for the vector commitments must be stored. After running for 100 years, a shareholder of ELSA consumes about 748 MB while a shareholder of LINCOS consumes about 559 MB of storage space.

References

1. Bayer, D., Haber, S., Stornetta, W.S.: Improving the efficiency and reliability of digital time-stamping. In: Capocelli, R., De Santis, A., Vaccaro, U. (eds.) Sequences II: Methods in Communication, Security, and Computer Science, pp. 329–334. Springer, New York (1993). https://doi.org/10.1007/978-1-4613-9323-8_24
2. Bitansky, N., et al.: The hunting of the snark. J. Cryptol. **30**(4), 989–1066 (2017). https://doi.org/10.1007/s00145-016-9241-9
3. Braun, J., et al.: Lincos: a storage system providing long-term integrity, authenticity, and confidentiality. In: Proceedings of the 2017 ACM on Asia Conference on Computer and Communications Security, ASIA CCS 2017, pp. 461–468. ACM, New York (2017)
4. Braun, J., Buchmann, J., Mullan, C., Wiesmaier, A.: Long term confidentiality: a survey. Des. Codes Cryptogr. **71**(3), 459–478 (2014)
5. Buchmann, J., Dahmen, E., Hülsing, A.: XMSS - a practical forward secure signature scheme based on minimal security assumptions. In: Yang, B.-Y. (ed.) PQCrypto 2011. LNCS, vol. 7071, pp. 117–129. Springer, Heidelberg (2011). https://doi.org/10.1007/978-3-642-25405-5_8
6. Buldas, A., Geihs, M., Buchmann, J.: Long-term secure commitments via extractable-binding commitments. In: Pieprzyk, J., Suriadi, S. (eds.) ACISP 2017. LNCS, vol. 10342, pp. 65–81. Springer, Cham (2017). https://doi.org/10.1007/978-3-319-60055-0_4
7. Catalano, D., Fiore, D.: Vector commitments and their applications. In: Kurosawa, K., Hanaoka, G. (eds.) PKC 2013. LNCS, vol. 7778, pp. 55–72. Springer, Heidelberg (2013). https://doi.org/10.1007/978-3-642-36362-7_5
8. Geihs, M., Buchmann, J.: Elsa: Efficient long-term secure storage of large datasets (full version). arXiv:1810.11888 (2018)
9. Geihs, M., Karvelas, N., Katzenbeisser, S., Buchmann, J.: Propyla: privacy preserving long-term secure storage. In: Proceedings of the 6th International Workshop on Security in Cloud Computing, SCC 2018, pp. 39–48. ACM, New York (2018). https://doi.org/10.1145/3201595.3201599
10. Gennaro, R., Micali, S.: Independent zero-knowledge sets. In: Bugliesi, M., Preneel, B., Sassone, V., Wegener, I. (eds.) ICALP 2006. LNCS, vol. 4052, pp. 34–45. Springer, Heidelberg (2006). https://doi.org/10.1007/11787006_4

11. Gisin, N., Ribordy, G., Tittel, W., Zbinden, H.: Quantum cryptography. Rev. Mod. Phys. **74**, 145–195 (2002)
12. Haber, S., Stornetta, W.S.: How to time-stamp a digital document. J. Cryptol. **3**(2), 99–111 (1991). https://doi.org/10.1007/BF00196791
13. Halevi, S., Micali, S.: Practical and provably-secure commitment schemes from collision-free hashing. In: Koblitz, N. (ed.) CRYPTO 1996. LNCS, vol. 1109, pp. 201–215. Springer, Heidelberg (1996). https://doi.org/10.1007/3-540-68697-5_16
14. Herzberg, A., Jarecki, S., Krawczyk, H., Yung, M.: Proactive secret sharing or: how to cope with perpetual leakage. In: Coppersmith, D. (ed.) CRYPTO 1995. LNCS, vol. 963, pp. 339–352. Springer, Heidelberg (1995). https://doi.org/10.1007/3-540-44750-4_27
15. Hofheinz, D.: Possibility and impossibility results for selective decommitments. J. Cryptol. **24**(3), 470–516 (2011). https://doi.org/10.1007/s00145-010-9066-x
16. Lenstra, A.K.: Key lengths. In: The Handbook of Information Security. Wiley, Hoboken (2004)
17. Lenstra, A.K., Verheul, E.R.: Selecting cryptographic key sizes. J. Cryptol. **14**(4), 255–293 (2001)
18. Merkle, R.C.: A certified digital signature. In: Brassard, G. (ed.) CRYPTO 1989. LNCS, vol. 435, pp. 218–238. Springer, New York (1990). https://doi.org/10.1007/0-387-34805-0_21
19. National Institute of Standards and Technology: FIPS 197: Announcing the advanced encryption standard (AES) (2001)
20. National Institute of Standards and Technology: FIPS PUB 180-4: Secure hash standard (SHS) (2015)
21. Rivest, R.L., Shamir, A., Adleman, L.: A method for obtaining digital signatures and public-key cryptosystems. Commun. ACM **21**(2), 120–126 (1978)
22. Shamir, A.: How to share a secret. Commun. ACM **22**(11), 612–613 (1979)
23. Shannon, C.E.: Communication theory of secrecy systems. Bell Syst. Tech. J. **28**(4), 656–715 (1949)
24. Shor, P.W.: Polynomial-time algorithms for prime factorization and discrete logarithms on a quantum computer. SIAM J. Comput. **26**(5), 1484–1509 (1997). https://doi.org/10.1137/S0097539795293172
25. Vigil, M.A.G., Buchmann, J.A., Cabarcas, D., Weinert, C., Wiesmaier, A.: Integrity, authenticity, non-repudiation, and proof of existence for long-term archiving: a survey. Comput. Secur. **50**, 16–32 (2015)
26. Weinert, C., Demirel, D., Vigil, M., Geihs, M., Buchmann, J.: Mops: a modular protection scheme for long-term storage. In: Proceedings of the 2017 ACM on Asia Conference on Computer and Communications Security, ASIA CCS 2017, pp. 436–448. ACM, New York (2017)

How to Block the Malicious Access to Android External Storage

Sisi Yuan[1,2,3], Yuewu Wang[2,3], Pingjian Wang[2,3(✉)], Lingguang Lei[2,3],
Quan Zhou[2,3], and Jun Li[4]

[1] School of Cyber Security, University of Chinese Academy of Sciences,
Beijing, China
[2] Institute of Information Engineering, CAS, Beijing, China
wangpingjian@iie.ac.cn
[3] Data Assurance and Communication Security Research Center, CAS,
Beijing, China
[4] Zhongxing Telecommunication Equipment Corporation, Shenzhen, China

Abstract. External storage (e.g., SD card) is an important component of the Android mobile terminals, commonly used for storing of the user information (including sensitive data such as photos). However, current protection mechanisms (e.g., the permission mechanism) on the external storage are somehow coarse-grained, where the external storage is controlled as a whole, which means all files on the external storage are accessible once the permission is assigned to an APP. This coarse-grained control weakness could be easily leveraged by the attackers. For example, the ransomware can obtain the access permission of the external storage and encrypt the files on external storage stealthily for ransom. In this paper, we introduce an Access Control List (ACL) mechanism to enforce the fine-grained control on the external storage. With ACL, the access control policy can be defined at the file granularity, and the access permissions will only be granted to legitimate APPs specified in a white list. First, we activate the Linux ACL mechanism on Android system and extend it to the Filesystem in Userspace (FUSE). Because the external storage is built on the FUSE filesystem, which is different from the traditional Linux filesystems (e.g., EXT4) and thus not supported by the traditional Linux ACL mechanism. Second, we introduce ACL-policy configuration interface in the Android framework, which enables the device owner and APP developers to set the fine-grained ACL access policies for their files on the external storage. Finally, we implement a prototype based on the Nexus 6 devices deployed Android 6.0.1 and Linux kernel 3.10.4, and evaluate it on the stability, effectiveness and performance. The results show our prototype system can effectively prevent illegal access to the files on the external storage with negligible performance overhead. As far as we know, this is the first work that can really enforce ACL access control on the external storage of Android.

Keywords: Access Control List · Android access control ·
External storage · Ransomware

© Springer Nature Switzerland AG 2019
K. Lee (Ed.): ICISC 2018, LNCS 11396, pp. 287–303, 2019.
https://doi.org/10.1007/978-3-030-12146-4_18

1   Introduction

External storage is indispensable in Android system. Data communications between Android and PC system are mainly via external storage. Furthermore, external storage is a cheap way to extend Android storage. Even today, APP developers are still willing to store large data on external storage and share those data with others.

However, Android security mechanism towards external storage is inadequate. Based on Linux kernel, Android inherits some features of the Linux architecture and mainly utilizes Linux discretionary access control (DAC) named user/group/other (UGO) to protect external storage. UGO enforces access control policies according to APP's Linux user ID (UID) and Linux group IDs (GIDs). Each APP will be assigned a unique Linux UID at install-time, and granted some Linux GIDs consistent with the permissions it requested. For external storage access, this kind of mechanism is rather coarse-grained. Device owner can only determine whether an APP could read or write all the files on external storage, while the range of accessible files is uncontrollable. SEAndroid is another important Android security mechanism, but this mandatory access control (MAC) mechanism cannot work well on external storage protection, because the policies of external storage access need to be changed frequently.

The lack of effectiveness protection on external storage makes it easy to become an object of attacks. In recent years, external storage has become one of ransomware's major targets. Since external storage is accessible after obtaining certain permissions, ransomware may encrypt all the files on external storage stealthily for ransom. This attack mode has been adopted by many ransomware such as Simplocker family [17] and DoubleLocker [8]. Moreover, malware may analyze the data on external storage, such as geographic information of photos, then user's sensitive information may be exposed [24].

If we can restrict the access to the files on external storage at the file-level, we can block the malicious access effectively while sharing files as we want. Linux ACL is such a mechanism that can support fine-grained access control policies on file access. It is hooked in the path of permission check after UGO mechanism, and cannot be bypassed by userspace code. For each file, the ACL can set different access permissions for different UIDs. The policies of ACL is richer than that of UGO, and more fine-grained access may be enforced by it. In addition, because every APP has a unique UID, ACL mechanism is suitable for Android.

Unfortunately, it is not simple to implement ACL access control on current Android external storage. Firstly, starting with Android 4.4, *FUSE* filesystem is adopted to realize access control for external storage. Unlike traditional Linux filesystems, *FUSE* does not support ACL well. Secondly, there are two types of external storage filesystem format in Android, that is, the built-in SD card is *EXT4* format and the removable SD card is mostly *VFAT* format. *VFAT* format does not support ACL at all. Thirdly, we have to make the policies of ACL only be modified by the specific subject.

In this paper, we overcome foregoing difficulties and present a complete solution to enforce ACL access control on Android external storage. We first activate the ACL mechanism in Android Linux kernel. Then, the ACL mechanism is extended to *FUSE* filesystem through introducing ACL features into *sdcard daemon* that in charge of *FUSE* based external storage access control. We also add hooks into *VFAT* filesystem to make it support ACL features. Finally, ACL-policy configuration interface is implemented to enable device owner and APP developers to set ACL policies of their files on external storage. The interface can effectively prevent ACL-policy tampering.

In summary, we make following contributions in this paper.

- We presented an ACL access control mechanism for Android external storage. Although ACL mechanism has been introduced into Linux kernel successfully to enforce fine-grained access control on system file, the *FUSE* used by Android external storage management does not support the ACL mechanism well. Our solution is the first work that implements the extension of ACL to *FUSE* based Android external storage. With the ACL mechanism, we can set fine-grained policies for every APP to block the potential malicious access to these files on external storage.
- We design and implement the ACL-policy configuration interface compatible with Android permission mechanism. The interface allows device owner and APP developers to create customizable ACL policies. For device owner, a system-level APP is provided, by using which the owner can modify the ACL policies. For APP developers, a set of APIs is provided to customize the ACL policies of files created by the APP. Android permission mechanism is used to assure that only authorized policies configuration operations are allowed.
- We develop a prototype system based on Nexus 6 devices with Android 6.0.1, and Linux kernel 3.10.4 to evaluate the effectiveness and efficiency. Experimental results demonstrate that our work may work well with negligible performance overhead.

The rest of paper is organized as follows. Section 2 describes the related background knowledge. Section 3 shows the design of our system. Section 4 introduces how we implement our prototype in detail. Section 5 presents the evaluation of our prototype system. Section 6 shows related work. In Sect. 7, we summarize our work.

2   Background

2.1   Access Control List

Access control list (ACL) is a kind of access control mechanism adopted by many systems. Linux has implemented complete ACL package [2]. Compared with Linux default filesystem access control mechanism-UGO, ACL can provide an additional and more flexible permission mechanism. UGO access control mechanism just uses 9 bits to represent subject's (that is, owner, group and others) permissions towards a certain file. Take *rwxrw-r-* for example, the file owner can read, write and execute it, the group that the file belongs to can read and write it, and others can only read it. UGO is more or less coarse-grained. Android introduced Permission mechanism based on UGO to implement fine-grained authority assignment at the APP level [28]. Permission mechanism can only make sure whether an APP can access external storage, and cannot cover access control for every file on Android external storage.

ACL allows us to give permissions for any user or group to any file. ACL policy is a white list of permissions attached to a file. Each entry in the list specifies what permissions a UID is granted when accessing the file. For example, an entry *user:BOB:rwx* means BOB can read, write and execute the file. ACL mechanism is very suitable for enforcing access control on Android APP, because every APP has a unique UID.

The ACL mechanism is compatible with the UGO mechanism. ACL is an optional mechanism after UGO mechanism. If ACL mechanism is enabled, a file access will be confronted with ACL check after passing UGO check. Thus, introducing ACL into Android will not affect existing Android security mechanisms.

2.2   FUSE Filesystem

Filesystem in Userspace (FUSE) [18] is a software module for Unix-like OSes that enables non-privileged user to create his own filesystem without modifying kernel code. This is achieved by running user's filesystem code in userspace while the *FUSE* module provides a "bridge" to the actual kernel-level filesystem.

Figure 1 shows how Android works with *FUSE* filesystem. As shown in Fig. 1, raw external storage devices are mounted as *EXT4* filesystem (built-in SD card) or *VFAT* filesystem (removable SD card). Android uses *FUSE* to wrap the raw external storage devices. Thus, any access to external storage has to go through *FUSE* first, and then uses userspace filesystem called *sdcard daemon* to access real filesystems.

It can be seen that external storage access in Android is actually done by *sdcard daemon*. *Sdcard daemon* masks the details of the actual filesystem of external storage. If we want to achieve functionality on *FUSE* based external storage, we need to enable certain feature on *sdcard daemon* too.

2.3   Access Control for External Storage in Android

In Android versions before 1.5, an APP was permitted to write and read the entire external storage freely. Since Android 1.6, an APP has to apply for some permissions statically to access external storage. From Android 4.4, a UGO-like permission management was adopted for external storage access [33]. All the files on external storage are set up separately with the UGO policy. Although Android 6.0 adopted the dynamic external storage access permission applying, granularity of access control is still very coarse.

As mentioned above, Android external storage permission management is fixed and coarse-grained. There is no way to set permissions of a certain file on

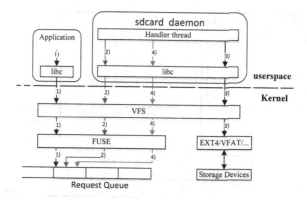

Fig. 1. Operation flow when APP accesses external storage.

external storage for a specific APP. As long as the device owner grants external storage related permissions to a third party APP, the APP can read and write all the public files on external storage as it wants.

3 System Design

3.1 Design Principles and Architecture Overview

Threat Model. Our work mainly aims at the attacks on external storage in Android framework. The prevailing forms of attacks are malwares. Malwares we discuss may take advantage of the inflexible coarse-grained permission management to utilize or damage the files on external storage, such as dig out user's sensitive information, encrypt files and modify files.

Assumption. In our work, we assume the Linux kernel of the Android platform is trustworthy. The attack will not destroy the security mechanisms of the Linux kernel. We also assume the Android security mechanisms, such as Android permission and sandbox, are effective and cannot be bypassed.

Goals. Our work is designed to provide ACL access control at the file-level for external storage of Android. Only when the certain permissions are granted to the UID attached to an APP, the APP is able to access the file on external storage in a way specified by the permissions.

Thus, we propose a complete solution that enables ACL access control on Android external storage so as to customize ACL policies for files on external storage. Our system consists of following three modules: ACL policy management module, ACL enabled module and ACL policy storage module, as shown in Fig. 2.

3.2 ACL Policy Management Module

ACL policy management module is designed to support ACL management. This module consists of three parts: shared library, system-level APP and APIs.

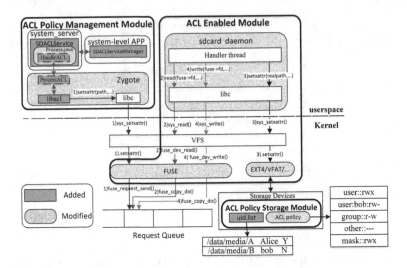

Fig. 2. Architecture overview of our system.

This module enriches the Android shared library. Since *POSIX ACL* [21] generally saves as extended attributes (so-called xattrs), ACL entries and xattrs need to convert each other. Android does not support that conversions, so this module imposes a shared library called libacl-devel [3] library into Android to support that transformation.

In addition, a system-level APP is designed to interact with device owner to customize ACL policies for the files on the external storage. Because most of the files on the external storage are owned by root, a system service called *SDACLService* attached with *signatureOrSystem*-level permission is designed to communicate with *Zygote* to manage ACL policies using *root* privileges. Then, our system-level APP uses this service to achieve ACL policies management.

Moreover, a set of APIs for APP developers is designed to customize ACL policies for APP's files on external storage. APP developers can use APIs to block any other APP's access to their APP's files that may leak user's information. These APIs include wrapper functions that perform the ACL-related operations provided by implanted libacl-devel library. The permission check embedded in Linux kernel will determine whether the APP has capabilities to set ACL policies on certain object.

3.3   ACL Policy Storage Module

ACL policies are stored in this module. Typical formats of external storage are *VFAT* and *EXT4*. If we enforce ACL access control for these two filesystems, we may need to store corresponding ACL policies in the filesystem. It is generally known that *EXT4* filesystem stores ACL information with the format of xattrs as a part of *inode*. Once ACL features of *EXT4* filesystem are enabled, we can utilize xattrs-related operations of *EXT4* to store ACL policies. As for *VFAT* filesystem, it does not support ACL features in essence, so additional ACL policies storage recorded in this module is necessary for this kind of filesystem to realize the ACL access control.

In addition, except for specific directories, the owner of the files on Android external storage is root, even if the files are created by an APP. In order to achieve the APP's ACL management towards its files on external storage, we need to change the owner of those files. Thus, we record the corresponding relationship between APP and its files in the *uid.list*, and use hooks to modify the owner of files on external storage. The entity recorded in *uid.list* contains three parts: the path of the file, the uid of file's creator, and effectiveness of the entity ("Y" for valid, "N" for invalid). For example, an entity */data/media/A Alice Y* means the file */data/media/A* is created by *Alice*, and *Y* means this entity is valid.

3.4   ACL Enabled Module

ACL enabled module is the central part of our work, since all permission setting, querying and checking have to base on it. Three major functions are included in this module.

Enable embedded ACL of Linux kernel. Any ACL-related operation generated in ACL policy management module will finally turn into a system call for Linux kernel. Linux kernel has embedded ACL supports disabled by Android for some reasons. So this module will enable the embedded *POSIX ACL* features.

Extend ACL features to *FUSE* filesystem. Android uses *FUSE* filesystem to wrap the real filesystems external storage based on, and uses a userspace daemon to operate real filesystems external storage based on. Since current Android does not implement the ACL or xattrs features for foregoing parts, this module will extend ACL features from *FUSE* filesystem to the filesystems external storage based on.

Enrich SEAndroid policies. Android has introduced a security mechanism based on SELinux [30], called SEAndroid [32], to enhance system security. Any additional function to Android system must be declared in the SEAndroid policies before being enabled. So, as we enhance the security module, we must synchronously add the corresponding SEAndroid policies.

4   Implementation

In this section, the implementation details of some critical components of our work are described.

4.1   Enabling ACL Features in Linux Kernel

First of all, we enable the embedded ACL features of Linux kernel. Although Android disables some features of the Linux kernel, the entire functions related to *POSIX ACL* still can be found in the source code of Linux kernel that Android uses. Thus, in our work, we modify the build configuration file of the Linux kernel corresponding to the Android Open Source Project (AOSP) [5] version we use, to enable embedded *POSIX ACL* features. Table 1 shows modified arguments.

Table 1. Configuration file modification

Argument name	Original value	Value after modification
CONFIG_EXT4_FS_POSIX_ACL	Not set	y
CONFIG_FS_POSIX_ACL	Not set	y
CONFIG_GENERIC_ACL	Not set	y

Then, we enable the ACL features for FUSE filesystem. To be specific, we add source code of xattr-related functions (e.g., *setxattr* and *getxattr*) into a new file called *xattr.c*. Then, we backport ACL-related functions (such as *setacl* and *getacl*) into a file called *acl.c* to handle ACL-related operations. With these two files, FUSE can call the ACL mechanism in kernel to complete ACL operations.

Furthermore, we enable ACL features on the real filesystems that external storage based on to handle the ACL-related operations sent by FUSE. After modifying the kernel build configuration, we enable the ACL features of *EXT4* filesystem. Because *VFAT* filesystem dose not support ACL at all, we have to introduce additional code into *VFAT* filesystem to make it support ACL features. The ACL related code is hooked in the control-flow of *VFAT* filesystem.

After making the above modifications, we recompile the modified kernel source code, and package it into the image. The new kernel image will support the ACL operations towards external storage.

4.2   Introducing ACL Features into Userspace Daemon

The *FUSE* filesystem uses a userspace daemon called *sdcard daemon* to handle the file operation requests acquired from FUSE Request Queue shown in Fig. 2. Therefore, while modifying Linux kernel, we need to enable ACL-related features in *sdcard daemon*, especially xattrs-related opertaions. Because *FUSE* will convert ACL-related requests to xattrs-related requests and transfer the requests to *sdcard daemon*.

We add xattrs-related operations into the source code of *sdcard daemon* called *sdcard.c* to make *sdcard daemon* handle xattrs-related requests from kernel *FUSE* filesystem. By tracking call flow, we found that it is *handle_fuse_request()* function in *sdcard.c* that handles *opcode* of *fuse_in_header* structure gotten from FUSE Request Queue and calls the corresponding function to complete the specific operation. Thus, we add extra opcodes and corresponding functions, as shown in Listing 1.1, to complete the specific operation.

Listing 1.1. Handle Additional Opcodes

```
1 switch opcode do
2     ......
3     case FUSE_GETXATTR
4         call handle_getxattr
5     case FUSE_SETXATTR
6         call handle_setxattr
```

To enforce ACL at the file level, the ownership of the files on external storage need to be various. However, as mentioned above, the default owner of the files on external storage is *root*. We use the *uid.list* stored on ACL policy storage module to make the owner of the files on external storage be specified. Following steps are adopted to complete this function.

(i) We add a hook into the *sdcard daemon*'s function that parses *packages.list*. Whenever FUSE based external storage is mounted or *packages.list* is changed, this function will be invoked. As List 1.2 shows, the added hook parses *uid.list* and stores the file's ownership information in the hashmap, if the entity is valid. If the entity is invalid, it will be removed from *uid.list*.

(ii) We add a judgment into permission-deriving function of *sdcard daemon*. As shown in List 1.3, the judgment will assign the uid gotten from one hashmap entity to the node.uid of the file, if the path of the file matches with that of the hashmap entity.

(iii) We also add modifications into functions, such as file creation and deletion in *sdcard daemon*, to make them dynamically modify the *uid.list* and the hashmap.

Listing 1.2. Parse the *uid.list*

```
1 for all item in uid.list do
2     if entity is invalid then
3         remove the entity from uid.list
4     else
5         hashmapPut(hashmap , entity.path , entity.uid)
6     end if
7 end for
```

Listing 1.3. Derive Permissions for Node

```
1 derive all attributes from the node.parent
2 if node.path exists in hashmap then
3     node.uid ← hashmapGet(hashmap , node.path)
4 end if
```

4.3 SEAndroid Configuration

SEAndroid is a mandatory access control mechanism, so-called MAC. In SEAndroid, each process and file is associated with a security context. When every process and file is attached with a security context, the system administrator can make the security access policies based on the security context, that is, decides what kind of process can access the given file.

Due to the lack of rules in SEAndroid configuration, system will block the functions we added. That is, *SDACLService* and *sdcard daemon* cannot operate files' attributes as well as read or write files on external storage, even after what we do above. To make what we added available, we add rules into SEAndroid configuration, showed in Table 2.

Table 2. Added SEAndroid configuration

File name	Policy
sdcardd.te	allow sdcardd storage_file:lnk_file {read write getattr}
zygote.te	allow zygote storage_file:lnk_file {read write} allow zygote fuse:file {getattr setattr}
service_context	SDACLService u:object_r:system_server_service:s0

5 Evaluation

In this section, we will evaluate our work. In order to evaluate our work, we compare our prototype system (named as "Modified" in the tables) with the unmodified system (named as "Original" in the tables) that based on same AOSP version and kernel version as ours. We build the AOSP images from version 6.0.1 for Nexus 6 devices, and build kernel images from version 3.10.4. Then, we evaluate the systems based on following metrics.

(i) Test the impact on system's stability.
(ii) Verify effectiveness of our work.
(iii) Evaluate performance overhead.

5.1 Stability

We downloaded 50 APPs that may read or write external storage from the Android official market [20] for this test. Then, we manually run those APPs in our system as well as in unmodified system. After properly granting permissions, those APPs can read and write external storage as their functionality design.

Thus, we believe that our system is rather stable and does not affect the legitimate operations towards external storage.

5.2 Effectiveness

Effectiveness Towards Malicious APP. To test effectiveness of our system, we develop a testing tool to simulate the behavior of ransomware Simplocker family, that is, the testing tool can stealthily encrypt the files on external storage.

Then, we apply the testing tool on both unmodified system and our system. In our system, we disable group privileges of external storage and set proper ACL policies for external storage before starting the testing tool. After running the testing tool, both of testing systems grant the runtime permissions towards external storage to the tool.

As a result, barely all files on external storage are encrypted without notifying the device owner in unmodified system. On the contrary, the testing tool do nothing harmful to external storage except for its own files in our system.

Effectiveness of System APP's Functionality. In this part, we will illustrate the effectiveness of the system APP (named as "ManagementAPP") mentioned in Sect. 3.2, as shown in Figs. 3 and 4.

Firstly, we show the case that no permission is granted to the given APP, as shown in Fig. 3. Above all, we select a folder named "Ringtones", as shown in Fig. 3(a), and click the button named "GRANT_PERMISSION". Then, in the permission-granting interface as shown in Fig. 3(b), we input "rootexplorer" the name of an APP (if no input, it means to set the permission for the group to which the file or folder belongs), and we do not select any permission. As shown in Fig. 3(c), the APP cannot access the data inside the selected folder because the APP does not have any permission towards the folder.

(a) Select What You Want to Set Permissions for

(b) No Permission is Granted to Given APP

(c) the Given APP Cannot Access

Fig. 3. No permission is granted.

Secondly, we show the case that all permissions are granted to the given APP, as shown in Fig. 4. The first two steps shown in Figs. 4(a) and (b) are similar to those shown in Figs. 3(a) and (b) above. As shown in Fig. 4(c), since the APP obtains permissions, the APP can access the data in the selected folder.

In conclusion, our prototype can effectively block the malicious access towards external storage.

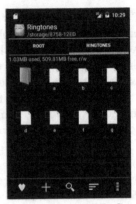

(a) Select What You Want to Set Permissions for

(b) All Permissions are Granted to the Given APP

(c) the Given APP Can Access

Fig. 4. All permissions are granted.

5.3   Performance Overhead

Overall Performance Overhead. We use three common overall performance testing sets to evaluate our work's overall performance overhead, that is, Quadrant, AnTuTu Benchmark [4] and Geekbench4 [19].

The scores and results of above benchmark tools are shown in Table 3. From the table, we can see that the overhead produced by our system in Quadrant Benchmark is no more than 5.26%, in AnTuTu Benchmark is no more than 5.6% and in GeekBench4 is less than 2.08%, all of which can be negligible.

So we reach a conclusion that our system has little impact on overall performance.

Table 3. Benchmark result

Testing item	Original	Modified	Diff.	Diff. (%)
Quadrant: CPU	46699	45680	1019	2.18
Quadrant: MEM	14715	14715	0	0.00
Quadrant: I/O	5585	5291	294	5.26
Quadrant: Total	13780	13545	235	1.71
Antutu: GPU	5265	5118	147	2.79
Antutu: MEM	29348	29320	28	0.10
Antutu: UX	21327	20132	1195	5.60
Geekbench4: Single-Core Score	1009	988	21	2.08
Geekbench4: Multi-Core Score	2837	2827	10	0.352

External Storage Performance Overhead. We use an external storage performance testing tool called A1 SD Bench [1] to evaluate the performance of external storage.

We performed A1 SD Bench 10 times on unmodified system and our system, then we calculated the average and standard deviation of the results. As results of A1 SD Bench shown in Table 4, we find out that additional performance overhead that our system introduces is acceptable.

Table 4. A1 SD bench

	Original		Modified	
	Mean (MB/s)	SD (MB/s)	Mean (MB/s)	SD (MB/s)
Random I/O: Read	27.629	3.411	26.284	3.802
Random I/O: Write	0.845	0.019	0.828	0.058
Accurate: Read	50.241	1.107	49.82	1.617
Accurate: Write	63.789	1.495	58.349	1.729
Long Time: Read	47.887	0.399	47.437	1.282
Long Time: Write	63.269	1.302	57.867	1.328

6   Related Work

This section shows the overview of the related work. There are plenty of security extensions proposed enhancing Android security, most of which are for the Android middleware layer or kernel layer. Section 6.1 introduces improvements that mainly resolved the access control problem at the middleware layer. Section 6.2 shows solutions that introduced MAC access control into Android. Section 6.3 presents some solutions for Android external storage.

6.1   Android Middleware Layer Improvements

TaintDroid [15] allows users to track and analyze flows of sensitive data and potentially identify suspicious APPs. Saint [27] enforces policies that leverage the relationship between the caller APP and the callee APP. Apex [26] allows users to accept a subset of the permissions declared by APPs. Kirin [16] alarms users when APP's declared permissions violate predefined policies. SPAC [36] scheme provides fine-grained permission enforcement at component level. [6,7,13,23,29] are based on context to enhance Android security. Cai et al. [12] modify both Android middleware layer and kernel layer to enforce ACL access control on Android platform, but they do not offer an effective access control scheme for Android external storage.

6.2 Android MAC Access Control Enhancements

There has been a lot of work to harden the Linux kernel with MAC based systems that exercise the principle of least privilege more strictly. There has been prior studies [25,31,32] on integrating and applying SELinux [30] in Android. Besides, TrustDroid [10] and XManDroid [9] provide MAC at both middleware layer and kernel layer, relying upon pathname-based security model of TOMOYO Linux [34]. And FlaskDroid [11] is a generic kernel and middleware MAC based architecture that can support multiple fine-grained security policies and use cases.

6.3 Android Improvements for External Storage

Currently, there are few improvements to enhance the security of Android external storage. Wang et al. [35] implemented an encryption filesystem on Android utilizing FUSE, and can only protect external storage from the attacks from outside the device. Our work can intercept internal APPs' illegal access to files on external storage. Do et al. [14] enforced filesystem permission on external SD card by reformatting it to the EXT4 filesystem format. Our work does not need that reformation. Liu et al. [24] carried out an empirical study on the data stored on external storage of Android. They presented several attacks based on these data and proposed a defense framework. The limitation of their solution is that it only works at framework/API layer. Huang et al. [22] introduced an external storage data sharing GID (ESDS-GID) into current Android security model. This work extends the GID used to manage external storage, but it cannot set flexible access control policies for a certain file on external storage as we did.

7 Conclusions and Future Work

In this paper, we introduced the Access Control List (ACL) for external storage. We allow APP developers and device owner to create customized ACL policies for resources they can manage on external storage. We present a customized system by modifying source code of framework layer, middleware layer and Linux kernel. Besides, we provide APIs for APP developers to enable them to use ACL policies to protect APP's resources on external storage, and offer a system-level APP for device owner to create customized ACL policies. Evaluation results suggest that the system can efficiently prevent the unauthorized APP from accessing external storage as well as offer a stable environment with negligible cost.

Since Android 8.0 Oreo, SDCardFS replaces FUSE and avoids extra round trip by being an in-kernel FAT32 emulation layer, and it is newly integrated into AOSP. Since SDCardFS was not the mainstream when we started our work, so our work aims at FUSE filesystem. In the future, we may introduce our work into SDCardFS and simplify the process of granting permission.

Acknowledgements. We would like to thank our anonymous reviewers for their valuable comments and suggestions. This work is supported by the National Key Research and Development Program of China under Grant No. 2016YFB0800102, the National Cryptography Development Fund under Award No. MMJJ20170215, and the Youth Innovation Promotion Association CAS.

References

1. A1 sd bench. http://a1dev.com/sd-bench/
2. Access control list. https://en.wikipedia.org/wiki/Access_control_list
3. Acl open source community. http://savannah.nongnu.org/projects/acl
4. Antutu benchmark. http://www.antutu.com/en/index.htm
5. Aosp. https://source.android.com/
6. Arena, V., Catania, V., Torre, G.L., Monteleone, S., Ricciato, F.: Securedroid: an android security framework extension for context-aware policy enforcement. In: 2013 International Conference on Privacy and Security in Mobile Systems, PRISMS 2013, Atlantic City, NJ, USA, 24–27 June 2013, pp. 1–8 (2013). https://doi.org/10.1109/PRISMS.2013.6927185
7. Bai, G., Gu, L., Feng, T., Guo, Y., Chen, X.: Context-aware usage control for android. In: Proceedings 6th International ICST Conference Security and Privacy in Communication Networks - SecureComm 2010, Singapore, 7–9 September 2010, pp. 326–343 (2010). https://doi.org/10.1007/978-3-642-16161-2_19
8. Beware of ransomware and high risks threats on android devices. https://www.symantec.com/connect/articles/beware-ransomware-and-high-risks-threats-android-devices
9. Bugiel, S., Davi, L., Dmitrienko, A., Fischer, T., Sadeghi, A., Shastry, B.: Towards taming privilege-escalation attacks on android. In: 19th Annual Network and Distributed System Security Symposium, NDSS 2012, San Diego, California, USA, 5–8 February 2012 (2012). http://www.internetsociety.org/towards-taming-privilege-escalation-attacks-android
10. Bugiel, S., Davi, L., Dmitrienko, A., Heuser, S., Sadeghi, A., Shastry, B.: Practical and lightweight domain isolation on android. In: SPSM 2011, Proceedings of the 1st ACM Workshop Security and Privacy in Smartphones and Mobile Devices, Co-located with CCS 2011, 17 October 2011, Chicago, pp. 51–62 (2011). https://doi.org/10.1145/2046614.2046624
11. Bugiel, S., Heuser, S., Sadeghi, A.: Flexible and fine-grained mandatory access control on android for diverse security and privacy policies. In: Proceedings of the 22th USENIX Security Symposium, Washington, DC, USA, 14–16 August 2013, pp. 131–146 (2013). https://www.usenix.org/conference/usenixsecurity13/technical-sessions/presentation/bugiel
12. Cai, X., Gu, X., Wang, Y., Zhou, Q., Cao, Z.: Enforcing ACL access control on android platform. In: Proceedings 20th International Conference Information Security - ISC 2017, Ho Chi Minh City, Vietnam, 22–24 November 2017, pp. 366–383 (2017). https://doi.org/10.1007/978-3-319-69659-1_20
13. Conti, M., Crispo, B., Fernandes, E., Zhauniarovich, Y.: Crêpe: a system for enforcing fine-grained context-related policies on android. IEEE Trans. Inf. Forensics Secur. **7**(5), 1426–1438 (2012). https://doi.org/10.1109/TIFS.2012.2204249

14. Do, Q., Martini, B., Choo, K.R.: Enforcing file system permissions on android external storage: android file system permissions (AFP) prototype and owncloud. In: 13th IEEE International Conference on Trust, Security and Privacy in Computing and Communications, TrustCom 2014, Beijing, China, 24–26 September 2014, pp. 949–954 (2014). https://doi.org/10.1109/TrustCom.2014.53

15. Enck, W., et al.: TaintDroid: an information flow tracking system for real-time privacy monitoring on smartphones. Commun. ACM **57**(3), 99–106 (2014). https://doi.org/10.1145/2494522

16. Enck, W., Ongtang, M., McDaniel, P.D.: On lightweight mobile phone application certification. In: Proceedings of the 2009 ACM Conference on Computer and Communications Security, CCS 2009, Chicago, Illinois, USA, 9–13 November 2009, pp. 235–245 (2009). https://doi.org/10.1145/1653662.1653691

17. Eset analyzes simplocker. https://www.welivesecurity.com/2014/06/04/simplocker/

18. Filesystem in userspace. https://en.wikipedia.org/wiki/Filesystem_in_Userspace

19. Geekbench4. http://www.geekbench.com

20. Google play. https://play.google.com/store

21. Grünbacher, A.: POSIX access control lists on linux. In: Proceedings of the FREENIX Track: 2003 USENIX Annual Technical Conference, San Antonio, Texas, USA, 9–14 June 2003, pp. 259–272 (2003). http://www.usenix.org/events/usenix03/tech/freenix03/gruenbacher.html

22. Huang, F., Wu, W., Yang, M., Luo, J.: A fine-grained permission control mechanism for external storage of android. In: 2016 IEEE International Conference on Systems, Man, and Cybernetics, SMC 2016, Budapest, Hungary, 9–12 October 2016, pp. 2911–2916 (2016). https://doi.org/10.1109/SMC.2016.7844682

23. Jung, C., Feth, D., Seise, C.: Context-aware policy enforcement for android. In: IEEE 7th International Conference on Software Security and Reliability, SERE 2013, Gaithersburg, MD, USA, 18–20 June 2013, pp. 40–49 (2013). https://doi.org/10.1109/SERE.2013.15

24. Liu, X., Zhou, Z., Diao, W., Li, Z., Zhang, K.: An empirical study on android for saving non-shared data on public storage. In: Proceedings 30th IFIP TC 11 International Conference, ICT Systems Security and Privacy Protection - SEC 2015, Hamburg, Germany, 26–28 May 2015, pp. 542–556 (2015). https://doi.org/10.1007/978-3-319-18467-8_36

25. Nakamura, Y., Sameshima, Y.: SELinux for consumer electronics devices. In: 2008 Proceedings of the Linux Symposium OLS, Ottawa, Ontario, Canada, 23–26 July 2008, pp. 125–134 (2008)

26. Nauman, M., Khan, S., Zhang, X.: Apex: extending android permission model and enforcement with user-defined runtime constraints. In: Proceedings of the 5th ACM Symposium on Information, Computer and Communications Security, ASIACCS 2010, Beijing, China, 13–16 April 2010, pp. 328–332 (2010). https://doi.org/10.1145/1755688.1755732

27. Ongtang, M., McLaughlin, S.E., Enck, W., McDaniel, P.D.: Semantically rich application-centric security in android. Secur. Commun. Netw. **5**(6), 658–673 (2012). https://doi.org/10.1002/sec.360

28. Permissions overview. https://developer.android.com/guide/topics/permissions

29. Roesner, F., Kohno, T., Moshchuk, A., Parno, B., Wang, H.J., Cowan, C.: User-driven access control: rethinking permission granting in modern operating systems. In: IEEE Symposium on Security and Privacy, SP 2012, San Francisco, California, USA, 21–23 May 2012, pp. 224–238 (2012). https://doi.org/10.1109/SP.2012.24

30. Security-enhanced linux. http://www.nsa.gov/research/selinux
31. Shabtai, A., Fledel, Y., Elovici, Y.: Securing android-powered mobile devices using selinux. IEEE Secur. Priv. 8(3), 36–44 (2010). https://doi.org/10.1109/MSP.2009.144
32. Smalley, S., Craig, R.: Security enhanced (SE) android: bringing flexible MAC to android. In: 20th Annual Network and Distributed System Security Symposium, NDSS 2013, San Diego, California, USA, 24–27 February 2013 (2013). https://www.ndss-symposium.org/ndss2013/
33. Storage. https://source.android.com/devices/storage/
34. Tomoyo linux home page. http://tomoyo.sourceforge.jp/
35. Wang, Z., Murmuria, R., Stavrou, A.: Implementing and optimizing an encryption filesystem on android. In: 13th IEEE International Conference on Mobile Data Management, MDM 2012, Bengaluru, India, 23–26 July 2012, pp. 52–62 (2012). https://doi.org/10.1109/MDM.2012.31
36. Wu, L., Du, X., Zhang, H.: An effective access control scheme for preventing permission leak in android. In: International Conference on Computing, Networking and Communications, ICNC 2015, Garden Grove, CA, USA, 16–19 February 2015, pp. 57–61 (2015). https://doi.org/10.1109/ICCNC.2015.7069315

A Novel Tamper Evident Single Database Information-Theoretic Private Information Retrieval for User Privacy Applications

Radhakrishna Bhat[(⊠)] and N. R. Sunitha

Department of Computer Science and Engineering,
Siddaganga Institute of Technology, Visvesvaraya Technological University,
B H Road, Tumakuru 572103, Karnataka, India
rsb567@gmail.com, nrsunithasit@gmail.com

Abstract. Providing *perfect privacy* to the user against analytics enabled *trusted-but-curious* type of database server during private information retrieval has gained major attention. The major problem with the existing *user privacy* preserving information retrieval methods is that either server has adopted its own privacy preserving policy (i.e., user privacy is guaranteed through the server privacy policy) or user has conveyed to use intractability assumption based user privacy preserving techniques. Due to this, user privacy is not completely assured till date. We have successfully constructed a perfect user privacy preserving information retrieval scheme in a single database setting called sitPIR using the concept of Private Information Retrieval (PIR). In the proposed scheme, the identically distributed $\mathcal{O}(5 \log N)$ bits query exhibit *perfect privacy* where N is the RSA composite. Note that the proposed scheme preserves *user privacy* (i.e., user interest) using an information-theoretic query against *the curious server* and preserves *data privacy* through $\mathcal{O}(o(n)+2 \log N)$ response bits against computationally bounded *intermediate adversary* using Quadratic Residuosity Assumption (QRA) where n is the database size. We have also extended the proposed scheme to a tamper-evident single database information-theoretic Private Block Retrieval (PBR) scheme called sitPBR.

Keywords: Private Information Retrieval · Information-theoretic · Perfect privacy · Quadratic residuosity · Private Block Retrieval · Tamper evident retrieval

1 Introduction

Consider a scenario where the user wants to retrieve a record of information (may be a single bit or block) from a single database server privately without revealing any information about the record retrieved.

© Springer Nature Switzerland AG 2019
K. Lee (Ed.): ICISC 2018, LNCS 11396, pp. 304–321, 2019.
https://doi.org/10.1007/978-3-030-12146-4_19

Private Information Retrieval (PIR) [10] is one of the user privacy preserving techniques and involves two participating entities: the *User* and *Server* in which the *User* wants to retrieve or read a bit from the *Server* without revealing his interest. Hence, PIR is the way of retrieving the required information through the database reference (the reference may be the index or the address of the information stored on the server) from the *Server* by hiding the reference. The primary concern in any PIR protocol is to hide the reference from the server along with reading the required bit from the database server. This concept was introduced in a replicated database setting and coined as private information retrieval by Chor et al. [10,12]. Private Block Retrieval (PBR) is the realistic version of PIR in which the user retrieves a block from the set of blocks maintained by the server.

If the PIR protocol involves a computationally bounded (or computationally intractable) database server entities then such scheme is considered as computationally bounded PIR (cPIR) in which the user privacy is preserved based on the well-defined cryptographic intractability assumption(s) based queries. If the PIR protocol involves non-colluding replicated database server entities then such scheme is considered as information-theoretic PIR (itPIR) in which the user privacy is preserved based on information-theoretically private queries.

Most of the existing computationally bounded PIR schemes including ϕ-hiding assumption based scheme proposed by Cachin et al. [6], Paillier's cryptosystem based scheme proposed by Chang [9], one-way function based scheme proposed by Chor and Gilboa [11], decision subgroup problem called ϕ-hiding assumption based scheme proposed by Gentry and Ramzan [14], the multi-query scheme introduced by Groth et al. [17], anonymity technique based scheme proposed by Ishai et al. [18], quadratic residuosity assumption based scheme introduced by Kushilevitz and Ostrovsky [19], one-way trapdoor permutation based scheme proposed by Kushilevitz and Ostrovsky [20], composite residuosity assumption based scheme presented by Lipmaa [22], Coding theory and Lattice assumption based PIR scheme is also presented by Aguilar-Melchor and Gaborit [24], trapdoor group based scheme presented by Jonathan and Andy [25], Lattice based scheme presented by Aguilar-Melchor et al. [1] and preprocessing based scheme presented by Canetti et al. [7] are all involved a single intractability assumption to preserve both *user privacy* and *data privacy*.

Several information-theoretic schemes [2–4,8,15,16,23] and PBR extension schemes [5,12,14,21,22] proposed in PIR environment are also suffering from the imbalance between the communication cost and resource utilization, involvement of additional techniques (like pre-processing, caching etc.) and multiple rounds, inability to achieve the non-trivial communication cost in a single database environment.

Problems With Existing PIR Schemes: There are two major problems in the existing single database PIR schemes as described below.

– Much attention is given on computationally protecting the privacy of the user using a well-defined intractability assumption that the database server is computationally bounded (i.e., the database server has limited computation capability). All such cryptographic intractability assumption based privacy preserving techniques fail to provide *user privacy* if the database server attains high computational power.

– All the single database PIR schemes rely on a single intractability assumption (like Phi-hiding, Lattice, QRA, Composite Residuosity and n-th residuosity, one-way Functions etc.) to preserve both *user privacy* (assuming that the curious server is computationally bounded) and *data privacy* (assuming that the intermediate adversary is computationally bounded). Intuitively, if the adversary is able to reveal any one of them, he/she will get the other without any extra effort.

In order to provide perfect user privacy preserving single database PIR solution, the best way is that all the successively generating queries must be *identically distributed*. Therefore, all the existing single database PIR schemes have clearly failed to generate the identically distributed queries due to the existence of underlying intractability assumption.

Perfect User Privacy Preserving PIR Solution: Conventionally, we use the term "server" for the database server, we use the term "privacy" for user privacy, we use the term "perfect privacy" for perfect user privacy unless and until externally stated. We also use the terms "perfect privacy" and "information-theoretic privacy" interchangeably.

We have constructed an information-theoretic query which takes identically distributed random input from the quadratic residuosity set $\mathbb{Z}_N^{+1}$ for preserving *user privacy* (Note that in the QRA based Kushilevitz and Ostrovsky [19] scheme, the query input for the interested bit is always drawn from quadratic non-residue. Therefore, all such randomly generated queries are not identically distributed.) and new QRA based recursive 2-bit encryptions called *pair-link encryption* (PLF) (which encrypts two bits and decrypts one bit at a time) for preserving the *data privacy*. Also, we have introduced new methods of selection and encryption of database bits called *criss-cross encryption* (CCE) and *snake-walk encryption* (SWE) using the pair-link encryption as the basic building block during PIR invocation.

With the aid of the proposed pair-link encryption and CCE/SWE methods, we have successfully constructed the perfect privacy preserving single database PIR scheme with the following results.

– All the random queries generated in the proposed scheme are identically distributed over $\mathbb{Z}_N^{+1}$ and hence, exhibit *information-theoretic* privacy over the trusted-but-curious server (i.e., *user privacy* is always guaranteed and independent of the security parameter). That is, an individual query or a randomly selected pair of queries gives no information (not even a partial information) about the user interest.

– The proposed scheme uses quadratic residuosity as the underlying *data privacy* primitive to preserve the communicating data over the intermediate adversary.

- The overall communication cost is $(o(n) + 2 \log N)$ where n is the database size, N is the RSA composite modulus. The communication cost can reach non-triviality (i.e., less than the database size) for all $c_0 > c$ and $n = 2^{c_0}$ where c is an integer constant.
- Inbuilt tamper evidence for the communicating data (from server to user) when the PBR version of the proposed scheme is used.
- The proposed scheme can easily be extended to oblivious transfer and computationally bounded PIR schemes.

Organization: All the necessary preliminaries and notations are described in Sect. 2. The proposed information-theoretic PIR scheme along with required building blocks, security proofs, the performance details and the PBR version are described in Sect. 3. Finally, the open problems are described along with the conclusion in Sect. 4.

2 Preliminaries and Notations

2.1 Notations

Let $[u] \triangleq \{1, 2, \cdots, u\}$, Let k denote the security parameter, $N \xleftarrow{R} \{0,1\}^k = pq$ be the RSA composite modulus where $p \equiv 3 \pmod 4$, $q \equiv 3 \pmod 4$, $\mathbb{Z}_N^{+1}$ denote the set of all elements with *Jacobi Symbol* (*JS*) 1. Let Q_R and $\overline{Q}_R$ denote the quadratic residue and quadratic non-residue sets with $JS = 1$ respectively. Let $<a, b>$ be two components set where $a \in \mathbb{Z}_N^{+1}$ and $b = \{i : i \in \{0, 1\}\}$.

2.2 Preliminaries

Quadratic Residuosity: For any element $a \in \mathbb{Z}_N^*$ if there exists an element b^2 congruent to a modulo N then a is called the quadratic residue otherwise quadratic non-residue modulo N. Intuitively, JS is equal to 1 for all elements that belongs to $\mathbb{Z}_N^{+1}$ and JS is equal to -1 for all elements that belongs to $\mathbb{Z}_N^{-1}$ where $JS(\cdot)$ is the *Jacobi Symbol* modulo N.

Quadratic Residuosity Predicate (QRP): $\forall a \in \mathbb{Z}_N^*$, QRP is a function to return a value (0 or 1) to indicate whether a is Q_R if $QRP_{p,q}(a) = 0$ or $\overline{Q}_R$ if $QRP_{p,q}(a) = 1$.

Quadratic Residuosity Assumption (QRA): For all $N \in \{0,1\}^k$, for all $\mathcal{R} \in \mathbb{Z}_N^{+1}$, for all probabilistic polynomial time intermediate adversary Ad, $PROB[Ad(N, \mathcal{R}) = QRP_N(\mathcal{R})] < p^{QR}$ where $p^{QR} = (1/2) + (1/k^c)$ and c is a constant.

Quadratic Residuosity Based Lossy Trapdoor Function of Freeman et al. [13] (LTDF): For all $\alpha \in \mathbb{Z}_N^*$, $s \in \overline{Q}_R$ and $r \in \mathbb{Z}_N^{-1}$, the lossy trapdoor function $\mathcal{T}: \mathbb{Z}_N^* \to \mathbb{Z}_N^*$ is $\mathcal{T} = (\alpha^2 \cdot r^{jx} \cdot s^{hx} \equiv z \pmod N))$ such that jx is equal to 1 if $JS(\alpha) = -1$ otherwise jx is equal to 0. The value of hx is equal to 1 if $\alpha > N/2$ otherwise hx is equal to 0. The respective inverse function is $\mathcal{T}^{-1} = (\sqrt{(z \cdot s^{-hx}) \cdot r^{-jx}} \equiv \alpha \pmod N))$. We use the alternative square root syntax as $\mathcal{T}^{-1} = (\sqrt[jx,hx]{z} \equiv \alpha \pmod N))$.

3   A Single Database Information-Theoretic Private Information Retrieval (sitPIR)

Definition 1 *(A Single database information-theoretic PIR (sitPIR)):* It is a 4-tuple (KG, QF, RC, IE) protocol that involves two communicating parties: user $\mathcal{U}_{pir}$ and server $\mathcal{S}_{pir}$ in which $\mathcal{S}_{pir}$ maintains n bit single dimensional matrix database $\mathcal{DB} = \{b_1, b_2, \cdots, b_n\}$. User $\mathcal{U}_{pir}$ requests the interested bit b_i, $i \in [n]$, privately from $\mathcal{S}_{pir}$ by generating information-theoretic query $\mathcal{Q}$ such that $\forall b_i, b_j \in \mathcal{DB}$, $i, j \in [n]$, any two random generated queries $\mathcal{Q}_i$ and $\mathcal{Q}_j$ exhibit same level of information-theoretic privacy equivalent to *perfect privacy* as described by Chor et al. [12] and the server $\mathcal{S}_{pir}$ in-turn generates the response $\mathcal{R}$ with the communication $\mathcal{O}(o(n) + 2 \log N)$. The setting consists of the following polynomial time algorithms.

1. *Key Generation* (KG): $\mathcal{U}_{pir}$ calculates RSA modulus $N \overset{R}{\leftarrow} \{0,1\}^k$. $\mathcal{U}_{pir}$ then generates (public, private) key pair $(pk, sk) \overset{R}{\leftarrow} KG(1^k)$ where $pk = (N, (\mathcal{PK}_1, \mathcal{PK}_2) \in \mathbb{Z}_N^{+1})$ and $sk = (p, q)$.
2. *Query Formulation* (QF): $\mathcal{U}_{pir}$ generates (pk, sk) from the key generation algorithm KG. $\mathcal{U}_{pir}$ then generates the perfect privacy preserving query as $\{(\mathcal{Q} = (\alpha, pk), sk) \leftarrow QF(1^k) : \alpha \overset{R}{\leftarrow} \mathbb{Z}_N^{+1}\}$ and keeps sk secret. Importantly, the random generation of the index (or reference) independent input α from either Q_R or $\overline{Q}_R$ always exhibits *perfect privacy* (Note: All the query input α randomly selected from $\mathbb{Z}_N^{+1}$ are always domain independent. Also, note that the "domain" here is the quadratic residuosity sets like Q_R and $\overline{Q}_R$).
3. *Response Creation* (RC): $\mathcal{S}_{pir}$ generates the response $\mathcal{R}$ using the query $\mathcal{Q}$ and the database $\mathcal{DB}$ as $\mathcal{R} \leftarrow RC(\mathcal{Q}, \mathcal{DB}, n, 1^k)$.
4. *Interest Extraction* (IE): Using the response $\mathcal{R}$ and the secret sk, $\mathcal{U}_{pir}$ extracts the required bit b_i, $i \in [n]$, as $b_i \leftarrow IE(\mathcal{R}, sk, n, 1^k)$.

Definition 2 *(Single database information-theoretic PIR (sitPIR)):* Let $\mathcal{DB} = \{b_1, b_2, \cdots, b_n\}$ be n bit server database. Let query formulation QF, response creation RC of Definition 1 be the Probabilistic Polynomial Time (PPT) algorithms and interest extraction IE be the deterministic polynomial time algorithm. We say that 4-tuple (KG, QF, RC, IE) protocol of Definition 1 is a single database perfect privacy PIR scheme or single database itPIR scheme if the following conditions hold.

1. *Perfect user privacy:* For any two identically distributed random queries $\mathcal{Q}_i$ and $\mathcal{Q}_j$, $i, j \in [n]$,
   $\mathrm{PROB}[(\mathcal{Q}_i, sk) \overset{R}{\leftarrow} QF(1^k) : Adv(n, \mathcal{Q}_i, 1^k) = 1] = \mathrm{PROB}[(\mathcal{Q}_j, sk) \overset{R}{\leftarrow} QF$ $(1^k) : Adv(n, \mathcal{Q}_j, 1^k) = 1]$
2. *Correctness:* $\forall z \in [n], \mathrm{PROB}[IE((\mathcal{R}, sk, n, 1^k) : \mathcal{R} \overset{R}{\leftarrow} RC(\mathcal{Q}_z, \mathcal{DB}, n, 1^k)$, $(\mathcal{Q}_z, sk) \overset{R}{\leftarrow} QF(1^k)) = b_z] = 1$
3. *Data privacy:* For any two randomly selected queries $\mathcal{Q}_1$ and $\mathcal{Q}_2$, for all security parameter k and for all ciphertexts $\mathcal{R}_1, \mathcal{R}_2$,

$$|\text{PROB}[(\mathcal{Q}_1, sk) \xleftarrow{R} QF(1^k), \mathcal{R}_1 \xleftarrow{R} RC(\mathcal{Q}_1, \mathcal{DB}, n, 1^k): Ad(n, \mathcal{R}_1, 1^k) = 1]$$
$$- \text{PROB}[(\mathcal{Q}_2, sk) \xleftarrow{R} QF(1^k), \mathcal{R}_2 \xleftarrow{R} RC(\mathcal{Q}_2, \mathcal{DB}, n, 1^k): Ad(n, \mathcal{R}_2, 1^k) = 1]|$$
$$< (p^{QR} + p^R + p^C)$$

where PROB$[\cdot]$ is the *privacy* revealing probability, $Adv(\cdot)$ is the trusted-but-curious server, $Ad(\cdot)$ is the polynomial time intermediate adversary, k is the security parameter, p^{QR} is the QRA probability, p^R is the single fair coin toss probability, p^C is some combination identification probability.

Let n bit 1-dimensional matrix database be $\mathcal{DB} = \{b_1, b_2, ..., b_n\}$. Consider $S_a, S_o \subseteq \mathcal{DB} \times \mathcal{DB}$ where S_a is viewed as the subset of ordered pairs $\{\mathcal{B}_1, \mathcal{B}_2, \cdots, \mathcal{B}_h\}$ where $\mathcal{B}_1 = (b_1, b_2)$, $\mathcal{B}_2 = (b_3, b_4)$, and so on till $\mathcal{B}_h = (b_{n-1}, b_n)$ and S_o is viewed as the subset of ordered pairs $\{\mathcal{B}_1', \mathcal{B}_2', \cdots, \mathcal{B}_{h-1}'\}$ where $\mathcal{B}_1' = (b_2, b_4)$, $\mathcal{B}_2' = (b_4, b_6)$, and so on till $\mathcal{B}_{h-1}' = (b_{n-2}, b_n)$ and $h = \frac{n}{2}$ is the total number of ordered pairs of S_a. Let the communication bit sets for S_a and S_o be $T_a = \{i : i \in \{0, 1\}\}$ and $T_o = \{j : j \in \{0, 1\}\}$ respectively with $|T_a| = \frac{n}{2} - 1$ and $|T_o| = \frac{n}{2} - 2$.

Note that the bits are arranged in S_a and S_o in such a way that second bit of each $\mathcal{B}_i$, $1 \leq i \leq h - 1$, is same as first bit of $\mathcal{B}_i'$. All the following subsections use this database (i.e., $\mathcal{DB}$) definition only.

3.1 Building Blocks

Communication Bit: It is a special bit (i.e., $hx \in \{0, 1\}$) used as a "trapdoor" to inverse the LTDF ($\mathcal{T}$) described in Sect. 2. If the input α of the LTDF ($\mathcal{T}$) of Sect. 2 is restricted to $\mathbb{Z}_N^{+1}$, then the value of "jx" of α is always zero. Then, the modified function would be $\mathcal{T} = (\alpha^2 \cdot r^0 \cdot s^{hx} \equiv z \pmod{N}) \Rightarrow \mathcal{T}' = (\alpha^2 \cdot s^{hx} \equiv z \pmod{N})$. Therefore, we have considered this modified function $\mathcal{T}'$ to define a reduced new communication bit function $\mathcal{MT}: \mathbb{Z}_N^{+1} \to \mathbb{Z}_N^{+1}$. For all $\alpha \in \mathbb{Z}_N^{+1}$, the communication bit function is

$$\mathcal{MT}(\alpha) = (\alpha^2 \cdot w^{\delta \% 2} \equiv z \pmod{N}) = <z, hx_\alpha> \tag{1}$$

where $\alpha \in \mathbb{Z}_N^{+1}$, $w \in \overline{Q}_R$, $\delta \in \{0, 1, 2\}$, the value of "hx_α" is considered as the "communication bit". The value of hx_α is 1 if $\alpha > N/2$ otherwise hx_α is 0. The respective inverse is $\mathcal{MT}^{-1}(z, hx_\alpha) = (^{0,hx}\sqrt{z \cdot (w^{\delta \% 2})^{-1}} \equiv \alpha \pmod{N})$.

QRA Based Single Bit Encryption (SBE): For all bit $b \in \{0, 1\}$, for all random $x, y, \mathcal{PK}_1, \mathcal{PK}_2 \in \mathbb{Z}_N^{+1}$ with $QRP(\mathcal{PK}_1) \neq QRP(\mathcal{PK}_2)$, for all random $\mathcal{PK}_3 \in \mathbb{Z}_N^{-1}$, the single bit encryption $\mathcal{E}_s(b, N, x, y, \mathcal{PK}_1, \mathcal{PK}_2, \mathcal{PK}_3)$ is given in Eq. 2. Each input $x, y \in \mathbb{Z}_N^{+1}$ consists of their respective jx, hx values as described in LTDF. There are four jx, hx combinations (listed in the first column of Eq. 2) possible for any $x, y \in \mathbb{Z}_N^{+1}$. Use the respective pair of equations to encrypt the bit b. For instance, if $jx_x = 0, hx_x = 0$ and $jx_y = 0, hx_y = 1$, then for all $b = 0$ use the pair of equations defined in second row and second column of Eq. 2; for all $b = 1$ use second row and third column of Eq. 2.

Table 1. Possible residuosity property combinations of ciphertext y when the input $x \in Q_R$.

a	b	$(x \cdot PK_l) \cdot PK_{l'} \equiv y \pmod{N}$
0	0	$(x \cdot PK_1) \cdot PK_1 \equiv y \in Q_R$
0	1	$(x \cdot PK_1) \cdot PK_2 \equiv y \in \overline{Q}_R$
1	0	$(x \cdot PK_2) \cdot PK_1 \equiv y \in \overline{Q}_R$
1	1	$(x \cdot PK_2) \cdot PK_2 \equiv y \in Q_R$

Table 2. Possible residuosity property combinations of ciphertext y when the input $x \in \overline{Q}_R$.

a	b	$(x \cdot PK_l) \cdot PK_{l'} \equiv y \pmod{N}$
0	0	$(x \cdot PK_2) \cdot PK_2 \equiv y \in Q_R$
0	1	$(x \cdot PK_2) \cdot PK_1 \equiv y \in Q_R$
1	0	$(x \cdot PK_1) \cdot PK_2 \equiv y \in \overline{Q}_R$
1	1	$(x \cdot PK_1) \cdot PK_1 \equiv y \in \overline{Q}_R$

$$
\mathcal{E}_s = \begin{cases}
\begin{array}{lll}
\mathbf{jx, hx} & \mathbf{If\ } b = 0 & \mathbf{If\ } b = 1 \\
0,0 & x^2 \cdot PK_1 \equiv c_1 & x^2 \cdot PK_1 \equiv c_1 \\
0,0 & y^2 \cdot PK_3 \equiv c_2 & y^2 \cdot PK_2 \equiv c_2
\end{array} \left. \right\} \text{if } x \leq N/2, y \leq N/2 \\[2em]
\begin{array}{lll}
0,0 & x^2 \cdot PK_1 \equiv c_1 & x^2 \cdot PK_3 \equiv c_1 \\
0,1 & y^2 \cdot PK_1 \equiv c_2 & y^2 \cdot PK_1 \equiv c_2
\end{array} \left. \right\} \text{if } x \leq N/2, y > N/2 \\[2em]
\begin{array}{lll}
0,1 & x^2 \cdot PK_3 \equiv c_1 & x^2 \cdot PK_2 \equiv c_1 \\
0,0 & y^2 \cdot PK_2 \equiv c_2 & y^2 \cdot PK_2 \equiv c_2
\end{array} \left. \right\} \text{if } x > N/2, y \leq N/2 \\[2em]
\begin{array}{lll}
0,1 & x^2 \cdot PK_2 \equiv c_1 & x^2 \cdot PK_2 \equiv c_1 \\
0,1 & y^2 \cdot PK_1 \equiv c_2 & y^2 \cdot PK_3 \equiv c_2
\end{array} \left. \right\} \text{if } x > N/2, y > N/2
\end{cases}
\tag{2}
$$

The decryption of $\mathcal{E}_s$ to get the bit b involves the identification of respective quadratic residuosity properties of the ciphertexts c_1 and c_2 as follows.

   Step-1: Identify $QRP(c_1)$ and $QRP(c_2)$. Output the respective b and (jx, hx) combinations of x, y.

   Step-2: Find x^2, y^2 using the respective public key inverses. Then, given x^2 and (jx_x, hx_x) values, identify x as described in Eq. 1. Similarly, given y^2 and (jx_y, hx_y) values, identify y as described in Eq. 1.

Axiom 1. *For all RSA composite $N = pq$ where $|p| = |q| = k$, $|Q_R| = |\overline{Q}_R| = 1/|\mathbb{Z}_N^{+1}|$.*

Pair-Link Encryption (PLE): It is newly constructed quadratic residuosity based encryption method which encrypts two bits at a time and its respective decryption function decrypts only a single bit (Note: PLE is analogous to logical "xor" operation).

   For any ordered bit pair $(a, b) \in \mathcal{B}_i$, $i \in [h]$, of S_a or $(a, b) \in \mathcal{B}'_j$, $j \in [h-1]$, of S_o of n bit database $\mathcal{DB}$ where $h = n/2$ and for all the random input $t \in Q_R$ and for all the public key $PK_1, PK_2 \in \mathbb{Z}_N^{+1}$ with $QRP(PK_1) \neq QRP(PK_2)$, the encryption function $\mathcal{E} : \mathbb{Z}_N^{+1} \to \mathbb{Z}_N^{+1}$ is given as

$$
\mathcal{E}((a, b), x = t^\rho, PK_1, PK_2) = ((x \cdot PK_l) \cdot PK_{l'} \equiv y \pmod{N})
\tag{3}
$$

where $l, l' \in [2]$ and $\rho \in \{1, 2\}$.

In order to understand this method, let us look at the encryption Tables 1 and 2. For instance, let us consider Table 1 and the input pair $a = 0$, $b = 1$ and the random input $x \in Q_R$. Let $\mathcal{PK}_l \in Q_R$, $\mathcal{PK}_2 \in \overline{Q}_R$. Now, the encryption of $a = 0$ and $b = 1$ is $\mathcal{E}((0,1), x, \mathcal{PK}_1, \mathcal{PK}_2) = ((x \cdot \mathcal{PK}_l) \cdot \mathcal{PK}_{l'} \equiv y \pmod{N})$ where $l = 1, l' = 2$ and the ciphertext y is always a quadratic non-residue. Similarly, the encryption of $a = 1$ and $b = 1$ is $\mathcal{E}((1,1), x, \mathcal{PK}_1, \mathcal{PK}_2) = ((x \cdot \mathcal{PK}_l) \cdot \mathcal{PK}_{l'} \equiv y \pmod{N})$ where $l = 2, l' = 2$ and the ciphertext y is always a quadratic residue and so on.

In order to find the decryption of PLE, one should know the value of the second bit b in advance (bit b acts as an inverse factor) and the quadratic residuosity properties of the ciphertext y and the input x. On identifying the residuosity property of the ciphertext y and given b (Note: there is some means to get the second bit b when this isolated encryption instance is used in combination with other encryption instances in CCE or SWE methods. At this point, assume that the second bit b is given), the decryption of PLE to get the first bit a is calculated as

$$\mathcal{E}^{-1}(y, b, \mathcal{SK}_1, \mathcal{SK}_2) = ((y \cdot \mathcal{SK}_l) \cdot \mathcal{SK}_{l'} \equiv x \pmod{N}) = <x, a> \qquad (4)$$

where $l, l' \in [2]$ and $\mathcal{PK}_l \cdot \mathcal{SK}_l \equiv \mathcal{PK}_{l'} \cdot \mathcal{SK}_{l'} \equiv 1 \pmod{N}$.

If the quadratic residuosity properties of the ciphertext y and the input x (Note: the quadratic residuosity property of a number should be calculated using the private key p, q) and the second bit b is known then the first bit a can easily be calculated. Initially, using the private key p, q, identify quadratic residuosity property of y. Then, given QRP(y) and b, identify the corresponding first bit a and the public key inverse combinations $\mathcal{SK}_l, \mathcal{SK}_{l'}$ for the unique quadratic residuosity property combinations of the ciphertext y (i.e., QRP(y)) and the input x (i.e., QRP(x)). Finally, using the identified public key inverse combinations $\mathcal{SK}_l, \mathcal{SK}_{l'}$, $l, l' \in [2]$, get back the input x. For example, if the pair $a = 0$, $b = 1$ is encrypted then the ciphertext y is always a quadratic non-residue i.e., QRP(y) $= 1$ and let the input x is a quadratic residue i.e., QRP(x) $= 0$. Now, it is clear that, for QRP(y) $= 1$, QRP(x) $= 0$ and $b = 1, a$ is equal to 0 and $l = 2, l' = 1$. Finally, get the input x as $(y \cdot \mathcal{SK}_2) \cdot \mathcal{SK}_1 \equiv x \pmod{N}$. Similarly, if the pair $a = 1$, $b = 1$ is encrypted then the ciphertext y is always a quadratic residue i.e., QRP(y) $= 0$ and let the input x is a quadratic residue i.e., QRP(x) $= 0$. Now, it is clear that, for QRP(y) $= 0$, QRP(x) $= 0$ and $b = 1, a$ is equal to 1 and $l = 1, l' = 1$. Finally, get the input x as $\mathcal{E}^{-1}(y, b, \mathcal{SK}_1, \mathcal{SK}_2) = (y \cdot \mathcal{SK}_1) \cdot \mathcal{SK}_1 \equiv x \pmod{N}$. Note that, for a single encryption-decryption instance and for a given $\rho \in \{1, 2\}$, use any one of the Tables 1 and 2.

Lemma 1. *Let $N \in \{0,1\}^k$ be the RSA composite. Let p^R be a single fair coin toss probability, p^C be a combination selection probability and p^{QR} be a QRA probability. For all given $\mathcal{E}((a,b), \cdot)$ where $a, b, \in \{0,1\}$, public key $(N, \mathcal{PK}_1 \in \mathbb{Z}_N^{+1}, \mathcal{PK}_2 \in \mathbb{Z}_N^{+1})$, for all probabilistic polynomial time intermediate adversary $Ad(\cdot)$, for any random number $x \in \mathbb{Z}_N^{+1}$, for all security parameter k,*

$$\text{PROB}[\mathcal{E}((a,b), x, \mathcal{PK}_1, \mathcal{PK}_2) : Ad(N, \mathcal{PK}_1, \mathcal{PK}_2, \mathcal{E}) = (a,b)] < (p^{QR} + p^R + p^C) \qquad (5)$$

Proof. Since the pair-link encryption described in Eq. 3 is based on quadratic residuosity, the adversary has minimum probability equivalent to $p^{QR} = (1/2 + p(k))$ where $p(k)$ is some inverse polynomial in k. Also, for any two equal plaintext bits, the pair-link function in Eq. 3 always produces with the same property ciphertext (Refer the Tables 1 and 2). By this approach, the intermediate adversary has additional $p^R = 1/2$ probability to get the correct plaintext along with p^{QR}. In addition to that, the encryption of Eq. 3 uses eight combination tables for any input $x \in \mathbb{Z}_N^{+1}$ and public key $\mathcal{PK}_1, \mathcal{PK}_2$. Hence, probability of getting the exact combination is $p^C = 1/8$. Therefore, the total success probability to know the exact plaintext bits (a, b) would always be less than $p^{QR} + p^R + p^C$.

Axiom 2. *For all $a, b \in \mathbb{Z}_N^{+1}$, the equation $ax \equiv b \pmod{N}$ always has a unique solution if $\gcd(a, N) = 1$.*

Lemma 2. *Every pair-link encryption described in Eq. 3 has unique solution.*

Proof. It is clear from the Axiom 2 that if the *gcd* of participating equation variables w.r.t the modulus is 1, then there exists a unique modular solution. Therefore, for every unique one-to-one mapping function $\mathcal{E}$ (as given in Eq. 3) there exists a unique inverse mapping function $\mathcal{E}^{-1}$ (as given in Eq. 4). In other words, for all the pair (a, b) and the random input $x \in \mathbb{Z}_N^{+1}$ and public key $\mathcal{PK}_1, \mathcal{PK}_2$, the encryption of Eq. 3 always maps to unique y since $\gcd(\mathcal{PK}_1, N) = 1$ and $\gcd(\mathcal{PK}_2, N) = 1$ and $p, q \equiv 3 \pmod{4}$. Similarly, for all the given ciphertext y and $\text{QRP}(x)$ and the second bit b, the decryption of Eq. 4 always gives the unique bit a and the input x. That means, for any given encryption table i.e., Tables 1 and 2, the encryption $\mathcal{E}$ always maps to the unique property ciphertext (though the value of the ciphertext may be different) and for any given corresponding decryption, the decryption $\mathcal{E}^{-1}$ always maps to the unique bit and the unique input. This implies that Eq. 3 has unique solution.

Connective Function ($\mathcal{C}$): Since the individual PLE described in Eq. 3 alone can encrypt only two bits at a time, connect any two successive pair-link encryptions with the aid of the communication bit function $\mathcal{MT}$ of Eq. 1 to encrypt two pair of bits (instead of a single pair). Therefore, either select the input bits $a, b, c, d \in \{0, 1\}$ where $(a, b) \in \mathcal{B}_i$ and $(c, d) \in \mathcal{B}_{i+1}$, with the condition $b \neq c$, $1 \leq i \leq (h - 1)$, from S_a or select the input bits $a, b, c, d \in \{0, 1\}$ where the ordered pairs $(a, b) \in \mathcal{B}'_j$ and $(c, d) \in \mathcal{B}'_{j+1}$, with the condition $b = c$, $1 \leq j \leq (h - 2)$, from S_o of the n bit database $\mathcal{DB}$.

For all bit pairs $(a, b), (c, d)$ as selected above and input x and the public key $\mathcal{PK}_1, \mathcal{PK}_2 \in \mathbb{Z}_N^{+1}$ with $\text{QRP}(\mathcal{PK}_1) \neq \text{QRP}(\mathcal{PK}_2)$ and for any two successive pair-link encryptions $\mathcal{E}$ and $\mathcal{F}$ as each described in Eq. 3, the connective function $\mathcal{C}: \mathbb{Z}_N^{+1} \rightarrow \mathbb{Z}_N^{+1}$ is given as

$$\mathcal{C}((a, b), (c, d), \mathcal{E}, \mathcal{F}) = \mathcal{F}((c, d), \mathcal{MT}(\mathcal{E}((a, b), x, \mathcal{PK}_1, \mathcal{PK}_2) = y) = <y^2, t>, \mathcal{PK}_1, \mathcal{PK}_2) = z$$

$$(6)$$

where $l, l' \in [2]$, t is equal to 1 if $y > N/2$ otherwise t is equal to 0. We treat this "t" as "communication bit" equivalent to the "hx" value described in communication bit function $\mathcal{MT}$.

In the connective function of Eq. 6, the preceding pair-link encryption $\mathcal{E}$ receives the ordered pair of bits (a, b), the input x, the public key $\mathcal{PK}_1, \mathcal{PK}_2$ with $\mathrm{QRP}(\mathcal{PK}_1) \neq \mathrm{QRP}(\mathcal{PK}_2)$ and encrypts (a, b) and generates the ciphertext y. Further, the communication bit function $\mathcal{MT}$ receives the ciphertext y as input and produces the ciphertext y^2 and the communication bit t where $t = 1$ if $y > N/2$ otherwise $t = 0$. Finally, the succeeding pair-link encryption $\mathcal{F}$ receives the ordered pair of bits (c, d), the ciphertext y^2 as input, the public key $\mathcal{PK}_1, \mathcal{PK}_2$ and encrypts (c, d) and generates the final ciphertext z. Therefore, any two successive pair-link encryptions connected using connective function $\mathcal{C}$ always encrypt four bits and produce one communication bit in between them.

Types of Connection: We define two types of successive pair-link encryption connections as follows.

- *Criss-cross*: In this, every successive ordered pairs $\mathcal{B}_i$ and $\mathcal{B}_{i+1}$, $1 \leq i \leq (h - 1)$, of S_a are encrypted using connective function. Similarly, every successive ordered pairs $\mathcal{B}'_j$ and $\mathcal{B}'_{j+1}$, $1 \leq j \leq (h - 2)$, of S_o are encrypted using connective function $(\mathcal{C})$.

- *Snake-walk*: Let $S_f \subseteq \mathcal{DB} \times \mathcal{DB}$ where S_f is viewed as a set of ordered pairs $\{\mathcal{B}''_1, \mathcal{B}''_2, \cdots, \mathcal{B}''_{h-1}\}$ where $\mathcal{B}''_1 = (b_2, b_3)$, $\mathcal{B}''_2 = (b_4, b_5)$, and so on till $\mathcal{B}''_{h-1} = (b_{n-2}, b_{n-1})$. In this, every successive pair $\mathcal{B}_i$ and $\mathcal{B}_{i+1}$, $1 \leq i \leq (h - 1)$, of S_a are encrypted using connective function. Similarly, every successive pairs $\mathcal{B}''_j$ and $\mathcal{B}''_{j+1}$, $1 \leq j \leq (h - 2)$, of S_f are encrypted using connective function $(\mathcal{C})$.

In order to decrypt the connective function of Eq. 6, the seconds bits d, b that were encrypted using $\mathcal{F}$ and $\mathcal{E}$ are essential along with the quadratic residuosity properties of z and x. How to get these second bits? or Who will provide these second bits?

By careful observation, in the *criss-cross* connection type, it is clear that the decryption of every connective function involving encryption of successive pairs $\mathcal{B}'_j$ and $\mathcal{B}'_{j+1}$, $1 \leq j \leq (h-2)$, of S_o provides the second bits for the decryption of every connective function involving encryption of successive pairs $\mathcal{B}_i$ and $\mathcal{B}_{i+1}$, $1 \leq i \leq (h - 1)$, of S_a (since every first bit of $\mathcal{B}'_j$ and $\mathcal{B}'_{j+1}$ are same as every second bit of $\mathcal{B}_i$ and $\mathcal{B}_{i+1}$). Similarly, in the *snake-walk* connection type, it is clear that the decryption of every connective function involving encryption of successive pairs $\mathcal{B}_i$ and $\mathcal{B}_{i+1}$, $1 \leq i \leq (h - 1)$, of S_a provides the second bits for the decryption of every connective function involving encryption of successive pairs $\mathcal{B}''_j$ and $\mathcal{B}''_{j+1}$, $1 \leq j \leq (h - 2)$, of S_f since every first bit of $\mathcal{B}_i$ and $\mathcal{B}_{i+1}$ are same as every second bit of $\mathcal{B}''_j$ and $\mathcal{B}''_{j+1}$.

Definition 3 *(Chain of successive connective function ($\mathcal{CHAIN}$)):* It is the chain of successive connective functions of the form $\mathcal{CHAIN}(N, S, \alpha, \mathcal{PK}_1, \mathcal{PK}_2) = ([\xrightarrow{\alpha} \mathcal{C}_1 \Rightarrow o_1] \xrightarrow{o_1} [\mathcal{C}_2 \Rightarrow o_2] \xrightarrow{o_2} \cdots \xrightarrow{o_{g-2}} [\mathcal{C}_{g-1} \Rightarrow o_{g-1}] \xrightarrow{o_{g-1}} [\mathcal{C}_g \Rightarrow o_g]) = <o_g, \mathcal{TB}>$ where $1 \leq g \leq h, h = n/2$, $(\alpha, \mathcal{PK}_1, \mathcal{PK}_2) \in \mathbb{Z}_N^{+1}$, $S \subseteq \mathcal{DB} \times \mathcal{DB}$ and

o_g is the final output ciphertext and $\mathcal{TB}$ is the communication bit set and each connective function $\mathcal{C}$ is drawn from Eq. 6. Let $\mathcal{CHAIN}^{-1}(o_g, \mathcal{TB}, (p,q))$ is the respective inverse chain.

Remark: For the CCE encryption, there are two concurrently executing chains $\mathcal{CHAIN}_1(N, S_a, \alpha, \mathcal{PK}_1, \mathcal{PK}_2)$ with $g = h-1$ and $\mathcal{CHAIN}_2(N, S_o, \alpha, \mathcal{PK}_1, \mathcal{PK}_2)$ with $g = h-2$ in which the subset S_a is encrypted using $\mathcal{CHAIN}_1$ and the subset S_o is encrypted using $\mathcal{CHAIN}_2$ resulting in the generation of g number of communication bits from each chain along with each chain ciphertexts. Similarly, for the SWE type of encryption, there are two concurrently executing chains $\mathcal{CHAIN}_1(N, S_a, \alpha, \mathcal{PK}_1, \mathcal{PK}_2)$ with $g = h$ and $\mathcal{CHAIN}_2(N, S_f, \alpha, \mathcal{PK}_1, \mathcal{PK}_2)$ with $g = h-1$ in which the subset S_a is encrypted using $\mathcal{CHAIN}_1$ and the subset S_f is encrypted using $\mathcal{CHAIN}_2$ resulting in the generation of g number of communication bits from each chain along with each chain ciphertexts.

3.2   Proposed sitPIR Scheme

In order to generate the response from the server, the main trick here is to execute two chain of successive connective functions (individual connective function $\mathcal{C}$ is described in Subsect. 3.1 and the chain of successive connective functions is described in Definition 3) in parallel on the database and produce the respective ciphertexts. Also, encrypt the last database bit using QRA based single bit encryption (SBE) of Sect. 3.1 and produce the final ciphertexts. Consider the criss-cross type of encryption for instance. The detailed description of the algorithms is as follows.

- **Query Generation (QG):** Let $N \xleftarrow{R} \{0,1\}^k$. User $\mathcal{U}_{pir}$ sends information-theoretic query $\mathcal{Q} = (\alpha, N, \mathcal{PK}_1, \mathcal{PK}_2, \mathcal{PK}_3)$ to the server where $\mathcal{PK}_1, \mathcal{PK}_2 \in \mathbb{Z}_N^{+1}$ with $\mathrm{QRP}(\mathcal{PK}_1) \neq \mathrm{QRP}(\mathcal{PK}_2)$, $\mathcal{PK}_3 \in \mathbb{Z}_N^{-1}$ and $\alpha \xleftarrow{R} \mathbb{Z}_N^{+1}$.
- **Response Creation (RC):** Server $\mathcal{S}_{pir}$ generates the response $\mathcal{R}$ consisting of two ciphertexts and two communication bit sets as follows. Initially, using the query $\mathcal{Q}$, server executes two parallel chain of successive connective functions $\mathcal{CHAIN}_1$ and $\mathcal{CHAIN}_2$ (each chain is described in Definition 3) on either using CCE or SWE type of the database and produces respective chain ciphertexts β_1, β_2 and respective communication bit sets T_a and T_o as follows. Consider CCE type for instance. All the ordered pairs of the subset S_a are encrypted using $\mathcal{CHAIN}_1$ as

$$
\begin{aligned}
\mathcal{CHAIN}_1(N, S_a, \alpha, \mathcal{PK}_1, \mathcal{PK}_2) &= \mathcal{C}_i(N, \mathcal{MT}(\mathcal{C}_{i-1}), \mathcal{PK}_1, \mathcal{PK}_2) \\
&= <\beta_1, (T_a = (t_1, \cdots, t_{i-1}))> \\
&= <\beta_1, T_a>
\end{aligned} \tag{7}
$$

where $\delta = \rho = 2$, $i \in [h,2]$, $h = n/2$, β_1 is the output ciphertext generated from $\mathcal{CHAIN}_1$, T_a is the communication bit set with $(h-1)$ number of communication bits and $\mathcal{C}_1((b_1, b_2), (b_3, b_4), \mathcal{E}_1, \mathcal{E}_2) = \mathcal{E}_2((b_3, b_4), \mathcal{MT}(\mathcal{E}_1((b_1, b_2), \alpha, \mathcal{PK}_1, \mathcal{PK}_2) = y) = <y^2, t_1>, \mathcal{PK}_1, \mathcal{PK}_2)$.

Similarly, all the ordered pairs of the subset S_o are encrypted using $CHAIN_2$ as

$$CHAIN_2(N, S_o, \alpha, PK_1, PK_2) = C_i(N, MI(C_{i-1}), PK_1, PK_2)$$
$$= <\beta_2, (T_o = (t'_1, \cdots, t'_{i-1}))> \qquad (8)$$
$$= <\beta_2, T_o>$$

where $i \in [h-1, 2]$, $h = n/2$, β_2 is the output ciphertext generated from $CHAIN_2$, T_o is the communication bit set with $(h-2)$ number of communication bits and $C_1((b_2, b_4), (b_4, b_6), \mathcal{E}_1, \mathcal{E}_2) = \mathcal{E}_2((b_4, b_6), MI(\mathcal{E}_1((b_2, b_4), \alpha, PK_1, PK_2) = y) = <y^2, t'_1>, PK_1, PK_2)$.

It is evident that both the chains $CHAIN_1$ and $CHAIN_2$ interlock the database bits (we call this type of encryption as "criss-cross encryption" and the cipher generated from it as criss-cross cipher alternative to substitution or transposition ciphers) and hence, all the ordered pairs of subsets S_a and S_o should be retrieved alternatively using the respective inverse chains $CHAIN_1^{-1}$ and $CHAIN_2^{-1}$. That means, every second bit of each ordered pair of S_a is encrypted as a first bit of each pair of S_o. Hence, during retrieval, it is impossible to retrieve the required bit(s) of the subset S_a or S_o alone without the aid of other inverse chain.

Further, the last bit b_n is encrypted using the single bit encryption SBE as $\mathcal{E}_s(b_n, N, \beta_1, \beta_2, PK_1, PK_2, PK_3) = (\gamma_1, \gamma_2)$. Finally, the PIR response $\mathcal{R}$ is generated as $\mathcal{R} = \{C_1 = (\gamma_1, T_a), C_2 = (\gamma_2, T_o)\}$. Therefore, for the whole database, there are two constant k size ciphertexts and $(2h-3)$ number of communication bits generated in total. This response $\mathcal{R}$ is sent back to the user.

- **Interest Extraction (IE):** Using the response $\mathcal{R}$ and the private key (p, q), user $\mathcal{U}_{pir}$ privately reads the required bit of the database DB as follows. Initially, using the ciphertext (γ_1, γ_2), find the last bit b_n as $\mathcal{E}_s(\gamma_1, \gamma_2) = b_n$. Since both the chains were adopted criss-cross encryption during response creation on the server, exact reverse order should be maintained to get the required bit using the obtained last bit b_n and chain specific ciphertexts β_1, β_2 and T_a, T_o.

It is intuitive that the last bit b_n of the database DB is always same as the second bit of $B_h \in S_a$ and $B'_{h-1} \in S_o$. Since both the chains $CHAIN_1$ and $CHAIN_2$ have adopted criss-cross encryption, it is also clear that the first bit of each $B'_i \in S_o$ is always equal to second bit of each $B_i \in S_a$ where $h-1 \geq i \geq 1$. Hence, find the first bits of $B_h \in S_a$ and $B'_{h-1} \in S_o$ by inverting respective chains $CHAIN_1$ and $CHAIN_2$ and continue the inverse process till the required bit of interest.

3.3   A Toy Example

Let us consider $N = 133$, $p = 19$, $q = 7$ and database $DB = \{1, 1, 0, 0, 1, 1, 1, 1\}$ where $|DB| = n = 8$. Therefore, $S_a = \{(1, 1), (0, 0), (1, 1), (1, 1)\}$ and $S_o = \{(1, 0), (0, 1), (1, 1)\}$. Let $\alpha = 25$, $PK_1 = 44$, $PK_2 = 48$, $PK_3 = 15$. Let us assume that the user is interested in b_5. An illustrative example is given in Table 3.

Table 3. An illustrative example of response creation (RC) and interest extraction (IE) algorithms of the proposed sitPIR scheme.

Response Creation (RC)		
Step-1		*Step-2*
$\mathcal{CHAIN}_1(N, S_a, \alpha, \mathcal{PK}_1, \mathcal{PK}_2)$	$\mathcal{CHAIN}_2(N, S_o, \alpha, \mathcal{PK}_1, \mathcal{PK}_2)$	$\mathcal{E}_s(\cdot)$
1 $\mathcal{E}(1, 1, 25, 133, 44, 48) = 11$ $\mathcal{MT}(11) = <121, 0>$	$\mathcal{E}(1, 0, 25, 133, 44, 48) = 132$ $\mathcal{MT}(132) = <1, 1>$	$92^2 \cdot 48 \equiv 90$ $102^2 \cdot 15 \equiv 51$
2 $\mathcal{E}(0, 0, 121, 133, 44, 48) = 43$ $\mathcal{MT}(43) = <120, 0>$	$\mathcal{E}(0, 1, 1, 133, 44, 48) = 117$ $\mathcal{MT}(117) = <123, 1>$	
3 $\mathcal{E}(1, 1, 120, 133, 44, 48) = 106$ $\mathcal{MT}(106) = <64, 1>$	$\mathcal{E}(1, 1, 123, 133, 44, 48) = 102$	
4 $\mathcal{E}(1, 1, 64, 133, 44, 48) = 92$		
$\beta_1 = 92, T_a = (0, 0, 1)$	$\beta_2 = 102, T_o = (1, 1)$	$\gamma_1 = 90, \gamma_2 = 51$
Therefore, $\mathcal{C}_1 = <90, (0, 0, 1)>$, $\mathcal{C}_2 = <51, (1, 1)>$		
Interest Extraction (IE)		
Step-2		*Step-1*
$\mathcal{CHAIN}_1^{-1}(\beta_1, T_a, p, q)$	$\mathcal{CHAIN}_2^{-1}(\beta_2, T_o, p, q)$	$\mathcal{E}_s^{-1}(\cdot)$
1 $\mathcal{E}^{-1}(1, 92, 19, 7, 130, 97) = <64, 1>$ $\mathcal{MT}^{-1}(64, 0, 1) = 106$	$\mathcal{E}^{-1}(1, 102, 19, 7, 130, 97) = <123, 1>$	$90 \in \overline{Q}_R$ $51 \in \mathbb{Z}_N^{-1}$
2 $\mathcal{E}^{-1}(1, 106, 19, 7, 130, 97) = <120, 1>$		
$b_5 = 1$		$b_n = 1$

3.4   Security Proofs

Lemma 3. *For all pair-link encryption ($\mathcal{E}$) described in Lemma 2, the input $x \in \mathbb{Z}_N^{+1}$ is identically distributed over Q_R and $\overline{Q}_R$ (or identically distributed over $\mathbb{Z}_N^{+1}$).*

Proof. Case-1: Let us consider $x \in Q_R$ and the public key $\mathcal{PK}_1, \mathcal{PK}_2 \in \mathbb{Z}_N^{+1}$ where $\mathcal{PK}_1$ and $\mathcal{PK}_2$ belong to different subsets either Q_R or $\overline{Q}_R$ (i.e., ($\mathcal{PK}_1 \in Q_R, \mathcal{PK}_2 \in \overline{Q}_R$) or ($\mathcal{PK}_1 \in \overline{Q}_R, \mathcal{PK}_2 \in Q_R$)) from Tables 1 and 2. For all the pair $a, b \in \{0, 1\}$ with $a = b$, $\mathcal{E}$ successfully generates a unique quadratic residue ciphertext $y \in Q_R$. Similarly, for all the pair $a, b \in \{0, 1\}$ with $a \neq b$, $\mathcal{E}$ successfully generates a unique quadratic non residue ciphertext $y \in \overline{Q}_R$.

Case-2: Let us consider $x \in \overline{Q}_R$ and the public key $\mathcal{PK}_1, \mathcal{PK}_2 \in \mathbb{Z}_N^{+1}$ where $\mathcal{PK}_1$ and $\mathcal{PK}_2$ belong to different subsets either Q_R or $\overline{Q}_R$ (i.e., ($\mathcal{PK}_1 \in Q_R, \mathcal{PK}_2 \in \overline{Q}_R$) or ($\mathcal{PK}_1 \in \overline{Q}_R, \mathcal{PK}_2 \in Q_R$)). For all the pair $a, b \in \{0, 1\}$ with $a = b$, $\mathcal{E}$ successfully generates a unique quadratic non residue ciphertext $y \in \overline{Q}_R$. Similarly, for all the pair $a, b \in \{0, 1\}$ with $a \neq b$, $\mathcal{E}$ successfully generates a unique quadratic residue ciphertext $y \in Q_R$.

Therefore, for all $x \in \mathbb{Z}_N^{+1}$ (whether $x \in Q_R$ or $x \in \overline{Q}_R$) and for all the input pair $a, b \in \{0, 1\}$, the ciphertext generated from $\mathcal{E}$ encryption function is always identically distributed over $\mathbb{Z}_N^{+1}$. Therefore, it is now intuitive that the input x drawn from $\mathbb{Z}_N^{+1}$ is "identically distributed" over Q_R and $\overline{Q}_R$.

Theorem 1. *Any two randomly generated PIR queries from the proposed sit-PIR scheme as described in Definition 1 are identically distributed and hence are information-theoretically indistinguishable.*

Proof. From the response creation algorithm RC of the proposed PIR scheme described in Sect. 3.2, it is clear that each response creation process involves the execution of two parallel chains of successive connective functions and input number to each connective function is always identically distributed over $\mathbb{Z}_N^{+1}$ as described in Lemma 3. Since the input of each $\mathcal{E}$ function is identically distributed over $\mathbb{Z}_N^{+1}$, any two randomly generated PIR queries $\mathcal{Q}_i$ and $\mathcal{Q}_j$, $i, j \in [n]$, with the respective inputs (say) $r \in \mathbb{Z}_N^{+1}$ and $s \in \mathbb{Z}_N^{+1}$ are always identically distributed. Since the queries $\mathcal{Q}_i$ and $\mathcal{Q}_j$ are identically distributed over $\mathbb{Z}_N^{+1}$,

$$\mathrm{PROB}[(\mathcal{Q}_i, sk) \xleftarrow{R} QF(1^k) : Adv(n, \mathcal{Q}_i, 1^k) = 1] = \mathrm{PROB}[(\mathcal{Q}_j, sk)$$
$$\xleftarrow{R} QF(1^k) : Adv(n, \mathcal{Q}_j, 1^k) = 1] \tag{9}$$

Hence any two randomly selected queries $\mathcal{Q}_i$ and $\mathcal{Q}_j$ from query generation algorithm are always independent to each other and consist of "identically distributed" input numbers.

If the queries are identically distributed, then the privacy leak through the mutual information is always zero. Therefore, let any two independent random variables X and Y be $[(\mathcal{Q}_i, sk) \xleftarrow{R} QF(1^k) : Adv(n, \mathcal{Q}_i, 1^k) = 1]$ and $[(\mathcal{Q}_j, sk) \xleftarrow{R} QF(1^k) : Adv(n, \mathcal{Q}_j, 1^k) = 1]$ respectively. Intuitively $\mathrm{PROB}(XY) = \mathrm{PROB}(X, Y) = \mathrm{PROB}(X) \cdot \mathrm{PROB}(Y) = \mathrm{PROB}(YX)$. Then the conditional distribution of X and Y is calculated as

$$\mathbf{PROB(X\,|\,Y)} = \frac{\mathrm{PROB}(XY)}{\mathrm{PROB}(Y)} = \frac{\mathrm{PROB}(X) \cdot \mathrm{PROB}(Y)}{\mathrm{PROB}(Y)} = \mathrm{PROB}(X)$$
$$\mathbf{PROB(Y\,|\,X)} = \frac{\mathrm{PROB}(YX)}{\mathrm{PROB}(X)} = \frac{\mathrm{PROB}(Y) \cdot \mathrm{PROB}(X)}{\mathrm{PROB}(X)} = \mathrm{PROB}(Y) \tag{10}$$

Then, the mutual information of X and Y is calculated as

$$\mathbf{I(X,Y)} = \sum_X \sum_Y \mathrm{PROB}(X,Y) \, log \, \frac{\mathrm{PROB}(X,Y)}{\mathrm{PROB}(X) \cdot \mathrm{PROB}(Y)} = 0 = \mathbf{I(X,Y)} \tag{11}$$

Intuitively, X and Y are information-theoretically indistinguishable. Therefore, all such queries exhibit *perfect privacy* i.e, leaks no information about the user interest on the curious server side.

Theorem 2. *For all the single database information-theoretically indistinguishable PIR (sitPIR) scheme defined in Definition 1, the server communication cost is always guaranteed to be $\mathcal{O}(o(n) + 2 \, log \, N)$ where $(2 \, log \, N)$ is the fixed size chain specific ciphertexts.*

Proof. By referring the Eqs. 7 and 8, it is clear that the PIR response creation involves two chain of successive connective functions $CHAIN_1$ and $CHAIN_2$.

There are $(h-1)$ number of connective functions used in $\mathcal{CHAIN}_1$ and each connective function generates one communication bit. Therefore, there are $(h-1)$ number of communication bits generated from $\mathcal{CHAIN}_1$ where $h = n/2$. Similarly, there are $(h-2)$ number of connective functions used in $\mathcal{CHAIN}_2$ and each connective function generates one communication bit. Therefore, there are $(h-2)$ number of communication bits generated from $\mathcal{CHAIN}_2$. In total, considering both $\mathcal{CHAIN}_1$ and $\mathcal{CHAIN}_2$, there are $(h-1)+(h-2) = 2h-3 \Rightarrow (2 \cdot n/2)-3 \Rightarrow (n-3)$ number of communication bits generated from the database which is clearly less than the database size i.e., $o(n)$. Also, there are two fixed $log\ N$ size chain ciphertexts β_1, β_2. The overall server communication would be $(n - 3 + 2\ log\ N)$ which is slightly greater than the trivial communication (without any optimization). But, the scheme will achieve non-trivial communication when $((o(n) + 2\ log\ N)/n) = 0$ for all $c_0 > c$ and $n = 2^{c_0}$ where c is an integer constant.

Correctness Proof: When the underlying standard quadratic residuosity product is correct and the LTDF function of Freeman et al. [13] is successfully invertible and the communication bit sets T_a, T_o and the ciphertexts set $\mathcal{R}$ sent from the server are unchanged during transmission, the proposed *sitPIR* scheme always generates the required bit of the database. Therefore, $\forall z \in [n]$, for all the security parameter k ($\forall k \in \mathbb{N}$), for all the RSA composite N, for the database $\mathcal{DB}$,

$$IE((\mathcal{R}, sk) : \mathcal{R} \leftarrow RC(\mathcal{Q}, \mathcal{DB}, n, 1^k), (\mathcal{Q}, sk) \xleftarrow{R} QF(1^k)) = b_z$$

Proof. By the Lemma 2, it is clear that each pair-link function of Eq. 3 has unique solution. That means, for all the given ciphertext, the inverse pair-link function always produces the unique plaintext. It is intuitive that each chain described in Definition 3 is composed of combination of pair-link functions described in Eq. 3 and LTDFs. Since the underlying components produces, unique solution, the chain also produces the unique solution. Additionally, the criss-cross encryption ensures that one chain ($\mathcal{CHAIN}_2$) supplies the required bits to the other chain ($\mathcal{CHAIN}_1$) during retrieval process. Therefore, for all the given ciphertext and communication bits, $\mathcal{CHAIN}_2$ always gives the unique plaintext bits of S_o. For all the given ciphertext and communication bits and the bits supplied by $\mathcal{CHAIN}_2$, $\mathcal{CHAIN}_1$ always gives the unique plaintext bits of S_a. Hence, for all the given response and the private key, the interest extraction (IE) algorithm always produces the required bit of interest.

3.5   Performance

PRIVACY: Since the proposed scheme generates information-theoretic queries, privacy is evenly distributed over $\mathbb{Z}_N^{+1}$. This information-theoretic query makes the curious server to achieve only fair coin toss probability to reveal user privacy. One of the greatest advantages of the proposed scheme is that the *data privacy* level can be adjusted from $(p^{QR} + p^R)$ to $(p^{QR} + p^R + p^C)$.

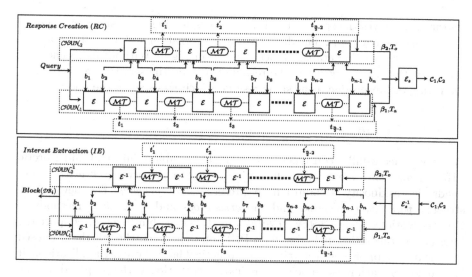

Fig. 1. A single block response creation (RC) and interest extraction (IE) algorithm execution for the proposed sitPBR scheme.

COMMUNICATION: For the given database of size n, the proposed scheme generates $\mathcal{O}(5 \log N)$ number of user query bits, generates $\mathcal{O}(n - 3 + 2 \log N)$ server response bits. If the caching is enabled by storing all the communication bits generated during response creation, then the succeeding PIR invocations generate only constant size response (i.e., $2 \log N$ bits). Also, note that the non-trivial communication can be achieved when $((o(n) + 2 \log N)/n) = 0$ for all $c_0 > c$ and $n = 2^{c_0}$ where c is an integer constant. For example, if $k = 2048$, the non-trivial communication can be achieved for all $c_0 = 18$. Similarly, if $k = 4096$, the non-trivial communication can be achieved for all $c_0 = 20$.

COMPUTATION: In the proposed scheme, server executes $\mathcal{O}((3n/2) - 1)$ number of modular multiplications from $\mathcal{CHAIN}_1$ function, $\mathcal{O}((3n/2) - 4)$ number of modular multiplications from $\mathcal{CHAIN}_2$ function and two modular multiplications. User executes minimum two modular multiplications and maximum $\mathcal{O}(3n - 5)$ + 2 number of modular inverse multiplications. In the *criss-cross* method, response creation can be executed with two parallel sub-processes (in which each sub-process executes each chain in parallel) and interest extraction cannot be assigned to sub-processes due to the dependency of one chain on the other. In the *snake-walk* method, both response creation and interest extraction processes can be assigned to two sub-processes.

3.6   A Single Database Information-Theoretic Private Block Retrieval (sitPBR)

The proposed sitPIR scheme is easily extended to sitPBR scheme as follows. Let a two dimensional matrix $n = uv$ bit database $\mathcal{D} = \{\mathcal{DB}_1, \cdots, \mathcal{DB}_u\}$ where $|\mathcal{DB}_i| = v$, $i \in [1, u]$. The QG algorithm generates identically distributed random queries $\{\mathcal{Q}_1, \cdots, \mathcal{Q}_u\}$ and the RC algorithm generates $\{\mathcal{R}_1, \cdots, \mathcal{R}_u\}$ responses. Finally, IE algorithm retrieves the specific block $j \in [u]$ by selecting the respective response $\mathcal{R}_j$ and private key (p, q). The detailed response creation and interest extraction execution for a single database block is given in Fig. 1. Without extra effort, it is evident that the integrity of the response sent by the server is verified when the IE algorithm produces the same residue which was sent in the query.

4   Conclusion

We have successfully constructed the single database information-theoretic PIR scheme using information-theoretic queries to preserve *user privacy* and quadratic residuosity assumption to preserve *data privacy*. The newly constructed *pair-link encryption* and the *criss-cross* and *snake-walk* methods of PIR encryptions using $\mathcal{CHAIN}_1$, $\mathcal{CHAIN}_2$ in RC algorithm together support the information-theoretic single database PIR solution. Even though the proposed scheme fully supports *perfect privacy*, for practical large database applications, it is required to reach reasonable communication cost. Hence, the proposed scheme is only the stepping stone and can further be modified to attain efficient communication cost using pre-processing techniques. There are several additional open problems like considering bandwidth utilization, robustness, fault-tolerance etc. in a single database information-theoretic PIR and among them the construction of communication efficient perfect privacy preserving single database PIR solution for privacy critical applications is still an open problem.

References

1. Aguilar-Melchor, C., Barrier, J., Fousse, L., Killijian, M.-O.: Xpir: private information retrieval for everyone. Cryptology ePrint Archive, Report 2014/1025 (2014). https://eprint.iacr.org/2014/1025
2. Beimel, A., Ishai, Y., Kushilevitz, E.: General constructions for information-theoretic private information retrieval. J. Comput. Syst. Sci. **71**(2), 213–247 (2005)
3. Beimel, A., Ishai, Y., Malkin, T.: Reducing the servers computation in private information retrieval: PIR with preprocessing. In: 20th Annual International Cryptology Conference Advances in Cryptology - CRYPTO 2000, Santa Barbara, California, USA, pp. 55–73 (2000)
4. Beimel, A., Stahl, Y.: Robust information-theoretic private information retrieval. J. Crypt. **20**(3), 295–321 (2007)
5. Chor, B., Gilboa, N., Naor, M.: Private information retrieval by keywords. Cryptology ePrint Archive, Report 1998/003 (1998). http://eprint.iacr.org/1998/003
6. Cachin, C., Micali, S., Stadler, M.: Computationally private information retrieval with polylogarithmic communication. In: Stern, J. (ed.) EUROCRYPT 1999. LNCS, vol. 1592, pp. 402–414. Springer, Heidelberg (1999). https://doi.org/10.1007/3-540-48910-X_28
7. Canetti, R., Holmgren, J., Richelson, S.: Towards doubly efficient private information retrieval. Cryptology ePrint Archive, Report 2017/568 (2017). https://eprint.iacr.org/2017/568
8. Chakrabarti, A., Shubina, A.: Nearly private information retrieval. In: Kučera, L., Kučera, A. (eds.) MFCS 2007. LNCS, vol. 4708, pp. 383–393. Springer, Heidelberg (2007). https://doi.org/10.1007/978-3-540-74456-6_35

9. Chang, Y.-C.: Single database private information retrieval with logarithmic communication. In: Wang, H., Pieprzyk, J., Varadharajan, V. (eds.) ACISP 2004. LNCS, vol. 3108, pp. 50–61. Springer, Heidelberg (2004). https://doi.org/10.1007/978-3-540-27800-9_5

10. Chor, B., Goldreich, O., Kushilevitz, E., Sudan, M.: Private information retrieval. In: Proceedings of the 36th FOCS, FOCS 1995, pp. 41–50. IEEE Computer Society (1995)

11. Chor, B., Gilboa, N.: Computationally private information retrieval (extended abstract). In: Proceedings of 29th STOC, STOC 1997, pp. 304–313. ACM (1997)

12. Chor, B., Kushilevitz, E., Goldreich, O., Sudan, M.: Private information retrieval. J. ACM **45**(6), 965–981 (1998)

13. Freeman, D.M., Goldreich, O., Kiltz, E., Rosen, A., Segev, G.: More constructions of lossy and correlation-secure trapdoor functions. Cryptology ePrint Archive, Report 2009/590 (2009). http://eprint.iacr.org/2009/590

14. Gentry, C., Ramzan, Z.: Single-database private information retrieval with constant communication rate. In: Caires, L., Italiano, G.F., Monteiro, L., Palamidessi, C., Yung, M. (eds.) ICALP 2005. LNCS, vol. 3580, pp. 803–815. Springer, Heidelberg (2005). https://doi.org/10.1007/11523468_65

15. Gertner, Y., Ishai, Y., Kushilevitz, E., Malkin, T.: Protecting data privacy in private information retrieval schemes. In STOC 1998, pp. 151–160. ACM (1998)

16. Goldberg, I.: Improving the robustness of private information retrieval. In: IEEE Symposium on Security and Privacy, pp. 131–148 (2007)

17. Groth, J., Kiayias, A., Lipmaa, H.: Multi-query computationally-private information retrieval with constant communication rate. In: Nguyen, P.Q., Pointcheval, D. (eds.) PKC 2010. LNCS, vol. 6056, pp. 107–123. Springer, Heidelberg (2010). https://doi.org/10.1007/978-3-642-13013-7_7

18. Ishai, Y., Kushilevitz, E., Ostrovsky, R., Sahai, A.: Cryptography from anonymity. In: Proceedings of 47th FOCS, FOCS 2006, pp. 239–248. IEEE Computer Society (2006)

19. Kushilevitz, E., Ostrovsky, R.: Replication is not needed: single database, computationally-private information retrieval. In: Proceedings of 38th FOCS, FOCS 1997, p. 364. IEEE Computer Society (1997)

20. Kushilevitz, E., Ostrovsky, R.: One-way trapdoor permutations are sufficient for non-trivial single-server private information retrieval. In: Preneel, B. (ed.) EUROCRYPT 2000. LNCS, vol. 1807, pp. 104–121. Springer, Heidelberg (2000). https://doi.org/10.1007/3-540-45539-6_9

21. Lipmaa, H.: An oblivious transfer protocol with log-squared communication. In: Zhou, J., Lopez, J., Deng, R.H., Bao, F. (eds.) ISC 2005. LNCS, vol. 3650, pp. 314–328. Springer, Heidelberg (2005). https://doi.org/10.1007/11556992_23

22. Lipmaa, H.: First CPIR protocol with data-dependent computation. In: Lee, D., Hong, S. (eds.) ICISC 2009. LNCS, vol. 5984, pp. 193–210. Springer, Heidelberg (2010). https://doi.org/10.1007/978-3-642-14423-3_14

23. Liu, T., Vaikuntanathan, V.: On basing private information retrieval on NP-hardness. In: Kushilevitz, E., Malkin, T. (eds.) TCC 2016. LNCS, vol. 9562, pp. 372–386. Springer, Heidelberg (2016). https://doi.org/10.1007/978-3-662-49096-9_16

24. Aguilar-Melchor, C., Gaborit, P.: A lattice-based computationally-efficient private information retrieval protocol (2007)

25. Trostle, J., Parrish, A.: Efficient computationally private information retrieval from anonymity or trapdoor groups. In: Burmester, M., Tsudik, G., Magliveras, S., Ilić, I. (eds.) ISC 2010. LNCS, vol. 6531, pp. 114–128. Springer, Heidelberg (2011). https://doi.org/10.1007/978-3-642-18178-8_10

Attacks and Software Security

Practical Algebraic Side-Channel Attacks Against ACORN

Alexandre Adomnicai[1,2(✉)] ⓘ, Laurent Masson[1] ⓘ,
and Jacques J. A. Fournier[3] ⓘ

[1] Trusted Objects, Aix-en-Provence, France
{a.adomnicai,l.masson}@trusted-objects.com
[2] Mines Saint-Étienne, CEA-Tech, Centre CMP, Gardanne, France
[3] Univ. Grenoble Alpes, CEA-LETI, DSYS, Grenoble, France
jacques.fournier@cea.fr

Abstract. The authenticated cipher ACORN is one of the two finalists of the CAESAR competition and is intended for lightweight applications. Because such use cases require protection against physical attacks, several works have been undertaken to achieve secure implementations. Although dedicated threshold and masked schemes have been proposed, no practical side-channel attack against ACORN has been published in the literature yet. It has been theoretically demonstrated that ACORN is vulnerable against differential power analysis but the feasibility of the attack has not been validated in a practical manner. This paper details the results obtained when putting the attack into practice against a software implementation running on a 32-bit micro-controller. Especially, these practical results led us to propose two optimizations of the reference attack: one that requires less knowledge of initial vectors and another one that is less prone to errors and requires fewer acquisitions.

Keywords: ACORN · Authenticated encryption ·
Side-channel attacks

1 Introduction

In January 2013, the competition for authenticated encryption: security, applicability, and robustness (CAESAR) has been launched with the objective to push for the adoption of authenticated encryption schemes that offer advantages over AES-GCM and are suitable for widespread adoption. In March 2018, the finalists for different use cases were announced. Among them, ACORN is still competing for lightweight applications. This category is defined by various criteria such as compactness of the implementation (in software and hardware), a low overhead for short messages and an intrinsic ability to protect against physical attacks. While several studies have been carried out in order to investigate the last point, most of them discuss the susceptibility of ACORN towards fault attacks [4,14,19]. The first work with regards to side-channel attacks has been

© Springer Nature Switzerland AG 2019
K. Lee (Ed.): ICISC 2018, LNCS 11396, pp. 325–340, 2019.
https://doi.org/10.1007/978-3-030-12146-4_20

recently published [5]. In this paper, the authors propose threshold implementations of some CAESAR candidates, including ACORN, in order to compare their ability to integrate countermeasures against differential power analysis (DPA) in hardware. To justify the need of such countermeasures, they apply the non-specific t-test [13] to each unprotected implementation on a Spartan 6 FPGA in order to detect the presence of leakages. Their results show that ACORN seems to be the most leakage resilient candidate in the unprotected setting and has the lowest area when implemented with countermeasures. The second work dealing with side-channel attacks follows the same approach by studying the integration of the masking countermeasure to the finalists ACORN and Ascon in software [1]. This latter also introduces the first theoretical DPA against ACORN but does not provide any practical results. Therefore, the only two available studies on the susceptibility of ACORN towards side-channel analysis only deal with leakage detection and theoretical attacks.

Our Contribution. Although leakage assessment methodologies give a good overview of the resilience of an implementation against side-channel attacks, it might not be sufficient to guarantee its security level [16]. Because such statistic tools are not meant to perform a key recovery, it is recommended to run additional tests (*e.g.* DPA) in order to assess the security of an implementation in an accurate manner. However, the only DPA against ACORN reported in the literature has not been validated in practice. To fill the gap, we run the attack described in [1] on a software implementation of ACORN on a 32-bit microcontroller. In addition to bringing information on the effectiveness of the attack and the difficulties that might be encountered in practice, our results allow us to introduce more efficient attack paths.

Outline. The rest of this paper is organised as follows. Section 2 briefly recalls the specification of ACORN and the principle of correlation eletromagnetic analysis. Section 3 recalls the theoretical attack against this algorithm and provides some missing elements in order to put it into practice. Subsequently, Sect. 4 details how the attack was applied in a practical manner and presents the results obtained. Section 5 introduces two optimized variants of the reference attack, each one having its own advantages. Finally, we summarise our main results and provide some perspectives in Sect. 6.

2   Preliminaries

2.1   ACORN

ACORN [18] is a stream cipher based authenticated encryption with associated data (AEAD) algorithm designed by Hongjun Wu. ACORN uses a 128-bit key, a 128-bit initialization vector (IV) and produces a 128-bit authentication tag. Its internal state is 293-bit long and consists of the concatenation of six LFSRs in addition to a 4-bit register, as shown in Fig. 1. We note S^i the state after i updates and S_j the jth bit of the state.

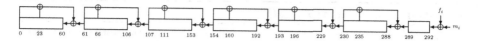

Fig. 1. The concatenation of 6 LFSRs in ACORN. f_i and m_i indicate the overall feedback bit and the message bit for the ith step, respectively.

ACORN relies on three main functions: an output keystream generation function, a nonlinear feedback function, and a state update function. The keystream generation function is defined by

$$\kappa(S) = S_{12} \oplus S_{154} \oplus Maj(S_{235}, S_{61}, S_{193}) \oplus Ch(S_{230}, S_{111}, S_{66}) \tag{1}$$

where $Maj(x, y, z) = (x \wedge y) \oplus (x \wedge z) \oplus (y \wedge z)$ and $Ch(x, y) = (x \wedge y) \oplus (\neg x \wedge z)$. The nonlinear feedback function is defined by

$$\varphi(S, k, ca, cb) = S_0 \oplus \neg S_{107} \oplus Maj(S_{244}, S_{23}, S_{160}) \oplus (ca \wedge S_{196}) \oplus (cb \wedge k). \tag{2}$$

The variables ca and cb allow to define different variants of the feedback function for the four phases of the cipher: initialization, additional data processing, encryption and tag generation. All of them rely on the state update function, defined in Algorithm 1, which is the core of ACORN.

Algorithm 1. StateUpdate(S^i, m_i, ca, cb)

$S^i_{289} \leftarrow S^i_{289} \oplus S^i_{235} \oplus S^i_{230}$ ▷ Update using six LFSRs
$S^i_{230} \leftarrow S^i_{230} \oplus S^i_{196} \oplus S^i_{193}$
$S^i_{193} \leftarrow S^i_{193} \oplus S^i_{160} \oplus S^i_{154}$
$S^i_{154} \leftarrow S^i_{154} \oplus S^i_{111} \oplus S^i_{107}$
$S^i_{107} \leftarrow S^i_{107} \oplus S^i_{66} \oplus S^i_{61}$
$S^i_{61} \leftarrow S^i_{61} \oplus S^i_{23} \oplus S^i_0$
$ks_i \leftarrow \kappa(S^i)$
$c_i \leftarrow ks_i \oplus m_i$ ▷ Encryption of the input
$f_i \leftarrow \varphi(S^i, ks_i, ca, cb)$ ▷ Nonlinear feedback bit generation
for j from 0 to 291 **do**
 $S^{i+1}_j \leftarrow S^i_{j+1}$ ▷ Shift the state
$S^{i+1}_{292} \leftarrow f_i \oplus m_i$ ▷ Injection of the input

---

Initialization. The initialization phase takes as input the encryption key and the IV. First, the entire state is initialized to zero. Then the cipher is run for 1792 steps as described in Algorithm 2.

Additional Data Processing. After the initialization step, the associated data is used to update the state. The cipher is run for at least 256 steps, even if there is no associated data to process.

Algorithm 2. AcornInit(S^0, K, IV)

$(S_0^0, ..., S_{292}^0) \leftarrow (0, ..., 0)$ ▷ Initialize the state to zero

for i from 0 to 127 **do**

    $S^{i+1} \leftarrow$ StateUpdate($S^i, K_i, 1, 1$)   ▷ Update the state with key bits as input

for i from 0 to 127 **do**

    $S^{129+i} \leftarrow$ StateUpdate($S^{128+i}, IV_i, 1, 1$)   ▷ Update the state with IV bits as input

$S^{257} \leftarrow$ StateUpdate($S^{256}, K_0 \oplus 1, 1, 1$)

for i from 1 to 1535 **do**

    $S^{257+i} \leftarrow$ StateUpdate($S^{256+i}, K_{i \bmod 128}, 1, 1$) ▷ Update the state with key bits as input

---

Encryption. At each step of the encryption, one bit from the plaintext is encrypted. The cipher is run for at least 256 steps, even if there is no plaintext to encrypt.

Finalization. At the end, an n-bit authentication tag is computed. The state is updated 768 times and the tag consists of the last n keystream bits generated.

2.2 Correlation Electromagnetic Analysis

Since the publication of DPA [9], it is common knowledge that the analysis of the power consumed by the execution of a cryptographic primitive might reveal information about the secret involved. A few years later, correlation power analysis (CPA) has been widely adopted over DPA as it requires fewer traces and has been shown to be more efficient [3]. The principle is to target a sensitive intermediate state of the algorithm which depends on a subpart of the key, and try to predict its value for all hypotheses. The function that defines the intermediate state from the known input and the subkey is called *selection function*. Then, to uncover the link between these predictions and the leakage measurements, the Pearson correlation coefficient between these two variables is computed using an appropriate leakage model. The Hamming weight (HW) and the Hamming distance (HD) models are the most commonly used models to simulate the leakage of a cryptographic device. For each subkey hypothesis, it results in a value between -1 and 1, indicating how much it correlates with the recorded values for every point in time. Finally, the hypothesis which matches with the real subkey should return a significantly higher coefficient than the other hypotheses. The procedure is described in details in Algorithm 3. This attack remains valid when analyzing electromagnetic emanations [6,11] instead of power consumption, since they are mainly due to the displacement of current through the rails of the metal layers. In this case, we refer to it as correlation electromagnetic analysis (CEMA).

Algorithm 3. CEMA(φ, $\mathcal{L}$, $D^{1\cdots n}$, $[a, b]$, $M^{1\cdots n}$)

Require: Selection function φ ; Leakage model $\mathcal{L}$; Data acquisitions $D^{1\cdots n}$; Interval of samples to consider $[a, b]$; Input messages $M^{1\cdots n}$

Ensure: subkey candidate $\bar{k}$

   for i from 1 to n **do**

      for k from 0 to $|\mathcal{K} - 1|$ **do** $\triangleright$ $\mathcal{K}$ denotes the key search space

         $H_k^i \leftarrow \mathcal{L}\left(\varphi(k, M^i)\right)$ $\triangleright$ Prediction of the intermediate state leakage

   for i from a to b **do** $\triangleright$ For each sample to consider

      for k from 0 to $|\mathcal{K} - 1|$ **do**

        $C_k^i \leftarrow \mathsf{Corr}\left(\left[H_k^1, \cdots, H_k^n\right], \left[D_i^0, \cdots, D_i^n\right]\right)$

 $C_{\bar{k}} \leftarrow \mathsf{max}(C)$ $\triangleright$ Most likely subkey among all samples in $[a, b]$

---

3 Reference Attack Against ACORN

3.1 Theoretical Basics

The attack introduced in [1] details how a DPA can be mounted against leakages caused by the calculation of $S_{292}^{i+1} \leftarrow f_i \oplus m_i$ when updating the state update during the initialization phase for $128 \leq i \leq 255$. More precisely, it assumes the knowledge of the input $m_i = IV_{i-128}$ and thus targets the feedback bits f_i. However, because feedback bits are defined by nonlinear combinations of several key bits, the attack does not lead to a direct key recovery but returns a system of Boolean equations to be solved. This kind of attack is called algebraic side-channel attack (ASCA) [12] and has already been applied to other stream ciphers such as Trivium and Grain [8].

In the case of ACORN, the state is first updated 128 times with the key. Especially, after the 128th initialization step, the state is as follows

$$\left(S_0^{128}, ..., S_{164}^{128}\right) = (0, ..., 0)$$

$$\left(S_{165}^{128}, ..., S_{198}^{128}\right) = (\neg K_0, ..., \neg K_{33})$$

$$\left(S_{199}^{128}, ..., S_{201}^{128}\right) = (K_{34} \oplus K_0, ..., K_{36} \oplus K_2)$$

$$\left(S_{202}^{128}, ..., S_{218}^{128}\right) = (\neg K_{37} \oplus K_3 \oplus K_0, ..., \neg K_{53} \oplus K_{19} \oplus K_{16})$$

$$\left(S_{219}^{128}, ..., S_{223}^{128}\right) = (K_{54} \oplus K_{20} \oplus K_{17} \oplus K_0, ..., K_{58} \oplus K_{24} \oplus K_{21} \oplus K_4)$$

$$\left(S_{224}^{128}, ..., S_{229}^{128}\right) = (\neg K_{59} \oplus K_{25} \oplus K_{22} \oplus K_5 \oplus K_0, ..., \neg K_{64} \oplus K_{30} \oplus K_{27} \oplus K_{10} \oplus K_5) \tag{3}$$

$$\left(S_{230}^{128}, ..., S_{261}^{128}\right) = (\neg K_{65} \oplus K_{11} \oplus K_6, ..., \neg K_{96} \oplus K_{42} \oplus K_{37})$$

$$\left(S_{262}^{128}, ..., S_{272}^{128}\right) = (K_{97} \oplus K_{43} \oplus K_{38} \oplus f_{97}, ..., K_{107} \oplus K_{53} \oplus K_{48} \oplus f_{107})$$

$$\left(S_{273}^{128}, ..., S_{288}^{128}\right) = (\neg K_{108} \oplus K_{54} \oplus K_{49} \oplus K_0 \oplus f_{108}, ..., \neg K_{123} \oplus K_{69} \oplus K_{64} \oplus K_{15} \oplus f_{123})$$

$$\left(S_{289}^{128}, ..., S_{292}^{128}\right) = (\neg K_{124} \oplus f_{124}, ..., \neg K_{127} \oplus f_{127})$$

where f_i defines the nonlinear feedback bit.

$$f_i = \begin{cases} 1 & \text{if } 0 \leq i \leq 96 \\ K_{i-97} & \text{if } 97 \leq i \leq 99 \\ (\neg K_{i-58}) \wedge (\neg K_{i-100}) \oplus K_{i-97} & \text{if } 100 \leq i \leq 111 \\ (K_{i-58} \oplus K_{i-112}) \wedge (\neg K_{i-100}) \oplus K_{i-97} & \text{if } 112 \leq i \leq 116 \\ \neg(K_{i-58} \oplus K_{i-112} \oplus K_{i-117}) \wedge (\neg K_{i-100}) \oplus K_{i-97} & \text{if } 117 \leq i \leq 127 \end{cases} \quad (4)$$

Then, the state is updated with the IV as input for the next 128 steps. As a result, one can run a DPA by targeting the result of the XOR between IV bits and feedback bits in order to get a system of Boolean equations to be solved. However, f_i is constant for a given key if and only if $i \leq 176$. Indeed, from $i = 177$, IV bits that have been injected into the internal state have been shifted to such an extent that they are involved in the computation of f_i. Therefore, the use of XOR as selection function is only possible from f_{128} to f_{176}, which results in a Boolean system $\mathscr{F}$ that depends on all key bits.

The task of recovering the key bits from $\mathscr{F}$ can be reduced to a variant of the Boolean satisfiability (SAT) problem, which decides whether a given propositional formula in conjunctive normal form (CNF) is satisfiable. As the CNF derived from $\mathscr{F}$ is satisfiable at least by the encryption key K, the purpose of the attack is to get all of the truth assignments of SAT. Because $\mathscr{F}$ defines a system of 49 equations with 128 unknowns, there are so many solutions that we were not able to determine the number of truth assignments by means of 600 core-hours. In order to reduce the number of solutions, it is possible to extend $\mathscr{F}$ by recovering the next feedback bits using more sophisticated selection functions.

Actually, the DPA should not target the next f_i themselves but their component parts that are IV-independent. The principle is to isolate the key bit combinations from the IV bits so that they can be recovered through a DPA and then be added to the Boolean system. As a result, each DPA against f_i for $i \geq 177$ adds $n + 1$ equations to the Boolean system where n is the number of IV bits involved in the calculation of f_i. The authors computed the value of each equation of $\mathscr{F}$ for a random key and investigated how many equations are necessary to return a single key hypothesis. Their experimentations led to the conclusion that $\mathscr{F}$ has a unique truth assignment only if it results from the leakage of at least the first 82 updates (*i.e.* from f_{128} to f_{209}). Therefore, ACORN is theoretically vulnerable to DPA and it is only necessary to have knowledge of the first 82 IV bits to recover the entire encryption key.

3.2   Remarks and Clarifications

Although [1] clearly exhibits how to proceed in order to isolate the key bit combinations when a single IV bit interferes in the recovery of f_i with $i \geq 177$, the case where multiple IV bits are involved is left to the reader as an exercise. Thanks to the distributive property of AND over XOR, f_i can be rewritten in terms of IV bits as shown in Table 1, where f_i' refers to f_i for the null IV.

Table 1. Intermediate bit $\beta_{i\in\mathcal{I}}$ to consider when running a DPA against $f_{128+i}\oplus IV_i$

$\mathcal{I}$	$\beta_{i\in\mathcal{I}}$
$[0,48]$	$f_{128+i}\oplus IV_i$
$[49,57]$	$f'_{128+i}\oplus\left(S_{160}^{128+i}\wedge IV_{i-49}\right)\oplus IV_i$
$[58,62]$	$f'_{128+i}\oplus\left(S_{160}^{128+i}\wedge IV_{i-49}\right)\oplus\left(S_{193}^{128+i}\wedge IV_{i-58}\right)\oplus IV_i$
$[63,96]$	$f'_{128+i}\oplus\left(S_{160}^{128+i}\wedge IV_{i-49}\right)\oplus\left(S_{193}^{128+i}\wedge IV_{i-58}\right)\oplus\left(S_{111}^{128+i}\wedge IV_{i-63}\right)\oplus IV_i$

It straightforwardly follows the definition of the selection function $\varphi_{i\in\mathcal{I}}$ to use for a given feedback bit index.

$$
\varphi_i:\begin{cases}
x\mapsto x\oplus IV_i & \text{if } 0\le i\le 48\\
(x,y)\mapsto x\oplus(y\wedge IV_{i-49})\oplus IV_i & \text{if } 49\le i\le 57\\
(x,y,z)\mapsto x\oplus(y\wedge IV_{i-49})\oplus(z\wedge IV_{i-58})\oplus IV_i & \text{if } 58\le i\le 62\\
(x,y,z,t)\mapsto x\oplus(y\wedge IV_{i-49})\oplus(z\wedge IV_{i-58})\oplus(t\wedge IV_{i-63})\oplus IV_i & \text{if } 63\le i\le 96
\end{cases}
\tag{5}
$$

Throughout this paper, $\mathscr{F}_\mathcal{I}$ refers to the system resulting from the leakage of $\beta_{i\in\mathcal{I}}$. In order to express the values of $\mathscr{F}_\mathcal{I}$ in terms of key bits, we implemented a software version of ACORN which operates on strings instead of numeric values (*i.e.* 'a' $\oplus$ 'b' = 'a ^ b'). For instance, $\mathscr{F}_{[0,81]}$ which should return a unique solution according to [1], is defined in Eq. 6.

$$
\mathscr{F}_{[0,81]}=\begin{cases}
(\neg K_{70}\oplus K_{11}\oplus K_{16})\wedge\neg K_{28}\oplus K_{31} & = f_{128}\\
\quad\vdots & \vdots\\
(\neg(K_{69}\oplus K_{10}\oplus K_{15})\wedge\neg K_{27}\oplus K_{30}\oplus K_{127}\oplus K_{68}\oplus K_9\oplus\cdots)\wedge\cdots & = f_{176}\\
(f_{128}\oplus K_{69}\oplus K_{10}\oplus K_{74}\oplus K_{20})\wedge S_{160}^{177}\oplus\neg(K_{61}\oplus K_2\oplus K_7)\oplus\cdots & = f'_{177}\\
\quad\vdots & \vdots\\
(f_{160}\oplus(\neg K_{44}\wedge\neg K_2)\oplus K_5\oplus K_{102}\oplus K_{43}\oplus K_{48})\wedge(\neg K_{60}\oplus\cdots)\oplus\cdots & = f'_{209}\\
\neg K_{44}\oplus K_7\oplus K_{10}\oplus K_5\oplus K_{11} & = S_{160}^{177}\\
\quad\vdots & \vdots\\
K_{76}\oplus K_{17}\oplus K_{22}\oplus K_{39}\oplus K_2\oplus K_{42}\oplus K_8\oplus K_{37}\oplus K_0\oplus K_3\oplus\cdots & = S_{160}^{209}\\
\neg K_{86}\oplus K_{27}\oplus K_{32}\oplus K_{49}\oplus K_{12}\oplus K_{14}\oplus K_{52}\oplus K_{15}\oplus K_{18} & = S_{193}^{186}\\
\quad\vdots & \vdots\\
(\neg K_{51}\wedge\neg K_9)\oplus K_{21}\oplus K_{109}\oplus K_{50}\oplus K_{55}\oplus K_1\oplus K_{72}\oplus K_{13}\oplus\cdots & = S_{193}^{209}\\
\neg K_9 & = S_{111}^{191}\\
\quad\vdots & \vdots\\
\neg K_{27} & = S_{111}^{209}
\end{cases}
\tag{6}
$$

Because the equations resulting from the recovery of S_{111}^{128+i} depend on a single key bit, they are especially useful to solve $\mathscr{F}_{[0,81]}$. As a result, it might be interesting to ignore leakages related to f_{128+i} for $i\in[0,62]$ and rather focus on $\mathscr{F}_{[63,96]}$.

4 From Theory to Practice

4.1 Targeted Implementation

Although ACORN is designed to process one bit per step, because its smallest LFSR is 37-bit long, up to 37 steps can be processed in parallel. Within the scope of the CAESAR contest, Hongjun Wu provided an optimized software implementation which processes 32 steps at once. In this way, each function defined in Sect. 2 should be seen as operating on 32-bit words instead of bits. Its implementation dedicates a 64-bit register to each LFSR. Although it consumes more memory than needed, since all LFSRs contains less than 64 bits, it increases the performances by saving some instructions in order to build the 32-bit working variables. An ARM assembly implementation of the state update function based on the same principle is provided in [1]. We chose to run the attack against this specific implementation, as it is the most appropriate for 32-bit platforms. The hard-coded 128-bit encryption key $K = $ 'Encryption key K' was used to encrypt and authenticate 5 000 messages, using random IVs.

4.2 Experimental Setup

All practical experiments presented below were done using a microcontroller equipped with an ARM Cortex-M3 running at 24 MHz. Note that the device under test does not embed any hardware countermeasure against side-channel attacks. A trigger signal was inserted at the beginning and the end of the initialization phase in order to guarantee a proper synchronization. EM emanations were measured using a Langer HF-U 5 near-field probe (30 MHz–3 GHz) combined with a Langer PA 303 BNC preamplifier providing a gain of 30 dB. The sampling acquisition was performed using a PicoScope 6404D sampled at 1 GS/s. We recorded the leakage from state updates where IV words are given as input, but also from five further ones in case they would also contain information to exploit. As shown in Fig. 2, state updates are clearly discernible and each of them are roughly made up of 10 000 samples.

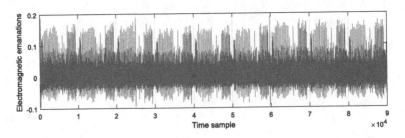

Fig. 2. Data acquisition of nine 32-bit state updates during the initialization phase

4.3 Practical Correlation Electromagnetic Analysis

Even if the targeted feedback bits are actually stored in 32-bit registers, it has been proven that one can compute a partial correlation of the entire variable in order to reduce the computational complexity [17]. Therefore we chose to apply the attack as defined above, in a mono-bit manner, using the Hamming weight leakage model and the Pearson's correlation coefficient as distinguisher. In order to precisely target leakages related to the insertion of f_{128+i} into the state, the window on which the attack is run depends on the feedback bit index. More precisely, an attack against $f_{128+i} \oplus IV_i$ is managed by executing

$$\mathsf{CEMA}\left(\varphi_i, \mathrm{HW}, D^{1\cdots5\,000}, \left[10\,000 \times \left\lceil\frac{i}{32}\right\rceil + 1, 10\,000 \times \left\lceil\frac{i+32}{32}\right\rceil\right], IV^{1\cdots5\,000}\right).$$

As suggested in the theoretical specification, we run the attack for i from 0 to 81. After assigning the CEMA results to the corresponding equations within $\mathscr{F}_{[0,81]}$, they are converted into CNF formulas using the bc2cnf tool [7] and finally given as input to the SAT solver CryptoMiniSat5 [15]. On the first try, it turns out that the input system is not satisfiable and therefore does not lead to a key recovery. Because this issue can be due to many factors (e.g. some erroneous CEMA results or ineffectiveness of some selection functions), we carried out investigations starting by visually examining the CEMA output for various feedback bits.

Figure 3 illustrates the points of interests (POI) for some of them. The first observation that can be made is that information leakage is not identical for all feedback bits. For instance, Fig. 3a shows three samples (5 913, 7 245 and 9 160) that might reveal information about f_{128} while Fig. 3b shows only two (5 913 and 9 329) regarding f_{157}.

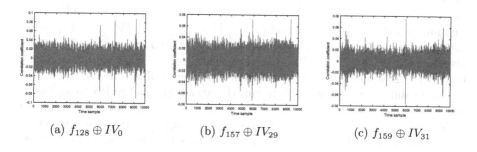

(a) $f_{128} \oplus IV_0$ (b) $f_{157} \oplus IV_{29}$ (c) $f_{159} \oplus IV_{31}$

Fig. 3. POI for several feedback bits during the first state update

Because the implementation processes 32 steps at once, all but the last four – stored in the 4-bit register – feedback bits have already been updated using S_{230} and S_{235} before being added to the state. For instance, after the state update of S^{128} with $IV_{0\cdots31}$ as input, the last 32 bits of S^{160} are as follows.

$$\left(S_{261}^{160}, \cdots, S_{288}^{160}\right) = \left(f_{128} \oplus S_{230}^{132} \oplus S_{235}^{132} \oplus IV_0, \cdots, f_{155} \oplus S_{230}^{159} \oplus S_{235}^{159} \oplus IV_{27}\right)$$
$$\left(S_{289}^{160}, \cdots, S_{292}^{160}\right) = (f_{156} \oplus IV_{28}, \cdots, f_{159} \oplus IV_{31}) \tag{7}$$

Especially, the implementation computes $(f_i \| \cdots \| f_{i+31}) \oplus (IV_i \| \cdots \| IV_{i+31})$ before finally updating its 28 most significant bits and adding it into the state. Therefore, $\varphi_{i \in [0,48]}$ should not only return a candidate for f_{128+i}, but also for $f_{128+i}^1 = f_{128+i} \oplus S_{230}^{132+i} \oplus S_{235}^{132+i}$ when $i < 28 \bmod 32$. Moreover, the implementation of the state update function is generic in the sense that the input is always encrypted using the keystream, even during the initialization phase. Although encryption is not necessary during this phase, it allows the use of the same code through all the authenticated encryption process. As a result, the selection function $\varphi_{i \in [0,48]}$ also targets $ks_i \oplus IV_i$ unintentionally.

These remarks highlight the first difficulty when putting the attack into practice. Because selection functions can lead to the recovery of several key bit combinations (keystream, feedback and updated feedback bits), an attacker has to associate each leakage to an intermediate value. Indeed, if the highest correlation coefficient is reached for the keystream bit but its value is assigned to the feedback bit equation, then the ASCA will fail because of an erroneous Boolean system. For instance, this scenario is depicted in Fig. 3c where the highest correlation peak is reached for the leakage of $ks_{159} \oplus IV_{31}$ instead of $f_{159} \oplus IV_{31}$. The methodology that was used to clearly identify each leakage is described below.

When targeting software implementations on load/store architectures, data transfers due to memory accesses are known to leak the most information compared to arithmetic and logic operations, which only occur between registers and are usually more difficult to exploit in practice [2,10]. Especially, on top of memory accesses that store the last 32 bits into the state as described in Eq. 7, the assembly code under test performs two additional store instructions that are likely to be critical. It consists of $(ks_i \| \cdots \| ks_{i+31}) \oplus (IV_i \| \cdots \| IV_{i+31})$ as computation of the ciphertext, and $(f_{128+i}^1 \| \cdots \| f_{155+i}^1) \oplus (IV_i \| \cdots \| IV_{i+27})$ as computation of the updated feedback word in a temporary register. Therefore, attacks using $\varphi_{i \in [0,48]}$ should lead to three leakages for $i < 28 \bmod 32$ and only two for $i \geq 28 \bmod 32$, which is consistent with the results from Fig. 3.

As a result, our first attempt to run the attack in practice led to an unsatisfiable system because some CEMA results did not match the expected key bit combinations. More generally, our investigations highlight that the theoretical DPA against ACORN as described in Sect. 3 does not necessarily apply to all unprotected implementations. However, a tweaked version of the attack can still be applied in order to deal with exploitable leakages on the device under test. We chose to ignore leakages related to ciphertext computations as they can be easily avoided during the initialization phase.

The required modifications affect some of the selection functions. Indeed, even if they remain valid when $i \geq 28 \bmod 32$, this is not the case anymore from $i = 54$ since S_{235}^{132+i} depends on IV_0. In this case, additional IV bits have to be considered. The tweaked selection functions are noted φ_i^1 and are defined in Eq. 8.

$$\varphi_i^1 = \begin{cases} \varphi_i & \text{if } i \geq 28 \bmod 32 \text{ or } i \leq 54 \\ \varphi_{i-54} \circ \varphi_i & \text{if } 54 \leq i \leq 58 \\ \varphi_{i-59} \circ \varphi_{i-54} \circ \varphi_i & \text{otherwise} \end{cases} \tag{8}$$

Of course, the Boolean system $\mathscr{F}$ has to be modified in order to be compliant with the intermediate bits defined by φ_i^1, and we refer to this variant as $\mathscr{F}^1$. As leakages related to keystream words are not taken into consideration, the attack is run on the last $3\,000$ samples of each state update window, by executing for i from 0 to 81,

$$\mathsf{CEMA}\left(\varphi_i^1, \mathsf{HW}, D^{1\cdots 5\,000}, \left[10\,000 \times \left\lceil \frac{i}{32} \right\rceil + 7\,000, 10\,000 \times \left\lceil \frac{i+32}{32} \right\rceil \right], IV^{1\cdots 5\,000}\right).$$

This time, the attack is successful as the resulting system $\mathscr{F}_{[0,81]}^1$ is satisfiable and returns the expected key as the unique solution. Moreover, unlike $\mathscr{F}_{[0,n]}$ that requires $n \geq 81$ to return a unique solution, $n \geq 78$ is enough for $\mathscr{F}_{[0,n]}^1$. For each kind of selection function, Fig. 4 shows a CEMA result and the maximum correlation coefficient reached for each key hypothesis, depending on the number of acquisitions.

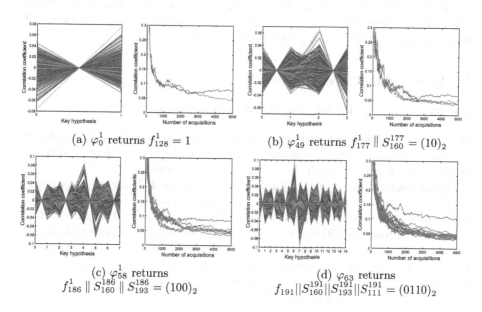

(a) φ_0^1 returns $f_{128}^1 = 1$

(b) φ_{49}^1 returns $f_{177}^1 \,\|\, S_{160}^{177} = (10)_2$

(c) φ_{58}^1 returns
$f_{186}^1 \,\|\, S_{160}^{186} \,\|\, S_{193}^{186} = (100)_2$

(d) φ_{63}^1 returns
$f_{191}^1 \|S_{160}^{191}\|S_{193}^{191}\|S_{111}^{191} = (0110)_2$

Fig. 4. Experimental results for the different selection functions

4.4 Lessons Learned

Several conclusions can be drawn from these experimentations. First, although the practical application does not exactly follows the theoretical specification, it still validates the reference attack in the sense that the evaluation of selection functions remains the same. Indeed, the XOR of additional IV bits does not change the way the hypothetical bits interact with each other as detailed in Eq. 9.

$$\varphi_{65}(x,y,z,t) = x \oplus (y \wedge IV_{16}) \oplus (z \oplus IV_7) \oplus (t \wedge IV_2) \oplus IV_{65}$$
$$\varphi_{65}^1(x,y,z,t) = x \oplus (y \wedge IV_{16}) \oplus (z \oplus IV_7) \oplus (t \wedge IV_2) \oplus (IV_{65} \oplus IV_{11} \oplus IV_6)$$

(9)

Second, since f_{128+i} has to be XORed with IV_i at some point, it might be tempting to perform this calculation during the encryption step in order to achieve a generic implementation of the state update function. However, as shown by our practical experiments, it can lead to additional leakages that an attacker could exploit. Therefore, we argue that the encryption computation should be removed during the initialization phase and that the XOR with IV_i should only occur once f_{128+i} has been entirely computed.

Third, some of the selection functions can be used to recover several intermediate values during the initialization phase (*e.g.* keystream and feedback bits for $\varphi_{i \in [0,48]}$). On the one hand, it introduces the fact that several variants of the attack can be defined depending on the leakage available to the attacker. On the other hand, it requires to clearly identify the points of interests in order to avoid misinterpretation of the results. Finally, some of the selection functions perform better than others. In the next section, we discuss how to take advantage of these results in order to propose more efficient attack paths.

5   Other Attack Variants

5.1   Minimizing the Knowledge of Initial Vectors

Because the knowledge of plaintexts (or IVs in the case of ACORN) is sometimes an unrealistic assumption in practice, it might be interesting to identify the most efficient attack path given the fewer input bits to consider.

In this case, focusing on $\varphi_{i \in [0,48]}$ is of great interest as knowledge of a single IV bit allows to recover several key bit combinations. Because ACORN is defined by the concatenation of six LFSRs, each feedback bit is updated six times before being thrown from the internal state. Therefore, regardless of a potential leakage related to the keystream computation, $\varphi_{i \in [0,48]}$ could theoretically be used to target up to seven key bit combinations: the feedback bit itself f_{128+i} and its six updated values, noted from f_{128+i}^1 to f_{128+i}^6 and defined in Eq. 10, which are computed just before being shifted in each LFSR.

$$f_i^1 = f_i \oplus S_{235}^{i+4} \oplus S_{230}^{i+4}$$
$$f_i^2 = f_i^1 \oplus S_{196}^{i+63} \oplus S_{193}^{i+63}$$
$$f_i^3 = f_i^2 \oplus S_{160}^{i+100} \oplus S_{154}^{i+100}$$
$$f_i^4 = f_i^3 \oplus S_{111}^{i+139} \oplus S_{107}^{i+139} \tag{10}$$
$$f_i^5 = f_i^4 \oplus S_{66}^{i+186} \oplus S_{61}^{i+186}$$
$$f_i^6 = f_i^5 \oplus S_{23}^{i+232} \oplus S_0^{i+232}$$

In order to investigate whether this statement is verified in practice, we ran attacks using $\varphi_{i \in [0,48]}$ on the same acquisitions but this time, on the entire window of 90 000 samples. Indeed, focusing on the state updates with IV as input only allows to exploit leakages related to the update of three LFSRs. Therefore, also considering five further 32-bit state updates gives access to leakages of all LFSRs' updates. As shown in Fig. 5, each state update leads to several leakages in time (usually two). In addition to the leakage produced by the final store instruction, we suspect that the other peak is due to a previous memory access that loads the state from RAM to registers. Note that a 32-bit shift does not necessarily imply an LFSR update and thus, several leakages may refer to the same intermediate value. Therefore, the window to consider when targeting a specific feedback bit depends on its index. For instance, Fig. 5 indicates for each state update the key bit combination targeted by φ_{29}.

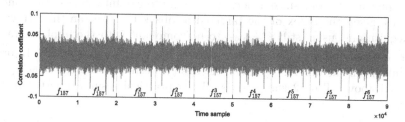

Fig. 5. Leakage in time from $\mathsf{CEMA}\left(\varphi_{29}, \mathrm{HW}, D^{1 \cdots 5\,000}, [1, 90\,000], IV^{1 \cdots 5\,000}\right)$

We tried to apply this attack to our acquisitions. Because the device under test does not leak $f_{128+i} \oplus IV_i$ when $i < 28 \bmod 32$, we were able to exploit six leakages in this case and seven otherwise. The resulting Boolean system is noted $\mathscr{F}_{[0,n]}^{1 \to 6}$ where n refers to the number of IV bits considered. Finally, the SAT solver returned the correct key hypothesis as the unique truth assignment for $n \geq 18$. However, the resulting system required more than one hour to be solved on a commonly available laptop while previous systems required less than a second. Indeed, the more an intermediate value is updated, the more terms are involved in its definition, significantly increasing the number of CNF clauses in the resulting system.

5.2   Maximizing the Practical Efficiency

In cases where the attacker has full knowledge of the IVs, other leakages should be preferred. As mentioned in Sect. 3.2, leakages related to f_{128+i} for $i \in [63, 96]$ are of particular interest as some components targeted by $\varphi_{i \in [63,96]}$ refer to single key bits, not combinations of them, allowing to simplify the Boolean system. Moreover, focusing on $\mathscr{F}_{[63,96]}$ brings additional benefits from a practical point of view.

First, it results from Fig. 4 that φ_{64} shows better results than the other selection functions. Especially, it seems that the more IV bits involved in the selection function, the more efficient it is. This can be explained by the fact that selection functions make IV bits interact with hypotheses through the bitwise AND operator, which is nonlinear. Nonlinearity is a valuable property for selection functions as it ensures a good distinguishability between the correct and incorrect hypotheses and reduces the risk of false positives in practice. As a result, $\varphi_{i \in [63,96]}$ requires fewer acquisitions than other selection functions for the correct hypothesis to stand out.

Second, attacking an intermediate bit using $\varphi_{i \in [63,96]}$ returns a result for four Boolean equations at once. This allows to build a meaningful system by targeting fewer intermediate values and thus, making this attack path less prone to errors. For instance, $\mathscr{F}_{[64,95]}$ is composed of $32 \times 4 = 128$ equations and has only six truth assignments. Another benefit from this variant is the fact that all the leakages take place during the same state update. Therefore, it is of great interest for 32-bit implementations since it does not require to carefully choose the window to attack given the index of the targeted feedback bit. We ran this attack on our acquisitions, using still $\varphi^1_{i \in [64,95]}$ to be compliant with the implementation under test. Solving $\mathscr{F}^1_{[64,95]}$ led to the correct key as the only solution. In order to highlight all the differences and subtleties between the different attack paths discussed above, Table 2 summarizes all the practical results reported in this paper.

Table 2. Summary and comparison of our practical experiments

	$\mathscr{F}^1_{[0,78]}$	$\mathscr{F}^1_{[64,95]}$	$\mathscr{F}^{1 \to 6}_{[0,18]}$
IV bits to consider	$IV_{i \in [0,78]}$	$IV_{i \in [0,95]}$	$IV_{i \in [0,18]}$
# of required acquisitions	$\geq 4\,000$	$\geq 2\,000$	$\geq 4\,000$
# of attacked bits	79	32	114
# of equations	148	128	114
# of CNF clauses	2165	1804	4251
Solving time (i5-6200U CPU)	0.05 s	0.04 s	87 min 17 s

6 Conclusion and Perspectives

The main objective of this paper was to validate the practical feasibility of side-channel attacks against ACORN. To do so, we first defined all selection functions required to put the attack introduced in [1] into practice. Because our experimental setup did not allow us to exploit some leakages required by the theoretical specification, we had to make some minor changes in the selection functions and thus in the resulting Boolean system, in order to achieve a successful attack against the 32-bit implementation under test. However, it does not call into question the reference attack as the results of the selection functions' evaluation remain valid in both cases. Among the different observations made during our experimentations, two of them allowed us to propose optimized variants of the attack. First, one of the selection functions can actually be used to recover several intermediate values, not just the feedback bit itself. It led to an attack that minimizes the number of IV bits to consider. On the device under test, we were able to recover the encryption key with only knowledge of 19 IV bits. Second, another selection function shows significantly better results as it provides a higher distinguishability of the correct hypothesis for fewer acquisitions. This observation led to an attack path that requires to target fewer intermediate values, and is therefore is less prone to errors. On the device under test, we were able to recover the encryption key by targeting 32 intermediate values with half as many acquisitions than other attack paths.

Further work should be undertaken on protected implementations in order to determine whether the selection functions discussed in this paper are efficient enough to deal with high-order side channel analyses. Moreover, the integration of countermeasures such as hiding and shuffling in the specific case of ACORN has not been studied yet and could be of great benefit as some selection functions require to clearly identify specific points of interests.

References

1. Adomnicai, A., Fournier, J.J., Masson, L.: Masking the lightweight authenticated ciphers ACORN and Ascon in software. In: Tiplea, F.L., Warinschi, B. (eds.) Cryptography and Information Security in the Balkans. Springer, Cham (2018). https://eprint.iacr.org/2018/708
2. Biryukov, A., Dinu, D., Großschädl, J.: Correlation power analysis of lightweight block ciphers: from theory to practice. In: Manulis, M., Sadeghi, A.-R., Schneider, S. (eds.) ACNS 2016. LNCS, vol. 9696, pp. 537–557. Springer, Cham (2016). https://doi.org/10.1007/978-3-319-39555-5_29
3. Brier, E., Clavier, C., Olivier, F.: Correlation power analysis with a leakage model. In: Joye, M., Quisquater, J.-J. (eds.) CHES 2004. LNCS, vol. 3156, pp. 16–29. Springer, Heidelberg (2004). https://doi.org/10.1007/978-3-540-28632-5_2
4. Dey, P., Rohit, R.S., Adhikari, A.: Full key recovery of acorn with a single fault. J. Inf. Secur. Appl. 29(C), 57–64 (2016). https://doi.org/10.1016/j.jisa.2016.03.003
5. Diehl, W., Abdulgadir, A., Farahmand, F., Kaps, J.P., Gaj, K.: Comparison of cost of protection against differential power analysis of selected authenticated ciphers. In: 2018 IEEE International Symposium on Hardware Oriented Security and Trust (HOST), pp. 147–152, April 2018. https://doi.org/10.1109/HST.2018.8383904

6. Gandolfi, K., Mourtel, C., Olivier, F.: Electromagnetic analysis: concrete results. In: Koç, Ç.K., Naccache, D., Paar, C. (eds.) CHES 2001. LNCS, vol. 2162, pp. 251–261. Springer, Heidelberg (2001). https://doi.org/10.1007/3-540-44709-1_21

7. Junttila, T.A., Niemelä, I.: Towards an efficient tableau method for boolean circuit satisfiability checking. In: Lloyd, J., et al. (eds.) CL 2000. LNCS (LNAI), vol. 1861, pp. 553–567. Springer, Heidelberg (2000). https://doi.org/10.1007/3-540-44957-4_37

8. Kazmi, A.R., Afzal, M., Amjad, M.F., Abbas, H., Yang, X.: Algebraic side channel attack on trivium and grain ciphers. IEEE Access 5, 23958–23968 (2017). https://doi.org/10.1109/ACCESS.2017.2766234

9. Kocher, P., Jaffe, J., Jun, B.: Differential power analysis. In: Wiener, M. (ed.) CRYPTO 1999. LNCS, vol. 1666, pp. 388–397. Springer, Heidelberg (1999). https://doi.org/10.1007/3-540-48405-1_25

10. McCann, D., Eder, K., Oswald, E.: Characterising and comparing the energy consumption of side channel attack countermeasures and lightweight cryptography on embedded devices. In: 2015 International Workshop on Secure Internet of Things (SIoT), pp. 65–71, September 2015. https://doi.org/10.1109/SIOT.2015.11

11. Quisquater, J.-J., Samyde, D.: Electromagnetic analysis (EMA): measures and counter-measures for smart cards. In: Attali, I., Jensen, T. (eds.) E-smart 2001. LNCS, vol. 2140, pp. 200–210. Springer, Heidelberg (2001). https://doi.org/10.1007/3-540-45418-7_17

12. Renauld, M., Standaert, F.-X.: Algebraic side-channel attacks. In: Bao, F., Yung, M., Lin, D., Jing, J. (eds.) Inscrypt 2009. LNCS, vol. 6151, pp. 393–410. Springer, Heidelberg (2010). https://doi.org/10.1007/978-3-642-16342-5_29

13. Schneider, T., Moradi, A.: Leakage assessment methodology. J. Cryptogr. Eng. 6(2), 85–99 (2016). https://doi.org/10.1007/s13389-016-0120-y

14. Siddhanti, A., Sarkar, S., Maitra, S., Chattopadhyay, A.: Differential fault attack on grain v1, ACORN v3 and lizard. In: Ali, S.S., Danger, J.-L., Eisenbarth, T. (eds.) SPACE 2017. LNCS, vol. 10662, pp. 247–263. Springer, Cham (2017). https://doi.org/10.1007/978-3-319-71501-8_14

15. Soos, M., Nohl, K., Castelluccia, C.: Extending SAT solvers to cryptographic problems. In: Kullmann, O. (ed.) SAT 2009. LNCS, vol. 5584, pp. 244–257. Springer, Heidelberg (2009). https://doi.org/10.1007/978-3-642-02777-2_24

16. Standaert, F.X.: How (not) to use welch's t-test in side-channel security evaluations. Cryptology ePrint Archive, Report 2017/138 (2017). https://eprint.iacr.org/2017/138

17. Tunstall, M., Hanley, N., McEvoy, R., Whelan, C., Murphy, C., Marnane, W.: Correlation Power Analysis of Large Word Sizes (2007). http://www.geocities.ws/mike.tunstall/papers/THMWMM.pdf

18. Wu, H.: ACORN: A Lightweight Authenticated Cipher (v3). Submission to the CAESAR competition (2016). https://competitions.cr.yp.to/round3/acornv3.pdf

19. Zhang, X., Feng, X., Lin, D.: Fault attack on ACORN v3. Comput. J. (2018). https://doi.org/10.1093/comjnl/bxy044

A Closer Look
at the Guo–Johansson–Stankovski Attack
Against QC-MDPC Codes

Tung Chou$^{(\boxtimes)}$, Yohei Maezawa, and Atsuko Miyaji

Graduate School of Engineering, Osaka University, 2-1 Yamadaoka, Suita,
Osaka 565-0871, Japan
blueprint@crypto.tw, maezawa@cy2sec.comm.eng.osaka-u.ac.jp,
miyaji@comm.eng.osaka-u.ac.jp

Abstract. In Asiacrypt 2016, Guo, Johansson, and Stankovski presented a reaction attack against QC-MDPC McEliece. In their attack, by observing the difference in failure rates for various sets Φ_d of error vectors, the attacker obtains the distances between 1's in the secret key and can thus recover the whole secret key. While the attack appears to be powerful, the paper only shows experiment results against the bit-flipping algorithm that uses precomputed thresholds, and the explanation of why the attack works does not seem to be convincing.

In this paper, we give some empirical evidence to show that the Guo–Johansson–Stankovski attack, to some extent, works independently of the way that the thresholds in the bit-flipping algorithm are chosen. Also, by viewing the bit-flipping algorithm as a variant of "statistical decoding", we point out why the explanation of the Guo–Johansson–Stankovski paper is not reasonable, identify some factors that can affect the failure rates, and show how the factors change for different Φ_d.

1 Introduction

In 1978, McEliece presented in his seminal paper [6] the first code-based public-key encryption system. The paper opens the area of code-based cryptography, which is considered as an important branch of the post-quantum cryptography today. The McEliece crpytosystem has stood firmly for 40 years and is thus considered rather confidence-inspiring. However, the public-key size (typically at the scale of 1 megabyte) makes it hard to deploy the scheme in some scenarios.

This work is partially supported by JSPS KAKENHI Grant (C) (JP15K00183), Microsoft Research Asia, CREST (JPMJCR1404) at Japan Science and Technology Agency, the Japan-Taiwan Collaborative Research Program at Japan Science and Technology Agency, and Project for Establishing a Nationwide Practical Education Network for IT Human Resources Development, Education Network for Practical Information Technologies. Permanent ID of this document: eac422391e669b6d7bbaf8d29c49d2ad. Date: 2018.11.2.

© Springer Nature Switzerland AG 2019
K. Lee (Ed.): ICISC 2018, LNCS 11396, pp. 341–353, 2019.
https://doi.org/10.1007/978-3-030-12146-4_21

In order to solve the problem of key size, in 2013, Misoczki, Tillich, Sendrier, and Barreto introduced the usage of QC-MDPC codes for McEliece [7]. Compared to the conventional McEliece system, QC-MDPC enjoys much smaller key sizes (typically at the scale of a few kilobytes). Despite the large advantage in key size, the decoding algorithm, the so-called "bit-flipping algorithm", is a probabilistic algorithm. Even worse, there is no satisfying way to evaluate the decoding failure rate when using the bit-flipping algorithm.

In 2016, Guo, Johansson, Stankovski presented in their paper [8] a reaction attack against QC-MDPC McEliece. In their attack, by observing the difference in failure rates for various sets Φ_d of error vectors, the attacker obtains the distances between 1's in the secret key and can thus recover the whole secret key. The Guo–Johansson–Stankovski attack appears to be quite effective as long as the decoding failures can be observed.

To show the effectiveness of the attack, [8] uses one specific variant of the bit-flipping algorithm: the variant that uses precomputed thresholds. As shown in some papers (e.g., [9]), there are many variants which performs better (in terms of decoding failure rate) than the one with precomputed thresholds. This invokes the natural questions: is the Guo–Johansson–Stankovski attack still effective when applied to other variants? Also, although some arguments are given in [8] to show why the attack works, the arguments are unfortunately not convincing to us.

In this paper, we first show that the Guo–Johansson–Stankovski attack works against a rather conservative variant of the bit-flipping algorithm. From the results, we conclude that there might be some factor that naturally causes different failure rates for different Φ_d's. We then discuss about one such factor and show the corresponding experiment results. Furthermore, we also discuss how to view the bit-flipping algorithm as statistical decoding. From such a viewpoint, it can be seen why the explanation in [8] does not seem to be reasonable, and it is shown in detail how the various factors which can affect the failure rates change between different Φ_d.

The organization of the paper is as follows. Section 2 gives a review on some basic concept related to the Guo–Johansson–Stankovski attack. Section 3 discusses about the effectiveness of Guo–Johansson–Stankovski attacks against different variants of the bit-flipping algorithm. Section 4 discusses how the bit-flipping algorithm can be viewed as statistical decoding and identifies two factors that can affect the failure rate. Section 5 tries to give a unified view of how Guo–Johansson–Stankovski attack works in the CPA case and the CCA case.

2   Preliminaries

In this section, we give a brief review on the basic concepts of QC-MDPC codes, the bit-flipping algorithm, and the Guo–Johansson–Stankovski attack.

2.1 QC-MDPC Codes

"MDPC" stands for "moderate-density-parity-check". As the name implies, an MDPC code is a linear code with a "moderate" number of non-zero entries in a parity-check matrix H. In some sense, MDPC codes are simply LDPC codes with H with sufficiently high density such that H cannot be easily recovered when used in code-based cryptography (which is a rather ambiguous definition).

For ease of discussion, in this paper it is assumed $H \in \mathbb{F}_2^{r \times n}$ where $n = 2r$, even though some parameter sets in [7] allow $n = 3r$ or $n = 4r$. H can be viewed as the concatenation of two square matrices, i.e.,

$$H = \left[H^{(0)} | H^{(1)} \right],$$

where $H^{(i)} \in \mathbb{F}_2^{r \times r}$. Each row of H contains a moderate number of 1's.

"QC" stands for "quasi-cyclic". Being quasi-cyclic means that each $H^{(i)}$ is cyclic. For ease of discussion, one may consider

$$H^{(k)}_{(i+1) \bmod r, (j+1) \bmod r} = H^{(k)}_{i,j},$$

even though [7] allows a row permutation on H. Note that being quasi-cyclic implies that H has a fixed row weight w. The following is a quasi-cyclic matrix with $r = 5$ and $w = 4$:

$$\begin{pmatrix} 1\,0\,1\,0\,0 & 0\,1\,0\,0\,1 \\ 0\,1\,0\,1\,0 & 1\,0\,1\,0\,0 \\ 0\,0\,1\,0\,1 & 0\,1\,0\,1\,0 \\ 1\,0\,0\,1\,0 & 0\,0\,1\,0\,1 \\ 0\,1\,0\,0\,1 & 1\,0\,0\,1\,0 \end{pmatrix}.$$

One can use QC-MPDC codes for the McEliece (as in [7]) or Niederreiter [10] (as in [11]) cryptosystem. One noticeable difference between the QC-MDPC McEliece/Niederreiter and traditional McEliece/Niederreiter is that there is no need to permute the columns to obtain the public keys; the public keys are just systematic generating matrices or systematic parity-check matrices. This allows us to maintain the quasi-cyclic structure and thus save the public-key size.

The number of errors a code is able to correct is often denoted as t. Since there is no good way to figure out the minimum distance for a given QC-MDPC code, t is usually merely an estimated value. In this paper, unless explicitly stated otherwise, we will consider the parameter set $(r, w, t) = (4801, 90, 84)$. This parameter set is evaluated to have a 80-bit security level in [7], and it is the one targeted by [8].

For the discussion in this paper, we reintroduce the concept of "distance spectrum" presented in [8]. The distance spectrum $D(v)$ for $v \in \mathbb{F}_2^r$ is defined as the multi-set that contains distances of 1's in the vector, where the distance is defined in a cyclic way:

$$D(v) = \{ \min(j - i, \ i - j + r) \mid v[i] = v[j] = 1, \ i < j \}.$$

The distance spectrum of a $r \times r$ cyclic matrix is defined to be the distance spectrum of any row of the matrix. We use $D(v)[d]$ to denote the multiplicity of $d \in \{1, \ldots, \lfloor r/2 \rfloor\}$ in the distance spectrum.

2.2   The Bit-Flipping Algorithm

As described in [7], the bit-flipping algorithm is a probabilistic, iterative algorithm for decoding LDPC codes. The algorithm takes a noisy codeword $y = c + e$ as input. In each iteration, some of the entries of (the current version of) the noisy codeword y' are considered to be more likely to be erroneous, and the bits are flipped to obtain a new (possibly) noisy codeword. In the simplest version of the algorithm, iterations are repeated until a codeword c' is reached. Our hope is that $c = c'$ so that decoding is successful.

Each iteration starts with computing the syndrome s of the current noisy codeword. Each entry s_i then indicates whether the noisy codeword satisfies the corresponding parity-check equation or not: if $s_i = 1$, the noisy codeword does not satisfy the parity check defined by the i-th row of H (denoted by H_i). The number of unsatisfied parity checks

$$u_j = |\{H_{i,j} \mid s_i = 1\}|$$

for each position of the n positions are then collected to form a vector u. The vector u serves as an indicator of how likely it is for the positions to be in error: the larger u_j is, the more likely the position is presumed to be in error. Then, y'_j is flipped if the corresponding u_j is considered to be "large enough". Note that a simple way to compute u is to sum up all H_i such that $s_i = 1$, where the summation is done in $\mathbb{Z}^n$.

Now the remaining problem is, which bits should be flipped given the vector u? In [7] two possibilities are given:

- Flip y'_i if $u_i \geq T_j$, where T_j is a precomputed threshold for the iteration j.
- Flip y'_i if $u_i \geq \max(u) - \delta$, where δ is a predefined small value ([7] proposed to use $\delta = 5$).

We note that, as shown in [9], there are also many other ways to set the thresholds. In particular, [9] proposed to modify y'_i in an "in-place" fashion; one can consider this as allowing to flip at most one y'_i in each iteration. In the remaining of this paper, we will focus on "out-of-place" decoders such as the ones given in [7], where in each iteration we flip all the y'_i with u_i greater or equal to the threshold.

2.3   The Guo–Johansson–Stankovski Attack

In [8], Guo, Johansson, and Stankovski presented an attack against the QC-MDPC McEliece scheme [7]. Their attack is a reaction attack: the attacker sends a bunch of ciphertexts to the private-key holder, and by observing whether the decodings are successful or not, the attacher is able to recover the secret key. They showed that the attack works in two settings, the "CPA case" and the "CCA case".

The CPA case essentially means that the sender is able to choose the error vector for each ciphertext. The attack works as follows.

1. For each distance d in $\{1,\ldots,\lfloor r/2 \rfloor\}$, generate a set Φ_d of weight-t error vectors. Each element $e \in \Phi_d$ has about $t/2$ pairs on 1's that are separated by distance d, and all the 1's lie in the first block of e.
2. Send the vectors in all Φ_d to the secret-key holder and observe the failure rate P_d for each Φ_d.
3. Generate a figure that shows the points (d, P_d). P_d will then form several non-overlapping groups, and each d will then be classified into one of the group based on P_d. The group with the highest failure rate will then contain all the distances with multiplicity 0, and the group with the second highest failure rate will contain distances with multiplicity 1, and so on.
4. With the multiplicities for each d, which essentially shows $D(H^{(0)})$, run [8, Algorithm 2] to obtain $H^{(0)}$. The algorithm essentially enumerates all candidates of $H^{(0)}$ that fit the distance spectrum. Once $H^{(0)}$ is obtained, $H^{(1)}$ can also be obtained easily.

In the CCA case, the sender does not have the ability to choose the error vector. One can consider that the error vectors are hash outputs. The attack works as follows.

1. Generate a bunch of ciphertext and let Φ be the corresponding set of error vectors. For each distance d in $\{1,\ldots,\lfloor r/2 \rfloor\}$, define Φ_d to be the set that contains all elements in Φ that has distance d; that is,

$$\Phi_d = \{e \in \Phi \mid d \in D(e^{(0)})\}$$

2. Send the vectors in all Φ and observe the failure rate P_d for each Φ_d.
3. Generate a figure that shows the points (d, P_d). P_d will then form several non-overlapping groups, and each d will then be classified into one of the group based on P_d. The group with the highest failure rate will then contain all the distances with multiplicity 0, and the group with the second highest failure rate will contain distances with multiplicity 1, and so on.
4. With the multiplicities for each d, which essentially shows $D(H^{(0)})$, run [8, Algorithm 2] to obtain $H^{(0)}$. Once $H^{(0)}$ is obtained, $H^{(1)}$ can also be obtained easily.

Note that, to demonstrate the effectiveness of the attack, [8] uses the bit flipping algorithm with precomputed thresholds (without specifying the actual thresholds). There is no evidence in [8] that the attack can work when the thresholds are chosen in other ways, e.g., when thresholds are set to be $\max(u) - \delta$. In [8] some arguments are given to show why the attack works, but as we will discuss in Sect. 4.2 the explanation has some flaws.

3   Effectiveness of the Guo–Johansson–Stankovski Attack

In [8], it is shown that the attack works when the thresholds are predefined fixed values. This naturally causes the doubt that whether the attack can really works when the thresholds are chosen in other ways (in particular, in more conservative ways). In this section, we try to give some empirical evidence and argue that the Guo–Johansson–Stankovski attack, to some extent, is independent of the way the thresholds are chosen.

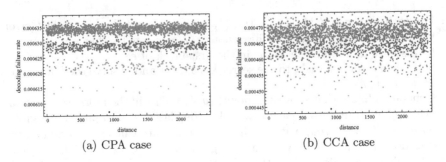

(a) CPA case   (b) CCA case

Fig. 1. Decoding failure rates for Φ_d's. For the CPA case $(n, r, w, t) = (9602, 4801, 90, 84)$. For the CCA case $(n, r, w, t) = (9602, 4801, 90, 90)$. For the CPA case $|\Phi_d| = 10^6$. For the CCA case we generate is $2.4 \cdot 10^9$ error vectors in total.

3.1 Experiment Results

To understand how the Guo–Johansson–Stankovski attack behaves when the thresholds are chosen in other ways, we consider the other option described in [7]: to use $\max(u) - \delta$ as the thresholds. In particular, we use $\delta = 0$, as this is more conservative than any $\delta > 0$. Using $\max(u)$ as thresholds is apparently the most conservative strategy among the "out-of-place" decoders. The results are in Fig. 1. Note that, for the CCA case, in order to increase the failure rate, we increase t to 90. As shown in the figure, the failure rate decreases as the multiplicity increases. Such phenomena has been shown in [8] for precomputed thresholds.

3.2 An Indicator of the Hardness of Decoding

The experiment result in the previous subsection causes the following questions to rise: can it be that the Guo–Johansson–Stankovski attack actually works independent of the thresholds? In other words, is there some reason that makes it intrinsically easier to decode vectors in Φ_d when the multiplicity of d gets larger? To answer the question, we would like to have an (possibly heuristic) indicator for the hardness of decoding when given in H and e. Our hope is that the indicator shows that Φ_d gets harder to decode as the multiplicity of d gets larger.

Recall that in each iteration of the bit-flipping algorithm, the r_i's with larger u_i's are flipped. As some non-erroneous positions r_i might have u_i that are greater than the threshold, it is possible that some non-erroneous positions are flipped. Roughly speaking, how much the u_i's for the erroneous positions and the non-erroneous positions are separated from each other determines how likely it is to distinguish the two cases. To quantify the idea, we thus consider the differences of the erroneous positions and the difference of non-erroneous positions. In other words, we define

$$\Delta_u = \sum_{e_i=1} u_i/t - \sum_{e_i=0} u_i/(n-t)$$

and use it as the indicator.

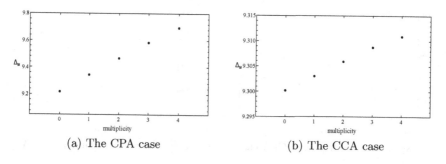

Fig. 2. Relationship between Δ_u and multiplicity.

The experiment results are given in Fig. 2. As shown in the figure, in both CPA and CCA cases, Δ_u increases as the multiplicity increases. Therefore, when Δ_u is considered, it seems that it should be easier to decode when the multiplicity increases. This matches our experiment results in the previous subsection and the results shown in [8].

4   A Deeper Look

In this section, we take a deeper look at the behaviour of the bit flipping algorithm to see what makes the difference in Δ_u. In particular, we will consider the bit-flipping algorithm in the view of "statistical decoding" and identify the factors that affect Δ_u.

4.1   Statistical Decoding

The statistical decoding algorithm described in [12] works as follows. Given a noisy codeword $y = c + e \in \mathbb{F}_2^n$ and a reasonably-large set $H_w \subset \mathbb{F}_2^n$ of weight-w vectors, the algorithm starts with computing

$$u = \sum_{h \in H_w,\ hy^T = 1} h \in \mathbb{Z}^n.$$

Then a set $I = \{i_1, \ldots, i_k\}$ is chosen such that $u_{i_1}, \ldots, u_{i_k}$ are the smallest entries in u. The set I is then considered as the "information set" (which means e_{i_j} are all 0), which can be used to decode y easily. Note that finding H_w can be a hard problem itself.

At this moment the reader should notice that statistical decoding is quite similar to the bit-flipping algorithm. Indeed, by letting H_w be the set of rows of the sparse parity-check matrix, the bit-flipping algorithm works essentially in the same way as statistical decoding. Therefore, the two algorithms can be considered to work in the same spirit: the only difference is that the bit-flipping

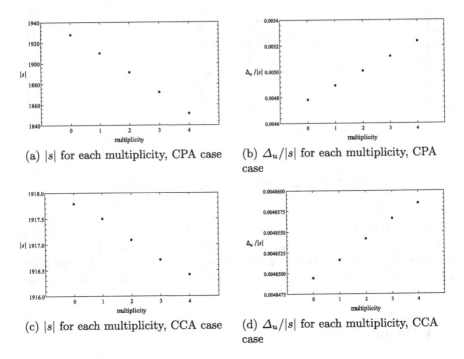

(a) $|s|$ for each multiplicity, CPA case

(b) $\Delta_u/|s|$ for each multiplicity, CPA case

(c) $|s|$ for each multiplicity, CCA case

(d) $\Delta_u/|s|$ for each multiplicity, CCA case

Fig. 3. The values for $|s|$ and $\Delta_u/|s|$ for each multiplicity.

algorithm works in an iterative way, and instead of the smallest entries in u the bit-flipping algorithm consider the largest entries in order to locate the positions in error.

In [12], it was discussed why such a simple algorithm actually works. For each h such that $hy^T = 1$, since the weight of h is only w, the non-zero entries are more likely to be in error. Let i be a nonzero position in h and $hy^T = 1$, let p_w^+ be the probability that $e_i = 1$ and q_w^+ be the probability that $e_i = 0$, [12] pointed out that, as long as $w < n/2$, we have $p_w^+ > q_w^+$. Such bias means that we can obtain a small amount of information about e for each h that satisfies $hy^T = 1$, and summing up all such h's results in the vector u where the erroneous positions tend to have larger values and the non-erroneous positions tend to have smaller values.

We can view such a bias in a different but equivalent way. For each h with $hy^T = 1$, the entries in u which correspond to the non-zero entries in h will be increases. The bias stated above indicates that, on average, the increase in the erroneous positions (u_i with $e_i = 1$) must be larger than the increase in the non-erroneous positions (u_i with $e_i = 0$). To be more precise, let

$$\ell = |\{j \mid h_j = e_j = 1\}|;$$

$\sum_{e_i=1} u_i/t$ would increase by ℓ/t, while $\sum_{e_i=0} u_i/(n-t)$ would increase by $(w - \ell)/(n-t)$. Taking $(n, r, w, t) = (9602, 4801, 90, 84)$ as example, this means that each h would create a difference of at least

$$1/84 - (90 - 1)/(9602 - 84) \approx 0.00255.$$

By summing up all such h's, a noticeable difference in $\sum_{e_i=1} u_i/t$ and $\sum_{e_i=0} u_i/(n - t)$ can be created.

From the view above, there are two important factors that would affect the result.

- The first factor is the size of H_w. It is apparently desirable to have a larger H_w so that we can make Δ_u larger. In the bit flipping algorithm, the size of H_w corresponds to the syndrome weight $|s|$, as each row of the parity-check matrix can be considered as h in statistical decoding.
- The second important factor is how much we can separate the two cases for each h, on average for each h with $hr^T = 1$. Note that the larger ℓ is, the more we can separate the two cases. From the view of the bit-flipping algorithm, this is simply $\Delta_u/|s|$.

We thus look at the relationship between $|s|$, $\Delta_u/|s|$ and the multiplicity; the result is shown in Fig. 3. As shown in the figure, when the multiplicity increases, $|s|$ decreases while $\Delta_u/|s|$ increases. Since Δ_u also increases with the multiplicity, it is clear that $\Delta_u/|s|$ increases at a faster rate than $|s|$ decreases. From the view of statistical decoding, when considering $|s|$ only, it seems that the failure rate should get higher as the multiplicity increases. However, the increase in $\Delta_u/|s|$ probably compensates for the decrease in $|s|$, and thus eventually we still have a lower failure rate for hight multiplicity. We note that similar observation on $|s|$ has been presented in [13], so we do not take the credit for this part.

The results in Figs. 2 and 3 do not depend on the thresholds of the bit-flipping algorithm. However, as the change in the distribution of u for different multiplicity is visible to the decoder, it seems possible to design the thresholds such that the failure rate increases as the multiplicity increases. We thus stress that the discussion above on the easiness of decoding with respect to $|s|$ and $\Delta_u/|s|$ does not take into account how the thresholds are determined.

4.2   Explanation of Guo–Johansson–Stankovski Paper

In [8], in addition to the description of the attack itself, the authors also tried to give some arguments about why the attack works. Similar to the discussion in the previous subsection, for each row (say row i) of H, we can define

$$\ell_i = |\{j \mid H_{i,j} = e_j = 1\}|.$$

Then ℓ_i indicates how many entries of u will be correctly changed (and thus how many will be wrongly changed), after adding H_i into u. Indeed, assuming that $s_i = 0$, $w - \ell_i$ entries would be correctly changed and ℓ_i would be wrongly changed; Assuming that $s_i = 1$, ℓ_i entries would be correctly changed and $w - \ell_i$ would be wrongly changed. One can thus obtain the following table (essentially [8, Table 2]).

ℓ_i	#(right change)	#(wrong change)
0	w	0
1	1	$w-1$
2	$w-2$	2
3	3	$w-3$
$\vdots$	$\vdots$	$\vdots$

[8] thus concludes that H_i's with an even ℓ_i help to decode, while H_i's with an odd ℓ_i gives a negative effect on decoding. [8] also concludes that it is desirable to have more H_i's with smaller, even ℓ_i.

To show the relationship of the argument above and the attack, they consider two cases in the CPA scenario:

– In CASE-0, the error vectors are from Φ_d where $d \notin D(H^{(0)})$, while
– in CASE-1, the error vectors are from Φ_d where $d \in D(H^{(0)})$.

Their experiment results are shown in the following table (essentially [8, Table 3]).

ℓ_i	CASE-0	CASE-1
0	0.4485	0.4534
1	0.3663	0.3602
≥ 2	0.1852	0.1864

As shown in the table, in CASE-1 the ratio of $\ell_i = 0$ increases and the ratio of $\ell_i = 1$ decreases. It is argued in [8] that both changes are in favor of decoding, and this is why we see a lower failure rate for larger multiplicity. It is not discussed in [8] the impact of $\ell_i \geq 2$.

From the perspective of statistical decoding, the explanation in [8] does not make sense. In particular, from the perspective of statistical decoding, the H_i's with odd ℓ_i's are the ones which help to decode, while those with even ℓ_i's do not help. As an extreme example, if all $\ell_i \in \{0,2\}$, then $u = 0$, which does not help to decode.

Nevertheless, we follow the approach in [8] to collect the ratios for all possible values for ℓ_i. The results are given in Table 1. The CASE-0 and CASE-1 in [8, Table 3] corresponds to multiplicity 0 and non-zero multiplicities. Also, in [8, Table 3] all the cases with $\ell \geq 2$ are considered as one case. Therefore Table 1 is much more detailed than [8, Table 3].

We note that it is possible to derive $|s|$ and $\Delta_u/|s|$ from Table 1. Let $T(m, \ell)$ be the entry of Table 1 for multiplicity m and $\ell_i = \ell$ (for one of the CPA and CCA case). Then it is easy to see that, for multiplicity m, $|s|$ is simply

$$(T(m,1) + T(m,3) + T(m,5) + T(m,7)) \cdot r,$$

while $\Delta_u/|s|$ is simply

$$\frac{T(m,1) \cdot \delta(1) + T(m,3) \cdot \delta(3) + T(m,5) \cdot \delta(5) + T(m,7) \cdot \delta(7)}{T(m,1) + T(m,3) + T(m,5) + T(m,7)},$$

where $\delta(i) = i/t - (w-i)/(n-t)$. It is probably not so easy to see directly how the $|s|$ changes when multiplicity increases. However, as the ratio of H_i with $\ell_i = 1$ decreases and all the ratio of H_i with $\ell_i = 3, 5, 7$ increases (and as $T(m,1)$ dominates $T(m,3), T(m,5), T(m,7)$), it is clear that $\Delta_u/|s|$ also increases.

Table 1. Relationship between ratios of ℓ_i and the multiplicity.

Case	mult.	$\ell_i=0$	1	2	3	4	5	6	7	8
CPA	0	0.450502	0.365833	0.141735	0.034886	0.006130	0.000819	0.000087	0.000007	0.000000
	1	0.454535	0.360821	0.140756	0.035992	0.006771	0.000994	0.000118	0.000012	0.000000
	2	0.458623	0.355647	0.139898	0.037070	0.007395	0.001171	0.000152	0.000016	0.000000
	3	0.462740	0.350338	0.139189	0.038161	0.008007	0.001354	0.000187	0.000022	0.000000
	4	0.466780	0.345376	0.138350	0.039124	0.008576	0.001527	0.000240	0.000029	0.000000
CCA	0	0.451770	0.362290	0.141901	0.036170	0.006754	0.000985	0.000117	0.000011	0.000001
	1	0.451843	0.362204	0.141877	0.036192	0.006767	0.000988	0.000118	0.000012	0.000001
	2	0.451934	0.362088	0.141858	0.036218	0.006779	0.000992	0.000118	0.000012	0.000001
	3	0.452020	0.361981	0.141838	0.036241	0.006792	0.000997	0.000119	0.000012	0.000001
	4	0.452088	0.361898	0.141819	0.036260	0.006802	0.001000	0.000120	0.000012	0.000001

5   A Unified View Between the CPA and CCA Cases

In [8], and also in the experiments for Figs. 1, 2, 3 and Table 1, we consider Φ_d with difference definitions in the CPA and the CCA case, but eventually we observe similar changes in the failure rates, Δ_u, $|s|$, and $\Delta_u/|s|$. This makes us wonder if there is a unified way to consider the results for the CPA and the CCA case. In other words, perhaps increasing the multiplicity of d causes some factor to change in a similar way for the CPA and CCA case, and the factor is what is really causing the change in Δ_u, $|s|$, and $\Delta_u/|s|$.

From the experiments for the CPA case, it appears that Δ_u increases, $|s|$ decreases, and $\Delta_u/|s|$ increases roughly linearly as the multiplicity in $D(H^{(0)})$ increases. In addition, as the syndrome can be considered as

$$h^{(0)}(x) \cdot e^{(0)}(x) + h^{(1)}(x) \cdot e^{(1)}(x)$$

where $h^{(i)}(x)$ and $e^{(i)}(x)$ are corresponding polynomials of $H^{(i)}$ and $e^{(i)}$ in $\mathbb{F}_2[x]/(x^r + 1)$ (as explained in [11]), it seems that the role of $e^{(0)}(x)$ and $h^{(0)}(x)$ are interchangeable. Therefore, it seems reasonable to assume that Δ_u would increase, $|s|$ would decrease, and $\Delta_u/|s|$ would increase roughly linearly as the multiplicity in $D(e^{(0)})$ increases. One evidence that supports this assumption is that we observe similar but much smaller changes in Δ_u, $|s|$, and $\Delta_u/|s|$ in the

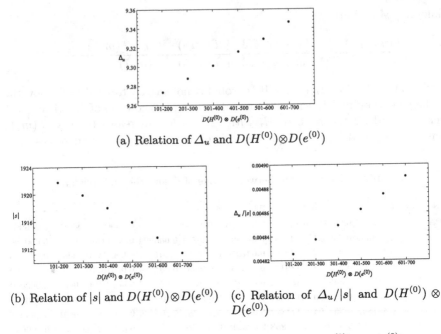

(a) Relation of Δ_u and $D(H^{(0)}) \otimes D(e^{(0)})$

(b) Relation of $|s|$ and $D(H^{(0)}) \otimes D(e^{(0)})$ (c) Relation of $\Delta_u/|s|$ and $D(H^{(0)}) \otimes D(e^{(0)})$

Fig. 4. Relationship between Δ_u, $|s|$, $\Delta_u/|s|$ and $D(H^{(0)}) \otimes D(e^{(0)})$

CCA case compared to the CPA, and in the CCA case the multiplicity of d in some $e \in \Phi_d$ is much smaller than that for the CPA case (at least $\lfloor t/2 \rfloor$).

Based on the discussion above, as increasing the multiplicity of d in $H^{(0)}$ and multiplicity in $e^{(0)}$ should both have help to increase the change in Δ_u, $|s|$, and $\Delta_u/|s|$, it seems reasonable to assume that the factors change linearly with

$$D(H^{(0)}) \otimes D(e^{(0)}) = \sum_i D(H^{(0)})[i] \cdot D(e^{(0)})[i].$$

Based on this assumption, we carried out experiments and present the results in Fig. 4. Interestingly, as shown in the figure, as $D(H^{(0)}) \otimes D(e^{(0)})$ increases, similar linear changes in Δ_u, $|s|$, and $\Delta_u/|s|$ can be observed as in Figs. 2 and 3.

The experiment results in [13, Fig. 3] might seem a bit similar to our results. We note that $|s|$, $\Delta_u/|s|$, and Δ_u are all threshold-independent, while numbers of iterations in [13, Fig. 3] are threshold-dependent.

References

1. Cheon, J.H., Takagi, T. (eds.): ASIACRYPT 2016. LNCS, vol. 10031. Springer, Berlin (2016). https://doi.org/10.1007/978-3-662-53887-6. ISBN 978-3-662-53886-9
2. Bertoni, G., Coron, J.-S. (eds.): CHES 2013. LNCS, vol. 8086. Springer, Berlin (2013). https://doi.org/10.1007/978-3-642-40349-1. ISBN 978-3-642-40348-4

3. Gierlichs, B., Poschmann, A.Y. (eds.): CHES 2016. LNCS, vol. 9813. Springer, Berlin (2016). https://doi.org/10.1007/978-3-662-53140-2. ISBN 978-3-662-53139-6
4. Batten, L.M., Safavi-Naini, R. (eds.): ACISP 2006. LNCS, vol. 4058. Springer, Berlin (2006). https://doi.org/10.1007/11780656. ISBN 3-540-35458-1
5. Lange, T., Steinwandt, R. (eds.): PQCrypto 2018. LNCS, vol. 10786. Springer, Berlin (2018). https://doi.org/10.1007/978-3-319-79063-3. ISBN 978-3-319-79062-6
6. McEliece, R.J.: A public-key cryptosystem based on algebraic coding theory, pp. 114–116. JPL DSN Progress Report (1978). http://ipnpr.jpl.nasa.gov/progress_report2/42-44/44N.PDF
7. Misoczki, R., Tillich, J.-P., Sendrier, N., Barreto, P.S.L.M.: MDPC-McEliece: new McEliece variants from moderate density parity-check codes. In: IEEE International Symposium on Information Theory, pp. 2069–2073 (2013). http://eprint.iacr.org/2012/409.pdf
8. Guo, Q., Johansson, T., Stankovski, P.: A key recovery attack on MDPC with CCA security using decoding errors. In: Cheon, J.H., Takagi, T. (eds.) ASIACRYPT 2016. LNCS, vol. 10031, pp. 789–815. Springer, Heidelberg (2016). https://doi.org/10.1007/978-3-662-53887-6_29
9. Heyse, S., von Maurich, I., Güneysu, T.: Smaller keys for code-based cryptography: QC-MDPC McEliece implementations on embedded devices. In: Bertoni, G., Coron, J.-S. (eds.) CHES 2013. LNCS, vol. 8086, pp. 273–292. Springer, Heidelberg (2013). https://doi.org/10.1007/978-3-642-40349-1_16. http://eprint.iacr.org/2015/425.pdf
10. Niederreiter, H.: Knapsack-type cryptosystems and algebraic coding theory. Probl. Control Inf. Theory **15**, 159–166 (1986)
11. Chou, T.: QcBits: constant-time small-key code-based cryptography. In: Gierlichs, B., Poschmann, A.Y. (eds.) CHES 2016. LNCS, vol. 9813, pp. 280–300. Springer, Heidelberg (2016). https://doi.org/10.1007/978-3-662-53140-2_14
12. Overbeck, R.: Statistical decoding revisited. In: Batten, L.M., Safavi-Naini, R. (eds.) ACISP 2006. LNCS, vol. 4058, pp. 283–294. Springer, Heidelberg (2006). https://doi.org/10.1007/11780656_24
13. Eaton, E., Lequesne, M., Parent, A., Sendrier, N.: QC-MDPC: a timing attack and a CCA2 KEM. In: Lange, T., Steinwandt, R. (eds.) PQCrypto 2018. LNCS, vol. 10786, pp. 47–76. Springer, Cham (2018). https://doi.org/10.1007/978-3-319-79063-3_3

Recurrent Neural Networks for Fuzz Testing Web Browsers

Martin Sablotny(✉)⬤, Bjørn Sand Jensen, and Chris W. Johnson

School of Computing Science, University of Glasgow, Glasgow, Scotland
m.sablotny.1@research.gla.ac.uk,
{bjorn.jensen,christopher.johnson}@glasgow.ac.uk

Abstract. Generation-based fuzzing is a software testing approach which is able to discover different types of bugs and vulnerabilities in software. It is, however, known to be very time consuming to design and fine tune classical fuzzers to achieve acceptable coverage, even for small-scale software systems. To address this issue, we investigate a machine learning-based approach to fuzz testing in which we outline a family of test-case generators based on Recurrent Neural Networks (RNNs) and train those on readily available datasets with a minimum of human fine tuning. The proposed generators do, in contrast to previous work, not rely on heuristic sampling strategies but principled sampling from the predictive distributions. We provide a detailed analysis to demonstrate the characteristics and efficacy of the proposed generators in a challenging web browser testing scenario. The empirical results show that the RNN-based generators are able to provide better coverage than a mutation based method and are able to discover paths not discovered by a classical fuzzer. Our results supplement findings in other domains suggesting that generation based fuzzing with RNNs is a viable route to better software quality conditioned on the use of a suitable model selection/analysis procedure.

Keywords: Software security · Fuzz testing · Browser security

1 Introduction

Fuzz testing has recently enjoyed increased popularity in theoretical and practical software testing. This can be primarily attributed to the apparent capability to trigger unintended behaviour in complex software systems, e.g. the summary of bugs found by American Fuzzy Lop (AFL) [28] and further evidenced by the use of fuzz testing in software companies like Microsoft and Google (e.g. through their open-source tool ClusterFuzz [12]) which shows success and applicability in many different domains. However, the standard approach of combining mutation on a set of input examples with an evolutionary approach has its limitation with increasing necessity of keywords and compliance to syntactic rules (e.g. HTML as considered in this work). Those problems can be tackled by generation-based

© Springer Nature Switzerland AG 2019
K. Lee (Ed.): ICISC 2018, LNCS 11396, pp. 354–370, 2019.
https://doi.org/10.1007/978-3-030-12146-4_22

fuzzers that are able to comply to those rules, use the correct keywords and generate novel inputs. Traditionally, the time needed to develop generation-based fuzzers is dependent on the input specification's complexity. For example it is less time consuming to develop a generator for a network protocol, which has a single field with three different possible values compared to implementing the File Transfer Protocol (FTP) [16] with it various fields and states. In addition, it is necessary to find the right balance between introduced errors and overall correctness to trigger code paths that lead to unintended behaviour.

The main bottleneck in the development of generation-based fuzzers is the need for a strict understanding and implementation of the input file format. Therefore, the potentially complex input specification has to be studied carefully to transfer it into a test case generator, which then needs to be fine tuned in order to find the right balance between correctness and introduced errors into the test cases. This implicit optimization process looks to maximize code coverage by generating test cases that deviate in certain areas from the given specification and therefore are capable of exercising different low-level execution paths. Thus, it is clear that methods which could automatically derive or lean the input specification would be able to speed up software testing by faster deployment of generation-based fuzzing techniques. This would potentially lead to an increase in software security and stability.

Learning an input specification (e.g. syntactic rules) is obviously not trivial, especially due to the long time dependencies input specifications can apply. Those dependencies have an direct impact on the possible outputs at a certain position and therefore have to be captured by a learning algorithm to produce specification adhering outputs. However, recent advancements in generative machine learning models ([2,3,6,26]) have demonstrated how machine learning models can be use to learn complex rules and distributions from examples and generate new examples from acquired knowledge.

These advancements have been previously explored for fuzz testing by Godefroid et. al. [11]. They demonstrated the use of deep neural networks to generate PDF-objects, which were used as input for a rendering engine. Those input files were able to trigger new instructions in the rendering engine. However, they focused on the tension between learning the correct input structure and fuzzing - or in other words, finding the balance between adhering to the learned specification and deviating from it. They did not provide an analysis of the learning process itself and gave no comparison to a naive mutation based baseline. In addition, they have not provided any information about the overlap between the baseline and their proposed sampling strategies. In order to use deep learning models during fuzz testing, it is important to see whether it is worth the development and training. Therefore, it is necessary to compare it with an easy to implement approach, like a naive mutation algorithm. The analysis of an existing overlap between different approaches also gives more insight into the model and sampling choice, since it is important to trigger as much new execution paths as possible during testing to find the ones that trigger unintended behaviour.

In this work, we investigate how Recurrent Neural Networks (RNNs) with different types of cells can be trained and used as a HTML-fuzzers. The models are trained on a dataset created by a generation based HTML-fuzzer, which allowed us to adjust the dataset size and complexity in a fast and systematic way. We use the models to generate new HTML-tags from the resulting probability distribution, which were used to form test cases. Those were executed with Firefox [19] to gather their code coverage data and compared to a baseline generated by the HTML-tags from the dataset and a naive mutated dataset. Thus, the contribution of the paper includes:

– A systematic and robust approach for training and evaluating recurrent neural networks with different types of cells for HTML fuzz testing.
– A procedure and metrics for model-selection and comparison of machine learning fuzzers against standard and a vanilla mutation-based methods including a similarity-based analysis.
– An extensive empirical evaluation on a web browser, demonstrating that learned fuzzers are able to outperform standard test methodologies.
– Open-source implementation and data available via Github[1].

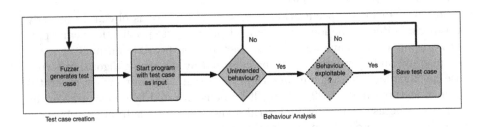

Fig. 1. Classic fuzzing workflow for finding security related flaws

2 Background

2.1 Fuzzing

Fuzz testing is a dynamic software testing approach, hereby dynamic means the software under test is actually executed in contrast to statically analysed. The goal of the fuzz test is to provoke unintended behaviour that was not detected in earlier testing stages, therefore software under test is executed with inputs created by a so-called fuzzer. Those inputs do not fully comply with the underlying input specification in order to find paths that lead to a state that triggers unintended behaviour. We adopt a broad definition of unintended behaviour, which makes it applicable for various kinds of software and devices [27]. For example,

---

[1] Code and data is available from https://github.com/susperius/icisc_rnnfuzz.

during fuzz testing desktop software, unintended behaviour can be the termination of a running process or even the possibility to take control over a process. Whereas during the test of a web application unintended behaviour might be defined as an information leak or the circumvention of access restriction both cases might happen due to a SQL-injection vulnerability, where arbitrary input is used as a valid SQL-statement.

As those examples highlight, a case of unintended behaviour becomes more severe if it could provide an attacker with an advantage. Here advantage can mean everything from accessing restricted information to taking over control of a device. In order to find those vulnerabilities fuzz testing is utilised. The general workflow during fuzz testing is shown in Fig. 1. The testing itself is split in two parts first the test case generation and secondly the behaviour analysis. In general, the creation of test cases during fuzzing can be divided into the two categories: mutation based and generation based [27], [8] and [20]. First mutation based fuzzing uses a valid input set and a mutation fuzzing in order to derive new test cases from the input set. This type of fuzzing can be implemented quickly if the input examples are available (e.g. JPG files). The main disadvantage is that test cases created by plain mutation based fuzzing are not able to quickly discover code paths deep in the call tree because many created test cases are filtered out in early program execution stages. A very prominent and successful example of this category is the aforementioned fuzzer AFL with its evolutionary mutation approach. Secondly, generation based fuzzing uses an approach where test cases are created from scratch, for example through grammar based creation. This method needs a lot of effort during studying the input structure and developing the generator but in general it is able to discover deeper lying code paths. However, a balance between complying to the rules and breaking them has to be found in order to provoke unintended behaviour in the target.

2.2 Recurrent Neural Networks

The input data for many software products is readily available on the internet (e.g. HTML, JPG, PNG) and deep learning algorithms have shown their performance in different use cases especially where they are trained on a large available dataset, for example text generation [26], program creation [3] and machine translation [2,6]. This led us to the use of a generative model for the test case creation during fuzz testing. In addition the structure of HTML and other input formats, where the actual character or byte is dependant on the previous positions in a sequence led to the use of RNNs.

RNNs are used to model sequential data, e.g. for text generation [26], language modelling and music prediction [21]. They use a hidden state as short term memory which carries information between time steps. The conventional RNN with input $\mathbf{x_t}$ is defined through a hidden state vector $\mathbf{h_t}$ and an output $\hat{\mathbf{y}}_t$ at time step t as follows

$$\mathbf{h_t} = f_h(\mathbf{x_t}, \mathbf{h_{t-1}}) \ , \ \hat{\mathbf{y}}_t = f_o(\mathbf{h_t}),$$

with f_h and f_o being the hidden transformation and output function respectively. Hereby, the input $\mathbf{x}_t$ can be a N-dimensional vector, representing the input structure, e.g. a single pixel's RGB values at position t.

As described by Hochreiter [13] and later by Bengio et al. [4], RNNs suffer from either the vanishing or exploding gradient problem. This means that the weight updates are becoming infinitesimal during training, which consumes a lot of time but does not lead to a better optimised network. Hochreiter and Schmidhuber introduced the concept of Long-Short Term Memory (LSTM) cells [14] RNNs using those cells do not suffer from the vanishing (exploding) gradient problem. LSTM cells use a hidden state, a candidate value and three gates namely a forget gate, an input gate and an output gate. The gates control how much information is forgotten, used from the input and controlling the flow into the new hidden state respectively. They are default feed forward neural networks and each have their own trainable parameters.

Another popular RNN cell, the Gated Recurrent Unit (GRU) was introduced by Cho et al. [6]. This unit only uses two gates, a reset and an update gate. Here the reset gate controls what information from the past hidden state is forgotten and the update gate controls the information flow into the new hidden state. This simpler model arguably makes it easier to train than a standard LSTM based model.

The capability to learn sequential structures, where dependencies to former inputs exist, is obviously an important characteristic when learning input format structures for test case generation. This is especially evident in for example HTML where there are long term dependencies between an opening-tag and the corresponding closing-tag.

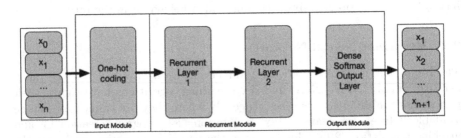

Fig. 2. Model overview for a stacked RNN with 2 recurrent layers (either LSTM or GRU)

3   Stacked RNN for HTML-Fuzzing

The basic concept of the model used in this work is shown in Fig. 2. The model consists of three modules. First, the input module, let $X = \{x_1, x_2, \ldots x_N\}$ be the sequence of input values with $x_t \in \mathbb{N}_0 \mid 1 \leq t \leq N$, where x_t is the natural number representing the character at position t in the input sequence. For example the character 'f' is at position t in the input sequence, its assigned number is 17 and $x_t = 17$.

The input module then takes such a x_t and transforms it into a one-hot coded vector $\hat{\mathbf{x}}_t \in \mathbb{R}^{\mathbf{I}}$ with $I = max(X) + 1$, the one is added to account for the zero. Let $\hat{\mathbf{x}}_t = (\hat{x}_1, \hat{x}_2, \ldots, \hat{x}_I)^{\intercal}$ then

$$\hat{x}_j = 0 \ \forall \ 1 \leq j \leq I \ : \ j \neq x_t \ \vee \ \hat{x}_j = 1 \ \Leftrightarrow \ j = x_t,$$

and for the former example character 'f' all $\hat{x}_j = 0$, except for $\hat{x}_1 7$, which equals 1. This conversion from integer values is necessary as interpret our input as categorical data (each character is its own category) and those categories are handled as features during the training process.

Secondly, the recurrent module consists of LSTM or GRU nodes as described in Sect. 2.2 with $s, l \in \mathbb{N}$ hereby s is the internal size of the nodes and l the amount of layers used, e.g. $l = 2$ for the LSTM based model shown Fig. 2. LSTM cells have demonstrated a high performance gain compared to the basic RNN approach as demonstrated by Chung et al. [7]. Gated Recurrent Units (GRUs) introduced by Cho et al. [6] perform similar to LSTM cells [7], however Jozefowicz et al. [17] have shown that LSTM cells perform better during XML modelling. We decided to evaluate the performance of both cells to analysis whether the XML modelling results are transferable to HTML modelling.

Finally, the output layer consists of a default feed forward network with I nodes. It takes the output of the last recurrent layer $\mathbf{h}_t^l \in \mathbb{R}^s$ as input value and after computing its output the *softmax* function is applied. The resulting $\hat{\mathbf{y}}_t$ provides the probability distribution for predicting the next value of the input sequence. The goal during training is to minimise the cross entropy loss function

$$\mathcal{L}(\Theta) = -\frac{1}{N} \sum_{i=1}^{N} \mathbf{y_i} \log(\hat{\mathbf{y}_i}) + (1 - \mathbf{y_i}) \log(1 - \hat{\mathbf{y}_i}),$$

where Θ denotes the model's parameters (i.e. a collection of $\mathbf{W}$'s and $\mathbf{b}$'s). In order to find a Θ that minimises the above loss $\mathcal{L}$ the ADAM [18] optimisation algorithm is applied. It is a gradient-based optimisation algorithm which only needs first order gradients and has a reduced memory footprint compared to other algorithms. Additionally, Dropout (30% dropout probability) [25] is used as regularisation.

4 Experiments

The following sections present the methodology that was used to validate our application of RNNs to generate test cases for fuzz testing of cyber security in complex systems.

The basic idea is to train the aforementioned neural networks with different depths on a large collection of HTML-tags. After training those models are used to generate HTML-tags directly using the probability distribution over characters given the sequence. The generated output is then used as input for a web browser. This browser is instrumented in order to gather the code coverage data during execution on a basic blocks basis. The collected code coverage data is then used to compare the models' performances with code coverage data collected by executing the dataset's HTML-tags and a naive mutation strategy performed on this HTML-tags.

4.1 Environmental Setup and Implementation

The model training took place on a Ubuntu 16.04 system equipped with a single NVIDIA GeForce 1080 Ti and a NVIDIA GeForce TITAN Xp, which shortens the necessary training time by utilising their parallel computational capabilities. The models were implemented using Google's TensorFlow framework [1] along with its Python bindings. This frameworks already provides the necessary cell types, optimisation algorithm and loss function for our model, which shortens the development time.

The code coverage data was collected on a Virtual Machine (VM) also running Ubuntu 16.04 and Firefox 57.0.1, which allows to run in so-called headless mode. In this mode Firefox does not display the graphical user interface, but it still renders the webpage. We also modified the standard configuration in order to disable internal services to avoid as much false code coverage data as possible. Furthermore safe mode was disabled, because during the automated code coverage collection Firefox was not closed correctly and therefore might tries to start in safe mode after just a few test cases. The use of the headless mode also saves time during the code coverage collection, which was collected by DynamoRIO's drcov tool (see Subsect. 4.4). The VM itself utilises 16 GB of RAM and a Solid State Disk. A VM was used to facilitate parallel data collection via cloning and deploying onto multiple host systems.

4.2 Data Set Generation

In order to provide a reproducible and controlled experiment, the training (and ground-truth) data set was generated by an existing HTML-fuzzer included in PyFuzz2 [24]. It provides a controllable generator thus ensuring less uncertainty about the variation within the training dataset in comparison to collecting a dataset from the Internet. Therefore, it was possible to control the complexity of the generated HTML on a per tag basis, whereas a collected set would have to be parsed and then filtered for unwanted HTML-tags to control the resulting dataset.

The pre-existing fuzzer was modified in order to avoid nesting of HTML tags, remove all Cascading Style Sheets and output exactly one HTML tag per line. Due to the restriction of not having nested HTML-tags some like *td* or *th* are excluded because they need an outer tag in this example *table*. Those restrictions were introduced to reduce to focus on the fundamental problem by reducing the overall data set complexity. This further reduced the necessary model complexity and effectively the time needed to train those models.

Listing 1.1 shows an excerpt from the data set used for training the models, which highlights the modification mentioned above. The created file consisted of 409,000 HTML-tags, which results in a total size of 36 MB.

4.3 Training

All models were trained to predict the input shifted by one on a per character basis. For example take "$< h2\ i$" from line 1 in Listing 1.1 as input sequence of length 5 then the label for that particular input sequence would be "$h2\ id$". The actual sequence length used during training was 150 characters and each model was trained for 50 epochs, which has shown sufficient for the models to converge. In order to train the models we used the previously mentioned ADAM [18] optimisation algorithm. The starting learning rate was set to 0.001 and halved every 10 epochs. The models were trained with a batch size of 512. The internal size of the LSTM and GRU cells was set to 256 for all models trained and the number of layers varied from 1 to 6. The weights of the layer were initialised by the Glorot uniform initializer [10]. So the weights are drawn from a uniform distribution in the interval $(-\frac{\sqrt{6}}{\sqrt{n_j+n_{j+1}}}, \frac{\sqrt{6}}{\sqrt{n_j+n_{j+1}}})$, with n_j being the internal size of layer j.

The first 30 MB of the data set were used for training and an additional generated 1MB for validation. All models were trained on 5 different training/validation splits repeated 3 times with different initialization (to mitigate extremely poor local minima) which results in a total of 90 trained models per cell type. The splits were chosen randomly without overlapping parts.

4.4 Data Collection

The code coverage data was collected by executing Firefox instrumented by DynamoRIO's *drcov* [9]. This tool gathers data about the executed basic blocks of the program under test. The collected code coverage data was parsed for uniquely executed basic blocks inside of Firefox's `libxul.so` library, which includes the whole web engine responsible for HTML rendering. It is possible to identify those basic blocks even when the process is restarted because the recorded data uses the offset of the basic block from the base address of the

```
1  <h2 id="id0" style="style" spellcheck="false" dir="rtl"
       title="eval(n1, $)"> 2e100 </h2>
2  <ul id="id3" style="style" translate="no"
       contenteditable="true" tabindex="4400000000">
       4400000000 </ul>
```

Listing 1.1. Example from the training set.

library in memory and this offset is always the same for a fixed version. Hereby a basic block is defined as a linear sequence of machine instructions with a single entry (branch target) and single exit (branch instruction).

All test cases consisted of a basic HTML-template with the HTML-tags inserted into the body tag. Initial experiments showed that executing the same test case multiple times returns different code coverage data. This is due to the other functions that are bundled into the `libxul.so` library, which are not part of the web engine itself. Those functions might for example only be executed after a number of restarts or in fixed time intervals. In order to identify the corresponding basic blocks the blank HTML-template was executed $1,024$ times and the resulting code coverage was store for later use.

The comparison baseline was established by using the HTML fuzzer to create $6 \times 16,384$ HTML-tags Each collection of $16,384$ HTML-tags was then used to create two datasets, one containing 64 files with 256 HTML-tags each and a second one with 128 files containing 128 HTML-tags. This resulted in twelve datasets.

In order to establish a second baseline for comparison, additional test sets were created by mutating the dataset test cases and collecting the code coverage from those. A simple mutation function was applied with a fixed chance that a position is replaced by a randomly chosen character (only characters that were already present in the dataset). The results were 20 additional test case sets, 10 sets consisting of 128 cases with 128 HTML-tags each and 10 sets consisting of 64 cases with 256 HTML-tags each, resulting in a total of $1,920$ additional cases. The replacement probability varied between 0.1% and 51.2%. This was done to ensure that there is difference and therefore an incentive to use a trained model for test case creation instead of implementing a naive mutation based approach.

For each trained model, a total $16,384$ HTML-tags were generated and then used to create two different sets of test cases. The first set used 128 HTML-tags per case, which resulted in 128 cases per model trained, whereas the second set used 256 HTML-tags per case, which resulted in 64 cases per model. This was done to analyse the impact of HTML-tags on code coverage and to observe the relationship with the model performance. The HTML-tags were generated by using the "<" character as starting input, sampling the next character from the resulting probability distribution, which was then used as new input. This was repeated until a "\n" (newline character) was sampled, since it marks the end of a HTML-tag.

Finally, the set difference between the collections of basic block sets from the test cases and the blank cases was computed to filter out the aforementioned irrelevant basic blocks.

```
1   <war id="id55804" scellcheck="false" tpalleaeck="false"
        class="style_class_0" title="50000000"> null</sab>
```

Listing 1.2. Example HTML-tag from a 1-layer LSTM model

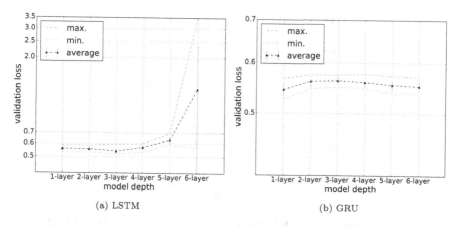

(a) LSTM   (b) GRU

Fig. 3. Average validation loss for models of different complexity (i.e. number of layers) models and dataset splits. Error-bars indicate the standard deviation.

4.5   Results

The training phase already showed a difference in behaviour between the two cell types. The LSTM based models showed a decrease in average validation loss and standard deviation up to three layers, as shown in Fig. 3a, with an increase afterwards. Especially, the 6-layer models show a large standard deviation and a huge increase in average validation loss compared to the other models This indicates that those models have too many parameters in order to be trained on our problem and training set. This behaviour is to be expected from a general machine learning perspective and since the training process is the same compared to other similar applications using generative neural networks, like generating text.

In contrast the training of the GRU based models showed a small increase from the 1-layer models to the 2-layers case, but a decrease afterwards with overall small differences in the standard deviation. This indicates that the GRU based models are either better suited to reproduced the input structure or do not reach the overall complexity of the 6-layer LSTM based model, which is also supported by comparing the trainable parameters of those models. The GRU based model has $2,276,971$ compared to $3,026,795$.

Overall, a small numeric difference in validation loss can lead to a big difference in the quality of the resulting HTML-tags. For example Listing 1.2 shows an

```
1   <p id="id38564" lang="mk">
      BBBBBBBBBBBBBBBBBBBBBBBBBBBBBBBBBBBBBBBBBBB</p>
2   <head id="id240801" sang="al" style="style" class="
      style_class_0" dir="rtl"> 7500000000</pre>
```

Listing 1.3. Example HTML-tag form a 3-layer LSTM model

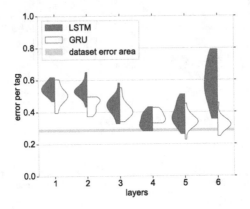

Fig. 4. Average error rate per HTML-tag generated by the LSTM and GRU based model in comparison to the datasets.

excerpt generated by a 1-layer LSTM model. It is barely recognisable as HTML and the model did not generate existing HTML-opening and closing tags and two of the generated HTML-attributes are misspelled in this particular example. In contrast to that Listing 1.3 shows two HTML-tags generated by a 3-layer LSTM model. Both use only existing HTML-tags, however the second one does not use the correct closing tag and misspelled one attribute name. Further evidence regarding the quality differences between the models of both cell types is provided by Fig. 4. It shows how the HTML error rate per tag follows the trend of the validation loss and highlights how small differences has a large effect on the HTML quality. The high spread of the 6-layer LSTM HTML error rate reflect the large standard deviation observed during training.

Test Cases with 128 HTML-Tags

In terms of code coverage performance the overall trend also follows the validation loss and standard deviation, where a smaller validation loss and standard deviation indicates a better performance. Figure 5a shows the total discovered basic blocks of both cell types per layer. It highlights that both types of 4-layer models and the GRU 5 and 6-layer models are able to discover basic blocks in the range of the datasets or even outperform it.

In addition, Fig. 6a shows the difference in number of basic blocks to the best performing dataset. It shows that all models were able to discover basic blocks not triggered by the dataset, with the 5-layer GRU models performing best on average. In comparison with the different mutation sets the maximum overlap reaches 90% with a mutation chance of 1.6%, which is not surprising because the same mutation set has an overlap of 87.6% with the best performing dataset, as also shown in Fig. 7. The best performing 5-layer GRU models have an overlap of 78% with the union of different mutation chances, highlighting the models ability to discover basic blocks, which can not be triggered by the naive mutation approach. The overall best performing models are also those with the largest overlap with the dataset.

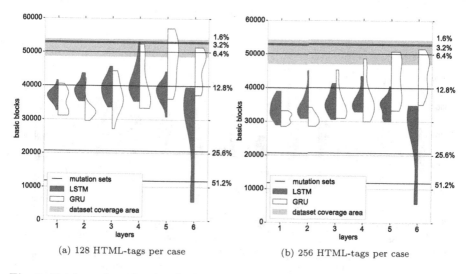

(a) 128 HTML-tags per case   (b) 256 HTML-tags per case

Fig. 5. Total number of uniquely discovered basic blocks on a per model basis. The dataset coverage area and the different mutation sets are included as baselines with the mutation probability indicated on the right vertical axis.

Test Cases with 256 HTML-Tags

The code coverage results for the test cases with 256 HTML-tags each showed a similar development, but a slightly lower overall performance, as shown in Figs. 5b and 6b. The lower overall performance was expected, because both runs basically use the same HTML-tags and only the number of inserted HTML-tags is different.

In terms of absolute basic blocks the 4-layer model was the best LSTM based model, however in this setting it did not reach the dataset coverage area. However, the 4-, 5- and 6-layer GRU based models were able to reach the dataset coverage area with the 6-layer model having the highest number of uniquely triggered basic blocks.

Considering the overlap with the mutation test cases the overall result is the same as in the 128 HTML-tags case. The best performing four layer models have an average overlap with the mutation sets of 74.6%. This shows that the 256 HTML-tags cases were also able to trigger new code paths in the web rendering engine.

5 Discussion

The results demonstrate that is is indeed possible to successfully train models and generate test HTML cases using the RNN based model. However, it is crucial to monitor this process to get robust results, e.g., the 6-layer LSTM model was not trainable in a reliable way. This may very well have been due to a lack of training data, or the high amount of parameters involved in the optimisation.

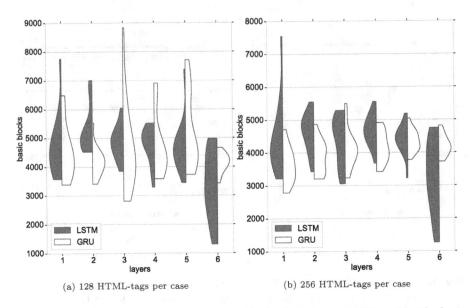

(a) 128 HTML-tags per case   (b) 256 HTML-tags per case

Fig. 6. Number of uniquely discovered basic blocks that were not triggered by the best performing dataset.

Once the models have been trained the results indicates that the average validation loss can be used as good initial selection criteria for choosing a good model for generation of test cases despite the implicit coupling with the code coverage metric. This is particularly interesting, since there is no code coverage data available during the model selection phase and covering as many code paths as possible during fuzz testing is important to discover software bugs. The results also have shown that the HTML error rate can be used to determine a good generative model and therefore augment the selection process. This is especially helpful, since the average validation loss and standard deviation alone might indicate a low difference between two models, see for example the Listings 1.2 and 1.3. The highest average validation loss difference between those models is ≤0.02, but the difference in the HTML error rate is 0.3. This means that the worst performing 1-layer LSTM model has twice as many error per tag than the best performing 3-layer LSTM model.

Overall the best performing models generated more valid HTML-tags than the other models, which leads to the use of existing HTML-tags. Those generated and generally valid HTML-tags are not always closed with right corresponding HTML-tag. This results in the best performing models building nested valid HTML-tags by accident, because those models use a valid opening HTML-tag, but do not generate the corresponding closing HTML-tag. However, this might still be generated at a later stage in the file. The assumed rendering behaviour and the creation of nested HTML-tags trigger code paths that have not been triggered by the baseline set, since in the baseline set every opened tag is closed with the corresponding closing tag in each line.

The similarity in terms of overlapping basic blocks (see Fig. 7a) between the LSTM models and the baseline set is lower than the overlap with the mutation sets and the models between each other in the 128 HTML-tag case. This might indicate that the models are not able to fully replicate the given input structure and therefore another model choice would be better suited to learn this structure or the provided training set was too small to capture the input structure with the chosen model architecture. For the GRU models the best performing models also show that the overlap with dataset is higher than the one with the mutation sets (see Fig. 7b). This further strengthens the assumption that a certain quality has to be reached by the models in order perform well.

Overall, we were able to demonstrate that especially GRU-based RNNs are capable of creating HTML-tags, which then can be used during fuzz testing a browser. Critically, the generated HTML test cases are also able to trigger a significant number of unique basic blocks, which were not reached by the dataset's baseline and the naive mutation approach.

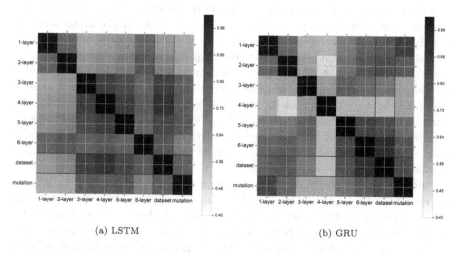

(a) LSTM   (b) GRU

Fig. 7. The similarity between all the models, the dataset and mutation-based fuzzer in terms of their overlapping basic blocks for test cases with 128 HTML-tags.

6 Related Work

The closest related work was done by Godefroid et al. [11]. They studied the achievable code coverage using a two layer stacked RNN to sample PDF-objects and focused on the effects the training duration has on this. The code coverage results they achieved were compared against a baseline, which was randomly selected from the training set. In contrast to that we used data not seen by the models during the training phase to establish our baseline for comparison. In addition, they analysed different approaches of creating test cases and compared those. They also highlighted an observed tension between learning and fuzzing and proposed an algorithm called SampleFuzz. This algorithm uses the lowest predicted probability, if the model's highest predicted probability is above a certain threshold value and a random coin toss is successful. Whereas our work studied a different input format, namely HTML, which is a more structure-reliant input format compared to PDF-objects. We also researched the effects of the model depth on the resulting code coverage. We were not able to observe the former described tension between learning and fuzzing. This might be connected to the relative large size of our training set or indicate that their models started to overfit to the training examples, thus requiring additional stochasticity to produce novel test cases. Regardless, we did not identify the need to introduce additional random values (e.g. through the use of SampleFuzz).

Other related works make use of the control and data flow during the execution in order to generate new test cases. Rawat et al. [23] utilise so-called evolutionary algorithms to derive new test cases. Whereas Höschele et al. [15] derives an input grammar from the collected execution information. Both approaches need direct access to the program under test to instrument it and to collect the necessary data. In contrary, our approach is able to learn the input structure directly from input examples, which shortens the design and learning process.

A different approach utilising code coverage and mutation-based fuzzing was presented by Böhme et al. [5]. They augmented AFL with Markov Chains in the mutation process. Their AFLFast called approach uses Markov Chains to determine the state transitions into new test inputs. They have shown that they shorten the time necessary for finding bugs in an ensemble of tested software. However, they have not provided any information on highly structure dependent input formats like HTML, which is described as a shortfall in the general AFL approach.

Another way of combining deep learning in order to find bugs in software was evaluated by Pradel et al. [22]. They used trained models in order classify potential buggy source code. Hereby they trained their models as individual classifiers for a certain bug category. In contrast to them we trained our models to generate inputs, which then can be used to trigger and observe bugs in software. Furthermore, their approach needs direct access to the source code, whereas we need access to enough input examples to train a RNN model.

7   Conclusion and Future Work

Our work provides evidence that it is possible to use a stacked RNN to generate HTML-tags in order generate novel test cases for fuzz testing a browser's rendering engine. The results also clearly show that the GRU based models are able to outperform LSTM ones even with less trainable parameters. Furthermore, the proposed evaluation procedure and similarity-based analysis demonstrates that the overlap in basic blocks between the dataset and the model generated test cases are very low on average. In addition, the overlap with the naively mutated sets is approximately 70% on average, which indicates that the trained networks

are able to discover new code paths formerly not discovered by the naive mutation approach with different mutation chances. This provides amble evidence
that RNNs can be trained and used as an effective HTML-fuzzer provided that
a suitable model-selection and analysis procedure is applied.

We are currently looking to extent the present work in least three ways:
Firstly, investigating more complex/suitable neural network models is necessary
to improve the overall quality of the generated HTML as other prevalent web
technologies, like JavaScript, cannot be used on broken HTML-tags. Secondly,
it is important to validate the generalisation of the current work on real-world
HTML-examples in contrast to the fuzzer generated training data considered
here. Lastly, we are exploring ways to utilise the gathered code coverage data
during the training process and rewarding the learning algorithm when discov-
ering unintended behaviour or new code paths. We speculate that this can be
achieved with the help of reinforcement learning to systematically trade-off the
model fit vs exploration.

Acknowledgements. We gratefully acknowledge the support of NVIDIA Corporation with the provision of the GeForce 1080 Ti and the GeForce TITAN Xp used for
this research. We also like to thank Chris Schneider from NVIDIA for his ongoing
interest in our research and his support.

References

1. Abadi, M., et al.: TensorFlow: large-scale machine learning on heterogeneous systems (2015). http://tensorflow.org/
2. Bahdanau, D., Cho, K., Bengio, Y.: Neural machine translation by jointly learning
   to align and translate. arXiv preprint arXiv:1409.0473 (2014)
3. Balog, M., Gaunt, A.L., Brockschmidt, M., Nowozin, S., Tarlow, D.: Deepcoder:
   learning to write programs. arXiv preprint arXiv:1611.01989 (2016)
4. Bengio, Y., Simard, P., Frasconi, P.: Learning long-term dependencies with gradient
   descent is difficult. IEEE Trans. Neural Netw. 5(2), 157–166 (1994)
5. Böhme, M., Pham, V., Roychoudhury, A.: Coverage-based Greybox Fuzzing as
   Markov Chain. IEEE Trans. Softw. Eng., 1 (2018). https://doi.org/10.1109/TSE.
   2017.2785841. ISSN 0098-5589
6. Cho, K., et al.: Learning phrase representations using RNN encoder-decoder for
   statistical machine translation. arXiv preprint arXiv:1406.1078 (2014)
7. Chung, J., Gulcehre, C., Cho, K., Bengio, Y.: Empirical evaluation of gated recurrent neural networks on sequence modeling. arXiv preprint arXiv:1412.3555 (2014)
8. DeMott, J.: The evolving art of fuzzing. DEF CON **14** (2006)
9. DynamoRIO: Dynamorio, June 2017. http://dynamorio.org/
10. Glorot, X., Bengio, Y.: Understanding the difficulty of training deep feedforward
    neural networks. In: Proceedings of The Thirteenth International Conference on
    Artificial Intelligence and Statistics, pp. 249–256 (2010)
11. Godefroid, P., Peleg, H., Singh, R.: Learn&fuzz: machine learning for input fuzzing.
    In: Automated Software Engineering (ASE 2017) (2017)
12. Google: Using clusterfuzz. http://dev.chromium.org/Home/chromium-security/
    bugs/using-clusterfuzz

13. Hochreiter, S.: Untersuchungen zu dynamischen neuronalen netzen. Diploma Technische Universität München **91** (1991)
14. Hochreiter, S., Schmidhuber, J.: Long short-term memory. Neural Comput. **9**(8), 1735–1780 (1997)
15. Höschele, M., Zeller, A.: Mining input grammars from dynamic taints. In: Proceedings of the 31st IEEE/ACM International Conference on Automated Software Engineering, pp. 720–725. ACM (2016)
16. Postel, J., Reynolds, J.: File transfer protocol. Technical report, October 1985. https://tools.ietf.org/html/rfc959
17. Jozefowicz, R., Zaremba, W., Sutskever, I.: An empirical exploration of recurrent network architectures. In: International Conference on Machine Learning, pp. 2342–2350 (2015)
18. Kingma, D., Ba, J.: Adam: a method for stochastic optimization. arXiv preprint arXiv:1412.6980 (2014)
19. Mozilla Corporation: Firefox, August 2018. https://www.mozilla.org/en-US/firefox/
20. Oehlert, P.: Violating assumptions with fuzzing. IEEE Secur. Priv. **3**(2), 58–62 (2005)
21. Pascanu, R., Gulcehre, C., Cho, K., Bengio, Y.: How to construct deep recurrent neural networks. arXiv preprint arXiv:1312.6026 (2013)
22. Pradel, M., Sen, K.: Deep learning to find bugs (2017)
23. Rawat, S., Jain, V., Kumar, A., Cojocar, L., Giuffrida, C., Bos, H.: Vuzzer: application-aware evolutionary fuzzing. In: Proceedings of the Network and Distributed System Security Symposium (NDSS) (2017)
24. Sablotny, M.: Pyfuzz2 - fuzzing framework (2017). https://github.com/susperius/PyFuzz2
25. Srivastava, N., Hinton, G., Krizhevsky, A., Sutskever, I., Salakhutdinov, R.: Dropout: a simple way to prevent neural networks from overfitting. J. Mach. Learn. Res. **15**(1), 1929–1958 (2014)
26. Sutskever, I., Martens, J., Hinton, G.E.: Generating text with recurrent neural networks. In: Proceedings of the 28th International Conference on Machine Learning (ICML 2011), pp. 1017–1024 (2011)
27. Sutton, M., Greene, A., Amini, P.: Fuzzing: Brute Force Vulnerability Discovery. Pearson Education (2007)
28. Zalewski, M.: American fuzzy lop (2017). http://lcamtuf.coredump.cx/afl/

Author Index

Printed in the United States
By Bookmasters